Mathematical Analysis and Applications

Part B

ADVANCES IN MATHEMATICS
SUPPLEMENTARY STUDIES, VOLUME 7B

ADVANCES IN
Mathematics
SUPPLEMENTARY STUDIES

Mathematical Analysis and Applications

ESSAYS DEDICATED TO LAURENT SCHWARTZ ON THE OCCASION OF HIS 65TH BIRTHDAY

Part B

ADVANCES IN MATHEMATICS
SUPPLEMENTARY STUDIES, VOLUME 7B

EDITED BY

Leopoldo Nachbin

Departamento de Matemática Pura
Universidade Federal do Rio de Janeiro
Rio de Janeiro
Brazil

and

Department of Mathematics
The University of Rochester
Rochester, New York

1981

ACADEMIC PRESS

A Subsidiary of Harcourt Brace Jovanovich, Publishers

New York London Toronto Sydney San Francisco

ACADEMIC PRESS, INC.
111 Fifth Avenue, New York, New York 10003

United Kingdom Edition published by
ACADEMIC PRESS, INC. (LONDON) LTD.
24/28 Oval Road, London NW1 7DX

Library of Congress Cataloging in Publication Data
Main entry under title:

Mathematical analysis and applications.

(Advances in mathematics. Supplementary studies; v. 7B)
English or French.
Includes bibliographical references.
1. Mathematical analysis--Addresses, essays, lectures. 2. Schwartz, Laurent. I. Nachbin, Leopoldo. II. Series.
QA300.M294 515 80-1780
ISBN 0-12-512802-9 (pt. B)

PRINTED IN THE UNITED STATES OF AMERICA

81 82 83 84 9 8 7 6 5 4 3 2 1

Contents

PART B

Hélices et quasi-hélices

Jean-Pierre Kahane

The Growth of Restrictions of Plurisubharmonic Functions

Christer O. Kiselman

Propagation des singularités des solutions d'équations aux derivées partielles non linéaires

B. Lascar

On the Cauchy–Kowalewski Theorem

Sigeru Mizohata

Factoring the Natural Injection $\iota^{(n)}: L_n^\infty \to L_n^1$ through Finite Dimensional Banach Spaces and Geometry of Finite Dimensional Unitary Ideals

A. Pelczynski and C. Schütt

De nouvelles caractérisations des ensembles de Sidon

Gilles Pisier

The Choice of Test Functions in Quantum Field Theory

A. S. Wightman

List of Contributors

Numbers in parentheses indicate the pages on which the authors' contributions begin.

JEAN-PIERRE KAHANE (417), Département de Mathématiques, Université de Paris-Sud, Orsay, France 91405

CHRISTER O. KISELMAN (435), Department of Mathematics, Uppsala University, Uppsala, Sweden S-752 38

B. LASCAR (455), Centre de Mathématiques, École Polytechnique, Palaiseau, France 91128

PETER D. LAX (483), Courant Institute of Mathematical Sciences, New York University, New York, New York 10012

J. LINDENSTRAUSS (489), Institute of Mathematics, Hebrew University, Jerusalem, Israel, and Department of Mathematics, Ohio State University, Columbus, Ohio

J. L. LIONS (499), Collège de France, Paris, France 75231

BERNARD MALGRANGE (513), Institut Fourier, Laboratoire de Mathématiques Pures associé au CNRS, Saint-Martin-d'Héres, France 38402

JOHN N. MATHER (531), Department of Mathematics, Princeton University, Princeton, New Jersey 08544

N. METROPOLIS (563), Los Alamos Scientific Laboratory, University of California, Los Alamos, New Mexico 87545

P. A. MEYER (577), Département de Mathématiques, Université de Strasbourg, Strasbourg, France 67084

YVES MEYER (603), Département de Mathématiques, Université de Paris-Sud, Orsay, France 91405

SIGERU MIZOHATA (617), Department of Mathematics, Kyoto University, Kitashirakawa, Kyoto, Japan

G. NICOLETTI (563), Istituto di Geometria, Università di Bologna, Bologna, Italia

A. PELCZYNSKI (653), Institute of Mathematics, Polish Academy of Sciences, Warsaw, Poland

RALPH S. PHILLIPS (483), Department of Mathematics, Stanford University, Stanford, California 94305

GILLES PISIER (685), Centre de Mathématiques, École Polytechnique, Palaiseau, France 91128

JOÃO B. PROLLA (727), Instituto de Matemática, Universidade Estadual de Campinas, Campinas, Brazil

GIAN-CARLO ROTA (563), Department of Mathematics, Massachusetts Institute of Technology, Cambridge, Massachusetts 02139, and Los Alamos Scientific Laboratory, University of California, Los Alamos, New Mexico

WALTER RUDIN (741), Department of Mathematics, University of Wisconsin, Madison, Wisconsin 53706

IRVING SEGAL† (749), The Institute for Advanced Study, Princeton, New Jersey 08544

C. SCHÜTT (653), Johannes Kepler Universität Linz, Mathematisch Institut, A-4040 Linz/Donau, Austria

C. STRICKER (577), Département de Mathématiques, Université de Strasbourg, Strasbourg, France 67084

L. TZAFRIRI (489), Institute of Mathematics, Hebrew University, Jerusalem, Israel, and Department of Mathematics, Ohio State University, Columbus, Ohio

M. VALDIVIA (759), Facultad de Matemáticas, Universidad de Valencia, Valencia, Spain

A. S. WIGHTMAN (769), Joseph Henry Laboratories of Physics, Princeton University, Princeton, New Jersey 08544

†Present address: Department of Mathematics, Massachusetts Institute of Technology, Cambridge, Massachusetts 02139

Contents of Part A

Part B

MATHEMATICAL ANALYSIS AND APPLICATIONS, PART B
ADVANCES IN MATHEMATICS SUPPLEMENTARY STUDIES, VOL. 7B

Hélices et quasi-hélices

JEAN-PIERRE KAHANE

Département de Mathématiques
Université de Paris-Sud
Orsay, France

EN HOMMAGE A LAURENT SCHWARTZ

Soit X un groupe abélien et $\mathscr{H}$ un espace de Hilbert. Une X-hélice dans $\mathscr{H}$ est une partie H de $\mathscr{H}$, paramétrée par X, telle que

$$\|h(y) - h(x)\| = \|h(y - x)\|, \qquad x \in X, \quad y \in Y,$$

h étant le paramétrage. A une isométrie près, une X-hélice dans $\mathscr{H}$ est bien définie par la fonction

$$\psi(x) = \|h(x)\|^2, \qquad x \in X,$$

qu'on appellera, suivant I. J. Schoenberg, fonction d'hélice (screw function).

Le théorème principal sur les hélices (théorème de Schoenberg) exprime l'équivalence des propositions suivantes, dans lesquelles ψ désigne une application de X dans $\mathbb{R}^+$:

(a) ψ est une fonction d'hélice

(b) $\psi(0) = 0$ et $\sum \psi(x_j - x_k)a_j a_k \leq 0$ quels que soient les $x_j \in X$ et les a_j réels tels que $\sum a_j = 0$ (sommes finies)

(c) $\psi(0) = 0$ et, pour tout $t > 0$, $\exp(-t\psi)$ est une fonction de type positif (Schoenberg [11–14]; von Neumann et Schoenberg [9]).

Par exemple, pour tout $\kappa \in]0, 1]$, la fonction $\psi(x) = |x|^{2\kappa}$ $(x \in \mathbb{R})$ est une fonction d'hélice (Schoenberg). Le cas $\kappa = \frac{1}{2}$ avait été étudié par W. A. Wilson (1935) [17]. L'hélice de Wilson est d'ailleurs réalisée par le processus de mouvement brownien ($\mathscr{H}$ étant un espace de Hilbert de variables gaussiennes centrées).

Nous présentons dans cet article deux études sans aucun point commun. Dans la première, nous considérons "l'hélice cubique", constituée par les points dont les coordonnées dans une base orthonormale sont 0 ou 1, et nous étudions, à la suite de K. Harzallah, les fonctions qui opèrent sur les fonctions de type négatif. Dans la seconde, nous répondons à un problème de P. Assouad et L. A. Shepp, qui consiste à construire une quasi-hélice de

ISBN 0-12-512802-9

Wilson dans un espace euclidien de dimension finie. Nous préciserons les questions et les résultats au début de chaque partie.

Première partie

Rappelons quelques définitions.

Une application ψ de X dans $\mathbb{R}^+$ qui vérifie l'une des propositions (a), (b), (c) du théorème de Schoenberg s'appelle fonction de type négatif. On donne quelquefois une définition plus large, applicable à des fonctions à valeurs complexes non nécessairement nulles en 0; une telle définition est utile en théorie des probabilités, mais serait sans intérêt ici.

Une fonction F à valeurs réelles définie sur la demi-droite ouverte $]0, \infty[$ est dite complètement monotone si elle vérifie l'une des conditions suivantes, qui sont équivalentes:

(a) F est de classe C^∞, et $(-1)^n F^{(n)} \geq 0$ pour $n = 0, 1, 2, \ldots$;

(b) $F(u) = \int_0^\infty e^{-ut}\, d\mu(t), \qquad d\mu \geq 0, \quad \int d\mu < \infty$ ou $= \infty$.

Une fonction F à valeurs réelles définie sur la demi droite fermée $[0, \infty[$ est dite complètement monotone si elle vérifie l'une des conditions équivalentes:

(a) F est continue sur $[0, \infty[$, et complètement monotone sur $]0, \infty[$;

(b) $F(u) = \int_0^\infty e^{-ut}\, d\mu(t), \qquad d\mu \geq 0, \quad \int d\mu < \infty.$

Une fonction G à valeurs réelles définie sur la demi-droite fermée $[0, \infty[$ s'appelle fonction de Bernstein nulle en 0 si elle vérifie l'une des conditions équivalentes:

(a) $G(0) = 0$ et, pour tout $t > 0$, la fonction $\exp(-tG)$ est complètement monotone sur $[0, \infty[$;

(b) $G(u) = bu + \int_0^\infty (1 - e^{-ut})\, d\nu(t), \qquad b \geq 0,$

$$d\nu \geq 0, \quad \text{et} \quad \int_0^\infty \frac{t}{1+t}\, d\nu(t) < \infty.$$

Nous dirons "fonctions de Bernstein" pour "fonctions de Bernstein nulles en 0".

Ces définitions et leurs équivalences se trouvent dans [3, 5, 16]. A partir des définitions il est très facile d'établir le théorème suivant.

Soit ψ une fonction de type négatif sur le groupe abélien X, F une fonction complètement monotone sur $[0, \infty[$, G une fonction de Bernstein sur $[0, \infty[$. Alors $F \circ \psi$ est une fonction de type positif et $G \circ \psi$ une fonction de type négatif sur X.

Voici la réciproque établie par Harzallah [6].

Soit X un groupe abélien localement compact infini, F et G deux fonctions continues à valeurs réelles définies sur $[0, \infty[$. *Si, pour toute fonction* ψ *de type négatif sur X,* $F \circ \psi$ *est de type positif, F est complètement monotone. Si, pour toute fonction* ψ *de type négatif sur X,* $G \circ \psi$ *est de type négatif, G est une fonction de Bernstein.*

Nous allons démontrer une réciproque plus précise dans le cas où X n'est pas compact.

THÉORÈME 1. *Soit X un groupe abélien localement compact et non compact. Il existe une fonction* ψ *de type négatif sur X ayant les propriétés suivantes:*

(a) *si F est une fonction continue définie sur* $[0, \infty[$ *et si* $F \circ \psi$ *est de type positif, F est complètement monotone*

(b) *si G est une fonction continue définie sur* $[0, \infty[$ *et si* $G \circ \psi$ *est de type négatif, G est une fonction de Bernstein.*

Il est immédiat que (b) est une conséquence de (a).

En vue d'établir le théorème 1 nous considérons d'abord le groupe Δ constitué par les suites $\delta = (\delta_1, \delta_2, \ldots, \delta_n, \ldots)$, $\delta_n = 0$ ou 1, $\sum \delta_n < \infty$ où l'addition est définie par

$$(\delta + \delta')_n \equiv \delta_n + \delta'_n \qquad (\mathrm{mod}\, 2).$$

Considérons la Δ-hélice H_c ("hélice cubique") définie par

$$h_c(\delta) = \sum \delta_n e_n \qquad \text{(somme finie)}$$

où $e_1, \ldots, e_n, \ldots$ est une base orthonormale de $\mathscr{H}$. La fonction d'hélice correspondante est

$$X(\delta) = \sum \delta_n^2 = \sum \delta_n = |\delta|$$

(définition de $|\delta|$). Nous allons déterminer les fonctions f et g telles que $f \circ \chi$ soit de type positif et $g \circ \chi$ de type négatif sur Δ.

THÉORÈME 2. *Pour que la fonction* $f \circ \chi$ *soit de type positif sur* Δ, *il faut et il suffit que*

$$f(n) = \int_{[-1,1]} y^n \, d\mu(y) \tag{1}$$

où $d\mu$ *est une mesure positive bornée. Pour que la fonction* $g \circ \chi$ *soit de type négatif sur* Δ, *il faut et il suffit que*

$$g(n) = kn + \int_{[-1,1]} (1 - y^n) \, d\mu(y) \tag{2}$$

où $k \geq 0$ *et* $d\mu$ *est une mesure positive, bornée sur* $[-1,0]$, *et intégrant* $1 - y$ *sur* $[0,1]$.

Pour la preuve du théorème 1, on a seulement besoin de la première moitié de la première partie du théorème 2.

Indiquons encore une application du théorème 2.

THÉORÈME 3. *Soit F une fonction continue sur* $[0, \infty[$. *Soit* φ *une fonction définie sur X à valeurs dans* $\mathbb{R}^+$, *paire* ($\varphi(x) = \varphi(-x)$), *nulle en* 0, *prenant des valeurs arbitrairement proches de* 0 *et différentes de* 0. *Si, pour tout entier* $m > 0$, *la fonction*

$$G(x_1, x_2, \ldots, x_m) = F(\varphi(x_1) + \varphi(x_2) + \cdots + \varphi(x_m))$$

est de type positif sur X^m, *F est complètement monotone.*

Le théorème 3 répond à une question que m'avait posée Jacques Deny, qui est à l'origine de cette étude. Dans le cas $X = \mathbb{R}$, $\varphi(x) = x^2$, on retrouve un théorème de Schoenberg [14].

Démonstration du théorème 2. Première partie. Supposons d'abord que $f \circ \chi$ soit de type positif sur Δ. Soit $D = D_m$ l'ensemble des points de Δ dont les coordonnées d'ordre $> m$ sont nulles: $\delta \in D \Leftrightarrow \forall n > m$, $\delta_n = 0$. D'après la caractérisation de Bochner on a

$$\sum f(|\zeta - \eta|) a_\zeta a_\eta \geq 0, \qquad \zeta, \eta \in D,$$

quel que soit le choix des a_ζ réels ($\zeta \in D$), ce qui s'écrit encore

$$\sum_{n=0}^{m} A_n f(n) \geq 0, \qquad A_n = \sum_{|\zeta - \eta| = n} a_\zeta a_\eta.$$

Soit μ un entier, $0 \leq \mu \leq m$; posons

$$\zeta' = (\zeta_1, \zeta_2, \ldots, \zeta_\mu, 0, 0, \ldots)$$

et $a_\zeta = (-1)^{|\zeta'|}$. Alors

$$a_\zeta a_\eta = (-1)^{|\zeta'| + |\eta'|} = (-1)^{|\zeta' - \eta'|}.$$

Quand η est fixé, $\zeta - \eta$ parcourt D quand ζ parcourt D. Donc

$$\sum_{\zeta,\, |\zeta - \eta| = n} a_\zeta a_\eta = \sum_{|\zeta| = n} (-1)^{|\zeta'|}.$$

Il s'ensuit que

$$A_n = 2^m \sum_{|\zeta| = \eta} (-1)^{|\zeta'|}, \qquad n = 0, 1, \ldots, m,$$

$$2^{-m} \sum_{n=0}^{m} A_n y^n = \sum_{\zeta \in D} (-1)^{|\zeta'|} y^{|\zeta|} = (1 - y)^\mu (1 + y)^{m - \mu}.$$

On a $\sum A_n f(n) \geq 0$ chaque fois que les A_n sont donnés par cette dernière formule, pour tout choix de μ et m ($0 \leq \mu \leq m$). Il en résulte bien (1): c'est essentiellement le théorème de Hausdorff–Bernstein sur les moments de Stieltjes (pour la preuve, voir, e.g. [7, pp. 58–59]).

Supposons maintenant que $f(n) = y^n$ avec $-1 \leq y \leq 1$. Alors

$$f(|\delta|) = y^{\delta_1} y^{\delta_2} \ldots y^{\delta_n} \ldots .$$

Chaque facteur y^{δ_n} est de type positif sur Δ (vérification immédiate) donc le produit l'est aussi. En intégrant par rapport à $d\mu$, on voit que $f(|\delta|)$ est de type positif lorsque f est donné par (1).

Deuxième partie. Si $g \circ \chi$ est de type négatif sur Δ, $\exp(-tg \circ \chi)$ est de type positif pour tout $t > 0$, donc

$$\exp(-tg(n)) = \int_{[-1,1]} y^n \, d\mu_t(y), \qquad d\mu_t \geq 0, \quad \int d\mu_t = 1$$

donc

$$g(n) = \lim_{t \to 0} \int_{[-1,1]} (1 - y^n) t^{-1} \, d\mu_t(y).$$

Il existe une suite de valeurs de t, tendant vers 0, telle que les mesures $(1 - y)t^{-1} \, d\mu_t(y)$ aient une limite faible. Désignons cette limite par $(1 - y) \, d\mu(y) + k\delta_1$ (δ_1 étant la mesure de Dirac au point 1). On obtient bien (2).

Inversement, si $g(n)$ est donné par (2), $g \circ \chi$ est une limite ponctuelle de fonctions de type négatif, donc c'est une fonction de type négatif.

Cela achève la démonstration du théorème 2.

Démonstration du théorème 1. Soit Γ le groupe dual de X. Rappelons qu'une suite (γ_j) de points de Γ est dite indépendante si, pour toute somme finie à coefficients entiers rationnels

$$\sum n_j \gamma_j = 0 \Leftrightarrow \forall j \; n_j \gamma_j = 0.$$

Il existe une suite (γ_j) indépendante tendant vers 0. Quitte à remplacer (γ_j) par une sous-suite, on peut supposer que l'une des deux conditions suivantes est satisfaite:

(α) l'ordre de γ_j est supérieur à $\pi 2^j$ (il peut être infini),
(β) l'ordre de γ_j est fini et fixe.

Quitte à remplacer (γ_j) par une sous-suite, on peut aussi supposer que la série $\sum |1 - \gamma_j(x)|$ converge uniformément sur tout compact de X. C'est ce que nous ferons désormais.

Supposons la condition (α) satisfaite. On va définir une suite (a_j) réelle bornée. On posera

$$h(x) = \sum a_j (1 - \gamma_j(x)) e_j,$$

paramétrage d'une X-hélice dans $\mathscr{H}$, dont la fonction d'hélice est

$$\psi(x) = \sum a_j^2 |1 - \gamma_j(x)|^2$$

Nous allons voir que, si les a_j sont bien choisis, cette fonction a les propriétés du théorème 1.

Définissons d'abord par induction une suite (x_k) d'éléments de X, telle que

$$\begin{aligned} |1 - \gamma_j(x_k)| &< 2^{-k} \quad \text{si} \quad j < k, \\ |1 + \gamma_k(x_k)| &< 2^{-k}. \end{aligned} \tag{3}$$

C'est possible à cause de l'indépendance des γ_j et de la condition (α). Définissons ensuite par induction une suite infinie $j(n)$ telle que

$$|1 - \gamma_{j(n)}(x_{j(m)})| < 2^{-n} \qquad \text{si} \quad n > m.$$

C'est possible parce que les γ_j tendent vers 0. Soit J l'ensemble des valeurs $\{j(n)\}$ ($n \in \mathbb{N}$). Décomposons J en une infinité d'ensembles infinis J_κ ($\kappa = 1, 2, \ldots$) et posons $a_j = 2^{-\kappa}$ quand $j \in J_\kappa$, $a_j = 0$ quand $j \notin J$.

Pour la fonction h définie ci-dessus, on a immédiatement

$$h(x_k) = 2a_k e_k + o(1), \qquad k \to \infty, \quad k \in J, \tag{4}$$

et, plus généralement, pour tout entier μ fixé,

$$h(x_{k_1} + x_{k_2} + \cdots + x_{k_\mu}) = 2a_{k_1}e_{k_1} + 2a_{k_2}e_{k_2} + \cdots + 2a_{k_\mu}x_{k_\mu} + o(1)$$

($k_j \in J$, k_j distincts, $\inf k_j \to \infty$).

Fixons un entier $\nu > 0$. Etant donné une partie K de J_1, de cardinal ν, considérons les 2^ν points de X de la forme

$$\sum_{k \in K} \delta_k x_k, \qquad \delta_k = 0 \text{ ou } 1,$$

que nous noterons s_δ. On a

$$h(s_\delta) = h\left(\sum_{k \in K} \delta_k x_k\right) = \sum_{k \in K} \delta_k e_k + o(1)$$

quand $\inf K \to \infty$. Supposons que $F \circ \psi$ est de type positif sur X. La condition de Bochner donne, pour tout choix des nombres complexes c_δ,

$$\sum_{\delta, \delta'} F(\psi(s_\delta - s_{\delta'}))c_\delta \bar{c}_{\delta'} \geq 0$$

soit

$$\sum_{\delta, \delta'} F(\|h(s_\delta) - h(s_{\delta'})\|^2)c_\delta \bar{c}_{\delta'} \geq 0.$$

En faisant tendre $\inf K$ vers l'infini, on obtient, compte tenu de la continuité de F,

$$\sum_{\delta, \delta'} F(|\delta - \delta'|)c_\delta \bar{c}_{\delta'} \geq 0, \qquad \left(|\delta - \delta'| = \sum_{k \in K} |\delta_k - \delta_{k'}|\right).$$

Comme v est arbitraire, il en résulte que la fonction $F(|\delta|)$ est de type positif sur le groupe Δ, donc, par le théorème 1, que

$$F(n) = \int_{[-1,1]} y^n \, d\mu(y)$$

d'où

$$F(2n) = \int_0^1 y^{2n}(d\mu(y) + d\mu(-y)) = \int_0^\infty e^{-2nt} \, dv_1(t),$$

dv_1 étant une mesure positive de masse totale $F(0)$.

Si on remplace ci-dessus J_1 par J_κ, on voit que la fonction $F(2^{-\kappa+1}|\delta|)$ est de type positif sur Δ, donc

$$F(2^{-\kappa+2}n) = \int_0^\infty e^{-2^{-\kappa+2}nt} \, dv_\kappa(t),$$

dv_κ étant une mesure positive de masse totale $F(0)$. Si dv est une limite faible d'une suite de telles mesures dv_κ, on a

$$F(u) = \int_0^\infty e^{-ut} \, dv(t)$$

quand u est un nombre binaire, et, par continuité, pour tout $u \geq 0$.

Donc φ a bien la propriété (a) du théorème 1. Nous avons déjà observé que la propriété (b) en résulte.

Si c'est la condition (β) qui est satisfaite au lieu de la condition (α), on répète exactement la construction des x_k est des a_j dans le cas où l'ordre N des γ_j est pair. Si l'ordre N est impair, les seules modifications à faire consistent à écrire

$$|\gamma_k(x_k) - \exp(2\pi i/N)| < 2^{-k}$$

au lieu de (3), puis

$$h(x_k) = (1 - \exp(2\pi i/N))a_k e_k + o(1)$$

au lieu de (4), et les remplacements de 2 par $1 - \exp(2\pi i/N)$ qui s'ensuivent.

Cela achève la démonstration du théorème 1.

Démonstration du théorème 3. Comme dans la démonstration du théorème 2, soit $D = D_m$ l'ensemble des points de Δ dont les coordonnées d'ordre $>m$ sont nulles. Si $\zeta = (\zeta_1, \zeta_2, \ldots \zeta_m, 0, 0, \ldots) \in D$, et $x \in X$, notons

$$y_\zeta = (\zeta_1 x, \zeta_2 x, \ldots, \zeta_m x) \in X^m.$$

L'hypothèse entraîne, par la condition de Bochner,

$$\sum G(y_\zeta - y_\eta) c_\zeta \bar{c}_\eta \geq 0 \qquad \zeta, \eta \in D; \quad c_\zeta \text{ complexes.}$$

Si l'on pose $\varphi(x) = \varphi(-x) = \varepsilon$,

$$G(y_\zeta - y_\eta) = F\left(\sum_j \varphi((\zeta_j - \eta_j)x)\right) = F(|\zeta - \eta|\varepsilon).$$

La condition de Bochner montre que la fonction $\delta \to F(|\delta|\varepsilon)$ est de type positif sur D, donc (m étant arbitraire) sur Δ. On conclut comme dans la démonstration du théorème 1 que F est complètement monotone.

Remarque. L'hypothèse du théorème 3 n'entraîne pas que φ est de type négatif. En effet, il est connu, et facile à voir, qu'il existe des fonctions de type positif $P > 0$, telles que $\sqrt{P}$ ne soit pas de type positif. (Sinon, pour toute fonction de type positif $P \geq 0$, et pour tout $t = 2^{-n}$, P^t serait de type positif, donc la fonction caractéristique de l'ensemble $\{x \mid P(x) > 0\}$ serait de type positif sur le groupe X discret, c'est-à-dire que l'ensemble $\{x \mid P(x) > 0\}$ serait un sous-groupe du groupe discret X.) En choisissant $\varphi = -\log(P/P(0))$ et $F(u) = e^{-u}$, l'hypothèse est satisfaite, mais $e^{-t\varphi}$ n'est pas de type positif pour $t = \frac{1}{2}$, donc φ n'est pas de type négatif.

Deuxième partie

La théorie des hélices concerne des plongements isométriques dans un espace de Hilbert. P. Assouad a systématiquement étudié les plongements lipschitziens d'un espace métrique X dans un espace euclidien $\mathscr{E}$ de dimension finie: ce sont les applications F de X dans $\mathscr{E}$ telles que

$$A < \|F(x) - F(y)\|/d(x, y) < B, \qquad x, y \in X,$$

$A > 0$ et $B > 0$ ne dépendant que de F [1]. Par exemple, la droite $X = \mathbb{R}$, munie de la distance $d(x, y) = |x - y|^{1/2}$, admet un plongement lipschitzien dans un espace euclidien $\mathscr{E}$ si et seulement si la dimension de $\mathscr{E}$ dépasse 2 [2]. Comme, munie de cette distance, la droite X admet un plongement isométrique dans un espace de Hilbert de dimension infinie (c'est l'hélice de Wilson, ou l'hélice brownienne), on peut se demander si, quand la dimension de $\mathscr{E}$ est grande, on peut choisir A et B voisins de 1: c'est une question de P. Assouad et L. A. Shepp, à laquelle répond le théorème suivant.

THÉORÈME. *Il existe une suite α_n tendant vers 0 telle que, pour tout $N \geq 3$, il existe un plongement de $\mathbb{R}$ dans $\mathbb{R}^N$ euclidien vérifiant*

$$1 - \alpha_N \leq \|F(x) - F(y)\|^2/|x - y| \leq 1 + \alpha_N. \tag{5}$$

On peut choisir $\alpha_N = O(1/N)$.

Nous nous bornerons à définir F sur $\mathbb{R}^+$, et à vérifier (1) quand x et y appartiennent à $\mathbb{R}^+$. On définira $F(0) = 0$. L'application F_1 de $\mathbb{R}$ dans $\mathbb{R}^{2N}$ définie par $F_1(x) = (F(x), 0)$ pour $x \geq 0$ et $F_1(x) = (0, F(-x))$ pour $x \leq 0$ vérifie (5).

D'après le théorème d'Assouad il suffit de construire F et d'établir (5) pour N grand. Nous nous bornerons au cas $N = 2^\nu$ $(\nu \geq 2)$. L'idée est alors de construire d'abord F sur $\mathbb{N}$, et d'établir (5) et

$$F(N^2 x) = NF(x) \tag{6}$$

quand x et y appartiennent à $\mathbb{N}$. On prolonge ensuite facilement F sur $\mathbb{R}^+$, de façon que (5) et (6) restent valables.

Dans la construction de F sur $\mathbb{N}$ une suite ε_n $(n \in \mathbb{N})$ à valeurs $+1$ ou -1 joue un rôle essentiel. Cette suite a des relations intéressantes avec les caractères du groupe $(\mathbb{Z}/2\mathbb{Z})^\nu$. Dans le cas $N = 2$ (cas exclu par l'énoncé) c'est une suite classique de Shapiro et Rudin; la construction de F, dans ce cas, donne la courbe plane de Peano–Paul Lévy. Dans le cas considéré par l'énoncé, c'est (à la notation près) une suite de Morse généralisée au sens de Christol, Kamae, Mendès-France et Rauzy [8]. La courbe dèfinie par F est une "quasi-hélice", qui est une sorte d'interpolation entre la courbe de Peano et l'hélice brownienne.

La suite ε_n. N est un entier. Nous appellerons *mot* une suite finie constituée de lettres $+$ et $-$. Si $M = (m_1 m_2 \ldots m_k)$ et $M' = (m'_1, m'_2 \ldots m'_l)$ sont deux mots, nous notons $M + M'$ le mot $(m_1 m_2 \ldots m_k m'_1 m'_2 \ldots m'_l)$ (la cancaténation de M et M'), $-M'$ le mot $(-m'_1, -m'_2, \ldots, -m'_l)$, et $M - M'$ le mot $M + (-M')$. Nous allons définir une suite de mots M^j $(j = 0, 1, \ldots)$.

$M^0 = (+ + \ldots +)$ est le mot de N lettres constitué uniquement de $+$. Pour $j = 0, 1, \ldots, M^j$ est un mot de $2^j N$ lettres qui est la somme de N mots de longueur 2^j que l'on appelle E_n^j $(n = 0, 1, \ldots, N - 1)$:

$$M^j = E_0^j + E_1^j + \cdots + E_{N-1}^j. \tag{7}$$

On obtient M^{j+1} à partir de M^j en posant

$$\begin{aligned} E_n^{j+1} &= E_{2n}^j + E_{2n+1}^j, \\ E_{n+N/2}^{j+1} &= E_{2n}^j - E_{2n+1}^j, \end{aligned} \qquad n = 0, 1, \ldots, \tfrac{1}{2}N - 1. \tag{8}$$

Ainsi la première moitié du mot M^{j+1} est M^j. La n-ième lettre de M^j, soit $\varepsilon_n(M^j)$, ne dépend pas de j dès que $n \leq 2^j N - 1$. On pose

$$\varepsilon_n = \varepsilon_n(M_j), \qquad n \leq 2^j N - 1. \tag{9}$$

Dans la suite, on identifiera librement $+$ avec $+1$, et $-$ avec -1.

Voici une autre définition équivalente de ε_n dans le cas $N = 2^\nu$. Soit

$$\ldots e_k(n) \ldots e_2(n) e_1(n) e_0(n)$$

le développement binaire de n ($e_j(n) = 0$ ou 1, $n = \sum 2^j e_j(n)$). Soit $\mu(n)$ le nombre de valeurs de j telles que $e_j(n) = e_{j+\nu} = 1$. On a

$$\varepsilon_n = (-1)^{\mu(n)}.$$

Nous laissons la vérification au lecteur. La suite $\frac{1}{2}(1 - \varepsilon_n) \equiv \mu(n) \pmod 2$ est une "suite de Morse généralisée" suivant [4].

Pour $N = 1$, $\frac{1}{2}(1 - \varepsilon_n)$ est la suite de Morse classique. Pour $N = 2$, ε_n est la suite de Shapiro (ou de Rudin–Shapiro). On trouve dans [4] des généralisations et des propriétés intéressantes de telles suites. Désormais nous nous limitons au cas $N = 2^\nu$, $\nu \geq 1$.

Posons $m = 0, 1, \ldots, N-1$, $n = jN + m$ ($j = 0, 1, \ldots$), et

$$\varepsilon_n = w_j(m). \tag{10}$$

La matrice carrée $(w_j(m))$ ($m = 0, 1, \ldots, N-1$; $j = 0, 1, \ldots, N-1$) se construit facilement comme produit tensoriel de ν matrices. Désignons la par W:

$$W = \begin{pmatrix} + & + \\ + & - \end{pmatrix} \otimes \begin{pmatrix} + & + \\ + & - \end{pmatrix} \otimes \cdots \otimes \begin{pmatrix} + & + \\ + & - \end{pmatrix}.$$

Par exemple, pour $N = 4$,

$$W = \left[\begin{array}{cc|cc} + & + & + & + \\ + & - & + & - \\ \hline + & + & - & - \\ + & - & - & + \end{array}\right].$$

La matrice W est donc symétrique: $w_j(m) = w_m(j)$. Si l'on identifie m et le développement binaire

$$e_{\nu-1}(m)e_{\nu-2}(m) \ldots e_0(m)$$

considéré comme élément du groupe $D^\nu = (\mathbb{Z}/2\mathbb{Z})^\nu$ (en d'autres termes, m est l'ordre lexicographique sur D^ν) les fonctions w_j sont les caractères de D^ν. Les caractères coordonnés sont $r_1, r_2, \ldots r_\nu$:

$$r_i(m) = e_{i-1}(m)$$

et l'on a $w_j = \prod r_i^{\alpha_i}$ si $j = \sum \alpha_i 2^{i-1}$ ($\alpha_i = 0$ ou 1). Il est commode de poser $r_0 = w_0 = 1$ (fonction définie sur D^ν). Les notations r_i et w_j rappellent les fonctions de Rademacher et les fonctions de Walsh, que l'on définit d'ordinaire sur $[0, 1]$ ou sur D^∞.

Pour chaque $\kappa = 1, 2, \ldots$, la matrice

$$W_\kappa = (w_{j+\kappa N}(m)), \qquad m = 0, 1, \ldots, N-1; \quad j = 0, 1, \ldots, N-1,$$

s'obtient à partir de $W = W_0$ de la manière suivante. Pour $\kappa = 0, 1, \ldots, N-1$,

$$w_{j+\kappa N}(m) = w_\kappa(j)w_j(m) = w_j(\kappa)w_j(m) = w_j(\kappa \circ m), \tag{11}$$

où $\kappa \circ m$ est le produit dans le groupe D^ν. Pour $\kappa \equiv \kappa' \pmod N$ et $0 \leq \kappa' < N$, on a

$$W_\kappa = \varepsilon_\kappa W_{\kappa'}. \tag{12}$$

Par exemple, dans le cas $N = 4$, on obtient les matrices figurées:

+	+	+	+	
+	−	+	−	W_0
+	+	−	−	
+	−	−	+	
+	+	+	+	
−	+	−	+	W_1
+	+	−	−	
−	+	+	−	
+	+	+	+	
+	−	+	−	W_2
−	−	+	+	
−	+	+	−	
+	+	+	+	
−	+	−	+	W_4
−	−	+	+	
+	−	−	+	
W_0				W_5
$-W_1$				W_6

DÉFINITION ET PROPRIÉTÉS DE F SUR $\mathbb{N}$. Soit $N = 2^\nu$, $\nu \geq 2$, et soit $u_0, u_1, \ldots, u_{N-1}$ la base canonique de $\mathbb{R}^N$. Définissons u_n $(n = N, N+1, \ldots)$

par la condition de périodicité: $u_{n+N} = u_n$. Posons

$$F(0) = 0, \qquad F(n+1) - F(n) = \varepsilon_n u_n, \qquad n = 0, 1, \ldots. \tag{13}$$

Il est clair que

$$\|F(n') - F(n)\|^2 = |n' - n| \quad \text{si} \quad |n' - n| \leq N. \tag{14}$$

Posons

$$U_n^j = F(2^j(n+1)) - F(2^j n).$$

A partir de (7), (8), (9) on vérifie que

$$\begin{aligned} U_n^{j+1} &= U_{2n}^j + U_{2n+1}^j, \\ U_{n+N/2}^{j+1} &= U_{2n}^j - U_{2n+1}^j, \end{aligned} \qquad n = 0, 1, \ldots, \tfrac{1}{2}N - 1; \tag{15}$$

$$U_{n+N}^j = \pm U_n^j; \tag{16}$$

$$\{2^{-j/2} U_n^j\}, \quad n = 0, 1, \ldots, N-1, \quad \text{est une base orthonormale;} \tag{17}$$

$$U_n^{j+2\nu} = N U_n^j. \tag{18}$$

[(18) exprime que les sommes des éléments des colonnes dans la matrice W_0 sont respectivement $N, 0, 0, \ldots, 0$]. Il résulte de (15), (16), (17) que

$$\|F(2^j n') - F(2^j n)\|^2 = 2^j |n' - n| \quad \text{si} \quad |n' - n| \leq N \tag{19}$$

ce qui généralise (14).

Supposons maintenant $2^j n \leq m < 2^j(n+1)$. Si $j \leq \nu$, la ligne brisée $F(2^j n)F(2^j n + 1)F(2^j n + 2) \ldots F(2^j(n+1))$ a ses côtés mutuellement orthogonaux, donc ses sommets sont sur la sphère ayant pour diamètre le segment joignant ses extrémités; par conséquent

$$\|F(m) - \tfrac{1}{2}(F(2^j n) + F(2^j(n+1)))\|^2 = 2^{j-2}. \tag{20}$$

Si $j > \nu$, désignons par m' et m'' les multiples de $2^{j-\nu}$ les plus voisins de m. On a encore (20) en remplaçant m par m' et par m'', et aussi

$$\|F(m) - \tfrac{1}{2}(F(m') + F(m''))\|^2 = 2^{j-\nu-2} \quad \text{si} \quad j \leq 2\nu. \tag{21}$$

Si $j > 2\nu$, on répète le procédé, en approchant m par les multiples de $2^{j-\kappa\nu}$ les plus voisins de m, tant que $\kappa\nu < j$, et on obtient

$$\begin{aligned} &\|F(m) - \tfrac{1}{2}(F(2^j n) + F(2^j(n+1)))\| \\ &\quad \leq 2^{j/2-1}(1 + 2^{-\nu/2} + \cdots + 2^{-\kappa\nu/2} + \cdots) \leq 2^{j/2}. \end{aligned} \tag{22}$$

Supposons $|m - m'| > N$, et soit j l'entier tel que

$$2^{j-1} N \leq |m - m'| < 2^j N \tag{23}$$

puis n et n' les entiers tels que

$$2^j n \leq m < 2^j(n+1), \qquad 2^j n' \leq m' < 2^j(n'+1). \tag{24}$$

Il résulte de (19) et (22), puis (23) et (24), que

$$\big|\|F(m) - F(m')\| - 2^{j/2}|n' - n|^{1/2}\big| \leq 4.2^{j/2},$$
$$\big|\|F(m) - F(m')\| - |m - m'|^{1/2}\big| \leq 5.2^{j/2} < 8N^{-1/2}|m - m'|^{1/2},$$
$$\left|\frac{\|F(m) - F(m')\|^2}{|m - m'|} - 1\right| \leq 8N^{-1/2}(2 + 8N^{-1/2}),$$

soit

$$1 - \alpha_N \leq \frac{\|F(m) - F(m')\|^2}{|m - m'|} \leq 1 + \alpha_N \tag{25}$$

avec $\alpha_N = O(N^{-1/2})$. Naturellement, (25) a lieu aussi si $|m - m'| \leq N$ [voir (14)].

D'autre part il résulte de (18) que

$$F(N^2 m) = NF(m) \tag{26}$$

pour tout entier m.

On verra plus loin comment améliorer l'estimation de α_N.

Définition et propriétés de F sur $\mathbb{R}^+$. Pour $x = pN^{-2q}$ (p et q entiers ≥ 0) on pose

$$F(x) = N^{-q}F(N^{2q}x)$$

ce qui a un sens d'après (26), et on vérifie d'après (25) que

$$1 - \alpha_N \leq \|F(x) - F(x')\|^2/|x - x'| \leq 1 + \alpha_N \tag{$*$}$$

quand x' est de la même forme que x. Ainsi F est uniformément continue sur un ensemble dense de $\mathbb{R}^+$. On peut la prolonger par continuité à $\mathbb{R}^+$, et ce prolongement est la fonction F cherchée. Elle vérifie à la fois ($*$), c'est-à-dire (5), et

$$F(N^2 x) = NF(x)$$

c'est-à-dire (6).

Amélioration de α_N. Le point de départ de l'estimation $\alpha_N = O(N^{-1/2})$ est (14). On obtient un meilleur résultat en étudiant de plus près $\|F(n) - F(n')\|^2$ quand $|n' - n| \leq N^2$. Soit lN, resp., $l'N$, le multiple de N immédiatement inférieur à n, resp., n'. On a

$$\|F(n) - F(n')\|^2 = \|F(lN) - F(l'N)\|^2 + \rho(n, n'),$$
$$\rho(n, n') = \|F(n) - F(lN) - F(n') + F(l'N)\|^2$$
$$+ 2(F(lN) - F(l'N), F(n) - F(lN) - F(n') + F(l'N)),$$
$$\rho(n, n') \leq 4N + 4 \sup_{\substack{\mu, \mu', n \\ |\mu - \mu'| \leq N}} |(F(\mu N) - F(\mu' N), F(n) - F(lN))|.$$

On va montrer que le second membre est $O(N)$.

Supposons d'abord que μ est un multiple de N, $\mu = \kappa N$, et $\mu' = \mu + k$, $0 < k \leq N$. Alors

$$F(\mu' N) - F(\mu N) = x_k(0)u_0 + x_k(1)u_1 + \cdots + x_k(N-1)u_{N-1}$$

où x_k est la somme des k premières lignes de la matrice W_κ. D'autre part

$$F(n) - F(lN) = w_l(0)u_0 + w_l(1)u_1 + \cdots + w_l(p-1)u_{p-1}, \qquad p = n - lN.$$

Il s'agit de majorer le module de

$$S = S_{\kappa,l,k,p} = w_l(0)x_k(0) + \cdots + w_l(p-1)x_k(p-1).$$

A cause de (12) on peut supposer $\kappa < N$, et on a d'après (11)

$$x_k(m) = \sum_{j=0}^{k-1} w_{j+\kappa N}(m) = \sum_{j=0}^{k-1} w_j(\kappa) w_j(m)$$

donc

$$S = \sum_{m=0}^{p-1} \sum_{j=0}^{k-1} w_j(\kappa) w_j(m) w_l(m).$$

Désignons par A_ν la borne supérieure de $|S|$ pour $0 \leq \kappa, l, k, p \leq N = 2^\nu$. Si $p \leq \frac{1}{2}N$ et $k \leq \frac{1}{2}N$, on a

$$|S| \leq A_{\nu-1}.$$

En effet, on peut supposer que $\kappa \leq \frac{1}{2}N$ et $l \leq \frac{1}{2}N$ parce que

$$S_{\kappa,l,k,p} = S_{\kappa,\, l+\frac{1}{2}N,\, k,\, p} = S_{\kappa+\frac{1}{2}N,\, l,\, k,\, p} = S_{\kappa+\frac{1}{2}N,\, l+\frac{1}{2}N,\, k,\, p},$$

et dans la première de ces sommes on peut considérer les w_j comme les caractères du groupe $D^{\nu-1}$ au lieu du groupe D^ν (voir la construction des matrices W). Si $p > \frac{1}{2}N$, resp., $k > \frac{1}{2}N$, on décompose la somme $\sum_{m=0}^{p-1}$, resp., $\sum_{j=0}^{k-1}$, sous la forme

$$\sum_{m=0}^{\frac{1}{2}N-1} + \sum_{m=N/2}^{p-1}, \quad \text{resp.,} \quad \sum_{j=0}^{N/2-1} + \sum_{j=N/2}^{k-1}.$$

Ainsi S se trouve décomposé en 2 ou 4 sommes. Pour la dernière somme, on se ramène au cas précédent et on a la majoration par $A_{\nu-1}$. Pour la ou les premières sommes, elles sont égales à 0 ou $N/2$, à cause des formules

$$\sum_{m=0}^{N/2-1} w_j(m)w_l(m) = \begin{cases} 0 & \text{si} \quad j \not\equiv l(N/2), \\ N/2 & \text{si} \quad j \equiv l(N/2) \end{cases}$$

et de la symétrie de la matrice W. On a donc

$$A_\nu \leq \tfrac{3}{2}N + A_{\nu-1} \leq \tfrac{3}{2}(N + \tfrac{1}{2}N + \tfrac{1}{4}N + \cdots) \leq 3N.$$

Dans l'hypothèse faite sur μ, on obtient donc $|S| \leq 3N$.

Sans hypothèse sur μ, on obtient, en intercalant un multiple de κN,

$$|(F(\mu N) - F(\mu' N), F(n) - F(lN))| \leq 6N$$

donc

$$\rho(n, n') \leq 28N,$$
$$|\|F(n) - F(n')\|^2 - |n - n'|| \leq 30N.$$

Si $\frac{1}{2}N^2 < |n - n'| \leq N^2$, on a donc

$$1 - \frac{60}{N} \leq \|F(n) - F(n')\|^2/|n' - n| \leq 1 + \frac{60}{N}.$$

En remplaçant les vecteurs u_n par les U_n^j, on obtient

$$1 - \frac{60}{N} \leq \|F(2^j n) - F(2^j n')\|^2/|2^j n - 2^j n'| \leq 1 + \frac{60}{N}. \tag{27}$$

Si enfin on donne deux entiers positifs quelconques, m et m', tels que $|m - m'| > N^2$, on choisit l'entier j de façon que

$$2^{j-1}N^2 < |m - m'| \leq 2^j N^2$$

puis n et n' de façon que

$$|2^j n - m| \leq 2^{j-1}, \qquad |2^j n' - m'| \leq 2^{j-1}.$$

On sait alors [voir (22)] que

$$\|F(m) - F(2^j n)\| \leq \tfrac{3}{2}2^{j/2}, \qquad \|F(m') - F(2^j n')\| \leq \tfrac{3}{2}2^{j/2}$$

ce qui, joint à (27), donne (25) avec $\alpha_N = O(1/N)$.

Cette estimation montre que la courbe paramétrée par F est presque une hélice quand N est grand. Pour les petites valeurs de N, on peut s'assurer que l'application F est bijective.

Le cas $N = 2$. Courbe de Peano–Paul Lévy. Pour $N = 2$, la suite ε_n définie par (9) est la suite de Shapiro, ou de Rudin [10] et Shapiro [15], qui intervient dans la théorie des séries trigonométriques. On prend pour u_0, u_1 deux vecteurs d'une base du plan euclidien, on pose $u_{2n} = u_0$, $u_{2n+1} = u_1$, et on définit une suite de points $F(n)$ par (13). On obtient de la même façon (19) et (26), et on peut définir F sur $\mathbb{R}^+$ par

$$F(p4^{-q}) = 2^{-q}F(p)$$

et par continuité. On voit facilement que la fonction obtenue vérifie

$$\|F(x) - F(x')\|^2 \leq |x - x'|, \qquad F(4x) = 2F(x),$$

mais ce n'est plus une application bijective. On vérifie que c'est la courbe de Peano–Paul Lévy. Elle couvre tout l'angle $\{x_0u_0 + x_1u_1, x_0 \geq x_1 \geq 0\} = \mathscr{A}$, et, pour tout ensemble measurable $E \subset \mathbb{R}^+$, la mesure superficielle de $F(E)$ est égale à la mesure linéaire de E. Pour le voir, il suffit d'engendrer la suite $F(n)$ par étapes à l'aide des transformations géométriques simples (translation et symétrie) qui expriment (8): pour $0 \leq n < 2^j$, les $F(n)$ sont tous les points à coordonnées entières d'un triangle T_j, les triangles T_j sont emboités et leur réunion est l'angle $\mathscr{A}$. Nous laissons les détails au lecteur.

De façon indépendante, M. Mendès-France et G. Tenenbaum ont découvert ce lien entre les suites de Rudin–Shapiro et les courbes de Peano, avec d'intéressants développements sur le pliage des papiers et la construction de courbes étranges [8].

Références

1. P. Assouad, Espaces métriques, plongements, facteurs, Thèse d'Etat, Orsay, 1977. Publications Mathematiques d'Orsay nº 223-77-69, Zbl 396-46035.
2. P. Assouad, *C. R. Acad. Sci. Paris* **290** (1980), 591–594.
3. C. Berg and G. Forst, "Potential Theory on Locally Compact Abelian Groups," Ergebnisse der Mathematik, Vol. 87, Springer-Verlag, Berlin and New York, 1975.
4. C. Christol, T. Kamae, M. Mendès-France, and G. Rauzy, Suites algébriques, automates et substitutions, *Bull. Soc. Math. France* **108** (1980).
5. J. Faraut and K. Harzallah, Distances hilbertiennes invariantes sur un espace homogène, *Ann. Inst. Fourier* **24** (1974), 171–217.
6. K. Harzallah, Fonctions opérant sur les fonctions définies négatives réelles, *C. R. Acad. Sci. Paris* **260** (1965), 6790–6793.
7. G. Lorentz, "Bernstein polynomials," Mathematical Expositions, No. 8, Univ. of Toronto Press, Toronto, 1953.
8. M. Mendès-France and G. Tenenbaum, Dimensions des courbes planes, papiers pliés et suites de Rudin-Shapiro, *Bull. Soc. Math. France* **109** (1981).
9. J. von Neumann and I. J. Schoenberg, Fourier integrals and metric geometry, *Trans. Amer. Math. Soc.* **50** (1941), 226–251.
10. W. Rudin, Some theorems on Fourier coefficients, *Proc. Amer. Math. Soc.* **10** (1959), 855–859.
11. I. J. Schoenberg, Remarks to Maurice Fréchet's article Sur la définition axiomatique d'une classe d'espaces distanciés vectoriellement applicables sur l'espace de Hilbert, *Ann. of Math.* **36** (1935), 724–732.

12. I. J. SCHOENBERG, On certain metric spaces arising from euclidean spaces . . . , *Ann. of Math.* **38** (1937), 787–793.
13. I. J. SCHOENBERG, Metric spaces and positive definite functions, *Trans. Amer. Math. Soc.* **44** (1938), 522–536.
14. I. J. SCHOENBERG, Metric spaces and completely monotone functions, *Ann. of Math.* **39** (1938), 811–841.
15. H. S. SHAPIRO, Extremal problems for polynomials and power series, Thesis, M.I.T., 1951.
16. D. WIDDER, "The Laplace Transforms," Princeton Univ. Press, Princeton, New Jersey, 1941.
17. W. A. WILSON, On certain types of continuous transformations of metric spaces. *Amer. J. Math* **57** (1935), 62–68.

MATHEMATICAL ANALYSIS AND APPLICATIONS, PART B
ADVANCES IN MATHEMATICS SUPPLEMENTARY STUDIES, VOL. 7B

The Growth of Restrictions of Plurisubharmonic Functions

CHRISTER O. KISELMAN

Department of Mathematics
Uppsala University
Uppsala, Sweden

TO LAURENT SCHWARTZ ON THE OCCASION OF HIS 65TH BIRTHDAY

Contents

1. INTRODUCTION

If $f:\mathbb{C}^2 \to \mathbb{C}$ is an entire function of two complex variables, then the growth at infinity of the partial function $y \mapsto f(x, y)$ is essentially the same for all values of x, except when x belongs to a certain small set. Many results generalizing this statement are known, with varying degree of generality and precision concerning the class of functions considered, the type of growth allowed, and the measure of smallness used for the exceptional set.

Our purpose here is to prove a result of this nature which is both general and precise (Theorem 2.1). It is general because it allows all kinds of growth (finite order, refined orders, infinite order) and even permits different types of growth in different directions; it is precise in that it describes the ideal generated by the exceptional sets (Theorem 2.5); and also in the sense that the condition (2.3) on the function which defines the set of exceptionally slow growth is necessary and sufficient (Theorem 2.4). We apply this result to the growth of restrictions of plurisubharmonic functions to linear or affine subspaces (Sections 4 and 5, respectively).

Let ω be an open set in $\mathbb{C}^n$. A subset E of ω is said to be *polar in* ω if there is a plurisubharmonic function in ω which is $-\infty$ on E but not identically $-\infty$ in any component of ω. We prove that the set of exceptionally slow growth is polar, or locally polar in the case of the Grassmannian.

ISBN 0-12-512802-9

Of the earlier results in this direction let us mention Sire [17], who considered entire functions of two variables of finite order, the exceptional set being of one-dimensional Hausdorff measure zero. Lelong [10] allowed entire functions of two variables of more general type of growth, the exceptional set being of capacity zero in $\mathbb{C}$. Hengartner [3] worked with entire or plurisubharmonic functions of n variables, refined orders, and his exceptional sets are negligible. Gauthier and Hengartner [2] considered restrictions of meromorphic and plurisubharmonic functions to subspaces, refined orders, and the set of m planes where the order is smaller is shown to be of Lebesgue measure zero in the Grassmannian manifold of all m planes.

Polar sets are negligible, of Lebesgue measure zero, of Newtonian capacity zero, and of Γ-capacity zero. Ronkin [15, Theorem 2.7.1] introduces a different kind of measure of smallness: E is said to satisfy condition A if $E \neq \mathbb{C}^n$ and every complex line which is not contained in E cuts E in a set of logarithmic capacity zero. Let us say for convenience that a subset of $\mathbb{C}^n$ belongs to A_1 if it is contained in a set satisfying condition A. It is obvious that every polar set is in A_1, and Ronkin proves that every set in A_1 is of Γ-capacity zero [15, Lemma 2.2.8.]. However, the polar sets are strictly contained in A_1, for neither A_1 nor the class of sets of Γ-capacity zero is invariant under biholomorphic maps. Indeed, it is easy to construct a set E which is of positive Γ-capacity and a biholomorphic map h such that $h(E)$ intersects every complex line in a finite set; consequently $h(E)$ satisfies condition A. Specifically, we may take

$$E = \{z \in \mathbb{C}^2;\ \operatorname{Im} z_1 = \operatorname{Re}(z_1 + z_2) = 0\}$$

and

$$h(z) = (z_1 - z_2^2, z_2), \qquad z \in \mathbb{C}^2;$$

this example, which was mentioned in a letter to L. I. Ronkin of January 10, 1973, is closely related to that of V. Šeinov in [15, p. 148].

A special case of Theorem 2.1 is obtained by considering the order of partial functions. Let v be plurisubharmonic in $\omega \times \mathbb{C}^m$, where ω is an open set in $\mathbb{C}^n$. There are various ways to define the order $\rho(x)$, $x \in \omega$, of the partial function $y \mapsto v(x, y)$, $y \in \mathbb{C}^m$ (see, e.g., [1, 9, 12]). It turns out that $-1/\rho^*$ is plurisubharmonic in ω, where $\rho^*(x) = \limsup \rho(x')$ as $x' \to x$. If it is known that ρ^* is constant in ω (and this is of course so if $\omega = \mathbb{C}^n$), then the set of points $x \in \omega$ with $\rho(x) < \rho^*$ is polar according to a theorem of Lelong. For results of this kind, using different definitions of $\rho(x)$, see [12, Theorem 6.6.2], [1, Theorem 1.1], and [9, Theorem 4.1]. For a related result on restrictions to subspaces, see Theorem 4.6 of the present paper. Ronkin [15, Theorems 2.7.1 and 3.2.1] proves that the exceptional set satisfies condition A if $\omega = \mathbb{C}^n$ (a subset of such a set is not necessarily of the same type). However, the function $-1/\rho^*$ can be any negative plurisubharmonic function in ω [9, Theorem 4.2], so ρ^* need not be constant if ω is bounded.

We shall use methods from convex analysis as presented, e.g., in Rockafellar [14] or Ioffe and Tihomirov [5]. The key result used in the proof is the minimum principle for plurisubharmonic functions established in [7]. For general properties of such functions we refer to Lelong [12] or Hörmander [4].

The main results of this paper were presented at the 7th Conference on Analytic Functions in Kozubnik, Poland, April 18–26, 1979, and at lectures given at the Stefan Banach International Mathematical Center, Warsaw, in May, 1979.

2. Growth of Partial Functions

In this section we present and prove the main theorem. We shall consider plurisubharmonic functions of two groups of variables, $x \in \mathbb{C}^n$ and $y \in \mathbb{C}^m$, and study the growth of the partial functions $y \mapsto v(x, y)$. As to notation we shall write $\mathrm{PSH}(\Omega)$ for the class of all plurisubharmonic functions in Ω, the constant $-\infty$ being permitted. If f is a numerical function on $\mathbb{R}^m$ (i.e., with values in $[-\infty, +\infty]$), we denote its Legendre transform by $\tilde{f}$:

$$\tilde{f}(\eta) = \sup_{y \in \mathbb{R}^m} (\langle y, \eta \rangle - f(y)), \qquad \eta \in \mathbb{R}^m. \tag{2.1}$$

Theorem 2.1. *Let ω be a finite-dimensional complex analytic manifold and $v \in \mathrm{PSH}(\omega \times \mathbb{C}^m)$ a function which is independent of $\mathrm{Im}\, y$, where y is the variable in $\mathbb{C}^m$. Let ω_0 and ω_1 be subsets of ω with $\omega_1 \Subset \omega_0 \Subset \omega$; assume that ω_0 is a domain (nonempty connected open set). Define*

$$v_k(y) = \sup_{x \in \omega_k} v(x, y), \qquad y \in \mathbb{R}^m, \qquad k = 0, 1, \tag{2.2}$$

and let v_2 be a real-valued function on $\mathbb{R}^m$. Assume:

$$\text{There is no constant} \quad c > 0 \quad \text{such that} \quad c\tilde{v}_0 + \tilde{v}_2 \leq (c+1)\tilde{v}_1. \tag{2.3}$$

Then the set

$$E = \{x \in \omega_0; \text{for all } y \in \mathbb{R}^m, v(x, y) \leq v_2(y)\} \tag{2.4}$$

is polar in ω_0.

The generality of the theorem is due to the fact that the scale of growth, which is defined by v_0 and v_1, is given by the function v itself. The assumption (2.3) can be heuristically rendered as saying "v_2 is considerably smaller than v_1 on the scale of growth defined by v_0 and v_1, at least as measured by some tangents." We may reformulate it in terms of the infimal convolution. Let us define, for $0 < \varepsilon < 1$,

$$f \square_\varepsilon g(y) = \inf_{y',y''} ((1-\varepsilon)f(y') + \varepsilon g(y''); (1-\varepsilon)y' + \varepsilon y'' = y), \qquad y \in \mathbb{R}^m.$$

This is meaningful except when f assumes the value $+\infty$ and g assumes the value $-\infty$ or vice versa. If $f < +\infty$ and $g < +\infty$ everywhere, then $f \square_\varepsilon g$ as well as $(1-\varepsilon)\tilde{f} + \varepsilon\tilde{g}$ are well defined and $(f \square_\varepsilon g)^\sim = (1-\varepsilon)\tilde{f} + \varepsilon\tilde{g}$. Thus an estimate $c\tilde{v}_0 + \tilde{v}_2 \le (c+1)\tilde{v}_1$ is equivalent to $v_0 \square_\varepsilon v_2 \ge v_1$. Hence (2.3) means that v_2 is so small that the inequality $v_0 \square_\varepsilon v_2 \ge v_1$ can hold for no $\varepsilon \in]0,1[$. (Note that $-\infty \le v_1 \le v_0 < +\infty$ and $-\infty < \tilde{v}_0 \le \tilde{v}_1 \le +\infty$ by hypothesis.)

LEMMA 2.2. *Let v and v_1 be as in Theorem* 2.1 *and denote by $\tilde{v}$ the partial Legendre transform of v with respect to the y variables, i.e.*,

$$\tilde{v}(x,\eta) = \sup_{y \in \mathbb{R}^m} (\langle y,\eta\rangle - v(x,y)), \qquad (x,\eta) \in \omega \times \mathbb{R}^m, \tag{2.5}$$

and define

$$q_1(\eta) = \inf_{x \in \omega_1} \tilde{v}(x,\eta), \qquad \eta \in \mathbb{R}^m.$$

Then $\tilde{v}_1$, defined by (2.1), *is equal to $\tilde{\tilde{q}}_1$.*

Proof. That $\tilde{v}_1 \le q_1$ follows from the general rule "$\sup_y \inf_x \le \inf_x \sup_y$." Hence, taking the transformation twice, we get $\tilde{v}_1 \le \tilde{\tilde{q}}_1$. On the other hand, $q_1(\eta) \le \tilde{v}(x,\eta)$ for every $x \in \omega_1$, giving $\tilde{q}_1(y) \ge v(x,y)$ for $x \in \omega_1$, $y \in \mathbb{R}^m$. Taking the supremum over all $x \in \omega_1$ we obtain $\tilde{q}_1(y) \ge v_1(y)$, hence $\tilde{\tilde{q}}_1 \le \tilde{v}_1$. Combining with the other inequality we have finally $\tilde{\tilde{q}}_1 = \tilde{v}_1$ as claimed. ■

It is easy to give examples where q_1 is not convex, thus $q_1(\eta) > \tilde{v}_1(\eta)$ for some η. Nevertheless, it turns out that we can substitute q_1 for $\tilde{v}_1$ in the condition (2.3):

LEMMA 2.3. *With $\tilde{v}_0, \tilde{v}_1, \tilde{v}_2$ and q_1 as given in Theorem* 2.1 *and Lemma* 2.2, *condition* (2.3) *is equivalent to*:

$$\text{There is no constant } c > 0 \text{ such that } c\tilde{v}_0 + \tilde{v}_2 \le (c+1)q_1. \tag{2.6}$$

Proof. We always have $q_1 \ge \tilde{v}_1$ so (2.6) implies (2.3). On the other hand, if (2.6) is not satisfied, then $c\tilde{v}_0 + \tilde{v}_2$ is a lower semicontinuous convex minorant of $(c+1)q_1$ for some $c > 0$ and since $(c+1)\tilde{v}_1$ is the largest such minorant by the previous lemma, we must have $c\tilde{v}_0 + \tilde{v}_2 \le (c+1)\tilde{v}_1$. Hence (2.3) is violated, i.e., we have proved that (2.3) implies (2.6). ■

Proof of Theorem 2.1. We keep the notation introduced in Lemma 2.2. We first choose points $\eta^j \in \mathbb{R}^m$ such that

$$j\tilde{v}_0(\eta^j) + \tilde{v}_2(\eta^j) > (j+1)q_1(\eta^j), \qquad j = 1, 2, \ldots;$$

this is possible in view of Lemma 2.3. We always have $-\infty < \tilde{v}_0 \leq \tilde{v}_1 \leq q_1 \leq +\infty$ so $q_1(\eta^j)$ and $\tilde{v}_0(\eta^j)$ must be finite then. We may thus rewrite the inequality as

$$q_1(\eta^j) - \tilde{v}_2(\eta^j) < j(\tilde{v}_0(\eta^j) - q_1(\eta^j)) \leq 0. \tag{2.7}$$

Note also that

$$\tilde{v}_0 - \tilde{v}_2 \leq \tilde{v}_1 - \tilde{v}_2 \leq q_1 - \tilde{v}_2,$$

whenever q_1 is finite. Now (2.7) and the remark just made makes it possible to find constants $\beta_j > 0$ such that

$$\beta_j(\tilde{v}_0(\eta^j) - \tilde{v}_2(\eta^j)) \to -\infty, \tag{2.8}$$

and

$$\beta_j(\tilde{v}_0(\eta^j) - q_1(\eta^j)) \to 0, \qquad \text{as} \quad j \to +\infty. \tag{2.9}$$

We now consider

$$\varphi_j(x) = \beta_j(\tilde{v}_0(\eta^j) - \tilde{v}(x, \eta^j)) \leq 0, \qquad x \in \omega_0.$$

We note that, in view of (2.8),

$$x \in E \quad \text{implies} \quad \varphi_j(x) \to -\infty. \tag{2.10}$$

Moreover, there exist points $x^j \in \omega_1$ such that $\tilde{v}(x^j, \eta^j)$ is close to $q_1(\eta^j)$, in fact so close that, in view of (2.9),

$$\varphi_j(x^j) \to 0 \qquad \text{as} \quad j \to +\infty. \tag{2.11}$$

The crucial property of φ_j is that it is plurisubharmonic in ω_0. This is a consequence of the "minimum principle" for plurisubharmonic functions established in [7, Theorem 2.2]. For the convenience of the reader we state the special case we need:

THE MINIMUM PRINCIPLE. *Let ω be a complex analytic manifold and $v \in \mathrm{PSH}(\omega \times \mathbb{C}^m)$ a function which is independent of* $\operatorname{Im} y$, $y \in \mathbb{C}^m$. *Then*

$$V(x) = \inf_{y \in \mathbb{C}^m} v(x, y), \qquad x \in \omega,$$

is plurisubharmonic in ω.

To apply this to our present situation we only have to note that the definition (2.5) of $\tilde{v}$ can be rewritten

$$-\tilde{v}(x, \eta) = \inf_{y \in \mathbb{C}^m} (v(x, y) - \langle \operatorname{Re} y, \eta \rangle), \qquad (x, \eta) \in \omega \times \mathbb{R}^m;$$

this shows that, for fixed $\eta \in \mathbb{R}^m$, $-\tilde{v}(x, \eta)$ is the infimum of a plurisubharmonic function of $(x, y) \in \omega \times \mathbb{C}^m$ which is, moreover, independent of Im y.

Let now $\varphi = \limsup \varphi_j$ and let φ^* denote the upper semicontinuous regularization of φ. We know that $\varphi^* \leq 0$ and claim that there is a point $a \in \omega_0$ such that $\varphi(a) = 0$. This is a standard result for subharmonic functions. The sequence (x^j) must have some accumulation point $x^0 \in \bar{\omega}_1 \subset \omega_0$ and since $\varphi_j(x^j) \to 0$ by (2.11) we deduce, writing $\psi_k = \sup_{j \geq k} \varphi_j$, that $\psi_k^*(x^0) = 0$. Now ψ_k^* is plurisubharmonic so this is only possible if ψ_k^* is identically zero in the domain ω_0. But $\psi_k = \psi_k^*$ almost everywhere in ω_0 (see, e.g., Lelong [12, Theorem 3.1.2]). Thus we also have $\varphi = \inf_k \psi_k = 0$ almost everywhere in ω_0, proving our claim that there is an $a \in \omega_0$ with $\varphi(a) = 0$.

Let now (j_k) be a sequence such that $\varphi_{j_k}(a) \to 0$ and define $\Phi = \sum \gamma_j \varphi_j$ where the constants γ_j are chosen so that $\gamma_j \geq 0, \sum \gamma_j = +\infty$ but $\sum \gamma_j \varphi_j(a) = \sum \gamma_{j_k} \varphi_{j_k}(a) > -\infty$. Since

$$\Phi = \lim_{m \to +\infty} \sum_{j=0}^{m} \gamma_j \varphi_j$$

is the limit of a decreasing family of plurisubharmonic functions, we have $\Phi \in \mathrm{PSH}(\omega_0)$. Because of our choice of γ_j, this function is not identically $-\infty$, in fact $\Phi(a) > -\infty$. However, at all points $x \in E$ we have $\Phi(x) = -\infty$ since $\varphi_j(x)$ tends to $-\infty$ for these points by (2.10). This shows that the set E defined by (2.4) is polar in ω_0 and completes the proof of Theorem 2.1.

The growth condition (2.3) is best possible:

THEOREM 2.4. *Let v_0 and v_1 be two convex functions $\mathbb{R}^m \to \mathbb{R}$ such that $v_1 \leq v_0$ and let v_2 be any real-valued function on $\mathbb{R}^m$ such that the Legendre transforms $\tilde{v}_k$ as defined by* (2.1) *violate* (2.3). *Moreover, let ω_0 be a domain in $\mathbb{C}^n$ which possesses a nonconstant bounded plurisubharmonic function and ω_3 any relatively compact nonempty open subset of ω_0. Then there exists a plurisubharmonic function v in $\omega_0 \times \mathbb{C}^m$ and an open set ω_1 with $\omega_3 \subset \omega_1 \Subset \omega_0$ such that*

$$\sup_{x \in \omega_1} v(x, y) = v_1(y), \qquad y \in \mathbb{R}^m, \tag{2.12}$$

$$\sup_{x \in \omega_0} v(x, y) \leq v_0(y), \qquad y \in \mathbb{R}^m, \tag{2.13}$$

and the exceptional set E, defined by (2.4), *contains ω_3.*

Proof. By assumption there is a number $c > 0$ such that

$$c\tilde{v}_0 + \tilde{v}_2 \leq (c+1)\tilde{v}_1 \quad \text{in } \mathbb{R}^m.$$

Define v_3 as the transform of $\sup(\tilde{v}_1, \tilde{v}_2)$. Thus $v_3 \leq v_1$ and $v_3 \leq v_2$. Clearly

$$c\tilde{v}_0 + \tilde{v}_3 \leq (c+1)\tilde{v}_1 \quad \text{in } \mathbb{R}^m. \tag{2.14}$$

Now define V on $[-\infty, 1/c[\, \times \mathbb{R}^m$ by

$$V(t,\eta) = \begin{cases} \tilde{v}_3(\eta) & \text{if} \quad -\infty \leq t \leq -1, \\ (1+t)\tilde{v}_1(\eta) - t\tilde{v}_3(\eta) & \text{if} \quad -1 < t < 0, \\ \tilde{v}_1(\eta) & \text{if} \quad t = 0, \\ ct\tilde{v}_0(\eta) + (1-ct)\tilde{v}_1(\eta) & \text{if} \quad 0 < t < 1/c. \end{cases}$$

Then V is a convex function in η for t fixed $< 1/c$; it is in fact a linear combination of the three convex functions $\tilde{v}_0$, $\tilde{v}_1$, and $\tilde{v}_3$ whose coefficients are nonnegative.

Moreover V is a concave decreasing function of t for fixed η. To see this consider first a point η such that $\tilde{v}_3(\eta) < +\infty$. Then $\tilde{v}_k(\eta) < +\infty$ for $k = 0, 1, 2$ and (2.14) says that

$$0 \geq \tilde{v}_1(\eta) - \tilde{v}_3(\eta) \geq c(\tilde{v}_0(\eta) - \tilde{v}_1(\eta)),$$

which means that the slope of V decreases when t increases. If, on the other hand, $\tilde{v}_3(\eta) = +\infty$, then (2.14) implies that $\tilde{v}_1(\eta) = +\infty$ and so $V(t,\eta) = +\infty$ for every $t < 1/c$.

Now the partial transform of V with respect to η,

$$\tilde{V}(t, y) = \sup_{\eta \in \mathbb{R}^m} (\langle \operatorname{Re} y, \eta \rangle - V(t,\eta)), \qquad (t, y) \in [-\infty, 1/c[\, \times \mathbb{C}^m,$$

is convex as a function of (t, y) and increasing in t. Define $t_0 = 1/c$, $t_1 = 0$ and $t_3 = -1$. When $t < t_k$ we have $V(t,\eta) \geq \tilde{v}_k(\eta)$ so we must also have $\tilde{V}(t, y) \leq v_k(y)$, $k = 0, 1, 3$; when $k = 1$ or 3 this holds even for $t \leq t_k$.

Now, finally, let $u \in \mathrm{PSH}(\omega_0)$ be a bounded nonconstant function. This means that

$$-\infty < \sup_{\omega_3} u < \sup_{\omega_0} u < +\infty$$

and we may as well assume that $\sup_{\omega_k} = t_k$, $k = 0, 3$. The function

$$v(x, y) = \tilde{V}(u(x), y), \qquad (x, y) \in \omega_0 \times \mathbb{C}^m,$$

is plurisubharmonic and satisfies $v(x, y) \leq \tilde{V}(t_k, y) \leq v_k(y)$ for $x \in \omega_k$, $k = 0, 3$. (Note that $u \leq t_3$ in ω_3 and $u < t_0$ in ω_0.) It is obvious now how we may choose an open set ω_1 with $\omega_3 \subset \omega_1 \Subset \omega_0$ and satisfying (2.12); note that $\tilde{V}$ is continuous and that $v(x, y) = v_1(y)$ whenever $u(x) = t_1 = 0$. (If ω_3 happens to be connected, ω_1 may be taken to be a domain.) The exceptional set is

$$\begin{aligned} E &= \{x \in \omega_0;\, v(x, y) \leq v_2(y) \text{ for all } y \in \mathbb{C}^m\} \\ &\supset \{x \in \omega_0;\, v(x, y) \leq v_3(y) \text{ for all } y \in \mathbb{C}^m\}. \end{aligned}$$

Hence E contains all points where $u(x) \leq t_3 = -1$. This proves the theorem. We remark that if v_1 grows so fast that $\tilde{v}_1 < +\infty$ everywhere, then $\sup_{t<t_0} \tilde{V}(t, y) = v_0(y)$; hence we get equality in (2.13). ∎

We shall also prove that every polar set is contained in an exceptional set. Thus Theorem 2.1 cannot be improved in this respect either.

THEOREM 2.5. *Let ω be a domain in $\mathbb{C}^n$, and ω_0 and ω_1 subdomains such that $\omega_1 \Subset \omega_0 \Subset \omega$. Let a nonvoid polar set E in ω be given. Finally, let two convex real-valued functions v_0 and v_1 be given on $\mathbb{R}^m$ such that $v_1 \le v_0 - \varepsilon$ for some $\varepsilon > 0$. Then there exists a function $v \in \mathrm{PSH}(\omega_0 \times \mathbb{C}^m)$ such that*

$$\sup_{x \in \omega_1} v(x, y) = v_1(y), \qquad y \in \mathbb{R}^m, \tag{2.15}$$

$$\sup_{x \in \omega_0} v(x, y) \le v_0(y), \qquad y \in \mathbb{R}^m, \tag{2.16}$$

and

$$E \cap \omega_0 \subset \{x \in \omega_0;\ v(x, y) = -\infty \text{ for all } y \in \mathbb{C}^m\}.$$

Proof. By hypothesis there exists a $u \in \mathrm{PSH}(\omega)$ such that $u(x) = -\infty$ for all $x \in E$ and such that

$$-\infty < \sup_{\omega_1} u < \sup_{\omega_0} u < \sup_{\omega} u \le +\infty.$$

Define $t_k = \sup_{\omega_k} u$, $k = 0, 1$, and

$$V(t,\eta) = \begin{cases} -\varepsilon \dfrac{t - t_1}{t_0 - t_1} + \tilde{v}_1(\eta) & \text{if } \ -\infty \le t \le t_1, \\[2ex] \dfrac{t - t_1}{t_0 - t_1}\tilde{v}_0(\eta) + \dfrac{t_0 - t}{t_0 - t_1}\tilde{v}_1(\eta) & \text{if } \ t_1 < t < t_0. \end{cases}$$

Then V is a concave decreasing function of $t \in [-\infty, t_0[$. In fact, if $\tilde{v}_1(\eta) = +\infty$, then $V(t,\eta) = +\infty$ for all $t \in [-\infty, t_0[$; if on the other hand $\tilde{v}_1(\eta) < +\infty$ then $\tilde{v}_0(\eta) < +\infty$ also and $V(t,\eta)$ is affine on the intervals $]-\infty, t_1]$ and $[t_1, t_0[$ with slope, respectively,

$$\frac{-\varepsilon}{t_0 - t_1} \qquad \text{and} \qquad \frac{\tilde{v}_0(\eta) - \tilde{v}_1(\eta)}{t_0 - t_1};$$

we have assumed that $-\varepsilon \ge \tilde{v}_0 - \tilde{v}_1$. For all $t \in [-\infty, t_0[$ we have $V(t,\eta) \ge \tilde{v}_0(\eta)$ so we get $\tilde{V}(t, y) \le v_0(y)$. Now $\tilde{V}(t, y)$ is convex in (t, y) and increasing in t, so $v(x, y) = \tilde{V}(u(x), y)$ is plurisubharmonic in $\omega_0 \times \mathbb{C}^m$ and satisfies (2.16). The estimate $V(t,\eta) \ge \tilde{v}_1(\eta)$ for $t \le t_1$ shows that $v(x, y) \le v_1(y)$ when $x \in \omega_1$ but since $V(t_1,\eta) = \tilde{v}_1(\eta)$ and $\tilde{V}$ is continuous in t we also see that equality holds in (2.15). To complete the proof we only have to note that $u(x) = -\infty$ for all $x \in E$ and that $\tilde{V}(-\infty, y) = -\infty$ by our construction. If $\tilde{v}_1$ happens to be finite everywhere, then $V(t,\eta)$ will tend to $\tilde{v}_0(\eta)$ as $t \to t_0$, $t < t_0$, so we will have equality in (2.16). ∎

3. FUNCTIONS OF SLOWLY VARYING GROWTH

If the function v_2 in Theorem 2.1 is very small in comparison with v_0 it may happen that condition (2.3) is satisfied regardless of ω_1. The next theorem describes a result of this kind. Then, in Theorem 3.2, we shall impose a condition on v which simplifies (2.3).

THEOREM 3.1. *Let ω, ω_0, v and v_0 be as in Theorem 2.1 and assume that a real-valued function v_2 on $\mathbb{R}^m$ is very small in comparison with v_0 in the sense that, for every ε with $0 < \varepsilon \leq 1$ and every number C, there is a point $y \in \mathbb{R}^m$ such that*

$$v_2(y) < \varepsilon^{-1} v_0(\varepsilon y) - C. \tag{3.1}$$

Then (2.3) holds for any domain $\omega_1 \Subset \omega_0$ and, consequently, the set E defined by (2.4) is polar in ω_0.

Proof. We may assume ω to be connected. To any two relatively compact subdomains ω_0 and ω_1 of ω there exist constants $a \geq 1$ and C such that

$$v_0(y) \leq a^{-1} v_1(ay) + C, \qquad y \in \mathbb{R}^m. \tag{3.2}$$

This result is essentially due to Lelong [12, Theorem 6.5.4]. See also the proof in Skoda [18, p. 105]. However, the inequality is proved there only for $m = 1$ and, more importantly, the factor a^{-1} on the right which is essential for us is omitted. The proof of the more precise inequality (3.2) is of course the same and depends on the fact that plurisubharmonic functions which are independent of the imaginary part of the variables are convex. To prove (3.2), let us assume that ω is an open neighborhood of the origin in $\mathbb{C}^n$ and define

$$w(z, y) = \sup_{|x| \leq |e^z|} v(x, y), \qquad (z, y) \in \Omega \times \mathbb{C}^m,$$

where Ω is a half-plane in $\mathbb{C}$. Since w is convex we have

$$w(z, y) \leq (1 - \lambda) w(z^0, y^0) + \lambda w(z^1, y^1)$$

with $z = (1 - \lambda) z^0 + \lambda z^1$ and similarly for y, $\lambda \in [0, 1]$. Write $a = 1/(1 - \lambda)$ and choose $y^1 = 0$ and z^1 so that ω contains the closed ball of radius $\exp|z^1|$; thus

$$w(z, y) \leq a^{-1} w(z^0, ay) + \lambda w(z^1, 0) = a^{-1} w(z^0, ay) + C.$$

Now this inequality means precisely that (3.2) holds with ω_0 as the ball of radius $|\exp z|$ and ω_1 as the ball of radius $|\exp z^0|$. The general case follows easily by induction.

Now (3.2) is equivalent to

$$\tilde{v}_0(\eta) \geq a^{-1}\tilde{v}_1(\eta) - C,$$

so if an estimate $c\tilde{v}_0 + \tilde{v}_2 \leq (c+1)\tilde{v}_1$ holds, we must have

$$c\tilde{v}_0 + \tilde{v}_2 \leq (c+1)\tilde{v}_1 \leq (c+1)a\tilde{v}_0 + (c+1)aC.$$

From this we deduce, by subtracting $c\tilde{v}_0$ at points where this function is real valued, that

$$\tilde{v}_2 \leq (ca + a - c)\tilde{v}_0 + (c+1)aC.$$

Taking the Legendre transform and putting $\varepsilon = (ca + a - c)^{-1}$ we get $0 < \varepsilon \leq 1$ and

$$v_2(y) \geq \tilde{\tilde{v}}_2(y) \geq \varepsilon^{-1} v_0(\varepsilon y) - (c+1)aC, \qquad y \in \mathbb{R}^m,$$

which contradicts the hypothesis and thus shows that (2.3) holds. ■

Let us agree to say that a function v defined on $\omega \times \mathbb{C}$ is *of slowly varying growth* if for any two domains $\omega_1 \subset \omega_0 \Subset \omega$ there is a constant b such that for all $t \in \mathbb{R}$,

$$\sup_{x \in \omega_0} \sup_{\operatorname{Re} y \leq t} v(x, y) \leq \sup_{x \in \omega_1} \sup_{\operatorname{Re} y \leq t + b} v(x, y).$$

This imposes no restriction on the growth of any single partial function $y \mapsto v(x, y)$, but implies that these functions do not change their growth too drastically from one domain to another. If w is defined on a vector bundle Γ over ω we may say it is of slowly varying growth if

$$v(x, y) = \sup_{z \in \Gamma_x,\ |z| \leq e^{\operatorname{Re} y}} w(x, z), \qquad (x, y) \in \omega \times \mathbb{C},$$

is of slowly varying growth in the sense already defined for some (or any) norm $|z|$ on the fibers Γ_x which depends continuously on x. We shall see in the next section that such functions appear naturally when we consider restrictions to subspaces.

For functions of slowly varying growth the condition (3.1) on v_2 can be very much relaxed:

THEOREM 3.2. *Let v be plurisubharmonic and of slowly varying growth on $\omega \times \mathbb{C}$ where ω is a finite-dimensional complex analytic manifold. Let ω_0 be a relatively compact subdomain of ω and define v_0 by* (2.2) *($m = 1$). Let v_2 be a given real-valued function on the real line which is small in comparison with v_0 in the sense that*

$$\textit{for every} \quad a \in \mathbb{R} \quad \textit{there exists} \quad y \in \mathbb{R} \quad \textit{such that} \quad v_2(y) < v_0(y - a). \tag{3.3}$$

Then (2.3) *holds for any domain $\omega_1 \Subset \omega_0$ and the set E defined by* (2.4) *is polar in ω_0.*

Proof. Let ω_1 be any relatively compact subdomain of ω_0 and define v_1 by (2.2). By hypothesis we have

$$v_0(y) \le v_1(y+b), \quad y \in \mathbb{R}^m,$$

and passing to the Legendre transforms we get

$$\tilde{v}_0(\eta) \ge \tilde{v}_1(\eta) - b\eta.$$

Hence an estimate $c\tilde{v}_0 + \tilde{v}_2 \le (c+1)\tilde{v}_1$ would imply

$$c\tilde{v}_0(\eta) + \tilde{v}_2(\eta) \le (c+1)\tilde{v}_0(\eta) + b(c+1)\eta,$$

from which we deduce, using that $-\infty < \tilde{v}_0 \le +\infty$,

$$\tilde{v}_2(\eta) \le \tilde{v}_0(\eta) + b(c+1)\eta,$$

and, taking the transform again,

$$v_2(y) \ge \tilde{\tilde{v}}_2(y) \ge v_0(y - b(c+1)),$$

which contradicts our hypothesis (3.3). Hence (2.3) must be true and we can apply Theorem 2.1. ■

4. Growth of Restrictions to Linear Subspaces

In this section we shall study the growth of restrictions of plurisubharmonic functions to linear subspaces of $\mathbb{C}^n$ and compare their growth with the overall growth. The exceptional set of subspaces where growth is considerably slower than usual now cannot be polar, of course, since the set of all subspaces forms a compact manifold. However, it turns out that the exceptional set is locally polar.

Theorem 4.1. *Let f be plurisubharmonic in $\mathbb{C}^n$ and let us measure its global growth by*

$$u_0(t) = \sup_{|x| \le e^t} f(x), \qquad t \in \mathbb{R}. \tag{4.1}$$

Let u_2 be a real-valued function on $\mathbb{R}$ which is considerably smaller than u_0 in the sense that

$$\text{for every} \quad a \in \mathbb{R} \quad \text{there exists} \quad t \in \mathbb{R} \quad \text{such that} \quad u_2(t) < u_0(t-a). \tag{4.2}$$

Denote by G_m the Grassmannian manifold of all m-dimensional subspaces of $\mathbb{C}^n$ for some m, $1 \le m \le n-1$. The set

$$E = \{X \in G_m;\, f(x) \le u_2(\log|x|) \text{ for all } x \in X\} \tag{4.3}$$

is then locally polar in G_m.

Gauthier and Hengartner [2] have proved a similar result with the exceptional set of Lebesgue measure zero, the scale of growth being the classical refined orders. Their proof relies on an earlier theorem of Hengartner [3] for the case $m = 1$ in which the exceptional set E is proved to be negligible.

In this paper the main interest is the growth of plurisubharmonic functions at infinity. Let us note, however, that Theorem 4.1 has nontrivial consequences also for the local behavior. For instance, we may take u_2 so that

$$\lim_{t\to-\infty}\frac{u_2(t)}{t} > \lim_{t\to-\infty}\frac{u_0(t)}{t},$$

and Theorem 4.1 then expresses the fact that $f|_X$ has density at $0 \in \mathbb{C}^n$ exceeding that of f only for a locally polar set of subspaces $X \in G_m$. In the case $m = 1$ this result is due to Lelong [13, p. 66]; for a discussion of the density, see also [8].

Proof of Theorem 4.1. Let Γ denote the canonical vector bundle over G_m, i.e., the set of all pairs (X, x) with $x \in X \in G_m$, with the usual vector bundle structure; the projection $\pi:\Gamma \to G_m$ being $\pi(X, x) = X$. (We may regard Γ as a subbundle of the trivial bundle $G_m \times \mathbb{C}^n$.) Given any element $X_0 \in G_m$ there exists an open neighborhood ω_0 of X_0 in G_m such that $\pi^{-1}(\omega_0)$ is isomorphic to $\omega_0 \times \mathbb{C}^m$; this means that there is a biholomorphic map

$$\omega_0 \times \mathbb{C}^m \to \pi^{-1}(\omega_0), \qquad (X, z) \mapsto (X, \varphi(X, z)),$$

where φ is holomorphic and defines a vector space isomorphism

$$\mathbb{C}^m \to X, \qquad z \mapsto \varphi(X, z),$$

for every fixed $X \in \omega_0$. By shrinking ω_0 if necessary we may assume that the norm $|z|_m$ of $z \in \mathbb{C}^m$ is equivalent to the norm $|\varphi(X, z)|_n$ in X induced by $\mathbb{C}^n$, uniformly for all $X \in \omega_0$, i.e.,

$$\alpha|z|_m \le |\varphi(X, z)|_n \le \beta|z|_m, \qquad (X, z) \in \omega_0 \times \mathbb{C}^m, \tag{4.4}$$

for some constants α and β, $0 < \alpha \le \beta < +\infty$.

The function whose growth we would like to study is

$$u(X, t) = \sup_{|x| \le |e^t|,\, x \in X} f(x), \qquad (X, t) \in G_m \times \mathbb{C};$$

note that it need not be plurisubharmonic on $G_m \times \mathbb{C}$. We therefore introduce

$$v(X, t) = \sup_{|z| \le |e^t|} g(X, z), \qquad (X, t) \in \omega_0 \times \mathbb{C}, \tag{4.5}$$

where

$$g(X, z) = f(\varphi(X, z)), \qquad (X, z) \in \omega_0 \times \mathbb{C}^m.$$

Since $\varphi:\omega_0 \times \mathbb{C}^m \to \mathbb{C}^n$ is holomorphic, g and v are plurisubharmonic. Now (4.4) implies, writing $a = \log\alpha$ and $b = \log\beta$ for convenience, that

$$u(X, t + a) \le v(X,t) \le u(X, t + b), \qquad (X,t) \in \omega_0 \times \mathbb{C}, \tag{4.6}$$

for each term is the supremum of $f(\varphi(X,z))$ when z varies in the balls, respectively,

$$|\varphi(X,z)|_n \le \alpha|e^t|, \qquad |z|_m \le |e^t|, \qquad |\varphi(X,z)|_n \le \beta|e^t|.$$

This means that u and v are comparable when it comes to an application of Theorem 3.2. Specifically, put

$$v_0(t) = \sup_{X \in \omega_0} v(X,t), \qquad t \in \mathbb{R},$$

and

$$v_2(t) = u_2(t + b), \qquad t \in \mathbb{R},$$

so that

$$\omega_0 \cap E = \{X \in \omega_0\,;\, f(x) \le u_2(\log|x|),\, x \in X\}$$
$$\subset E' = \{X \in \omega_0;\, v(X,t) \le v_2(t),\, t \in \mathbb{R}\}.$$

To prove that E is locally polar it suffices to prove that E' is polar and to do so we shall apply Theorem 3.2. We shall first check that v is of slowly varying growth. This is a simple geometric fact which we formulate as a separate lemma:

LEMMA 4.2. *The function v, defined by* (4.5), *is of slowly varying growth on* $\omega_0 \times \mathbb{C}$.

Proof. Let ω_1 be any subdomain of ω_0 and take $X_1 \in \omega_1$. The union Ω of all $X \in \omega_1$ is a neighborhood of $X_1 \cap S$, where S denotes the unit sphere in $\mathbb{C}^n$. By compactness there must exist an $\varepsilon > 0$ such that $\varepsilon B + X_1 \cap S \subset \Omega$, where we have written B for the unit ball. By homogeneity,

$$e^t B + X_1 \cap e^t\varepsilon^{-1} S \subset \Omega \cap e^t(1 + 1/\varepsilon)B.$$

Now, by the maximum principle, $f(x)$ is majorized by the supremum of f over $x + X_1 \cap e^t\varepsilon^{-1}S$, so if $|x| \le e^t$ we obtain the following estimate, using (4.6):

$$f(x) \le \sup_{y \in \Omega \cap e^t(1+1/\varepsilon)B} f(y) \le \sup_{X \in \omega_1} u(X, t + \log(1 + 1/\varepsilon))$$
$$\le \sup_{X \in \omega_1} v(X, t + \log(1 + 1/\varepsilon) - a). \tag{4.7}$$

Since this holds for all $x \in \mathbb{C}^n$ with $|x| \le e^t$ we finally see, using (4.6) again, that

$$\sup_{X \in \omega_0} v(X,t) = v_0(t) \le u_0(t + b) \le \sup_{X \in \omega_1} v(X, t + \log(1 + 1/\varepsilon) - a + b),$$

which shows that v is of slowly varying growth. This completes the proof of the lemma.

Next we shall check the condition (3.3) on v_2 and v_0; it turns out that it is equivalent to the condition (4.2) on u_2 and u_0. Indeed, we have

$$u_0(t + a - \log(1 + 1/\varepsilon)) \leq v_0(t) \leq u_0(t + b),$$

where the left inequality is obtained from (4.7) and the right one from (4.6). Thus everything is adapted for an application of Theorem 3.2, and Theorem 4.1 is finally proved. ∎

We should remark that the proof of Theorem 4.1 is considerably simpler in the case $m = 1$, i.e., for the growth of f along complex lines, G_m being the $(n - 1)$-dimensional projective space.

We shall now see that the condition on u_2 and u_0 in Theorem 4.1 is sharp.

THEOREM 4.3. *Let u_0 and u_2 be two real-valued functions on $\mathbb{R}$ which violate condition* (4.2). *Assume that u_0 is convex and increasing. Then there exists $f \in \mathrm{PSH}(\mathbb{C}^n)$ such that* (4.1) *holds and the set E is a neighborhood of an arbitrary element X_0 of G_m, $1 \leq m \leq n - 1$.*

Proof. Put $f(x) = u_0(\log d(x, X_0))$, $x \in \mathbb{C}^n$, where

$$d(x, X_0) = \inf_{y \in X_0} |x - y|$$

is the distance between x and the subspace X_0. Since d is a seminorm, $f \in \mathrm{PSH}(\mathbb{C}^n)$. We have assumed that $\dim X_0 < n$ so (4.1) holds.

The set ω_ε of all $X \in G_m$ such that

$$\sup_{x \in X,\, |x| = 1} d(x, X_0) \leq \varepsilon$$

is a neighborhood of X_0 for every $\varepsilon > 0$. We claim that E contains ω_ε if ε is small enough. Indeed, if (4.2) is not satisfied there is a number a such that

$$u_0(t - a) \leq u_2(t), \qquad t \in \mathbb{R},$$

and this implies that, for any $x \in X \in \omega_\varepsilon$,

$$f(x) \leq u_0(\log|x| + \log\varepsilon) \leq u_2(\log|x| + \log\varepsilon + a);$$

hence E contains ω_ε if $\log\varepsilon + a \leq 0$. ∎

We now ask, in analogy with Theorem 2.5, whether every locally polar set in G_m can occur as an exceptional set in Theorem 4.1. It is easy to see that the answer must be in the negative if $2 \leq m \leq n - 1$. In fact, if E is defined by (4.3), then the set

$$E_1 = \{Y \in G_1;\, Y \subset X \in E \text{ for some } X \in G_m\} \tag{4.8}$$

must be locally polar in G_1, provided u_0 and u_2 are as in Theorem 4.1. And this condition is not fulfilled by every locally polar set in G_m if $m \geq 2$. Indeed, take E as the set of all m planes containing a fixed nonzero vector in $\mathbb{C}^n$. Then E is locally polar if $1 \leq m \leq n - 1$. But the corresponding set E_1 is all of G_1 if $m \geq 2$. It turns out that the local polarity of E_1 is both necessary and sufficient:

THEOREM 4.4. *If a set $E \subset G_m$ can occur as an exceptional set in Theorem* 4.1, *then the set E_1 as defined by* (4.8) *is locally polar in G_1* $(1 \leq m \leq n-1)$. *Conversely, let $E \subset G_m$ be such that E_1 is locally polar in G_1, and let $u_0: \mathbb{R} \to \mathbb{R}$ be convex and increasing. Then there exists a function $f \in \mathrm{PSH}(\mathbb{C}^n)$ such that*

$$\sup_{|x| \leq e^t} f(x) = u_0(t), \qquad t \in \mathbb{R},$$

and

$$E \subset \{X \in G_m;\, f(x) \leq u_0(-\infty) \text{ for all } x \in X\}.$$

Thus E is contained in the exceptional set defined by u_2 if u_0 is not constant and we take u_2 as any real constant $\geq u_0(-\infty) \geq -\infty$.

Proof. The first part was already proved by the remarks preceding the statement of the theorem.

To prove the sufficiency of the condition, assume that E is given in G_m such that E_1 is locally polar in G_1 and take a function $\varphi \in \mathrm{PSH}(\mathbb{C}^n)$ such that $\varphi(sx) = \log|s| + \varphi(x)$ for every $(s, x) \in \mathbb{C} \times \mathbb{C}^n$, and $\varphi(x) = -\infty$ for every nonzero $x \in \mathbb{C}^n$ such that $\mathbb{C}x \in E_1$, but φ is not $-\infty$ identically. The existence of such a function is guaranteed by the next lemma. Define

$$f(x) = u_0\Big(\varphi(x) - \sup_{|y| \leq 1} \varphi(y)\Big),$$

where, of course, we have to agree to set $u_0(-\infty) = \lim_{t \to -\infty} u_0(t)$. Then f is plurisubharmonic in $\mathbb{C}^n$ and

$$\sup_{|x| \leq e^t} f(x) = u_0\Big(\sup_{|x| \leq e^t} \varphi(x) - \sup_{|y| \leq 1} \varphi(y)\Big) = u_0(t);$$

moreover $f(x) = u_0(-\infty)$ if $x \in X \in E$ for some $X \in G_m$. ■

The proof of Theorem 4.4 will be complete when we have established the following result:

LEMMA 4.5. *If E is locally polar in G_1, the projective space of dimension $n-1$, then there exists $\varphi \in \mathrm{PSH}(\mathbb{C}^n)$, not identically $-\infty$, such that $\varphi(x) = -\infty$ when the complex line defined by $x \in \mathbb{C}^n \setminus \{0\}$ belongs to E and with the homogeneity property $\varphi(sx) = \log|s| + \varphi(x)$, $(s, x) \in \mathbb{C} \times \mathbb{C}^n$.*

Proof. Let E' be a bounded, locally polar set in $\mathbb{C}^{n-1}$. According to Josefson's theorem [6], E' is globally polar, and by Siciak's theorem [16, Theorem 3.10, p. 125], the function taking $-\infty$ at the points of E' can be taken to grow at most like $\log|x|$, say,

$$\psi(x) \leq \log^+|x|, \qquad x \in \mathbb{C}^{n-1},$$

for some $\psi \in \mathrm{PSH}(\mathbb{C}^{n-1})$, not identically $-\infty$. Define $\varphi(x) = \log|x_n| + \psi(x_1/x_n, \ldots, x_{n-1}/x_n)$ when $x \in \mathbb{C}^n$, $x_n \neq 0$. Then φ is locally bounded above near every point in the hyperplane $x_n = 0$ so by a classical theorem, φ can be extended to all of $\mathbb{C}^n$ (see Lelong [11, Theorem 4, p. 35]). Clearly $\varphi(x) = -\infty$ if $x_n \neq 0$ and $(x_1/x_n, \ldots, x_{n-1}/x_n) \in E'$, i.e., $\mathbb{C}x \in E \subset G_1$ where E is the subset of G_1 defined by $E' \times \{1\} \subset \mathbb{C}^n$.

Now any locally polar set E in G_1 can be represented as the union of n sets similar to the one we have considered. Specifically, we may define E^k as the set of $\mathbb{C}x \in E$ such that $|x_k| = \sup_j |x_j|$, and then E^n is of the type already considered. For each E^k we get a plurisubharmonic function φ^k; the mean value φ of $\varphi^1, \ldots, \varphi^n$ has all desired properties. ■

We shall finally give a form of Theorem 4.1 which uses the language of relative order. If $h: \mathbb{R} \to \mathbb{R}$ is a given function, let us consider numbers α, $0 < \alpha < +\infty$, such that $f(x) \leq \alpha^{-1} h(\alpha \log|x|)$ when $|x|$ is large enough. The infimum ρ, $0 \leq \rho \leq +\infty$, of all such numbers α will be called the *h-order of* f. (For a discussion of this variant of the classical definition, see [9].)

THEOREM 4.6. *Let $f \in \mathrm{PSH}(\mathbb{C}^n)$ be of finite h-order ρ for some convex increasing function $h: \mathbb{R} \to \mathbb{R}$, which grows faster than any linear function. Then the set of m planes through the origin on which f is of h-order less than ρ is locally polar in G_m, provided $1 \leq m \leq n-1$.*

Proof. Let α_k be a strictly increasing sequence of positive numbers which tends to ρ. For every $X \in G_m$ such that $f\,|\,X$ is of h-order less than ρ there is a number k such that

$$f(x) \leq \sup(h_{\alpha_k}(\log|x|), k), \qquad x \in X,$$

where we have defined $h_\alpha(t) = \alpha^{-1} h(\alpha t)$. Hence it is enough to prove that the exceptional set E, defined as in Theorem 4.1, is locally polar when we choose

$$u_2(t) = \sup(h_{\alpha_k}(t), k), \qquad t \in \mathbb{R},$$

the set of m planes where the h-order is less than ρ being a denumerable union of these. To verify the hypotheses of Theorem 4.1 we only have to prove that there is no estimate $u_0(t) \leq u_2(t + a)$, $t \in \mathbb{R}$. We know that f is not of h-order $\leq \alpha_{k+1}$; hence there exists a sequence (t_j) of numbers tending to $+\infty$ such that

$$u_0(t_j) > h_{\alpha_{k+1}}(t_j).$$

By the following lemma, $h_{\alpha_{k+1}}(t) \geq h_{\alpha_k}(t+a)$ holds for all large t, a being fixed. Thus $u_0(t_j) > u_2(t_j + a)$ for large j, allowing an application of Theorem 4.1. ■

LEMMA 4.7. *With h as in Theorem 4.6, fix real numbers α, β, and a such that $0 < \alpha < \beta$. Then $h_\beta(t) \geq h_\alpha(t+a)$ for all large t.*

Proof. The desired conclusion is equivalent to

$$\tilde{h}_\beta(\tau) \leq \tilde{h}_\alpha(\tau) - a\tau \qquad \text{for } \tau \text{ large enough.}$$

Now $\tilde{h}_\alpha(\tau) = \alpha^{-1}\tilde{h}(\tau)$ so this, in turn, is equivalent to:

$$(\alpha^{-1} - \beta^{-1})\tilde{h}(\tau) \geq a\tau \qquad \text{for } \tau \text{ large.} \tag{4.9}$$

Since $\tilde{h}$ is a convex function which increases faster than any linear function, h being real valued, (4.9) holds. ■

5. GROWTH OF RESTRICTIONS TO AFFINE SUBSPACES

In this final section we shall consider restrictions to affine subspaces. The reasoning closely parallels that of Section 4, so we shall be rather brief. As in that section, G_m stands for the manifold of all m-dimensional vector subspaces of $\mathbb{C}^n$.

THEOREM 5.1. *Let $f \in \mathrm{PSH}(\mathbb{C}^n)$ and define*

$$u_0(t) = \sup_{|x| \leq e^t} f(x).$$

Assume that $u_2 : \mathbb{R} \to \mathbb{R}$ grows considerably slower than u_0 in the sense that

$$\textit{for every } a \in \mathbb{R} \textit{ there is a } t \in \mathbb{R} \textit{ such that } \sup(a, u_2(t)) < u_0(t-a). \tag{5.1}$$

Then the set

$$E = \{(x, X) \in \mathbb{C}^n \times G_m;\, f(x+y) \leq u_2(\log|y|) \textit{ for all } y \in X\}$$

is locally polar in $\mathbb{C}^n \times G_m$, $1 \leq m \leq n-1$.

Sketch of proof. Let us introduce the function

$$u(x, X, t) = \sup_y (f(x+y);\, y \in X \textit{ and } |y| \leq e^{\mathrm{Re}\, t}), \qquad (x, X, t) \in \mathbb{C}^n \times G_m \times \mathbb{C}.$$

As can be seen from simple examples, u is not necessarily of slowly varying growth in $(\mathbb{C}^n \times G_m) \times \mathbb{C}$. However, trouble appears only when $\mathrm{Re}\, t$ is small compared with $|x|$. Thus, let us fix a bounded open set Ω in $\mathbb{C}^n$ and let ω_1 be any relatively compact domain in $\Omega \times G_m$. In complete analogy with Lemma 4.2 it is seen that for some real number b,

$$u(x, X, t) \leq \sup_{(x', X') \in \omega_1} u(x', X', b+t) \qquad \text{for all} \quad t \geq 0 \quad \text{and all} \quad (x, X) \in \Omega \times G_m.$$

Let $a = \sup_{x \in \Omega, \, |y| \leq 1} f(x + y)$. Then $u(x, X, t) \leq a$ when $(x, X) \in \Omega \times G_m$, $t < 0$, so that

$$\sup(a, u(x, X, t)) \leq \sup_{(x', X') \in \omega_1} \sup(a, u(x', X', b + t))$$
$$\text{for all} \quad (x, X, t) \in \Omega \times G_m \times \mathbb{C}.$$

Note that b depends on ω_1 and Ω but that a is independent of ω_1. This shows that $\sup(a, u)$ is of slowly varying growth in $(\Omega \times G_m) \times \mathbb{C}$. We may thus apply Theorem 3.2 to $\sup(a, v)$, where v is obtained from u by a local trivialization as in the proof of Theorem 4.1; we have strengthened the hypothesis (5.1) accordingly. We conclude that the set of all $(x, X) \in E$ such that $x \in \Omega$ is locally polar in $\Omega \times G_m$; since Ω can be chosen to contain any prescribed point in $\mathbb{C}^n$ we are done. ■

Let us now consider the manifold A_m of all m-dimensional affine subspaces of $\mathbb{C}^n$. Every $Y \in A_m$ has a representation $Y = x + X$ for some $(x, X) \in \mathbb{C}^n \times G_m$. Let us say that f *grows at most like* u_2 *in* Y if there is a constant k such that

$$f(x + y) \leq u_2(k + \log^+|y|) \qquad \text{for all} \quad y \in X; \tag{5.2}$$

here $(x, X) \in \mathbb{C}^n \times G_m$ are chosen so that $x + X = Y$. If $x' + X$ is an other representation of Y we deduce from (5.2) that

$$f(x' + y) \leq u_2(k + \log^+|x' - x + y|) \leq u_2(k' + \log^+|y|)$$

for some other constant k' which depends on k, x, and x'. Thus the definition does not depend on the choice of representation for the affine subspace Y.

THEOREM 5.2. *Let f, u_0, and u_2 be as in Theorem* 5.1. *Then the set of all m-dimensional affine subspaces Y such that f grows at most like u_2 in Y is locally polar in A_m.*

Proof. Define

$$E_k = \{(x, X) \in \mathbb{C}^n \times G_m; f(x + y) \leq u_2(k + \log^+|y|) \text{ for all } y \in X\}.$$

The function $v_2(t) = u_2(k + t^+)$ satisfies (5.1). Thus an application of Theorem 5.1 yields that E_k is locally polar. Taking the union over all k we obtain that

$$E = \{(x, X) \in \mathbb{C}^n \times G_m; f \text{ grows at most like } u_2 \text{ in } x + X\}$$

is locally polar. What about

$$E' = \{Y \in A_m;\, f \text{ grows at most like } u_2 \text{ in } Y\}?$$

Let $\varphi:\mathbb{C}^n \times G_m \to A_m$ denote the map $(x, X) \mapsto x + X$. We shall prove that, locally, we have enough sections of φ: If $Y_0 = x_0 + X_0$ is an element of A_m and g a plurisubharmonic function defined in a connected neighborhood of (x_0, X_0) in $\mathbb{C}^n \times G_m$ which is $-\infty$ on E without being $-\infty$ identically, then we shall find a holomorphic section ψ of φ such that $g \circ \psi$ is not identically $-\infty$ (that ψ is a section implies that $g \circ \psi$ is $-\infty$ on E'). To construct ψ take a point (x'_0, X'_0) close to (x_0, X_0) where g is finite; it is possible to find (x'_0, X'_0) such that $x'_0 - x_0 \notin X_0$. Hence there exists a supplement $X_1 \in G_{n-m}$ to X_0 containing $x'_0 - x_0$. Write $Y_1 = x_0 + X_1 \in A_{n-m}$ and consider the restriction $\varphi|_{Y_1 \times G_m}$; its inverse ψ maps a neighborhood of Y_0 onto a neighborhood of (x_0, X_0) relative to $Y_1 \times G_m$. Now by our construction $x'_0 \in Y_1$ so $g \circ \psi(x'_0 + X'_0) = g(x'_0, X'_0) > -\infty$. This shows that E' is locally polar in A_m.

Acknowledgment

I am grateful to Mikael Ragstedt for valuable comments to the manuscript.

References

1. S. Ju. Favorov, O funkcijah klassa B i ih primenenii v teorii meromorfnyh funkciĭ mnogih peremennyh, *Teor. Funkciĭ Funkcional. Anal. i Priložen.* **20** (1974), 150–160.
2. P. M. Gauthier and W. Hengartner, Traces des fonctions méromorphes de plusieurs variables complexes, *in* "Rencontre sur l'analyse complexe à plusieurs variables et les systèmes surdéterminés," Presses Univ. de Montréal, 1975.
3. W. Hengartner, Famille des traces sur les droites complexes d'une fonction plurisousharmonique ou entière dans $\mathbb{C}^n$, *Comment. Math. Helv.* **43** (1968), 358–377.
4. L. Hörmander, "An Introduction to Complex Analysis in Several Variables," North-Holland Publ., Amsterdam, 1973.
5. A. D. Ioffe and V. M. Tihomirov, "Theory of Extremal Problems," North-Holland Publ., Amsterdam, 1979.
6. B. Josefson, On the equivalence between locally polar and globally polar sets for plurisubharmonic functions in $\mathbb{C}^n$, *Ark. Mat.* **16** (1978), 109–115.
7. C. O. Kiselman, The partial Legendre transformation for plurisubharmonic functions, *Invent. Math.* **49** (1978), 137–148.
8. C. O. Kiselman, Densité des fonctions plurisousharmoniques, *Bull. Soc. Math. France* **107** (1979), 295–304.
9. C. O. Kiselman, The use of conjugate convex functions in complex analysis, Banach Center Publication Series, Warsaw, to appear.
10. P. Lelong, Sur quelques problèmes de la théorie des fonctions de deux variables complexes, *Ann. Sci. École Norm. Sup.* **58** (1941), 83–176.

11. P. Lelong, "Fonctions plurisousharmoniques et formes différentielles positives," Gordon & Breach, New York, 1968.
12. P. Lelong, "Fonctionnelles analytiques et fonctions entières (n variables)," Cours d'été 1967, Presses Univ. de Montréal, Montréal, 1968.
13. P. Lelong, Plurisubharmonic functions in topological vector spaces: polar sets and problems of measure, *in* "Proceeding on Infinite Dimensional Holomorphy," Lecture Notes in Mathematics, No. 364, pp. 58–68, Springer-Verlag, Berlin and New York, 1974.
14. R. T. Rockafellar, "Convex Analysis," Princeton Univ. Press, Princeton, New Jersey, 1970.
15. L. I. Ronkin, "Vvedenie v teoriju celyh funkciĭ mnogih peremennyh," Nauka, Moscow, 1971.
16. J. Siciak, Extremal plurisubharmonic functions in $\mathbb{C}^n$, *Proc. 1st Finnish–Polish Summer School Complex Anal., Podlesice, 1977*, pp. 115–152, Univ. of Łódź, 1978.
17. J. Sire, Sur les fonctions entières de deux variables d'ordre apparent total fini. *Rend. Circ. Mat. Palermo* **31** (1911), 1–91.
18. H. Skoda, Fibrés holomorphes à base et à fibre de Stein, *Invent. Math.* **43** (1977), 97–107.

MATHEMATICAL ANALYSIS AND APPLICATIONS, PART B
ADVANCES IN MATHEMATICS SUPPLEMENTARY STUDIES, VOL. 7B

Propagation des singularités des solutions d'équations aux derivées partielles non linéaires

B. LASCAR[†]

Centre de Mathématiques
École Polytechnique
Palaiseau, France

EN HOMMAGE À LAURENT SCHWARTZ

Cet article est consacré à la propagation des singularités pour des équations aux dérivées partielles non linéaires; on y trouve les résultats annoncés dans [8]. Ce travail est constitué de deux parties.

La première construit une solution singulière du problème de Cauchy relatif à l'équation des ondes non linéaires, qui contredit le type de propagation classique (c'est-à-dire comme la partie linéaire) en dimension d'espace $n - 1 \geq 3$. Reed avait montré dans [11] que pour $n = 2$ la propagation des singularités se passe comme dans le cas linéaire. Bony a prouvé [1] que les singularités d'ordre H^s, s assez petit devant la régularité des données se propagent de la façon classique. Citons également les travaux de Rauch [10]. Notre exemple analyse le mécanisme de la propagation due aux termes non linéaires.

La seconde partie est consacrée à l'étude de la régularité microlocale d'équations aux dérivées partielles quasi linéaires elliptiques. On montre qu'il y a hypoellipticité microlocale dans le cone convexe propre (s'il existe) engendré par les singularités du second membre et que ce résultat est dans le cas le meilleur possible.

ENONCES DES RESULTATS

THÉORÈME 1. *Si $n \geq 4$, on peut construire $u \in H^{m_0}(\mathbb{R}^n)$ solution du problème de Cauchy dans ω où $\square u - u^2 \in C^\infty$, $(D_0 u, u)|_{x_0=0} = (u_0, u_1)$, telle que $m_0 > n/2$,*[‡]

$$\mathrm{WF}(u_0, u_1) = \{(-t_0 e_1, \xi_1 e_1), \xi_1 > 0\} \cup \{(-t_0 e_2, \xi_2 e_2), \xi_2 > 0\} \cup \{t_0 e_3, \xi_3 e_3), \xi_3 > 0\},$$

c'est à dire trois points sur la sphère cotangente $[(e_1, e_2, \ldots, e_n)$ base canonique de $\mathbb{R}^n]$; et telle que

$$\mathrm{WF}(u) \supset S_0 \cup C_+(S_0 \cap \operatorname{car} \square)$$

[†] Laboratoire associé au C.N.R.S. n° 169.

[‡] Voir [3] pour la dèfinition de WF(u).

ISBN 0-12-512802-9

où

$$S_0 = \left\{\left(x_0 = t_0, x' = 0, \xi' = \sum_{i=1}^{3} \xi_i e_i, \xi_0 = \sum_{i=1}^{3} \xi_i \varepsilon_i\right), \xi_i > 0\right\}$$

$$\varepsilon_1 = \varepsilon_2 = -\varepsilon_3 = 1.$$

C_+ est le flot bicaractéristique de $\square$ pris à temps croissant.

On a donc prouvé: que le WF ne reste pas dans les caractéristiques, que le support singulier n'est pas constitué de droites dirigées par les singularités des données initiales, que le support singulier pas plus que le WF n'est constitué de droites complètes.

THÉORÈME 2. *Soit* $u \in H^{\sigma+m}(\omega)$ *avec* $\sigma > n/2$.

$$\text{Soit } Pu = \sum_{|\alpha|=m} a_\alpha(x, D^\gamma u) D^\alpha u(x) + F(x, D^\gamma u)$$

une équation quasi linéaire où les $a_\alpha(x, U)$ *et* $F(x, U)$ *sont holomorphes en* U *sur* $\mathbb{C}^N$ *(ne dépendent des dérivées jusqu'à l'ordre* $m - 1$*). On suppose que* Γ *est un cône convexe fermé propre de* $T^*\omega\backslash 0$ *et que*

$$\left|\sum_{|\alpha|=m} a_\alpha(x, D_u^\gamma)\xi^\alpha\right| \geq c|\xi|^m \quad \forall x \in \omega, \quad \xi \in \mathbb{R}^n\backslash 0.$$

Alors $\mathrm{WF}_s(Pu) \subset \Gamma$ *entraîne* $\mathrm{WF}_{s+m}(u) \subset \Gamma$.

I. Construction de la solution et analyse sommaire

On construit d'abord une solution u de l'équation $\square u - u^2 \in C^\infty(\omega)$, u ayant des traces prescrites. Soit

$$\tilde{E}' u(x_0, \xi') = \int_0^{x_0} \frac{\sin(x_0 - s)|\xi'|}{2|\xi'|} \chi(\xi')\tilde{u}(s, \xi')\, ds$$

où on a noté par $\tilde{}$ la transformation de Fourier partielle, E' opère de $L^2(I, \mathscr{S}'(\mathbb{R}^{n-1}))$ dans $L^2(I, \mathscr{S}'(\mathbb{R}^{n-1}))$ si $I =]-T, T[$. Soit $\widetilde{Fu}(x_0, \xi') = e^{ix_0|\xi'|}\hat{u}(\xi')$ et $\widetilde{Gu}(x_0, \xi') = e^{-ix_0|\xi'|}\hat{u}(\xi')$. On va résoudre l'équation $u = E'(u^2) + Fu_1 + Gu_2 = \varphi(u)$ pour u_1 et $u_2 \in H^{m_0}(\mathbb{R}^{n-1})$ où $m_0 > n/2$, par la méthode de point fixe. Soit $\omega = I \times \mathbb{R}^{n-1}$ on définit

$$\|u\|_{s,T,\tau} = \inf(\|\tilde{u}\|_{s,\tau}\ \tilde{u}|_\omega = u) \qquad \text{pour} \quad u \in H^s(\omega), \quad \tilde{u} \in H^s(\mathbb{R}^n),$$

$$\|\tilde{u}\|_{s,\tau}^2 = \int \tau^{2s}|F\tilde{u}|^2\,(1 + |\xi'|^2 + |\xi_0|^2/\tau^2)^s\, d\xi \qquad \text{pour} \quad \tau \geq 1$$

et donc $\|u\|_{s,T,\tau} \leq \|u\|_{s',T,\tau}$ si $s \leq s'$.

Si $s \in \mathbb{N}$ on a

$$\|\varphi(x_0/2T)u\|_{s,\tau} \leq C_{s,\varphi}\|u\|_{s,\tau} \qquad \text{si} \quad T\tau \geq 1, \quad \varphi \in C_0^\infty(\mathbb{R}).$$

Si $u \in H^s(\mathbb{R}^n)$, $s \in \mathbb{N}$ on a

$$\sum_{j=0}^{s+1} \int_{-T}^{T} \|D_{x_0}^j E'u\|_{s-j+1}^2 \tau^{2(s-j)}\, dx_0 \leq C(T^2 + \tau^{-2}) \sum_{j=0}^{s} \int_{-T}^{T} \|D_{x_0}^j u\|_{s-j}^2 \tau^{2(s-j)}\, dx_0;$$

si $u \in H^s(\mathbb{R}^n)$ on a donc $\|\psi E'(\varphi u)\|_{s+1,\tau}^2 \leq C_s(1 + T^2\tau^2)\|u\|_{s,\tau}^2$ si $\psi = \tilde{\psi}(x_0/2T)$, $\varphi = \tilde{\varphi}(x_0/2T)$, $T\tau \geq 1$. On en déduit que

$$\|\psi E'(\varphi u)\|_{s,\tau}^2 \leq C(\tau^{-2} + T^2)\|u\|_{s,\tau}^2 \qquad \text{pour} \quad u \in H^s(\mathbb{R}^n), \quad s \in \mathbb{N}.$$

Par un argument d'interpolation on obtient la même inégalité lorsque $s \in \mathbb{R}^+$. On en déduit donc que $\|E'(u)\|_{s,T,\tau}^2 \leq C(\tau^{-2} + T^2)\|u\|_{s,T,\tau}^2$ pour $u \in H^s(\omega)$, $s \in \mathbb{R}^+$.

Par ailleurs si $s > n/2$, $\|uv\|_{s,\tau} \leq C\tau^{-(s-1/2)}\|u\|_{s,\tau}\|v\|_{s,\tau}$ et la même inégalité dans $H^s(\omega)$ pour la norme $\|\ \|_{s,T,\tau}$. Si $u \in H^s(\mathbb{R}^{n-1})$, $\|Fu|_\omega\|_{s,T,\tau} \leq C\tau^s T^{1/2}\|u\|_s$ si $T\tau \geq 1$, $s \in \mathbb{R}^+$ à l'aide aussi d'un argument d'interpolation.

On en déduit que

$$\|\varphi(u)\|_{m_0,T,\tau} \leq C[\tau^{m_0}T^{1/2}(\|u_1\|_{m_0} + \|u_2\|_{m_0}) + \tau^{-(m_0-\frac{1}{2})}(\tau^{-1} + T)\|u\|_{m_0,T,\tau}^2],$$
$$\|\varphi(u) - \varphi(v)\|_{m_0,T,\tau} \leq C\|u - v\|_{m_0,T,\tau}\tau^{-(m_0-\frac{1}{2})}(\tau^{-1} + T)(\|u\|_{m_0,T,\tau} + \|v\|_{m_0,T,\tau}).$$

L'application du théorème au point fixe donne une solution dans $T < \frac{1}{8}[\|u_1\|_{m_0} + \|u_2\|_{m_0}]$. Comme $m_0 > n/2$, $u \in C^0([-T, T] \times \mathbb{R}^n)$ puis $u \in C^0(I, L^2)$.

Si Λ désigne une partie conique convexe fermée propre de $\mathbb{R}^{n-1}$, on note $l_m(\Lambda) = \{u \in H^m(\mathbb{R}^{n-1}),\ \hat{u}(\xi)$ est nulle p.s. hors de $\Lambda\}$. Si u_1 et u_2 ont leurs supports dans Λ; $u \in l_{m_0}(\Lambda)$.

On montre alors que $u \in C^\infty(I, \mathscr{S}')$ et plus précisément que $u \in C^j(I, l_{m_j}(\Lambda))$ ce qui résulte aisément des arguments suivants: si

$$C_m^j = \{v \in C^j(I, l_{m_j}(\Lambda)),\ v \in C^l(I, l_{m_l}(\Lambda)),\ 0 \leq l \leq j\}.$$

F et G appliquent C_m^j dans C_m^j, $E'C_m^j$ dans C_m^{j+1}, et enfin que l'application $(u, v) \in l_m(\Lambda) \times l_{m'}(\Lambda) \mapsto uv \in l_{m''}(\Lambda)$ est bilinéaire continue si $m'' = m_- + (m' - \frac{1}{2}n)_- - 0$, $m_- = \inf(m, 0)$.

On va prouver maintenant que le $\mathrm{WF}_s(u)$ ne rencontre pas le normal à $x_0 = 0$ $\forall s$.

Soit $u_0 = Fu_1 + Gu_2$, $u = u_0 + v$; $v = E'(u^2)$. Notons que $\Box\, v - u^2 = \mathrm{op}((1 - \chi(\xi'))u(x_0; \cdot)) \in C^\infty(\omega)$ car $u \in C^\infty(I, \mathscr{S}')$; $u(x_0, \xi')$ est nul hors de Λ. Soit $s \in \mathbb{R}$, supposons que

$$\mathrm{WF}_{s-2}(v) \subset \omega \times \{(\xi_0, \xi') \mid \xi' \in \Lambda \text{ et } |\xi_0| \leq C|\xi'|\}$$

on en déduit que

$$\left.\begin{matrix}\mathrm{WF}_{s-2}(v^2) \\ \mathrm{WF}_{s-2}(u_0 v)\end{matrix}\right\} \subset \omega \times \{(\xi_0, \xi') \mid \xi' \in \Lambda \text{ et } |\xi_0| \leq C'|\xi'|\}$$

car $|\xi' + \eta'| \geq c(|\xi'| + |\eta'|)$ pour $\xi' \in \Lambda$, $\eta' \in \Lambda$. Et donc

$$\mathrm{WF}_s(v) \subset \text{car } \Box \cup [\mathrm{WF}_{s-2}(v^2) \cup \mathrm{WF}_{s-2}(u_0 v) \cup \mathrm{WF}_{s-2}(u_0^2)].$$

Soit $\varphi \in C_0$, $\alpha \in C_0^\infty$, $\alpha \equiv 1$ sur le support de φ,

$$\widehat{\varphi v}(\xi) = \int \hat{\varphi}(\xi - \eta)\widehat{\alpha v}(\eta)\, d\eta.$$

Comme $v \in lm_0$

$$|\widehat{\alpha(x')}v(\xi')| \leq C\|v\|_{m_0}(1 + |\xi'|)^{-m_0}.$$

Puisque $v(\cdot, \xi') \in l_{m_0}(\Lambda)$, on a uniformément au voisinage d'un point $\xi'^0 \notin \Lambda$:

$$|\widehat{\alpha(x')}v(\xi')| \leq C\|v\|_{m_0}\left(\int_\Lambda (1 + |\xi' - \eta'|)^{-N}(1 + |\eta'|)^{-2m_0}\, d\eta'\right)^{1/2}$$

$$\leq C_N(1 + |\xi'|)^{-N}\|v\|_{m_0}.$$

Soit $\alpha = \alpha(x_0)\alpha(x')$, $v(x_0, \xi') \in C^j(I, l_{m_j})$, on a

$$\mathscr{F}[\alpha v](\xi_0, \xi') = \int e^{-ix_0\xi_0}\widehat{\alpha(x')}v(x_0, \xi')\alpha(x_0)\, dx_0$$

et donc pour:

$$\xi_0' \notin \Lambda, \qquad \xi' \in V(\xi_0'), \qquad |\xi_0^p\widehat{\alpha v}(\xi)| \leq C_N(1 + |\xi'|)^{-N} \sup_{x_0 \in I,\, j \leq p} \|d_{x_0}^j v\|_{m_p}.$$

Donc si $0 \neq \xi'^0 \notin \Lambda$, $\exists V(\xi'^0)$ tel que $\forall N$, $\exists C_N$

$$|\widehat{\alpha v}(\xi)| \leq C_N(1 + |\xi|)^{-N} \qquad \text{pour} \quad \xi' \in V(\xi'^0).$$

On en déduit que

$$\mathrm{WF}(v) \subset \{(x, \xi) \mid x \in \omega,\ \xi' \in \Lambda \cup \{0\}\}.$$

Donc

$$\mathrm{WF}_s(v) \subset \{(x, \xi) \mid x \in \omega,\ \xi' \in \Lambda \text{ et } |\xi_0| \leq C|\xi'|\} \quad \forall s \in \mathbb{R}.$$

On déduira aisément les propriétés de propagation du WF_s de E', de celles de l'opérateur

$$Eu = \int_0^{x_0} \frac{\sin(x_0 - s)|\xi'|}{2|\xi'|}\tilde{u}\, ds$$

pour les $u \in C^\infty(I, \mathscr{S}')$ par les propriétés d'unicité des solutions de $\Box$.

Citons enfin, une proposition largement utilisée dans la suite, voir [8].

PROPOSITION. *Si* u_1(resp., u_2) $\in H^{s_1}$(resp., H^{s_2}) *avec* $s_1 > n/2$, $s_2 > n/2$. *Si* $0 \notin \mathrm{WF}_s(u_1) + \mathrm{WF}_s(u_2)$ *on a*:

$$— \mathrm{WF}_s(u_1 u_2) \subset \mathrm{WF}_s(u_1) \cup \mathrm{WF}_s(u_2) \cup [\mathrm{WF}_s(u_1) + \mathrm{WF}_s(u_2)],$$
$$— \mathrm{WF}_s(u_1 u_2) \subset \mathrm{WF}_s(u_1) \cup \mathrm{WF}_s(u_2) \text{ si } s < s_1 + s_2 - \tfrac{1}{2}n.$$

On trouvera une démonstration dans Rauch [10].

II. ANALYSE DES SINGULARITES

On a écrit u solution de $\square u = u^2 + C^\infty$ sous la forme: $u = u_0 + v$ où $\square v = (u_0 + v)^2 + C^\infty$, v traces nulles. Soit $v = E'((u_0 + v)^2)$. On pose $w = w_0 + w_1$

$$w_0 = 2u_0 E'(u_0^2),$$
$$w_1 = v^2 + 2u_0 E'(v^2 + 2u_0 v) + u_0^2,$$

et

$$w_0 + w_1 = (v + u_0)^2,$$
$$u = u_0 + E'(w).$$

On choisit

$$u_0 = Fu_1 + Fu_2 + Gu_3,\ \tilde{u}_j = Fu_j,\ \widehat{Fu_j}(x_0, \xi') = g_j(\xi')\exp(i\varepsilon_j(x_0 - t_0)|\xi'|)$$

où

$$\varepsilon_1 = \varepsilon_2 = -\varepsilon_3 = 1,\ t_0 > 0.$$

On définit $I_\delta^m(\Sigma)$ pour un cône $\Sigma = \mathbb{R}^+ e_1$ (voir Guillemin [6]) comme la classe des fonctions $g(\xi)$ sur $\mathbb{R}^{n-1}$ satisfaisant aux inégalités:

$$|D_{\xi_1}^{\alpha_1} D_{\tilde{\xi}}^{\tilde{\alpha}} g(\xi)| \leq C(1 + |\tilde{\xi}|\,|\xi_1|^{-\delta})^{-N}(1 + |\xi|)^{m - \delta|\tilde{\alpha}| - |\alpha_1|}$$

et supportée par une zone $\xi_1 \geq C$, si $\xi = \tilde{\xi} + \xi_1 e_1$.

On supposera que les $g_i \in I_\delta^{m_i}(\mathbb{R}^+ e_i)$ si $(e_1, e_2, e_3) \ni$ base canonique de $\mathbb{R}^{n-1}$ et qu'elles sont en particulier de la forme

$$g_1(\xi_1, \tilde{\xi}) = g_1^0(\xi_1) X(\tilde{\xi}/|\xi_1|^\delta), \qquad X \in C_0^\infty(\mathbb{R}^{n-2}), \quad X \geq \tfrac{1}{2}$$

dans la boule de rayon $\lambda/2$ et à support dans la boule de rayon λ. g_1^0 est un symbole elliptique de degré m_1 sur $\mathbb{R}^+$. Le choix

$$m_i = -\tfrac{1}{2} - m_0 - (n-2)(\delta/2) - \bar{\varepsilon}, \bar{\varepsilon}, > 0,$$

assure que $u_i \in H^{m_0}(\mathbb{R}^{n-1})$.

$$\text{Si} \quad g \in I_\delta^m(\mathbb{R}^+ e_1) \qquad \text{alors} \quad \mathrm{WF}(g) \subset \{(0, \xi_1 e_1) \text{ où } \xi_1 > 0\}.$$

On prouve que la distribution

$$u(x') = \int e^{i(x'\xi' - t_0|\xi'|)} g(\xi')\, d\xi'$$

vèrifie

$$\mathrm{WF}(u) \subset \{(x' = t_0 e_1, \xi' = \xi_1 e_1), \xi_1 > 0\}$$

Donc

$$\mathrm{WF}(\tilde{u}_i) \subset \{(x_0, x' = (x_0 - t_0)\varepsilon_i e_i, \xi' = \xi_i e_i, \xi_0 = |\xi_i| \varepsilon_i), \xi_i > 0\}$$

On posera donc:

$$\begin{aligned}
D_1^+ &= \{(x_0, x' = (x_0 - t_0)e_1, \xi' = \xi_1 e_1, \xi_0 = \xi_1), \xi_1 > 0\}, \\
D_2^+ &= \{(x_0, x' = -t_0 e_2 + x_0 e_2, \xi' = \xi_2 e_2, \xi_0 = -\xi_2), \xi_2 > 0\} \\
D_3^+ &= \{(x_0, x' = -(x_0 - t_0)e_3, \xi' = \xi_3 e_3, \xi_0 = -\xi_3), \xi_3 > 0\}.
\end{aligned}$$

D_i^+ sont les flots sortants respectivement des singularités des u_i. On rappelle que si $f \in C^\infty(I, \mathscr{S}'(\mathbb{R}^{n-1}))$ est telle que $\mathrm{WF}(f) \cap N_{x_0=0} = \varnothing$ où $N_{x_0=0}$ est le normal à $x_0 = 0$, on a

$$\mathrm{WF}_{s+1}(E'(f)) \subset \mathrm{WF}_s(f) \cup C(\mathrm{WF}_s(f)) \cup (C_0 \circ R_0)(\mathrm{WF}_s(f)) = C'(\mathrm{WF}_s(f)),$$

où C est la relation bicaractéristique. C_0 est le flot sortant, R_0 est la relation bicaractéristique de l'opérateur de restriction à $x_0 = 0$. Tandis que

$$\mathrm{WF}_{s+2}(E'f) \subset \mathrm{car}\, \square \cup \mathrm{WF}_s(f) \qquad \text{si} \quad f \in C^\infty(I, \mathscr{S}'(\mathbb{R}^{n-1})).$$

On introduit de nouveaux ensembles:

$$\begin{aligned}
D_i^- &= \{(x_0, -\varepsilon_i(x_0 + t_0)e_i, \xi_i e_i, -\varepsilon_i \xi_i), \xi_i > 0\}, & i &= 1, 2, 3. \\
L_i &= \{(0, -\varepsilon_i t_0 e_i, \xi_i e_i, \tau_i), \xi_i > 0, \tau_i \in \mathbb{R}\}, & i &= 1, 2, 3, \\
S_{ij}^+ &= \{(t_0, 0, \xi_i e_i + \xi_j e_j, \varepsilon_i \xi_i - \varepsilon_j \xi_j); \xi_i, \xi_j > 0\}, & i &< j, \\
S_{ij}^- &= \{(-t_0, 0, \xi_i e_i + \xi_j e_j, -\varepsilon_i \xi_i + \varepsilon_j \xi_j); \xi_i, \xi_j > 0\}.
\end{aligned}$$

Enfin

$$S_0^+ = \left\{\left(t_0, 0, \sum_{i=1}^3 \xi_i e_i, \sum_{i=1}^3 \xi_i \varepsilon_i\right), \xi_i > 0\right\}$$

$$S_0^- = \left\{\left(-t_0, 0, \sum_{i=1}^3 \xi_i e_i, -\sum_{i=1}^3 \xi_i \varepsilon_i\right), \xi_i > 0\right\}$$

et $C_0^+ = C(S_0^+)$ et $C_0^- = C(S_0^-)$.

On note les relations suivantes:

$D_i^{\pm} + D_j^{\pm} = S_{ij}^{\pm}, \qquad i \neq j;$
$D_i^{+} + D_j^{-} = \varnothing \qquad \text{si} \quad i \neq j \quad \text{et} \quad D_i^{+} + D_i^{-} = L_i;$
$D_i^{\pm} + D_i^{\pm} = D_i^{\pm}, \qquad D_i^{\pm} + L_j = \varnothing \qquad \text{si} \quad i \neq j, \quad D_i^{\pm} + L_i = L_i;$
$L_i + L_j = \varnothing \qquad \text{si} \quad i \neq j, \quad L_i + L_i = L_i;$
$D_i^{+} + S_{kl}^{+} = S_{kl}^{+} \qquad \text{si} \quad i \in \{k, l\}; \qquad D_i^{+} + S_{k,l}^{+} = S_0^{+} \qquad \text{si} \quad i \notin \{k, l\};$
$D_i^{-} + S_{kl}^{+} = D_i^{+} + S_{kl}^{-} = \varnothing; \qquad L_i + S_{kl}^{\pm} = \varnothing;$
$S_{ij}^{\pm} + S_{kl}^{\pm} = S_0^{\pm} \qquad \text{si} \quad \{i, j, k, l\} \supset \{1, 2, 3\}, \qquad \text{sinon} \quad S_{ij}^{\pm} + S_{ij}^{\pm} = S_{ij}^{\pm};$
$S_{ij}^{+} + S_{kl}^{-} = \varnothing.$

On a donc:

$$\mathrm{WF}(u_0) \subset \bigcup_{i=1\ldots3} D_i^{+} \qquad \text{et} \qquad \mathrm{WF}(u_0^2) \subset \bigcup_{i=1\ldots3} D_i^{+} \cup \bigcup_{i,j} S_{ij}^{+}.$$

car $$C'(D_i^{+}) = D_i^{+} \cup D_i^{-}.$$

Proposition 2.1. *Si* $\sigma < \inf(m_0 + 4, 2m_0 - (\delta/2)(n - 2) + 3 + \bar{\varepsilon})$, $\sigma > 2 + \frac{1}{2}n$ *on a*

$$\mathrm{WF}_\sigma(w_1) \subset \tilde{C} = \bigcup_{i=1\ldots3} D_i^{+} \cup D_i^{-} \cup \bigcup_{i<j} S_{ij}^{+} \cup S_{ij}^{-} \cup \bigcup_{i=1\ldots3} L_i.$$

Démonstration. Faisons remarquer d'abord que $C'(\tilde{C}) = \tilde{C}$ car $C(L_i) = D_i^{+} \cup D_i^{-} = (C_0 \circ R_0)(L_i)$, tandis que $C(S_{ij}^{\pm}) = \varnothing$ car les éléments de $S_{ij}^{\pm}$ ne sont pas caractéristiques.

$$\mathrm{WF}_{m_0+2}(v) \subset C'[\mathrm{WF}_{m_0+1}(u_0^2) \cup \mathrm{WF}_{m_0+1}(u_0)] \subset \bigcup_{i=1\ldots3} D_i^{+} \cup D_i^{-} \cup \bigcup_{i<j} S_{ij}^{+}$$

car $v \in H^{m_0+1}$, de plus $\mathrm{WF}_{m_0+2}(v) \subset \mathrm{car}\,\square$; donc

$$\mathrm{WF}_{m_0+2}(v) \subset \bigcup_{i=1\ldots3} D_i^{+} \cup D_i^{-}$$

et donc:

$$\mathrm{WF}_{m_0+2}(v^2), \mathrm{WF}_{m_0+2}(u_0 v) \subset \bigcup D_i^{+} \cup D_i^{-} \cup \bigcup S_{ij}^{+} \cup \bigcup L_i \cup \bigcup S_{ij}^{-} = \tilde{C}$$

donc $\mathrm{WF}_{m_0+3}(v) \subset \tilde{C}$. Donc pour $\sigma \leq m_0 + 4$, $\mathrm{WF}_{\sigma-1}(v) \subset \tilde{C}$ et $\mathrm{WF}_{\sigma-1}(v) \subset \mathrm{car}\,\square \cup \mathrm{WF}_{\sigma-3}(u_0^2)$. Le lemme suivant prouve que $\mathrm{WF}_{\sigma-3}(u_0^2) \subset \bigcup_{i=1\ldots3} D_i^{+}$ ce qui prouve que $\mathrm{WF}_{\sigma-1}(v) \subset \bigcup D_i^{+} \cup D_i^{-} \cup L_i$.

Lemme. *En un point de* S_{ij}^{+} u_0^2 *est de classe* H^s *si* $s < 2m_0 - (\delta/2)(n - 2) + \bar{\varepsilon}$.

Démonstration du lemme. Soit par exemple $(t_0, 0, \xi_1^0 e_1 + \xi_2^0 e_2, \xi_1^0 + \xi_2^0) \in S_{1,2}^+$. On calcule $F(\xi) = \widehat{\varphi F u_1 F u_2}(\xi)$ sous la forme:

$$\widehat{\varphi F u_1 F u_2}(\xi) = \int \varphi(x_0 + t_0, x') e^{-it_0\xi_0 - ix_0(\xi_0 - |\eta'| - |\lambda'|) - i(\xi' - \eta' - \lambda')}$$
$$\times\, g_1(\eta') g_2(\lambda')\, dx\, d\eta'\, d\lambda'.$$

Par une méthode que l'on détaillera plus loin dans un cas voisin, lorsque ξ reste voisin de ξ^0, on se réduit à

$$\varphi(0) \int_{\xi_0 = |\eta'| + |\lambda'|,\ \xi' = \eta' + \lambda'} X(\ ) g_1(\eta') g_2(\lambda')\, d\sigma + \text{autres terms.}$$

Et enfin lorsque ξ reste dans un voisinage conique assez petit de ξ^0

$$|F(\xi)| \leq C(1 + |\xi|)^{m_1 + m_2 + \delta(n-2)}.$$

Donc

$$\mathrm{WF}_{\sigma-1}(v^2),\ \mathrm{WF}_{\sigma-1}(u_0 v) \subset \tilde{C} \qquad \text{et} \qquad \mathrm{WF}_\sigma(v) \subset \tilde{C}$$

d'où

$$\mathrm{WF}_\sigma(v^2) \subset \tilde{C} \cup S_0^+ \cup S_0^-.$$

On va montrer que en fait $S_0^+ \cup S_0^-$ n'est pas dans $\mathrm{WF}_\sigma(v^2)$. Soit par exemple S_0^+, et $\mu_0 = (t_0, 0, \xi^0) \in S_0^+$. Si $\varepsilon > 0$ les ensembles

$$U_i = \{\xi \,|\, |\xi' - \xi_i e_i| < 2\varepsilon|\xi|\}$$

et

$$V_{ij} = \{\xi \,|\, |\xi_0^2 - |\xi'|^2| > 2\varepsilon^2|\xi|^2,\ |\xi - (\xi_i e_i + \xi_j e_j)| < \varepsilon|\xi|,\ |\xi_i| > \varepsilon|\xi|,\ |\xi_j| > \varepsilon|\xi|\}$$

constituant un recouvrement ouvert de $\mathrm{pr}_2(\mathrm{WF}_\sigma(v))$.

On en déduit une partition pseudo différentielle:

$$\varphi v = v_1 + \sum_i q_i(x, D) v + \sum_{i,j} p_{ij}(x, D) v$$

où $v_1 \in H^\sigma$, con supp $q_i \subset U_i$, con supp $p_{ij} \subset V_{ij}$.

On peut avoir choisi ε assez petit pour que $\xi^\circ \notin \bigcup U_i \cup \bigcup V_{ij}$, $\xi^0 \notin U_i + U_j$. Donc $\varphi^2 v^2 = v_1(\varphi v) + \sum q_i v q_j v + \sum p_{ij} v p_{kl} v + \sum q_i v p_{kl} v$, $(t_0, 0, \xi^0) \notin \mathrm{WF}_\sigma(v_1 \varphi v)$ ni dans $\mathrm{WF}_\sigma(q_i v q_j v)$, par ailleurs $p_{ij} v \in H^{\sigma - 1}$ donc $p_{ij} v p_{kl} v \in H^{2(\sigma-1) - \frac{1}{2}n - 0}$ en μ_0, de même $q_i v p_{kl} v \in H^{\sigma - 1 + m_0 + 1 + \frac{1}{2}n - 0}$ en μ_0. Donc $\mathrm{WF}_\sigma(v^2) \subset \tilde{C}$.

On va prouver maintenant que $\mathrm{WF}_\sigma(u_0 E'(v^2)) \subset \tilde{C}$. A priori

$$\mathrm{WF}_\sigma(u_0 E'(v^2)) \subset \tilde{C} \cup S_0^- \qquad \text{si} \quad p(x, D) \equiv 1$$

au voisinage de $(t_0, 0, \xi_2^0 e_2 + \xi_3^0 e_3, \xi_2^0 - \xi_3^0)$ on obtient clairement que $\mu_0 \notin \mathrm{WF}_\sigma(u_0(1 - p(x, D))E'(v^2))$ car $\mathrm{WF}_\sigma(E'(v^2)) \subset \tilde{C}$. Par ailleurs $(p(x, D)E')v^2 \in$

H^{m_0+4} puisque $\mathrm{WF}_{m_0+2}(v^2) \subset \mathrm{WF}_{m_0+2}(v) \subset \bigcup_i D_i^+ \cup D_i^-$. On peut écrire $\varphi u_0 = q(x, D)u_0 + u_0'$ où $u_0' \in C_0^\infty$ tandis que la transformée de Fourier de $q(x, D)u_0$ a son support dans un cône fermé de $\mathbb{R}^n \backslash 0$ qui ne contient pas ξ^0. Si on remarque que pour $\varphi \in C_0^\infty$, $\int |\widehat{\varphi u_0}(\xi)|(1 + |\xi|)^{m_0+1-\frac{1}{2}n} d\xi < +\infty$ on obtient que

$$q(x, D)u_0(p(x, D)E'v^2) \in H^{m_0+4+\frac{1}{2}-\frac{1}{2}n}$$

en μ_0,

$$\text{tandis que} \quad \mu_0 \notin \mathrm{WF}_\sigma(u_0' p(x, D)E'v^2).$$

Pour le terme $u_0 E'(u_0 v)$ la méthode est exactement la même mais cette fois-ci on a seulement $p(x, D)E'(u_0 v) \in H^{m_0+3/2}$ car $\mathrm{WF}_{m_0+3/2}(u_0 v) \subset \bigcup D_i^+ \cup D_i^-$. Soit $\mathrm{WF}_\sigma(u_0 E'(u_0 v)) \subset \tilde{C}$. On a donc obtenu $\mathrm{WF}_\sigma(w_1) \subset \tilde{C}$.

PROPOSITION 2.2. *Pour un choix convenable de* $(m_0, \sigma, \delta, \bar{\varepsilon})$, $S_0^+ \subset \mathrm{WF}_\sigma(w)$.

Démonstration. Notons tout d'abord que a priori

$$\mathrm{WF}(u_0 E'(u_0^2)) \subset \bigcup S_{ij}^+ \cup \bigcup L_i \cup S_0^+.$$

Il suffit donc de voir que $S_0^+ \cap \mathrm{WF}_\sigma(w_0) \neq \varnothing$.

On écrit $u_0 = \tilde{u}_1 + \tilde{u}_2 + \tilde{u}_3$ où $\mathrm{WF}(\tilde{u}_1) \subset D_i^+$, $\tilde{u}_i = Fu_i$, $1 \leq i \leq 2$, $\tilde{u}_3 = Gu_3$. Considérons le produit $u_0' = Fu_1 E'(Fu_2 Gu_3)$ si $\varphi = \alpha\beta \in C_0^\infty(\mathbb{R}^n)$

$$\widehat{\varphi u_0'}(\xi) = \int \hat{\alpha}(\xi_0 - \zeta_0 - |\eta'|, \xi' - \eta' - \zeta')\hat{u}_1(\eta')\widehat{\beta E'(\tilde{u}_2 \tilde{u}_3)}(\zeta)\, d\eta'\, d\zeta,$$

tandis qu'on exprime

$$\widehat{\beta E'(f)}(\zeta) = \int e^{-ix\zeta}\beta(x)e^{ix'\lambda'} \int_0^{x_0} \tilde{f}(s, \lambda') \times \left[\frac{e^{i(x_0-s)|\lambda'|} - e^{-i(x_0-s)|\lambda'|}}{2i|\lambda'|}\right] x(\lambda')\, ds\, d\lambda'\, dx.$$

α et β ont leurs supports au voisinage du point $(t_0, 0)$. Avec $f = h(t)Fu_2 Gu_3$ où $h \in C_0^\infty(\mathbb{R}^+)$ vaut 1 au voisinage de t_0: en effet

$$\mathrm{WF}\, E'((1 - h)Fu_2 Gu_3) \subset D_2^\pm \cup D_3^\pm.$$

Donc

$$\widehat{\beta E'(f)}(\zeta) = \int k(s)e^{-is(\zeta_0-\lambda_0)}\, ds \iint_0^{+\infty} \beta(x_0 + s, x') \times \left[\frac{e^{ix_0|\lambda'|} - e^{-ix_0|\lambda'|}}{2i|\lambda'|}\right] x(\lambda')\hat{f}(\lambda')e^{-ix'(\zeta'-\lambda')-ix_0\zeta_0}\, dx_0\, dx'\, d\lambda.$$

où $k \in C_0(\mathbb{R}^+)$ vaut 1 au voisinage du support de h. On pose donc

$$K(\zeta, \lambda) = \int k(s) e^{-is(\zeta_0 - \lambda_0)} ds \iint_0^{+\infty} \beta(x_0 + s, x') \times \left[\frac{e^{ix_0(|\lambda'| - \zeta_0)} - e^{-ix_0(|\lambda'| + \zeta_0)}}{2i|\lambda'|} \right] x(\lambda') e^{-ix'(\zeta' - \lambda')} dx$$

de plus

$$\hat{f}(\lambda) = \int h(t) e^{-it(\lambda_0 - |\lambda' - \theta'| + |\theta'|)} \hat{u}_2(\lambda' - \theta') \hat{u}_3(\theta') d\theta' dt.$$

Donc

$$F(\xi) = \varphi \widehat{F u_1 E'}(f)(\xi) = \int \hat{\alpha}(\xi_0 - \zeta_0 - |\eta'|, \xi' - \eta' - \zeta') e^{-it(\lambda_0 - |\theta' - \lambda'| + |\theta'|)} h(t) K(\zeta, \theta) \hat{u}_1(\eta') \hat{u}_2(\lambda' - \theta') \hat{u}_3(\theta') d\eta' d\zeta d\theta' d\lambda dt.$$

Soit

$$F(\xi) = \int e^{-ix_0(\xi_0 - |\eta'| - |\lambda'| + |\theta'|)} \left[\int_0^{+\infty} e^{-is(\xi_0 - |\eta'|)} \left[\frac{e^{is|\theta' + \lambda'|} - e^{-is|\theta' + \lambda'|}}{2i|\theta' + \lambda'|} \right] \times x(\lambda' + \theta') \varphi(x_0 + s, x') h(x_0) ds \right] \times e^{-ix'(\xi' - \eta' - \lambda' - \theta')} \hat{u}_1(\eta') \hat{u}_2(\lambda') \hat{u}_3(\theta') d\eta' d\lambda' d\theta' dx.$$

On a posé

$$\hat{u}_i(\zeta') = g_i(\zeta') e^{-i\varepsilon_i t_0 |\zeta'|}, \qquad 1 \leq i \leq 3,$$

d'où

$$F(\xi) = \int e^{-it_0\xi_0 - ix_0(\xi_0 - |\eta'| - |\lambda'| + |\theta'|) - ix'(\xi' - \eta' - \lambda' - \theta')} \times \Phi(x, \xi_0, \eta', \lambda', \theta') g_1(\eta') g_2(\lambda') g_3(\theta') dx d\eta' d\lambda' d\theta'$$

où

$$\Phi(x, \xi_0, \eta', \lambda', \theta') = h(x_0 + t_0) \int_0^{+\infty} e^{-is(\xi_0 - |\eta'|)} \times \left[\frac{e^{is|\theta' + \lambda'|} - e^{-is|\theta' + \lambda'|}}{2i|\theta' + \lambda'|} \right] x(\theta' + \lambda') \varphi(x_0 + t_0 + s, x') ds,$$

$$\Phi = \Phi_+(\ ) - \Phi_-(\ ).$$

Il est clair qu'une intégrale de la forme $\int_0^{+\infty} f(s, \tau) e^{-is\tau} ds$ où $s \to f(s, \tau)$ est C^∞ à support compact, tandis que $(s, \tau) \to f(s, \tau)$ est un symbole par rapport à τ, est également un symbole en τ. On pose $\varphi(x,\xi,\eta',\theta',\lambda') = x_0(\xi_0 - |\eta'| - |\lambda'| + |\theta'|) + x'(\xi' - \eta' - \lambda' - \theta')$ de sorte que

$$F(\xi) = e^{-it_0\xi_0} \int e^{-i\varphi}\Phi(x,\xi_0,\eta',\lambda',\theta')g_1(\eta')g_2(\lambda')g_3(\theta')\,dx\,d\eta'\,d\lambda'\,d\theta'.$$

$$\nabla_x\varphi = (\xi_0 - |\eta'| + |\theta'| - |\lambda'|, \xi' - \eta' - \lambda' - \theta').$$

Si $|\eta' + \lambda' + \theta'| \geq 2|\xi'|$ ou $|\eta' + \theta' + \lambda'| \leq \frac{1}{2}|\xi'|$ on aura $|\nabla_{x'}\varphi| \geq c(|\xi| + |\eta'| + |\lambda'| + |\theta'|)$ puisque $\xi' \neq 0$ dans un voisinage conique d'un point μ_0 fixé de $C_1 + C_1 + C_2 \backslash (C_1 \cup C_2)$, et que sur les supports coniques des g_i on a $|\eta' + \lambda' + \theta'| \approx |\eta'| + |\lambda'| + |\theta'|$ si on a pris chaque g_i a support compact conique assez près de e_i, la contribution de cette zone:

$$F^{\mathrm{II}}(\xi)e^{it_0\xi_0} = \int (1 - \zeta)\left(\frac{\eta' + \lambda' + \theta'}{|\xi'|}\right)e^{-i\varphi}\Phi(\quad)g_1g_2g_3\,dx\,d\eta'\,d\lambda'\,d\theta'$$

et donc à décroissance rapide en ξ. L'autre contribution est

$$F^{\mathrm{I}}(\xi)e^{it_0\xi_0} = \int \zeta\left(\frac{\eta' + \lambda' + \theta'}{|\xi'|}\right)e^{-i\varphi}\Phi(\quad)g_1g_2g_3\,dx\,d\eta'\,d\theta'\,d\lambda'$$

qui se réduit aisément à

$$F^{\mathrm{III}}(\xi)e^{it_0\xi_0} = \int \zeta\left(\frac{\eta' + \lambda' + \theta'}{|\xi'|}\right)\sigma(\varphi'_x/|\xi'|)e^{-i\varphi}\Phi g_1g_2g_3\,dx\,d\eta'\,d\lambda'\,d\theta'.$$

où le support de $\sigma(t)$ est dans $|t| \leq \varepsilon$. De sorte que l'on a sur le support de l'intégrande $|\eta'| + |\varphi'_{x'}| \geq |\varphi'_{x'} + \eta'| \geq c_1(|\xi'| + |\lambda'| + |\theta'|)$ si $|\xi'|$ est assez grand puisque alors $|\xi' - (\theta' + \lambda')| \geq c_1(|\xi'| + |\lambda'| + |\theta'|)$ et donc $|\eta'| \geq c'_1|\xi'|$ (si ε est assez petit); donc $|\eta'| \approx |\lambda'| \approx^{\mathrm{B}} |\theta'| \approx |\xi'|$. Dans ces conditions $|\lambda' + \theta'| \pm (|\lambda'| - |\theta'|) \geq c(|\lambda'| + |\theta'|)$ puisque sur les supports coniques de g_2 et g_3 $|\lambda'\theta' + |\lambda'||\theta'|| \geq c|\lambda'||\theta'|$. Ce qui fait que

$$|\xi_0 - |\eta'| \pm |\lambda' + \theta'|| = |\varphi'_{x_0} + (|\lambda'| - |\theta'|) \pm |\lambda' + \theta'|| \geq c'|\xi| - \varepsilon|\xi| \geq c''|\xi|;$$

il en résulte que dans la zone considérée la fonction $\Phi(x,\xi_0,\eta',\lambda',\theta')$ est un symbole de degé-2 dans les variables $(x,\xi_0,\eta',\lambda',\theta')$. Il en résulte que la fonction $\zeta((\eta' + \lambda' + \theta')/|\xi'|)\sigma(\varphi'_x/|\xi'|)\Phi$ est un symbole de degré-2 dans les variables $(x,\xi,\eta',\lambda',\theta')$ dans une zone où $|\eta'| \approx \cdots \approx |\xi|$. On veut obtenir un équivalent asymptotique de $F^{\mathrm{III}}(\xi)e^{it_0\xi_0}$, on fait remarquer que les diverses formes (voir [4] et [7]) du théorème de la phase stationnaire ne conviennent pas ici. On pose

$$I(a,g_1,g_2,g_3)(\xi) = \int a(x,\xi,\eta',\lambda',\theta')e^{-i\varphi}g_1(\eta')g_2(\lambda')g_3(\theta')\,dx\,d\eta'\,d\lambda'\,d\theta',$$

où $a(x,\ldots)$ est un symbole à support compact en x de degré μ et dont le support conique est contenu dans une région $|\eta'| \approx \cdots \approx |\xi|$; les $g_i \in I^{m_i}_\delta(\sum_i)$

ont un support conique voisin de $e_i\mathbb{R}_*^+$. On appelle $\mathscr{S}^{\mu m_1 m_2 m_3}$ la classe de ces intégrales. On note que

$$x_0 = \left(\frac{\partial\varphi}{\partial\theta_2} - \frac{\partial\varphi}{\partial\lambda_2}\right)\left(\frac{\theta_2}{|\theta'|} + \frac{\lambda_2}{|\lambda'|}\right)^{-1},$$

$$x' = -\frac{\partial\varphi}{\partial\theta'} + \left(\frac{\partial\varphi}{\partial\theta_2} - \frac{\partial\varphi}{\partial\lambda_2}\right)\left(\frac{\theta_2}{|\theta'|} + \frac{\lambda_2}{|\lambda'|}\right)^{-1}\frac{\theta'}{|\theta'|}.$$

La méthode va donc consister à développer a sur $x = 0$ par la formule de Taylor. Soit:

$$a(x,\xi,\eta',\lambda',\theta') = [a(0,\xi,\eta',\lambda',\theta') + x\,.\,a_1(x,\xi,\eta',\lambda',\theta')]X(x) \quad \text{où} \quad X \in C_0^\infty(\mathbb{R}^n).$$

Soit $f: U_1 \times U_2 \times U_3 \to \mathbb{R}^n$ définie par $(|\eta'| + |\lambda'| - |\theta'|, \eta' + \lambda' + \theta')$ dans des ouverts coniques assez voisins des supports des $g_i f$ est une submersion si $\lambda_2/|\lambda'| - \eta_2/|\eta'| > 0$ dans U. La mesure de surface sur $f^{-1}(\xi_0, \xi')$ est

$$\sigma_\xi = \left|\frac{\lambda_2(\xi,\hat{\lambda}',\theta')}{|\lambda'(\xi,\hat{\lambda}',\theta')|} - \frac{\eta_2(\xi,\hat{\lambda}',\theta')}{|\eta'(\xi,\hat{\lambda}',\theta')|}\right|^{-1} d\hat{\lambda}'\,d\theta'$$

où $\lambda'(\xi,\hat{\lambda}',\theta')$ et $\eta'(\xi,\hat{\lambda}',\theta')$ sont obtenues comme fonctions réciproques de: $\xi_0 = |\eta'| + |\lambda'| - |\theta'|$, $\xi' = \eta' + \lambda' + \theta'$, $\hat{\lambda}' = \lambda' - \lambda_2 e_2, \theta'$ $f(V)$ est ouvert. Le premier terme donne:

$$I^1 = \int_{f(V)} \hat{X}(\xi - \zeta)\,d\zeta \int_{f^{-1}(\zeta)} a(0,\xi,\eta',\lambda',\theta')g_1 g_2 g_3\,d\sigma_\zeta$$

ce qui fait que l'on pose

$$\begin{aligned} H(\xi,\zeta) &= \int_{f^{-1}(\zeta)} a(0,\xi,\eta',\lambda',\theta')g_1 g_2 g_3\,d\sigma_\zeta \\ &= \int a(0,\xi,\zeta' - \lambda'(\ ) - \theta', \lambda'(\ ), \theta')g_1 g_2 g_3 \mathscr{T}(\zeta,\hat{\lambda}',\theta')\,d\hat{\lambda}'\,d\theta' \end{aligned}$$

lorsque ζ est dans un voisinage conique assez petit de ξ.

On fait remarquer que dans les mêmes conditions

$$\left|\left(\frac{\partial}{\partial\zeta}\right)^\lambda H(\xi,\zeta)\right| \le C(1 + |\zeta|)^{\mu + m_1 + m_2 + m_3 + 2n - 3 - \delta|\lambda|}$$

où μ est le degré de a, car les g_j sont seulement des symboles du type δ. Donc

$$\begin{aligned} I^1 = \sum_{|\lambda| < N} (\ ) \int \hat{X}(\xi - \zeta)H^{(\lambda)}(\xi,\xi)\frac{(\xi - \zeta)^\lambda}{\lambda!}\,d\zeta \\ + \iint_0^1 (\ )\hat{X}(\xi - \zeta)H^N(\xi, \xi + s(\xi - \zeta))(\xi - \zeta)^N\,d\zeta\,ds \end{aligned}$$

et donc modulo $O(|\xi|^{-\infty})$,

$$I^1 \equiv \int_{f^{-1}(\xi)} a(0,\xi,\eta',\lambda',\theta')g_1g_2g_3\,d\sigma_\xi,$$

tandis que

$$I^2(a,g_1,g_2,g_3)(\xi) = \int \sum_{j\,\text{finie}} c_j(x,\xi,\eta',\lambda',\theta')\frac{\partial\varphi}{\partial\alpha_j}e^{-i\varphi}g_1g_2g_3\,dx'\,d\eta'\,d\lambda'\,d\theta',$$

où les c_j sont également de degré μ; on intègre par parties et on obtient une somme finie de termes appartenant à $\mathscr{S}^{\mu-1,m_1m_2,m_3}$ où à l'un des $\mathscr{S}^{\mu,m_1-\delta,m_2,m_3}$, itérant le procédé et négligeant un $O(|\xi|^{-\infty})$ on a un nombre fini d'intégrales sur la surface $f^{-1}(\xi)$ et un terme $\sum I(a^N,g_1^N,g_2^N,g_3^N)$ où la somme des degrés peut être rendue arbitrairement petite.

On exprime:

$$a(0,\xi,\eta',\lambda',\theta')|_{f^{-1}(\xi)} = \frac{1}{2}\frac{X(\lambda'+\theta')\varphi(t_0,0)}{(\xi_0-|\eta'|+|\theta'+\lambda'|)(\xi_0-|\eta'|-|\theta'+\lambda'|)}$$
$$+ \text{ symbole de degré } -3.$$

Chaque $g_i(\eta') = q_i(\eta_i)h(\hat{\eta}'|\eta_i|^{-\delta})$ où $\eta' = \hat{\eta}' + \eta_i e_i$ $1 \le i \le 3$, où q_i est un symbole elliptique sur $\mathbb{R}^+$ de degré m_i. Utilisant un argument de fonctions réciproques on voit que lorsque ξ décrit un voisinage conique assez petit de μ_0, $\forall\varepsilon > 0$, $\exists\varepsilon_0 > 0$ tel que:

$$|\theta' - \xi_3 e_3| \le \varepsilon_0|\xi|^\delta, \qquad |\hat{\lambda}'| \le \varepsilon_0|\xi|^\delta, \qquad |\xi''| \le \varepsilon_0|\xi|^\delta$$

et

$$\big|\xi_0 + |\xi_3| - |\xi_1| - |\xi_2|\big| \le \varepsilon_0|\xi|^\delta$$

entraînent

$$|\lambda_2(\xi,\hat{\lambda}',\theta') - \xi_2| \le \varepsilon|\xi|^\delta \qquad \text{et} \qquad |\eta'(\xi,\hat{\lambda}',\theta') - \xi_1 e_1| \le \varepsilon|\xi|^\delta.$$

Dans ces conditions (ε_0 assez petit) $|\theta_3| \ge c|\xi|$, $|\eta_1(\ )| \ge c|\xi|$, $|\lambda_2(\ )| \ge c|\xi|$ tandis que $|\hat{\lambda}'| \le \varepsilon_0|\xi|^\delta$, $|\hat{\eta}'| \le \varepsilon|\xi|^\delta$, $|\hat{\theta}'| \le \varepsilon_0|\xi|^\delta$; $g_1(\eta')g_2(\theta')g_3(\lambda') \ge c|\xi|^{m_1+m_2+m_3}$ au moins si $|\xi|$ est assez grand.

Alors

$$\left|\int_{f^{-1}(\xi)} a_0(0,\xi,\eta',\lambda',\theta')g_1g_2g_3\,d\sigma_\xi\right|$$
$$\ge c|\xi|^{m_1+m_2+m_3-2+\delta(2n-3)}1_{|\xi''|+|\xi_0+|\xi_3|-|\xi_1|-|\xi_2||\le\varepsilon_0|\xi|^\delta}$$

dans un voisinage conique de μ_0, $|\xi|$ assez grand.

En sens inverse, on note que

$$\left| \int_{f^{-1}(\xi)} a(\xi, \eta', \lambda', \theta') g_1 g_2 g_3 \, d\sigma_\xi \right| \leq C |\xi|^{m_1 + m_2 + m_3 + \mu + \delta(2m-3)}.$$

La contribution en μ_0 des termes du produit $u_0 E'(u_0^2)$ se réduit seulement à $(Fu)E'(FuGu_3)$, $(Fu_2)E'(Fu_1 Gu_3)$ et $(Gu_3)E'(Fu_1 Fu_2)$ des parties principales $a_0(\ )|_{f^{-1}(\xi)}$ respectivement égales à:

$$-\frac{1}{2}\frac{1}{(\lambda'.\theta' + |\lambda'||\theta'|)}, \qquad -\frac{1}{2(\eta'.\theta' + |\eta'||\theta'|)}, \qquad \frac{1}{2(|\lambda'||\eta'| - \lambda'.\eta')}$$

dont la somme est $\xi_0/|\eta'||\theta'||\lambda'| + O(|\xi|^{-2-(1-\delta)})$; si en $\mu_0 \xi_0 \neq 0$ on a un terme elliptique de degré -2. On peut donc conclure que si en $\mu_0 \in \tilde{C}$ et $\xi_0 \neq 0$, $\mu_0 \in \mathrm{WF}_\sigma(u_0 E'(\mu_0^2))$ si $\sigma + m_1 + m_2 + m_3 - 2 + \delta(2n-3) + \frac{1}{2}\delta(n-3) + \frac{3}{2} \geq 0$. En effet un voisinage conique quelconque de μ_0 contient toujours un ensemble de la forme:

$$|\xi''| + |\xi_0 + |\xi_3| - |\xi_1| - |\xi_2|| \leq \varepsilon |\xi'|^\delta \qquad \text{et} \qquad |\xi'/|\xi'| - \mu'^0/|\mu'^0|| \leq \varepsilon_1$$

si on choisit ε_1 et ε assez petits.

Les m_i doivent être choisis de facon que $(n-2)\frac{1}{2}\delta + m_i + m_0 + \frac{1}{2} < 0$ pour que $u_i \in H^{m_0}(\mathbb{R}^{n-1})$ ie $m_i = -(n-2)\frac{1}{2}\delta - \frac{1}{2} - m_0 - \bar{\varepsilon}$ donc $\sigma \geq 3m_0 - \delta n + \frac{3}{2}\delta + 2 + 3\bar{\varepsilon}$, on choisira donc m_0 de sorte que

$$2(m_0 - \tfrac{1}{2}n) + n(1-\delta) < 2 - \tfrac{3}{2}\delta - 3\bar{\varepsilon}$$

(ce qui est possible si δ est assez près de 1) et enfin $\bar{\varepsilon}$ assez petit. Les points μ_0 où $\xi_0 = 0$ sont obtenus dans l'adhérence. Ce qui achévera la preuve.

Cependant, ce sont les singularités de $u = u_0 + E'(w)$ qui nous importent. Il est clair qu'un point $\mu_0 \in \mathrm{WF}_\sigma(w)$ est alors dans $\mathrm{WF}_{\sigma+2}(E'w)$ et donc de u. Si $\mu_0 \in S_0^+ \cap \operatorname{car} \square$ on veut savoir si $\mu_0 \in \mathrm{WF}_{\sigma+1}(E'w)$ et également si la bicaractéristique à temps croissant $C_+(\mu_0)$ issue de μ_0 est dans le WF de $E'(w)$. C'est ce que lon va discuter. On étudiera donc $\widehat{\varphi E'(w_0)}(\xi)$ où φ a son support au voisinage d'un point $(\tilde{x}_0, \tilde{x} = [\tilde{x}_0 - t_0]\xi'^0/|\xi'^0|)$ où $\xi'^0 = \sum_{i=1}^3 \lambda_i^0 e_i$ où $\lambda_i^0 > 0$ et $\lambda_1^0 \lambda_2^0 - \lambda_3^0(\lambda_1^0 + \lambda_2^0) = 0$, $\tilde{x}_0 \geq t_0$; tandis que ξ est dans un voisinage conique de $(\xi_0', |\xi_0'|)$.

On introduit encore $h(x) \in C_0^\infty(\mathbb{R}_+^n)$ égale à 1 au voisinage de $(t_0, 0)$,

$$\mathrm{WF}((1-h)w_0) \subset \bigcup_{i,j} S_{ij}^+ \cup \bigcup_i L_i \cup D_1^+ \cup D_1^-.$$

Et donc $C_+(\mu_0) \cap \mathrm{WF}\, \mathrm{E}'((1-h)w_0) = \varnothing$. On exprime donc comme plus haut

$$\widehat{\varphi E'(hw_0)}(\xi) = \int \hat{\varphi}(\xi - \zeta) \widehat{hw_0}(\zeta) \times \left[\int_0^{+\infty} e^{-is\zeta_0} \left(\frac{e^{is|\zeta'|} - e^{-is|\zeta'|}}{2|\zeta'|} \right) X(\zeta') \rho(s)\, ds \right] d\zeta$$

où $\rho \in C_0^\infty(\mathbb{R})$ a son support au voisinage de $\tilde{x}_0 - t_0$, et est égale à 1 au voisinage de ce point.

Si $\tilde{x}_0 > t_0$ on prend soin d'avoir choisi les supports de φ et h de sorte que $\rho \in C_0^\infty(\mathbb{R}^+)$. Soit $\tilde{x}_0 > t_0$. On restreint l'intégrale en ζ à un voisinage conique assez petit de ξ^0. (pour ξ voisin de ξ^0) dans lequel on a vu que

$$\widehat{hw_0}(\zeta) = e^{-it_0\zeta_0} \int_{f^{-1}(\zeta)} \frac{\zeta_0}{|\lambda'||\eta'||\theta'|} g_1(\eta')g_2(\lambda')g_3'(\theta')\,d\lambda'\,d\theta'\,d\eta' + O(|\zeta|^{m_1+m_2+m_3+\delta(2n-3)-2-\varepsilon})$$

donc

$$|\widehat{\varphi E'}(hw_0)(\xi)| \leq C(1 + |\xi_0 - |\xi'||)^{-N}(1 + |\xi|)^{m_1+m_2+m_3+\delta(2n-3)-3}$$

dans un voisinage conique de ξ^0. Ce qui fait que $\varphi E'(hw_0) \in H^{\alpha+1}$ micro-localement si $\alpha < 3m_0 - \frac{1}{2}n(1 + \delta) + 3\bar{\varepsilon} + 4$. Ce qui sera en général le cas de σ. Dans le cas où $\tilde{x}_0 = t_0$, on obtient seulement la majoration:

$$|\widehat{\varphi E'}(hw_0)(\xi)| \leq C(1 + |\xi_0 - |\xi'||)^{-1}(1 + |\xi|)^{m_1+m_2+m_3+\delta(2n-3)-3}$$

ce qui mène encore à la même conclusion. On a prouvé:

PROPOSITION 2.3. *Si* $\alpha < 3m_0 - \frac{1}{2}n(1 + \delta) + 4 + 3\bar{\varepsilon}$, $E'(w) \in H^{\alpha+1}$ *le long des bicaractéristiques issues des points de* $S_0^+ \cap \operatorname{car} \square$.

Il faut donc poursuivre l'analyse des singularités de w à des ordres H^s plus grands. On suppose que $\sigma < 3m_0 - \frac{1}{2}n(1 + \delta) + 4 + 3\bar{\varepsilon}$. On considère $\hat{\sigma} < \sigma$ qui vérifie $\hat{\sigma} < m_0 + \frac{7}{2}$, $1 + \hat{\sigma} > 3m_0 - \frac{1}{2}n(1 + \delta) + 4 + 3\bar{\varepsilon}$, et $\hat{\sigma} < \sigma + \frac{3}{2}\delta - 1$, $\hat{\sigma} + 1 < 3m_0 - n + \frac{9}{2} + 2\bar{\varepsilon}$. On prouve alors:

PROPOSITION 2.4. *Si* $\mu_0 \in S_0^+ \cap \operatorname{car} \square$ *les fonctions* $E'(v^2)$ *et* $E'(u_0E'(v^2))$ *sont au moins de classe* $H^{\hat{\sigma}+2}$ *le long de la bicaractéristique* $C_+(\mu_0)$.

Démonstration. (1) Considérons d'abord le terme $E'(v^2)$: on a vu plus haut que $\mathrm{WF}_\sigma(w_1) \subset \tilde{C}$ donc $\mathrm{WF}_{\sigma+1}(E'w_1) \subset \tilde{C}$ tandis que

$$\mathrm{WF}_{\sigma+1}(E'(w_0)) \subset \bigcup S_{ij}^+ \cup \bigcup_i L_i \cup D_i^+ \cup D_i^- \cup C(S_0^+).$$

mais on a vu en que $C(S_0^+) \cap \mathrm{WF}_{\sigma+1}(E'(w_0)) = \varnothing$. On a vu cependant que $w_0 \in H^{\sigma-1/2}$ sur S_0^+; ce qui fait que

$$\mathrm{WF}_{\sigma+1}(E'w_0) \subset \bigcup S_{ij}^+ \cup \bigcup L_i \cup D_i^+ \cup D_i^-.$$

Soit $\mathrm{WF}_{\sigma+1}(v) \subset \tilde{C}$.

Et on a vu en Proposition 2.1 que $\mathrm{WF}_\sigma(v^2) \subset \tilde{C}$. On écrit $v = E'(u_0^2 + v^2 + 2u_0v)$. Soit

$$\xi_\pm^0 = \left(\sum_{i=1}^3 \xi_i^0 e_i, \pm \sum_{i=1}^3 \varepsilon_i|\xi_i^0|\right), \qquad \xi_i^0 > 0, \qquad \mu_0^\pm = ((\pm t_0, 0), \xi_\pm^0),$$

et

$$\varphi(x) \in C_0^\infty(\mathbb{R}^n)$$

à support près de $(\pm t_0, 0)$. On construit comme dans la proposition 2.1 une partition de l'unité sur le $(\mathrm{WF}_{\sigma+1}(\varphi v)$ par $p(x, D) = \sum_{i=1}^{3} q_i(x, D) + \sum_{i<j} p_{ij}(x, D)$ où supp con q_i (resp., $p_{ij}) \subset V_i$ (resp., V_{ij}) et $\mu_0^\pm \notin V_i + V_j$, ni à aucun des V_i et V_{ij}, et $(1 - p(x, D))(\varphi v) \in H^{\sigma+1}$. $\mathrm{WF}_{\sigma+1}((1 - p)vv) \subset \mathrm{WF}_{\sigma+1}(v)$ Les termes $p_{ij}(x, D)vp_{kl}(x, D)v \in H^\lambda$ en $\mu_0^\pm$ pour $\lambda \leq \inf(\sigma + 1,\ 2(\sigma - 1) - \frac{1}{2}n - 0)$ ne contribuent pas au $\mathrm{WF}_{\sigma+1}$ car $\sigma > 3 + \frac{1}{2}n$. Il reste donc seulement les termes $(q_i v)(p_{kl} v)$ que l'on écrit sous la forme $(q_i v)(p_{kl} E')(v^2) + 2(q_i v)(p_{kl} E')(u_0 v) + (q_i v)(p_{kl} E')(u_0^2)$. On fait remarquer que $\mathrm{WF}_{m_0+2}(v^2) \subset \bigcup_i D_i^+ \cup D_i^-$ puisque $v \in H^{m_0+1}$ et que $\mathrm{WF}_{m_0+2}(v) \subset \bigcup D_i^+ \cup D_i^-$. Donc $(p_{kl} E')(v^2) \in H^{m_0+4}$, $q_i v \in H^{m_0+1}$ on en déduit que

$$\mathrm{WF}_{\hat\sigma+1}(q_i v(p_{kl} E')(v^2)) \subset \tilde{C}$$

si $\hat\sigma < m_0 + 4$. Et $\mathrm{WF}_{\hat\sigma+2}(E'(q_i v(p_{kl} E')(v^2)) \subset \hat{C}$ et

$$\mathrm{WF}_{\hat\sigma+2}(E'(v(1 - p)v)) \subset \tilde{C}, \qquad C_\pm(\mu_0^\pm) \cap \mathrm{WF}_{\hat\sigma+2}(E'(q_i v q_j v)) = \varnothing.$$

Pour le terme $q_i v(p_{kl} E')(u_0 v)$; on fait remarquer que $\mathrm{WF}_{m_0+3/2}(u_0 v) \subset \bigcup_i D_i^+ \cup D_i^-$ puis que $\mathrm{WF}_{\sigma-1}(u_0 v) \subset \tilde{C}$ pour obtenir de même que

$$\mathrm{WF}_{\hat\sigma+1}(q_i v(p_{kl} E')(u_0 v)) \subset \tilde{C}$$

sous l'hypothèse $\hat\sigma < m_0 + \frac{7}{2}$. Soit $\mathrm{WF}_{\hat\sigma+2}(E'(q_i v(p_{kl} E')(u_0 v))) \subset \tilde{C}$. L'étude du terme $E'(q_i v(p_{kl} E')(u_0^2))$ au voisinage du point $\mu \in C_+(\mu_0)$ fait l'objet des développements suivants: on pose $\lambda = (p_{kl} E')(u_0^2)$, $\mu = q_i v$, pour ξ voisin de ξ^0 et φ à support près de $\tilde{x}$:

$$\widehat{\varphi E'}(\lambda\mu)(\xi) = \int \hat\varphi(\xi - \zeta)\hat\lambda(\zeta - \eta)\hat\mu(\eta)$$
$$\times \left[\int_0^{+\infty} e^{-is\zeta_0}\left(\frac{e^{is|\zeta'|} - e^{-is|\zeta'|}}{2|\zeta'|}\right) X(\zeta')\rho(s)\, ds\right] d\zeta\, d\eta.$$

Par ailleurs on note que

$$|\hat\lambda(\xi)| \leq C(1 + |\xi|)^{m_1 + m_2 - 2 + \delta(n-2)}$$

et est supporté par V_{kl}, $\hat\mu$ est intégrable avec le poids $(1 + |\eta|)^{m_0 + 1 - (n/2) - 0}$ tandis que le terme entre crochets et majoré par $(1 + |\zeta'|)^{-1}(1 + |\zeta_0 - |\zeta'||)^{-1}$ pour ζ voisin de ξ. On restreint l'intégrale en ζ à un voisinage conique de ξ^0 assez petit pour qu'on y ait:

$$|\zeta - \eta| \geq c|\zeta| \qquad \text{et} \qquad |\eta| \geq c|\zeta|$$

pour $\zeta - \eta \in V_{kl}$, $\eta \in V_i$. L'intégrale sur la région complémentaire étant

$O(|\xi|^{-\infty})$. Donc

$$(1+|\xi|)^{\hat{\sigma}+2}|\widehat{\varphi E'}(\lambda\mu)(\xi) \leq O(|\xi|^{-\infty}) + C\int(1+|\xi-\zeta|)^{-N}(1+|\zeta_0-|\zeta'||)^{-1}$$
$$\times(1+|\zeta'-\eta'|)^{\hat{\sigma}+1-(m_0-\frac{1}{2}n+1-0)+m_1+m_2-2+\delta(n-2)}$$
$$(1+|\eta|)^{m_0-\frac{1}{2}n+1-0}|\hat{\mu}(\eta)|\,dn\,d\zeta$$

$$\int_{\xi\in V(\xi^0)}(1+|\xi|)^{2(\hat{\sigma}+2)}|\varphi E'(\lambda\mu)|^2\,d\xi < +\infty \qquad \text{si} \quad \hat{\sigma} < 3m_0 + \tfrac{7}{2} - n + 2\bar{\varepsilon}.$$

Au total $\mathrm{WF}_{\hat{\sigma}+2}(E'(v^2)) \subset \tilde{C}$.

(2) Le terme $u_0E'(v^2)$ s'étudie à l'ordre $\sigma+1$ comme on l'a fait à l'ordre σ dans la proposition 2.1, en utilisant cette fois que $\mathrm{WF}_{\sigma+1}(E'(v^2)) \subset \tilde{C}$ la conclusion est que $\mathrm{WF}_{\hat{\sigma}+1}(u_0E'(v^2)) \subset \tilde{C}$ si $\hat{\sigma} \leq m_0 + \frac{7}{2}$, ce qui est le cas. L'étude du terme $E'(u_0E'(u_0v))$ est un peu plus longue.

PROPOSITION 2.5. *$E'(u_0E'(u_0v)) \in H^{\hat{\sigma}+2}$ le long de la bicaractéristique $C_+(\mu_0)$ issue d'un point $\mu_0 \in S_0^+ \cap \operatorname{car}\square$.*

Démonstration On étudie un terme: $E'(Fu_1E'(Fu_2v))$. Soit $\mu = (\tilde{x}_0, \tilde{x} = (\tilde{x}_0 - t_0)\xi'^0/|\xi'^0|; \xi^0) \in C_+(\mu_0)$; $\varphi(x) \in C_0^\infty(\mathbb{R}^n)$ à support au voisinage de $(\tilde{x}_0, \tilde{x})$. Soit $p(x,D)$ est un opérateur pseudo différentiel supporté près de $(t_0, 0, \xi_2^0e_2 + \xi_3^0e_3, \xi_2^0 - \xi_3^0)$. Il résulte des Propositions 2.1 et 2.3 que $\mathrm{WF}_\sigma(Fu_2E'(v^2)) \subset \tilde{C}$, que $\mathrm{WF}_\sigma(Fu_2E'(u_0v)) \subset \tilde{C}$ et que

$$\mathrm{WF}_{\sigma+1}(E'(Fu_2E'(u_0^2)) \subset \bigcup S_i^+ \cup \bigcup_i L_i \cup D_i^+ \cup D_i^-.$$

Donc $\mathrm{WF}_{\sigma+1}(E'(Fu_2v)) \subset \tilde{C}$. $\mathrm{WF}_{\sigma+1}(Fu_1(1-p)E'(Fu_2v)) \subset \tilde{C} \cup S_0^+ \cup S_0^-$ mais

$$\mu_0 \notin \mathrm{WF}_{\sigma+1}(Fu_1(1-p)E'(Fu_2v))$$

donc

$$C_+(\mu_0) \cap \mathrm{WF}_{\sigma+2}(E'[Fu_1(1-p)E'(Fu_2v)]) = \varnothing.$$

De même $\mu_0 \notin \mathrm{WF}_{\sigma+1}(Fu_1(1-p)E'(Fu_1v))$ mais

$$\mathrm{WF}_{\sigma+1}(pE'(Fu_1v)) \subset \bigcup_i D_i^+ \cup D_i^- \cup L_i \cup S_{12}^+ \cup S_{13}^+$$

et donc $\mathrm{WF}_{\sigma+1}(Fu_1(pE')(Fu_1v)) \not\ni \mu_0$. Et donc

$$C_+(\mu_0) \cap \mathrm{WF}_{\sigma+2}(Fu_1E'(Fu_2v)) = \varnothing.$$

Soit $q(x,D)$ un opérateur pseudo différentiel supporté au voisinage de $(0, t_0; e_3, -1)$. Il est clair que

$$\mathrm{WF}_{\sigma+1}(p(x,D)E'(Fu_2(1-q)v)) \subset \mathrm{WF}_{\sigma-1}(Fu_2(1-q)v)$$

qui en $(t_0, 0)$ ne contient que $\{(\xi_1 e_1 + \xi_2 e_2, \xi_1 - \xi_2), \xi_i \geq 0, \xi \neq 0\}$ et donc $\mu_0 \notin \mathrm{WF}_{\sigma+1}(Fu_1 p(x,D)E'(Fu_2(1-q)v))$ et

$$C_+(\mu_0) \cap \mathrm{WF}_{\sigma+2}(Fu_1(pE')(Fu_2(1-q)v)) = \varnothing.$$

Il reste donc $E'(Fu_1(p(x,D)E')(Fu_2 qv))$. On pose $\mu = q(x,D)v$; on suppose que $q(x,D) = q(D)q_1(x)$. Notant que

$$\begin{aligned}\widehat{p(x,D)E'(\alpha Fu_2\mu)}(\xi) = p(\xi) \int \hat{\psi}(\xi-\zeta)\widehat{\alpha F}u_2(\zeta-\eta)\hat{\mu}(\eta) \\ \times \left(\int_0^{+\infty} e^{-is\zeta_0}\left(\frac{e^{-is|\zeta'|} - e^{is|\zeta'|}}{2|\zeta'|}\right) X(\zeta')\rho(s)\,ds\right] d\zeta\, d\eta\end{aligned}$$

si $p(x,D) = X(x)p(D)\psi(x)$, $\alpha \in C_0^\infty(\mathbb{R}^m)$ a son support près de t_0; tandis que $\rho(s) \in C_0^\infty(\mathbb{R})$ est égale à 1 au voisinage de 0. $p(\xi)$ (pour $|\xi|$ grand) est une fonction homogène supportée par un voisinage conique de $(\xi_2^0 e_2 + \xi_3^0 e_3, \xi_2^0 - \xi_3^0)$ $\psi(x) \in C_0^\infty(\mathbb{R}^m)$ et $X(x)$ ont leur support au voisinage de $(t_0, 0)$. Ce qui permet de calculer

$$\begin{aligned}F(\xi) &= Fu_1\widehat{(p(x,D)E')}(\alpha Fu_2 q(x,D)v)(\xi) \\ &= \int \hat{X}(\xi_0 - \zeta_0 - |\eta'|, \xi' - \zeta' - \eta')\hat{\mu}_1(\eta')p(\zeta) \\ &\quad \times \hat{\psi}(\zeta_0 - \theta_0, \zeta' - \theta' - \kappa')\hat{\alpha}(\theta_0 - \kappa_0 - |\theta'|)\hat{\mu}_2(\theta')\hat{\mu}(\kappa) \\ &\quad \times \left[\int_0^{+\infty} e^{-is\theta_0}\left[\frac{e^{-is|\theta'+\kappa'|} - e^{is|\theta'+\kappa'|}}{2|\theta'+\kappa'|}\right] p\, d\eta X(\theta'+\kappa')\right] d\xi\, d\eta'\, d\kappa\, d\theta.\end{aligned}$$

$F(\xi)$ s'exprime aussi sous la forme

$$\begin{aligned}F(\xi) = \int e^{-ix_0(\xi_0-\zeta_0-|\eta'|)-ix'(\xi'-\zeta'-\eta')-it_0|\theta'|-it_0|\eta'|-iz'(\zeta'-\theta'-\kappa')-iz_0(\zeta_0-\kappa_0-|\theta'|)} \\ \times p(\zeta)X(x)\alpha(z_0)g_1(\eta')g_2(\theta')\hat{\mu}(\kappa)\left[\int_0^{+\infty} e^{-is\zeta_0}\left[\frac{e^{-is|\theta'+\kappa'|} - e^{is|\theta'+\kappa'|}}{2|\theta'+\kappa'|}\right]\right. \\ \left. \times X(\theta'+\kappa')\psi(z_0+s, z)\,ds\right] dx\, dz\, d\eta'\, d\theta'\, d\zeta\, d\kappa.\end{aligned}$$

Si on pose

$$\begin{aligned}\varphi(x,z,\eta',\theta',\kappa',\zeta) = x'(\xi'-\eta'-\zeta') + x_0(\xi_0-\zeta_0-|\eta'|) + z_0(\zeta_0-\kappa_0-|\theta'|) \\ + z'(\zeta'-\theta'-\kappa') - t_0\kappa_0,\end{aligned}$$

$$\begin{aligned}F(\xi) = e^{-it_0\xi_0}\int e^{-i\varphi}p(\zeta)g_1(\eta')g_2(\theta')\hat{\mu}(\kappa) \\ \times \Phi(x, z, \zeta_0, \theta'+\kappa')\,dx\,dz\,d\mu'\,d\theta'\,d\zeta\,d\kappa\end{aligned}$$

où

$$\Phi(x, z, \xi_0, \xi') = \alpha(z_0 + t_0) \int_0^{+\infty} e^{-is\xi_0} \left[\frac{e^{-is|\xi'|} - e^{is|\xi'|}}{2|\xi'|} \right]$$
$$\times X(\xi')\psi(z_0 + t_0 + s, z')X(x_0 + t_0, x') \, ds.$$

Comme plus haut on se débarasse des zones $|\eta' + \theta' + \kappa'| \geq 2|\xi'|$ ou $|\eta' + \theta' + \kappa'| \leq \frac{1}{2}|\xi'|$ dans lesquelles on a

$$c(|\xi'| + |\eta'| + |\theta'| + |\kappa'|) \leq |\xi'| + |\eta' + \theta' + \kappa'| \leq |\varphi'_{x'}| + |\varphi'_{z'}|$$

et intégrant par parties avec l'opérateur ${}^tL$ où

$$L = (1 + |\zeta_0 - |k'| - |\theta'||^2 + |\zeta' - \theta' - \kappa'|^2)^{-N}(1 - \Delta_z)^N$$
$$\times \left[(|\varphi'_{x'}|^2 + |\varphi'_{z'}|^2)^{-1} \left(\varphi'_{x'} \frac{\partial}{\partial x'} + \varphi'_{z'} \frac{\partial}{\partial z'} \right) \right].$$

On est ramené enfin à la situation:

$$F^{\mathrm{II}}(\xi) = e^{-it_0\xi_0} \int e^{-i\varphi} \sigma((\varphi'_{x'}, \varphi'_{z'})|\xi'|^{-1})\zeta((\eta' + \theta' + \kappa')|\xi'|^{-1})$$
$$\times p(\zeta)g_1(\eta')g_2(\theta')\hat{\mu}(\kappa)\Phi(x, z, \zeta_0, \theta' + \kappa') \, dx \, dz \, d\eta' \, d\theta' \, d\zeta \, d\kappa$$

sur le support de l'intégrande on a: pour $|\xi'|$ assez grand et àcondition d'avoir choisi le support de $\hat{\mu}$ assez près $\sum_3$, $|\kappa'|, |\eta'|, |\theta'| \approx |\xi'|$ si σ a son support assez près de zéro; et enfin $|\zeta| \leq c|\xi'|$ et $|\kappa_0| \leq c|\xi'|$, $|\theta' + \kappa'| \geq c'|\xi'|$.

Enfin $|\zeta_0 \pm |\theta' + \kappa'|| \geq c|\xi'|$ et $|\zeta'| \geq c_1|\xi'|$. Raisonnant comme plus haut, on va étudier dans un voisinage V_0 de ξ^0 une intégrale:

$$\mathscr{T}(a, g_1, g_2, p, \mu)(\xi) = \int a(x, z, \eta', \theta', \kappa, \zeta, \xi)e^{-i\varphi}p(\zeta)$$
$$\times g_1(\eta')g_2(\theta')\hat{\mu}(\kappa) \, dx \, dz \, d\eta' \, d\theta' \, d\zeta \, d\kappa.$$

Où $g_1 \in I_\delta^{m_1}(\sum_1)$, $g_2 \in I_\delta^{m_2}(\sum_2)$, supp, con $\hat{\mu} \subset U_3$, supp con $g_i \subset U_i$ $i = 1, 2$, supp con $p \subset V$, $a \in S^\mu_{\text{compact}}$, $p \in S^0(\mathbb{R}^m)$; et sur le support de a on a: $|\kappa| \sim |\eta| \sim |\theta| \sim |\xi| \sim |\zeta|$. On développe a sur $x = z$ et donc $a(x, z, \ldots) = [a(x, x, \ldots) + (z - x)a_1(x, z, \ldots)]X(x - z)$. Comme $z - x = \varphi'\zeta$, on peut intégrer par parties et écrire

$$\mathscr{T}(a, g_1, g_2, p, \mu) = \sum_{\text{finie}} \mathscr{T}(a^j, g_1, g_2, p^j, \mu) + \int a(x, x, \eta', \ldots)e^{-i\psi}$$
$$\times \hat{X}(\zeta_0 - \kappa_0 - |\theta'|, \zeta' - \theta' - \kappa')g_1g_2p\hat{\mu} \, dx \, d\eta' \, d\theta' \, d\kappa \, d\zeta.$$

où dans la lére somme l'un des termes a ou p a vu son degré diminuer.

$$\psi(x, \eta', \theta', \kappa) = x'(\xi' - \eta' - \theta' - \kappa') + x_0(\xi_0 - |\eta'| - |\theta'| - \kappa_0) - t_0\kappa_0.$$

Comme $|\theta' + \kappa'| \geq c|\xi'|$, la seconde intégrale est égale à

$$\int a(x, x, \eta', \theta', \kappa, \kappa_0 + |\theta'|, \kappa' + \theta', \xi) p(\kappa_0 + |\theta'|, \theta' + \kappa') \times e^{i\psi} g_1 g_2 \hat{\mu}\, dx\, d\eta'\, d\theta'\, d\kappa + O(|\xi|^{-\infty}).$$

Si $b(x, \eta', \theta', \kappa, \xi) = a(x, x, \eta', \ldots) p(\ldots)$ on a obtenu:

$$\mathscr{T}(a, g_1, g_2, p, \mu) = \sum_{\text{finie}} \mathscr{T}(a^j, g_1, g_2, p^j, \mu) + \int b(x, \eta', \theta', \kappa, \xi) e^{i\psi} g_1 g_2 \hat{\mu}\, dx\, d\eta'\, d\theta'\, d\kappa + O(|\xi|^{-\infty}).$$

Soit $I(b, g_1, g_2) = \int b(x, \eta', \theta', \kappa, \xi) e^{i\psi} g_1 g_2 \hat{\mu}\, dx\, d\eta'\, d\theta'\, d\kappa$; $b \in S^\mu$, sur le support de b on a $|\eta'| \sim |\theta'| \sim |\kappa| \sim |\xi|$. On va encore pouvoir réduire sur $x = 0$ car x est engendré par les $\partial\psi/\partial\theta'$ et $\partial\psi/\partial\eta'$; on a un C^∞ difféomorphisme homogène de

$$\mathscr{U} = \{(\eta', \theta', \kappa, \xi) \,|\, \eta' \in U_1, \theta' \in U_2, \kappa \in U_3, \xi \in V_0\} \overset{\Phi}{\mapsto} (\eta' + \theta' + \kappa', (\theta' - \theta_2 e_2 + (|\theta'| + |\eta'| + \kappa_0) e_2, \kappa, \xi)$$

dont le jacobien $\mathscr{T}(\Phi)(\eta', \theta', \kappa, \xi) = |(\theta_2/|\theta'|) + (\eta_2/|\eta'|)| \geq c$ dans $\mathscr{U}$ Soit $\Phi = (f(\eta', \theta', \kappa), \kappa, \xi)$.

On pourra de plus supposer que supp con b $\Subset \mathscr{U}$. Ceci permet d'écrire

$$I(b, g_1, g_2) = \sum_{\text{finie}} I(b^j, g_1^j, g_2^j) + \int_{f^{-1}(\xi)} b|_{x=0} e^{it_0\kappa_0} g_1 g_2 \hat{\mu}(\kappa)\, d\sigma_\xi + O(|\xi|^{-\infty}).$$

Le second terme s'écrit sous la forme:

$$H(\xi) = \int b|_{x=0} e^{it_0\kappa_0} g_1(\xi' - \kappa' - \theta'(\xi, \kappa, \hat{\theta}')) g_2(\theta'(\xi, \kappa, \hat{\theta}')) \times \hat{\mu}(\kappa)((\mathscr{T} \circ \Phi^{-1})(\xi, \hat{\theta}', \kappa, \xi))^{-1}\, d\hat{\theta}'\, d\kappa.$$

une telle intégrale se majore par:

$$|H(\xi)| \leq C\left(\int |\hat{\mu}(\kappa)|(1 + |\kappa|)^\alpha\, d\kappa\right)(1 + |\xi|)^{m_1 m_2 + \mu + \delta(n-2) - \alpha}$$

pour $\xi \in V(\xi^0)$, $|\xi|$ assez grand.

On pourra choisir $\alpha = m_0 - \frac{1}{2}n + 1 - 0$. On en déduit une majoration analogue pour $F(\xi)$ avec $\mu = -2$. Ce qui nous intéresse en fait est le terme

$$\mathscr{F}(\varphi E'(Fu_1(p(x, D)E')(\alpha F u_1 q(x, D)v)))(\xi) = G(\xi)$$

soit

$$G(\xi) = \int \hat{\varphi}(\xi - \zeta)F(\zeta)\left[\int_0^{+\infty} e^{-is\zeta_0}\left(\frac{e^{is|\zeta'|} - e^{-is|\zeta'|}}{2|\zeta'|}\right)X(\zeta')\rho_1(s)\,ds\right]d\zeta.$$

modulo $O(|\xi|^{-\infty})$ on restreint l'intégrale en ζ à $V(\xi^\circ)$, soit pour $\xi \in V_1(\xi^\circ)$

$$|G(\xi)|(1 + |\xi|)^{-[m_1+m_2-2+\delta(n-2)-(m_0-(n/2)+1-0)-1]}$$
$$\leq C\int(1 + |\xi - \zeta|)^{-N}(1 + |\zeta_0 - |\zeta'||)^{-1}\,d\zeta,$$

$$\int_{\xi\in V_1(\xi^0)} |G(\xi)|^2(1 + |\xi|)^{2t}\,d\xi < +\infty \qquad \text{si} \quad t < m_0 + 2(m_0 - \tfrac{1}{2}n) + \tfrac{11}{2} + 2\bar{\varepsilon},$$

$t = \hat{\sigma} + 2$ donc si $\hat{\sigma} < m_0 + 2(m_0 - \frac{1}{2}n) + \frac{7}{2} + 2\bar{\varepsilon}$.

Notons que les conditions envisagées sur $\hat{\sigma}$ sont compatibles si: $3(m_0 - \frac{1}{2}n) + \frac{1}{2}n(1 - \delta) + 3\bar{\varepsilon} < \frac{1}{2}\,(\delta > \frac{2}{3})$.

Il s'agit de voir que la singularité apparaît effectivement pour $E'(w_0)$ au point $\mu \in C_+(\mu_0)\backslash\mu_0$. Soit

$$G(\xi) = \widehat{\varphi E'}(hw_0)(\xi) = \int_{|\xi-\zeta|\leq\varepsilon|\xi|} \hat{\varphi}(\xi - \zeta)e^{-it_0\xi_0}H(\zeta)$$
$$\times\left(\int_{-\infty}^{+\infty} e^{-is(\zeta_0-|\zeta'|)}\frac{\rho(s)}{|\zeta'|}\,ds\right)d\zeta$$
$$+ O((1 + |\xi_0 - |\xi'||)^{-N}(1 + |\xi'|)^{m_1+m_2+m_3+\delta(2n-3)-3-\delta}).$$

Lorsque $\zeta \in V_1(\xi^0)$, $|\xi|$ grand, $\zeta \in V(\xi^\circ)$ dans lequel $H(\zeta)$ est un symbole de degré $m_1 + m_2 + m_3 + \delta(2n - 3) - 2$ et de type δ. D'où

$$G(\xi) = \left[\int \sigma(\xi - \zeta\,|\,|\xi'|)e^{-it_0\zeta_0}\hat{\rho}(\zeta_0 - |\zeta'|)\hat{\varphi}(\xi - \zeta)\,d\zeta/|\zeta'|\right]H(\xi)$$
$$+ O((1 + |\xi_0 - |\xi'||)^{-N}(1 + |\xi'|)^{m_1+m_2+m_3+\delta(2n-3)-3-\delta}).$$

Le terme entre crochets s'exprime par:

$$|\xi|^{-1}\int\varphi(s + t_0, s\xi'/|\xi'|)\rho(s)e^{-is(\xi_0-|\xi'|)-it_0\xi_0}\,ds + O(|\xi|^{-2})$$

par la phase stationnaire.

Le module du 1er terme $G_1(\xi)$ domine donc; $(1 + |\xi|)^{\Sigma m_i - 3 + \delta(2n-3)}$ dans $|\xi'/|\xi'| - \xi'^0/|\xi'^0|| \leq \varepsilon_1, |\xi''| + |\xi_0 + |\xi_3| - |\xi_1| - |\xi_2|| \leq \varepsilon|\xi'|^\delta$, et $|\xi_0 - |\xi'|| \leq \varepsilon_0$; donc dans un voisinage conique arbitraire de ξ^0 l'intégrale de $|G_1(\xi)|^2(1 + |\xi|)^{2s}$ va diverger si $s + \sum m_i + \delta(2n - 3) - 3 + [(n - 3)\delta + 2]/2 \geq 0$ soit si $s \geq 3m_0 + 3\bar{\varepsilon} - n\delta + \frac{3}{2}\delta + \frac{7}{2}$ tandis que l'intégrale relative au second terme converge si $s < 3m_0 - \frac{1}{2}n(1 + \delta) + \zeta + \delta + 3\bar{\varepsilon}$. On choisira $s = \hat{\sigma} + 2$, $m_0, \delta, \bar{\varepsilon}$ de façon à ce que toutes ces conditions soient remplies. Ceci achève la preuve du théorème 1.

III. Regularite microlocale de solutions d'équations aux derivées partielles quasi linéaires elliptiques

On montre d'abord le résultat suivant:

Proposition 3.1. *Soit ω ouvert de $\mathbb{R}^n$. $\mu \in H^{\sigma+m-1}(\omega)$, $\sigma > n/2$. Soit $F(x, U)$ une fonctcon telle que $F(x, U)$ est hollomorphe $\forall x \in \omega$ dans un ouvert Ω qui est voisinage de $\{D^\gamma u(x))_{|\gamma| \leq m-1}, x \in \omega\}$; $D_x^\alpha F(x, U)$ est continue sur $\omega \times \Omega$. On peut définir alors $F(x_1 D^\gamma u(x)) \forall x \in \omega$. Si Γ est un cône fermé convexe de $T^*\omega \backslash 0$, $\mathrm{WF}_{\sigma+m-1}(u) \subset \Gamma$ entraîne $\mathrm{WF}_\sigma(F(x, D^\gamma u)) \subset \Gamma$.*

Démonstration. On utilise le lemme suivant. On note

$$\Gamma_{x_0} = \{\xi \in \mathbb{R}^n \backslash 0 \,|\, (x_0, \xi) \in \Gamma];$$

Γ_{x_0} est un cône convexe fermé propre de $\mathbb{R}^n \backslash 0$.

Lemme. *Si $\xi_0 \notin \Gamma_{x_0}$, il existe un voisinage V de x_0 et un voisinage fermé convexe conique propre L de $k = pr_2(\Gamma \cap \pi^{-1}(V))$ tel que $\xi_0 \notin \bar{L}$.*

Démonstration. On prouve d'abord que le cône Γ_{x_0} possède un voisinage ouvert L' conique convexe propre tel que $\xi_0 \notin \bar{L}'$. On note

$$\Gamma_{x_0}^0 = \{n \in \mathbb{R}^n \,|\, nx \leq 0 \ \forall x \in \Gamma_{x_0}\};$$

Γ_x^0 est un cône convexe fermé de $\mathbb{R}^n$. $\mathrm{int}(\Gamma_x^0) = \{n \in \mathbb{R}^n \backslash 0 \,|\, nx < 0 \text{ si } x \in \Gamma_{x_0}\}$ en effet, soit $n \in \mathrm{int}(\Gamma_x^0)$ s'il existe $x \in \Gamma_{x_0}$ tel que $nx = 0$ dans un voisinage de n on a un vecteur de la forme $\tilde{n} = n + \varepsilon x$ et donc $\tilde{n}x > 0$ soit $\tilde{n} \notin \Gamma_{x_0}^0$ ce que est absurde.

L'hypothèse Γ_{x_0} est un cône convexe fermé propre implique que $\Gamma_{x_0}^0$ sépare les points de $\mathbb{R}^n$ donc $\Gamma_{x_0}^0$ engendre $\mathbb{R}^n$, il y a donc une base de $\mathbb{R}^n$ formée de vecteurs de $\Gamma_{x_0}^0$ donc $\mathrm{int}(\Gamma_{x_0}^0) \neq \emptyset$. $\Gamma_{x_0}^0$ étant convexe on en déduit alors que $\overline{\mathrm{int}(\Gamma_{x_0}^0)} = \Gamma_{x_0}^0$.

L'hypothèse $\xi_0 \notin \Gamma_{x_0}$ signifie exactement que $\Gamma_{x_0}^0 \cap \{n \in \mathbb{R}^n \backslash 0 \,|\, \xi_0 n > 0\} \neq \emptyset$ donc il existe $n_0 \in \mathrm{int}(\Gamma_{x_0}^0)$ tel que $\xi_0 n_0 > 0$.

On désigne alors par L_0' une partie fermée conique contenant un voisinage conique de n_0 et contenue dans $\mathrm{int}(\Gamma_{x_0}^0)$. Soit $L' = \{x \in \mathbb{R}^n \backslash 0 \,|\, \forall n \in L_0', xn < 0\}$ Il est clair que L' est un ouvert conique convexe. $L = \bar{L}' \backslash 0$ est une partie convexe fermée de $\mathbb{R}^n \backslash 0$ contenue dans $\{x \in \mathbb{R}^n \backslash 0 \,|\, \forall n \in L_0', xn \leq 0\}$ qui est un convexe propre car $\mathrm{int}(L_0') \neq \emptyset$. L_0' contient une base de $\mathbb{R}^n$. $\Gamma_{x_0} \subset L'$

car $x \in \Gamma_{x_0}$ et $n \in L'_0 \subset \text{int}(\Gamma^0_{x_0})$ entrainent $nx < 0$. $\xi_0 \notin L$ car $\xi_0 n_0 > 0$ et $n_0 \in L'_0$.

On montre maintenant qu'il y a un voisinage V de x_0 tel que $(x, \xi) \in \Gamma$ et $x \in V$ entraînent $\xi \in L'$. Sinon on a une suite $x_n \to x_0$ $(x_n, \xi_n) \in \Gamma$, $\xi_n \to \tilde{\xi}$, $\xi_n \notin L'$, $(x_0, \tilde{\xi}) \in \Gamma$, donc $\tilde{\xi} \notin L'$ ce qui contredit le fait que $\Gamma_{x_0} \subset L' \subset L$.

On construit à l'aide du lemme un voisinage V de x_0 et L un voisinage fermé convexe conique propre de $K = pr_2(\Gamma \cap \pi^{-1}(V))$ tel que $\xi_0 \notin L$. On déduit de l'hypothèse que

$$\forall K' \subset\subset \Omega \quad \forall K \subset\subset \omega \quad \exists C \qquad \sup_{\substack{|\alpha| \leq m + [\sigma] \\ x \in K,\, U \in K'}} \left| \left(\frac{\partial}{\partial U} \right)^{\lambda} D^{\alpha}_x F(x, U) \right| \leq C^{\lambda} \lambda!$$

Il n'est pas restrictif de supposer que $u \in \xi'(\mathbb{R}^n)$. Les $u_\varepsilon = u * \varphi_\varepsilon \in C^\infty_0(\mathbb{R}^n)$ (ε petit). On écrit

$$F(x, D^\gamma u(x)) = \sum_{|\lambda| = 0}^{+\infty} \left(\frac{1}{\lambda!} \right) \left(\frac{\partial}{\partial U} \right)^{\lambda} F(x, D^\gamma u_\varepsilon(x))\, (D^\gamma u(x) - D^\gamma u_\varepsilon(x))^\lambda.$$

La convergence étant uniforme sur toute partie compacte pour ε assez petit car: pour $K \subset\subset \omega$, $\sup_{x \in \mathbb{R}^n,\, |\gamma| \leq m-1} |D^\gamma u_\varepsilon(x) - D^\gamma u(x)| \to 0$ quand $\varepsilon \to 0$ et $\{(D^\gamma u(x), x \in K, |\gamma| \leq m - 1\} \subset\subset \Omega$.

Et donc les $(D^\gamma u_\varepsilon(x))_{|\gamma| \leq m-1}$ décrivent pour $x \in K$, $\varepsilon \leq \varepsilon_0$ une partie compacte de Ω.

Soit $\bar{L}_0$ un voisinage conique de ξ_0 qui ne rencontre pas L, $\alpha(x) \in C^\infty_0(K_0)$, K_0 assez petit pour que $K_0 \subset \mathring{V}$ (V a été déterminé plus haut). On va montrer que

$$\int_{L_0} |\alpha(x) F(x, D^\gamma u(x))(\xi)|^2 (1 + |\xi|)^{2s}\, d\xi < +\infty.$$

or

$$\int_{L_0} \left| \widehat{\alpha(x) \left(\frac{\partial}{\partial U} \right)^{\lambda} F(x, D^\gamma u_\varepsilon)(D^\gamma u(x) - D^\gamma u_\varepsilon(x))^\lambda}(\xi) \right|^2 (1 + |\xi|)^{2s}\, d\xi$$

$$\leq C_1^\lambda \sup_{x \in K,\, |\alpha| \leq N} \left| D^\alpha_x \left(\frac{\partial}{\partial U} \right)^{\lambda} F(x, D^\gamma u_\varepsilon(x)) \right|$$

$$\times \left(\sum_{|\gamma| \leq m-1} \int_{L'} |\widehat{\varphi_\gamma(u - u_\varepsilon)}(\xi)|^2 (1 + |\xi|)^{2(s+m-1)}\, d\xi + \|\varphi_\gamma(u - u_\varepsilon)\|_\sigma \right)^{|\lambda|/2}.$$

$L_{es}\varphi_\gamma \in C^\infty_0(V)$,

$$\sup_{|\alpha| \leq N,\, x \in K} \left| D^\alpha_x \left(\frac{\partial}{\partial U} \right)^{\lambda} F(x, D^\gamma u_\varepsilon(x)) \right| \leq \varepsilon^{-N} C_2^{|\lambda|} \lambda!$$

Par contre pour $L' \Subset K^c$,

$$\int_{L'} |\varphi_\gamma(u - u_\varepsilon)(\xi)|^2 (1 + |\xi|)^{2(s+m-1)}\, d\xi + \|u - u_\varepsilon\|_\sigma^2$$

tend vers zéro quand $\varepsilon \to 0$.

Donc pour ε fixé assez petit, les seconds membres convergent et on a donc prouvé la proposition.

On peut maintenant prouver:

THÉORÈME 2. *Soit $u \in H^{\sigma+m}(\omega)$ avec $\sigma > n/2$. Soit*

$$Pu = \sum_{|\alpha|=m} a_\alpha(x, D^\gamma u) D^\alpha u(x) + F(x, D^\gamma u)$$

une équation quasi linéaire, oū les $a_\alpha(x, U)$ et $F(x, U)$ sont holomorphes en U sur $\mathbb{C}^N$ (i.e., dépendent des dérivées jusqu'a l'orde $m - 1$). On suppose que Γ est un cône convexe fermé propre de $T^*\omega \backslash 0$ et que

$$\left| \sum_{|\alpha|=m} a_\alpha(x, D^\gamma u(x)) \xi^\alpha \right| \geq c|\xi|^m \quad \forall x \in \omega, \quad \forall \xi \in \mathbb{R}^m \backslash 0.$$

Alors $\mathrm{WF}_s(Pu) \subset \Gamma$ *entraîne* $\mathrm{WF}_{s+m}(u) \subset \Gamma$.

Démonstration. On suppose que $\mathrm{WF}_{s+m-1}(u) \subset \Gamma$. Soit $(x_0, \xi^\circ) \in \Gamma^c$, on détermine un voisinage ω_0 de x_0, L_1 un voisinage compact conique de ξ°, L un voisinage convexe compact de $K = pr_2(\Gamma \cap \omega_0 \times \mathbb{R}^m \backslash 0)$ tel que $\omega_0 \times L_1 \subset \omega_0 \times L^c$. Soit $p(\xi)$ homogène telle que $\mathrm{Supp}(1 - p) \subset L'$, où $\omega_0 \times L' \subset \Gamma^c$. Soit V_0 et L_0 tels que $\bar{V}_0 \subset \omega_0$ et $L_0 \subset \mathring{L}_1$. Soit $u_0 \in C_0^\infty(\omega)$ telle que $\sup_{x \in \omega', |\gamma| \leq m-1} |D^\gamma u(x) - D^\gamma u_0(x)|$ est assez petit pour que l'on ait:

$$\left| \sum_{|\alpha|=m} a_\alpha(x, D^\gamma u_0(x)) \xi^\alpha \right| \geq C|\xi|^m, \qquad \text{si} \quad x \in \omega', \bar{\omega}' \Subset \omega; \qquad u_0 = u * \varphi_{\varepsilon_0}$$

On désigne par Q_0 une paramétrixe de l'opérateur $P_0 = \sum_{|\alpha|=m} a_\alpha(x, D^\gamma u_0(x)) \xi^\alpha$ telle que $Q_0 P_0 = \mathrm{op}(X) + R_0$, $\mathrm{supp}\, X \subset \omega_0 \times L_1$ et $\mathrm{supp}\, q_0 \subset \omega_0 \times L_1$. R_0 de degré $-\infty$. Posant $P_m(x, u) \cdot v = \sum_{|\alpha|=m} a_\alpha(x, D^\gamma u) D^\alpha v$, $u_\varepsilon(x) = \alpha(x)\varphi_\varepsilon * (gu) \in C_0^\infty(V_0)$, où $g \in C_0^\infty(V_0)$, $\hat{\varphi}_\varepsilon(\xi) \in \mathscr{S}$, $\hat{\varphi}_\varepsilon(\xi) = \zeta(\varepsilon\xi)\varphi(\xi)$, où $\zeta \in C_0^\infty$, $\varphi \equiv 1$ sur $L_0' \cap \{|\xi| \geq 1\}$, $\mathrm{supp}\, \varphi \subset L_0$. Soit $u_\varepsilon = X_\varepsilon(u)$,

$$Q_0[P_m(x, u) X_\varepsilon u] = Q_0 X_\varepsilon(Pu) - Q_0 X_\varepsilon F(x, D^\gamma u) + Q_0[P_m(x, u), X_\varepsilon] u.$$

Soit

$$X u_\varepsilon + R_0 u_\varepsilon + \sum_{|\alpha|=m} Q_0([a_\alpha(x, D^\gamma u) - a_\alpha(x, D^\gamma u_0)] D^\gamma u_\varepsilon)$$
$$= Q_0 X_\varepsilon(Pu) - Q_0 X_\varepsilon F(x, D^\gamma u) + Q_0[P_m(x, u), X_\varepsilon] u.$$

Posant $w = (a_\alpha(x, D^\gamma u) - a_\alpha(x, D^\gamma u_0))D^\alpha u_\varepsilon$ si $h(x, D)$ est un opérateur pseudo différentiel supporté par $\omega_0 \times L_1$

$$\|h(x, D)w\|_s^2 \leq C\left(\int_{L'} |\widehat{\varphi(a_\alpha(x, D^\gamma u) - a_\alpha(x, D^\gamma u_0))}(\xi)|^2 (\ )^s \, d\xi \right.$$

$$\left. + \|\varphi(a_\alpha(\ ) - a_\alpha(\ ))\|_\sigma^2 \right)^{\frac{1}{2}} \left(\|u_\varepsilon\|_{\sigma+m}^2 + \int_{L'} |\hat{u}_\varepsilon(\xi)|^2 (\ )^{s+m} \, d\xi \right)^{\frac{1}{2}}$$

On sait que $\|Q_0 f\|_{s+m} \leq C\|f\|_s + C_{\varepsilon_0, N}\|f\|_{-N}$ où C est indépendant de ε_0.

L'intégrale $\int_{L'} |\hat{u}_\varepsilon(\xi)|^2 (\ )^{s+m} \, d\xi$ reste bornée par

$$\|\Psi(gu)\|_{s+m-1}^2 + \int_{L_0} |\hat{u}_\varepsilon|^2 (\ )^{s+m} \, d\xi$$

oû $\psi \equiv 1$ sur le support de φ et $\operatorname{supp} \psi \subset L_0$.

Donc si on a choisi ε_0 assez petit

$$\|Q_0(a_\alpha(x, D^\gamma u) - a_\alpha(x, D^\gamma u_0))D^\alpha u_\varepsilon\|_{s+m}^2 \leq \frac{1}{2}\left(\|u_\varepsilon\|_{\sigma+m}^2 + \int_{L_0} |\hat{u}_\varepsilon|^2 (\ )^{s+m} \, d\xi \right)$$

$$+ C(\|u\|_{\sigma+m-1}^2 + \|\psi(gu)\|_{\sigma+m-1}^2)$$

Par ailleurs

$$\|X\mu_\varepsilon\|_{s+m}^2 \geq \int_{L_0} |\hat{u}_\varepsilon(\xi)|^2 (\ )^{s+m} \, d\xi - C\|u_\varepsilon\|_{s+m-1}^2.$$

Il résulte de l'hypothèse que $\|Q_0 X_\varepsilon(P_u)\|_{s+m}$ et $\|Q_0 X_\varepsilon F(x, D^\gamma u)\|_{s+m}$ restent bornés quand $\varepsilon \to 0$.

Si on écrit

$$[P_m(x, u), X_\varepsilon] = \sum_{|\alpha|=m} [a_\alpha(x, D^\gamma u), X_\varepsilon]D^\alpha - a_\alpha(x, D^\gamma u)[X_\varepsilon, D^\alpha],$$

il suffit d'appliquer le lemme suivant à $[v, X(x, D)]$ ōu $v = a_\alpha(x, D_u^\gamma) \in H_{\Gamma^c}^s \cap H^{\sigma+m}$ poùr conclure.

LEMME. *Soit* $v \in H^{\sigma+1} \cap H_{\Gamma_c}^\sigma$, $u \in H^\sigma \cap H_{\Gamma_c}^\sigma$ *ou* Γ *est un cône convexe fermé de* $T^*\omega \backslash 0$ $(\sigma > n/2)$. *Si* $X(x, \xi)$ *est le symbole d'un o.p.d. de degré* 0, supp $X \subset V \times L_0$ *où* V *voisinage de* x_0 *et* L_0 *voisinage compact conique de* ξ^0 *sont tels qu'il existe un voisinage conique convexe compact* L *de* $K = \mathrm{pr}_2(\Gamma \cap V_0 \times \mathbb{R}^n \backslash 0)$ *tel que* $L_0 \subset L_c$. *Alors* $vX(x, D)u - X(x, D)(vu) \in H_{\Gamma_c}^{s+1}$

Démonstration. On exprime

$$vXu(x) - X(vu)(x) = A(u, v)(x) = \iint_0^1 X'_\xi(x, \xi + t\eta)\eta\hat{v}(\eta)\hat{u}(\xi)e^{ix(\xi+\eta)} \, d\xi \, d\eta \, dt.$$

Soit

$$\tilde{X}(\zeta, t, \xi, \eta) = \mathscr{F}(X'_\xi(x, \xi + t\eta)\eta)(\zeta)$$

on a les estimations:

$$|\tilde{X}(\zeta - \xi, t, \xi, \eta)| \leq C(1 + |\zeta - \xi|)^{-N}(1 + |\xi - \eta + t\eta|)^{-1}|\eta| 1_{L_0}(\xi - \eta + t\eta).$$

Soit $p(\xi) \equiv 1$ au voisinage de K, à support dans L.

$$\hat{v}(\eta) = p(\eta)\hat{v}(\eta) + (1 - p)(\eta)\hat{v}(\eta),$$
$$\hat{u}(\xi - \eta) = p(\xi - \eta)\hat{u}(\xi - \eta) + (1 - p)(\xi - \eta)\hat{u}(\xi - \eta).$$

Avec $$p(\eta)p(\xi - \eta)1_{L_0}(\xi - \eta + t\eta) = 0.$$

$$\widehat{A_1(u, v)}(\zeta) = \iint_0^1 \tilde{X}(\zeta - \xi, t, \xi, \eta)p(\eta)\hat{v}(\eta)(1 - p)(\xi - \eta)\hat{u}(\xi - \eta)\, d\xi\, d\eta\, dt$$

s'écrit comme somme $A'_1 + A''_1$ selon que $|\eta| \leq \frac{1}{2}|\xi - \eta|$ ou que $|\eta| > \frac{1}{2}|\xi - \eta|$ soit

$$\int_{L_0} |A'_1(u, v)(\xi)|^2(\ )^{2(s+1)}\, d\xi \leq C\left(\int |\eta|\, |\hat{v}(\eta)|\, d\eta\right)^2 \left(\int |(1 - p)(\xi)|^2 |\hat{v}(\xi)|^2 (\ )^{2s}\, d\xi\right)$$

et

$$A''_1(u, v)(\zeta) = \int_{|\xi - \eta| < 2|\eta|} p(\eta)\hat{v}(\eta)(1 - p)\hat{u}(\xi - \eta)(\hat{X}(\zeta - \xi, \xi - \eta) - \hat{X}(\zeta - \xi, \xi))\, d\xi\, d\eta$$

d'où

$$\int_{L_0} |A''_1(u, v)(\zeta)|^2(\ )^{2(s+1)}\, d\zeta \leq C\left(\int (1 + |\eta|)p(\eta)|\hat{v}(\eta)|\, d\eta\right)^2 \times \int |1 - p(\xi)|^2 |\hat{u}(\xi)|^2 (\ )^{2s}\, d\xi.$$

Soit

$$A_2(u_1, v)(\xi) = \int \hat{v}(\eta)(1 - p)(\eta)p(\xi - \eta)\hat{u}(\xi - \eta) \times (\hat{X}(\zeta - \xi, \xi - \eta) - \hat{X}(\zeta - \xi, \xi))\, d\xi\, d\eta,$$

$$\int_{L_0} |A_2(u, v)|^2(\ )^{2(s+1)}\, d\zeta \leq C \int |\hat{v}(\eta)|^2 |1 - p(\eta)|^2 (\ )^{2(s+1)}\, d\eta \left(\int p(\xi)\hat{u}(\xi)\, d\xi\right)^2,$$

$$A_3(u, v)(\xi) = \iint_0^1 (1 - p(\eta))\hat{v}(\eta)p(\xi - \eta)\hat{u}(\xi - \eta) \times \tilde{X}(\zeta - \xi, t, \xi, \eta)\, dt\, d\xi\, d\eta,$$

s'écrit comme A_1 comme somme de A'_3 et A''_3 qui vérifient respectivement:

$$\int_{L_0} |A'_3(u,v)(\zeta)|^2(\ )^{2(s+1)}\,d\zeta \leq C\int |1-p(\xi)|^2|\hat{u}(\xi)|^2(\ )^{2s}\,d\xi \left(\int |\eta|\,|\hat{v}(\eta)|\,d\eta\right)^2$$

et

$$\int_{L_0} |A''_3(u,v)|^2(\ )^{2(s+1)}\,d\zeta$$

$$\leq C\int |1-p(\eta)|^2|\hat{v}(\eta)|^2(\ )^{2(s+1)}\,d\eta \left(\int |1-p(\xi)|\,|\hat{u}(\xi)|\,d\xi\right)^2.$$

Ce sui achève la preuve du lemme.

Mentionnons pour finir un exemple qui prouve que le résultat du théorème 2 est optimal.

EXEMPLE. Soit

$$u_1(x) = \int e^{ix\xi} g_1(\xi)\,d\xi \qquad \text{avec} \quad g_1 \in I^m_\delta(\Sigma_1),$$

$g_1(\xi) = g(\xi_1)h(\hat{\xi}_1|\xi_1|^{-\delta})$ de sorte que

$$\mathrm{WF}(u_1) = \{(x=0, \xi = \xi_1 e_1), \xi_1 > 0\}.$$

De même $u_2 = \int e^{ix\xi} g_2(\xi)\,d\xi$ avec

$$\mathrm{WF}(u_2) = \{(x=0, \xi = \xi_2 e_2), \xi_2 > 0\}.$$

On choisit pour ω un voisinage de zéro assez petit pour que $x \in \omega$ entraîne que $u_1(x)$ et $u_2(x)$ sont proches de $\mathbb{R}^+_*$ [on note que $u_1(0)$ et $u_2(0) \in \mathbb{R}^+_*$]. On pourra prendre pour Ω un ouvert de $\mathbb{C}\setminus\{0\}$ dans lequel la détermination principale de l'argument est continue.

L'équation $\frac{1}{4}\Delta u\, u^{-3/2} - \frac{1}{2}u^{-1/2}\sum_{i=1}^n (D_{x_i}u)^2 + u^{-1/2} = (1-\Delta)(u_1+u_2) = f$ a pour solution la fonction $u = (u_1+u_2)^2$ on sait que

$$\mathrm{WF}_s(u) = \{(x=0, \xi = \xi_1 e_1 + \xi_2 e_2), \xi_i \geq 0, \xi \neq 0\}$$

pour s assez grand tandis qu'on a toujours $\mathrm{WF}_s(f) = D_1 \cup D_2$.

REFERENCES

1. J. M. BONY, Opérateurs paradifférentiels, *Colloq. e.d.p., St. Cast, 1979.*
2. J. M. BONY, Singularités d'e;d;p; non linéaires, Séminaire d'Orsay, 1978.
3. J. J. DUISTERMAAT, Cours professé au Courant Institute, 1973.

4. J. J. Duistermaat, Oscillatory intégrals, Lagrange immersions and unfolding of singularities, *Comm. Pure Appl. Math.* **27** (1974), 207–281.
5. J. J. Duistermaat and L. Hörmander, *Acta Math.* **128** (1972),
6. V. Guillemin, Symplectic spinors, *Colloq. C.N.R.S., Aix en Provence, 1975.*
7. L. Hörmander, *Acta Math.* **127** (1971),
8. B. Lascar, Singularités d'e.d.p. non linéaires, *C. R. Acad. Sci. Sér. A–B* **287** (1978),
9. J. Rauch, *Colloq. St. Cast, 1979.*
10. J. Rauch, *J. Math. Pures Appl.* (1979),
11. M. Reed, *Comm. Partial Differential Equations* **3** (1978),

MATHEMATICAL ANALYSIS AND APPLICATIONS, PART B
ADVANCES IN MATHEMATICS SUPPLEMENTARY STUDIES, VOL. 7B

A Local Paley–Wiener Theorem for the Radon Transform in Real Hyperbolic Spaces

PETER D. LAX

Courant Institute of Mathematical Sciences
New York University
New York, New York

AND

RALPH S. PHILLIPS†

Department of Mathematics
Stanford University
Stanford, California

DEDICATED TO LAURENT SCHWARTZ

The Paley–Wiener theorem proved in this paper is of the following type: Given that f vanishes near a point β_0 of the infinite plane bounding a hyperbolic space and that the Radon transform vanishes on the set of all horospheres of radius $r \leqq b$ which are tangent at points at infinity close to β_0 then f vanishes in the interior of all such horospheres. The local character of this result is rather surprising since what one expects by analogy with the Euclidean case is that if f is rapidly decreasing and if its Radon transform vanishes on all horospheres which do not meet a given ball B, then f vanishes in the exterior of B. This was proved by Helgason [1, Lemma 8.1 and Remark p. 473]. More recently Lax and Phillips [2] have proved an in-between result which holds if f is merely square integrable. In this case if the Radon transform of f vanishes on all horospheres in the exterior of a fixed horosphere Σ then f vanishes in the exterior of Σ.

We denote the points of the n-dimensional hyperbolic space Π by $w = (x, y)$ where $x \in \mathbb{R}^{n-1}$ and $y > 0$. The Riemann metric is

$$ds^2 = \frac{dx^2 + dy^2}{y^2}; \tag{1}$$

† The work of both authors was supported in part by the National Science Foundation, the first author under Grant No. MCS-76-07039 and the second under Grant No. MCS-77-04908A01.

ISBN 0-12-512802-9

and throughout we shall denote non-Euclidean volume and surface elements by dV and dS. In this geometry the points at infinity lie in the X plane and are denoted by $\beta \in R^{n-1}$. The horospheres are Euclidean spheres tangent to the X plane. We denote by $\xi(r, \beta)$ the horosphere tangent to the X plane at β with Euclidean radius r; it is described by the relation

$$|x - \beta|^2 + (y - r)^2 = r^2. \tag{2}$$

Finally we define the Radon transform $\hat{f}$ of a function f in Π by integrating over horospheres:

$$\hat{f}(r, \beta) = \int_{\xi(r,\beta)} f(w)\, dS, \tag{3}$$

where dS is a non-Euclidean surface element over ξ.

The precise statement of our result is

THEOREM. *If $f \in C(\Pi)$ vanishes near β_0 and if for some $\delta > 0$ and $b > 0$,*

$$\hat{f}(r, \beta) = 0$$

for all $r < b$ and $|\beta - \beta_0| < \delta$, then $f(w)$ vanishes for all w interior to the horospheres $[\xi(b, \beta); |\beta - \beta_0| < \delta]$.

Proof. We begin by introducing the family of functions: Φ is all C real-valued functions of one variable $\varphi(r)$ defined on $(0, b]$, vanishing near b but unconstrained at $r = 0$. We note that Φ is invariant under differentiation and multiplication by r^{-1}; that is,

$$\Phi = [\varphi'; \varphi \in \Phi], \tag{4$'$}$$

$$\Phi = [r^{-1}\varphi(r); \varphi \in \Phi]. \tag{4$''$}$$

As can be seen from (2) if w lies on the horosphere $\xi(r, \beta)$, then

$$r = (|x - \beta|^2 + y^2)/2y,$$

so that

$$\varphi(r) = \varphi\left(\frac{|x - \beta|^2 + y^2}{2y}\right) \tag{5}$$

is constant on horospheres. With this substitution in mind, we now define

$$I(\beta) = \int_{H(b,\beta)} \varphi(r) f(w)\, dV, \tag{6}$$

where dV is non-Euclidean volume; clearly for f vanishing near β, $I(\beta)$ is defined for every φ in Φ. Keeping β fixed but varying r, the volume element in $H(b, \beta)$, the interior of the horosphere $\xi(b, \beta)$ can be written as

$$dV = dS\, dr/r.$$

Hence (6) can be rewritten as

$$I(\beta) = \int_0^b \varphi(r) \left(\int_{\xi(r,\beta)} f(w)\, dS \right) \frac{dr}{r} = \int_0^b \varphi(r) \hat{f}(r, \beta) \frac{dr}{r}. \tag{7}$$

It follows that if f satisfies the hypothesis of the theorem, then

$$I(\beta) \equiv 0 \tag{8}$$

for $|\beta - \beta_0| < \delta$ and all $\varphi \in \Phi$.

The rest of the proof is divided into three steps:

Step 1. For all φ in Φ and all integer multi-indices $\alpha = (\alpha_1, \ldots, \alpha_{n-1})$,

$$\int_{H(b,\beta)} \varphi(r) \left(\frac{\beta - x}{y} \right)^{\alpha} f(w)\, dV \equiv 0 \qquad \text{when} \quad |\beta - \beta_0| < \delta. \tag{9}$$

It is convenient to denote by ε_j the index with one in the jth place and zeros elsewhere. We now proceed by induction, supposing that (9) is true for all α with $|\alpha| \leq m$. Take $|\alpha| = m$; if $\alpha_1 = 0$, then differentiating (9) with respect to β_1 yields after integration by parts

$$\int \varphi'(r) \left(\frac{\beta - x}{y} \right)^{\alpha + \varepsilon_1} f(w)\, dV = 0 \qquad \text{when} \quad |\beta - \beta_0| < \delta.$$

Since by (4)′, the φ's fill out Φ, the result holds for $\alpha + \varepsilon_1$.

If $\alpha_1 > 0$, then

$$\begin{aligned} 0 &= \partial_{\beta_i} \int \varphi(r) \left(\frac{\beta - x}{y} \right)^{\alpha} f(w)\, dV \\ &= \int \left[\varphi'(r) \left(\frac{\beta - x}{y} \right)^{\alpha + \varepsilon_1} + \frac{\alpha_1 \varphi(r)}{y} \left(\frac{\beta - x}{y} \right)^{\alpha - \varepsilon_1} \right] f(w)\, dV. \end{aligned} \tag{10$_1$}$$

Setting $\alpha^j = \alpha - \varepsilon_1 + \varepsilon_j$; then since $|\alpha^j| = m$ the induction assumption implies

$$\begin{aligned} 0 &= \partial_{\beta_j} \int \varphi(r) \left(\frac{\beta - x}{y} \right)^{\alpha j} f(w)\, dV \\ &= \int \left[\varphi'(r) \left(\frac{\beta - x}{y} \right)^{\alpha j + \varepsilon_j} + \frac{(\alpha_j + 1)\varphi(r)}{y} \left(\frac{\beta - x}{y} \right)^{\alpha - \varepsilon_1} \right] f(w)\, dV. \end{aligned} \tag{10$_j$}$$

The relation (2) can be rewritten in the form

$$|\beta - x|^2/y^2 = (2r/y) - 1.$$

Adding the relations $(10)_j$ for $j = 1, \ldots, n-1$ and making the above substitution we get

$$0 = \int \left(\frac{\beta - x}{y}\right)^{\alpha - \varepsilon_1} \left[\varphi'(r)\left(\frac{2r}{y} - 1\right) + \varphi(r)\frac{|\alpha| + n - 2}{y}\right] f(w)\,dV.$$

The induction assumption allows us to ignore the factor 1 of φ'. Moreover given any ψ in Φ we can find a φ in Φ such that

$$2r\varphi' + k\varphi = \psi, \qquad k = |\alpha| + n - 2,$$

namely,

$$\varphi = -r^{-k/2} \int_r^\infty s^{k/2} \frac{\psi(s)}{2s}\,ds.$$

Thus it follows that for all $\psi \in \Phi$,

$$\int \psi(r)\left(\frac{\beta - x}{y}\right)^{\alpha - \varepsilon_1} \frac{1}{y} f(w)\,dV = 0 \qquad \text{if} \quad \|\beta - \beta_0\| < \delta. \tag{11}$$

Inserting this into $(10)_j$ we see that

$$\int \varphi'(r)\left(\frac{\beta - x}{y}\right)^{\alpha + \varepsilon_1} f(w)\,dV = 0 \qquad \text{for} \quad |\beta - \beta_0| < \delta.$$

In view of (4)′, this completes the induction step on the first index and the other indices can be treated similarly.

Step 2. For all $\varphi \in \Phi$ and all n-indices (α, γ)

$$\int \varphi(r)\left(\frac{\beta - x}{y}\right)^{\alpha} \frac{1}{y^\gamma} f(w)\,dV = 0, \qquad |\beta - \beta_0| < \delta. \tag{12}$$

To this end we rewrite (2) as

$$\frac{1}{y} = \frac{1}{2r}\left[\frac{|\beta - x|^2}{y^2} + 1\right]. \tag{13}$$

Substituting this into the left member in (12) and expanding the right-hand side of (13) by the binomial theorem we obtain terms of the form

$$\int \frac{\varphi(r)}{r^\gamma}\left(\frac{\beta - x}{y}\right)^{\alpha'} f(w)\,dV$$

all of which vanish since $\varphi(r)/r^\gamma$ belongs to Φ. This proves (12).

Step 3. To obtain the assertion of the theorem we approximate

$$\varphi_0(r) = \begin{cases} 1 & r < b, \\ 0, & r \geq b, \end{cases}$$

pointwise by functions in Φ and obtain

$$\int_{H(b,\beta_0)} \left(\frac{(x-\beta_0)}{y}\right)^{\alpha} \frac{1}{y^{\gamma}} f(w)\, dV = 0,$$

where $H(b, \beta_0)$ = interior of $\xi(b, \beta_0)$. By assumption f vanishes near β_0 so that the support of f is contained in a truncated sphere of the form

$$H_1(b, \beta_0) = H(b, \beta_0) \cap [y \geq y_0 > 0].$$

The Stone–Weierstrass theorem shows that monomials of the form $[(x - \beta_0)/y]^{\alpha}(1/y^{\gamma})$ span the continuous functions on $H_1(b, \beta_0)$ in the uniform topology. It follows that $f(w) \equiv 0$ on $H(b, \beta_0)$. The same argument works for all $H(b, \beta)$ with $|\beta - \beta_0| < \delta$. This completes the proof of the theorem. ∎

We note that the above proof also works for any f in $L_1(\Pi)$ which vanishes near β_0.

References

1. S. Helgason, The surjectivity of invariant differential operators on symmetric spaces I, *Ann. of Math.* **98** (1973), 451–479.
2. P. D. Lax and R. S. Phillips, Translation representations for the solution of the non-Euclidean wave equation, *Comm. Pure Appl. Math.* **32** (1979), 617–667.

MATHEMATICAL ANALYSIS AND APPLICATIONS, PART B
ADVANCES IN MATHEMATICS SUPPLEMENTARY STUDIES, VOL. 7B

On the Isomorphic Classification of Injective Banach Lattices

J. LINDENSTRAUSS† AND L. TZAFRIRI‡

Institute of Mathematics
Hebrew University
Jerusalem, Israel
and
Department of Mathematics
Ohio State University
Columbus, Ohio

DEDICATED TO PROFESSOR LAURENT SCHWARTZ ON THE OCCASION OF HIS 65TH BIRTHDAY

A Banach lattice X is called injective if it is complemented by a positive projection in every Banach lattice Y containing it as a sublattice. This notion seems to have been discussed first in the literature by Lotz [5]. Lotz observed that, for every injective lattice X, there is a number λ such that, whenever X is a sublattice of a Banach lattice Y, there is a positive projection of norm $\leq \lambda$ from Y onto X. We call such a lattice X a λ-injective lattice. It is easily seen that X is a λ-injective lattice if and only if, for every Banach lattice W, sublattice V of W and every positive operator T from V into X there is a positive extension $\hat{T}$ of T from W into X with $\|\hat{T}\| \leq \lambda \|T\|$. In [5] Lotz observed that the class of 1-injective lattices includes besides the spaces $C(K)$ with K extremally disconnected (i.e., the so called $\mathscr{P}_1$ spaces) also all the $L_1(\mu)$ spaces and that this class is closed with respect to direct sums in the sense of l_∞.

A detailed study of 1-injective Banach lattices was undertaken by Cartwright [1] and Haydon [2]. Cartwright characterized the 1-injective lattices by some order intersection properties and proved that the finite-dimensional 1-injective lattices are exactly those which are order isometric to lattices of the form $(\sum_{j=1}^{k} \oplus l_1^{n_j})_\infty$ (i.e., direct sums in the sense of l_∞ of finite dimensional L_1 spaces). R. Haydon, in a thorough study, obtained a representation theorem for general 1-injective lattices. This representation, which involves the use of vector bundles, is quite complicated and shall not be reproduced here.

† Research supported in part by NSF Grant MCS 78-02194.
‡ Research supported in part by NSF Grant MCS 79-03042.

ISBN 0-12-512802-9

The tools used by Cartwright and Haydon are of a purely isometric nature and give no information on the structure of λ-injective lattices for $\lambda > 1$. The main natural question concerning such lattices is whether every λ-injective lattice is order isomorphic to a 1-injective lattice. The main result of the present note is a positive solution to the local version of this problem. We show that there exists a function $f(\lambda)$ so that every finite-dimensional λ-injective lattice is order isomorphic, with an isomorphism constant $\leq f(\lambda)$, to one of the lattices $(\sum_{j=1}^{k} \oplus l_1^{n_j})_\infty$. From this local result it follows easily that any discrete injective lattice is order isomorphic to a 1-injective lattice. The general case remains however open. We just mention that after the completion of the work reported here Haydon communicated to us a simple argument (due to himself and P.Mangheri) which gives an affirmative answer in another special case; the only order continuous injective lattices are those order isomorphic to $L_1(\mu)$ spaces.

Our solution to the local version of the λ-injective problem implies also a result of some interest concerning unconditional bases (answering a question posed to us by W. B. Johnson). An unconditional basic sequence $\{e_i\}_{i=1}^\infty$ is said to be *block injective* if, for every unconditional basic sequence $\{x_j\}_{j=1}^\infty$ and every block basis $\{f_i\}_{i=1}^\infty$ of $\{x_j\}_{j=1}^\infty$ which is equivalent to $\{e_i\}_{i=1}^\infty$, there is a bounded projection from $[x_j]_{j=1}^\infty$ onto $[f_i]_{i=1}^\infty$. We shall prove below that the only normalized block injective unconditional bases are (up to equivalence and a permutation) the unit vector bases of the following six spaces

$$c_0,\ l_1,\ c_0 \oplus l_1,\ \left(\sum_{n=1}^{\infty} \oplus l_1^n\right)_0,\ l_1 \oplus \left(\sum_{n=1}^{\infty} \oplus l_1^n\right)_0,\ \text{and}\ \left(\sum_{n=1}^{\infty} \oplus l_1\right)_0.$$

We follow the notation of [3, 4]. We shall work with lattices over the real field. However, all the results and the proofs carry over trivially to the complex case as well.

We pass to the local λ-injective problem. We find it convenient to break the solution into two steps. In the first step we prove that finite-dimensional λ-injective lattices are order isomorphic (with constant $\leq 4\lambda^2$) to lattices in which the norm is given by

$$\left\| \sum_{i=1}^{n} \alpha_i e_i \right\| = \sup_{1 \leq j \leq m} \sum_{i \in \sigma_j} |\alpha_i|,$$

for a suitable collection $\{\sigma_j\}_{j=1}^m$ of subsets of $\{1,2,\ldots,n\}$. In the second (and main) step we show that the sets $\{\sigma_j\}_{j=1}^m$ can be replaced by disjoint ones.

LEMMA 1. *Let X be an n-dimensional λ-injective lattice and let $\{e_i\}_{i=1}^n$ be its normalized atoms. Then there exist subsets $\{\sigma_j\}_{j=1}^m$ of $\{1,\ldots,n\}$ so that, for* *every choice of scalars* $\{\alpha_i\}_{i=1}^n$,

$$\frac{1}{2\lambda} \sup_{1 \le j \le m} \sum_{i \in \sigma_j} |\alpha_i| \le \left\| \sum_{i=1}^n \alpha_i e_i \right\| \le 2\lambda \sup_{1 \le j \le m} \sum_{i \in \sigma_j} |\alpha_i|.$$

Proof. There is no loss of generality in assuming that there is a finite set of positive functionals $\{\varphi_j\}_{j=1}^m$ of norm one in X^* so that $\|x\| = \sup_{1 \le j \le m} \varphi_j(|x|)$, for every $x \in X$. Indeed, for every $\varepsilon > 0$, there is such a set $\{\varphi_j\}_{j=1}^m$ so that $\|x\| = \sup_{1 \le j \le m} \varphi_j(|x|) \ge \|x\|/(1 + \varepsilon)$. Thus, by replacing λ with $\lambda(1 + \varepsilon) = \lambda'$, the general case reduces to the special case. Since X is finite dimensional and $\varepsilon > 0$ is arbitrary, a compactness argument shows that the general case holds also if $\lambda' = \lambda$.

Let Z be the direct sum in the sense of l_∞ of m copies of l_1^n. The natural unit vectors of Z are denoted by $\{u_{i,j}\}_{i=1,\, j=1}^{n,m}$. We have thus, for every choice of $\beta_{i,j}$,

$$\left\| \sum_{i,j} \beta_{i,j} u_{i,j} \right\| = \sup_{1 \le j \le m} \sum_{i=1}^n |\beta_{i,j}|.$$

The map $T : X \to Z$ defined by $Te_i = \sum_{j=1}^m \varphi_j(e_i) u_{i,j}$, $1 \le i \le n$, is a positive isometry from X onto a sublattice of Z. Since X is λ-injective there is a positive projection P from Z onto TX with $\|P\| \le \lambda$. By replacing P by its diagonal with respect to $\{Te_i\}_{i=1}^n$ (in the sense of [3, p. 20]), we may assume without loss of generality that, for every i and j, there is a $c_{i,j} \ge 0$ so that $Pu_{i,j} = c_{i,j} Te_i$.

For every $1 \le i \le n$ let $\eta_i = \{j, \varphi_j(e_i) > 1/2\lambda\}$ and

$$v_i = \sum_{j \in \eta_i} \varphi_j(e_i) u_{i,j}.$$

Clearly,

$$\left\| Te_i - v_i \right\| = \left\| \sum_{j \notin \eta_i} \varphi_j(e_i) u_{i,j} \right\| = \sup_{j \notin \eta_i} |\varphi_j(e_i)| \le 1/2\lambda.$$

Hence,

$$\|Pv_i\| \ge \|Te_i\| - \|P\| \, \|Te_i - v_i\| \ge \tfrac{1}{2}.$$

and thus $Pv_i = c_i Te_i$, for some $c_i \ge \frac{1}{2}$. For every j let now $\sigma_j = \{i; \varphi_j(e_i) > 1/2\lambda\} = \{i; j \in \eta_i\}$. By using the fact that the unconditionality constant of $\{u_{i,j}\}$ is one, we get, for every choice of scalars $\{\alpha_i\}_{i=1}^n$, that

$$\sup_{1 \le j \le m} \sum_{i \in \sigma_j} |\alpha_i|/2\lambda \le \left\| \sum_{i=1}^n \alpha_i v_i \right\| \le \left\| \sum_{i=1}^n \alpha_i e_i \right\| = \left\| \sum_{i=1}^n \alpha_i Te_i \right\|$$

$$\le 2 \left\| \sum_{i=1}^n \alpha_i P v_i \right\| \le 2\|P\| \left\| \sum_{i=1}^n \alpha_i v_i \right\| \le 2\lambda \sup_{1 \le j \le m} \sum_{i \in \sigma_j} |\alpha_i|. \quad \blacksquare$$

LEMMA 2. *Let X be a λ-injective lattice of dimension n with atoms $\{e_i\}_{i=1}^n$. Assume also that the norm in X is given by*

$$\left\|\sum_{i=1}^n \alpha_i e_i\right\| = \sup_{1\le j\le m} \sum_{i\in\sigma_j} |\alpha_i|,$$

for some collection $\{\sigma_j\}_{j=1}^m$ of subsets of $\{1,2,\ldots,n\}$. Then there exists a partition of $\{1,2,\ldots,n\}$ into disjoint sets $\{\tau_s\}_{s=1}^k$ so that

$$\sup_{1\le s\le k} \sum_{i\in\tau_s} |\alpha_i|/8\lambda \le \left\|\sum_{i=1}^n \alpha_i e_i\right\| \le 32\lambda^3 \sup_{1\le s\le k} \sum_{i\in\tau_s} |\alpha_i|,$$

for every choice of $\{\alpha_i\}_{i=1}^n$.

Proof. For every $1\le i\le n$ let $\eta_i = \{j; i\in\sigma_j\} \subset \{1,\ldots,m\}$. As in the proof of Lemma 1, let $Z = (\sum_{j=1}^m \oplus l_1^n)_\infty$ and let $\{u_{i,j}\}_{i=1,\,j=1}^{n,m}$ be its unit vectors. By our assumption, the map $T:X\to Z$ defined by $Te_i = \sum_{j\in\eta_i} u_{i,j}$ defines a positive isometry from X onto a sublattice of Z. Let Q be a positive projection of norm $\le\lambda$ from Z onto TX. By replacing Q, if necessary, by its diagonal, we may assume without loss of generality that

$$Qz = \sum_{i=1}^n y_i^*(z)Te_i; \qquad z\in Z,$$

where $y_i^* \in Z^*$ has the same support as Te_i. Since y_i^* is a positive functional and $y_i^*(Te_i) = 1$ we deduce that $\|y_i^*\| = 1$ for every i. Since $\operatorname{span}\{u_{i,j}\}_{j=1}^m$ is, for every i, isometric to l_∞^m we can consider each y_i^* also as an element of $(l_\infty^m)^*$. For future reference we note that, for every choice of positive $\{\beta_i\}_{i=1}^n$, we have

$$\left\|\sum_{i=1}^n \beta_i e_i\right\|_X = \left\|\sum_{i=1}^n \beta_i \chi_{\eta_i}\right\|_\infty.$$

In order to define the decomposition of $\{1,2,\ldots,n\}$ into disjoint subsets we first introduce the notion of "friendship" between integers. We say that i_1 and i_2 are *friends* if

$$y_{i_1}^*(\chi_{\eta_{i_2}}) \ge 1/8\lambda \qquad \text{and} \qquad y_{i_2}^*(\chi_{\eta_{i_1}}) \ge 1/8\lambda.$$

[here we consider the y_i^* as elements in $(l_\infty^m)^*$]. Note that this relation is symmetric but not necessarily transitive.

Set $i(1) = 1$ and let τ_1 be the set of all integers $\le n$ which are friends with $i(1)$. Let $i(2)$ be the first integer not belonging to τ_1 and let τ_2 be the set of all the friends of $i(2)$ which are not already friends of $i(1)$. We continue in an obvious way and obtain a decomposition of $\{1,2,\ldots,n\}$ into disjoint sets $\{\tau_s\}_{s=1}^k$. All the elements in τ_s are friends with $i(s)\in\tau_s$. If $s_1\ne s_2$ then

$i(s_1)$ and $i(s_2)$ are not friends. We are now going to prove the inequalities appearing in the statement of the lemma.

The left-hand inequality is easy:

$$\left\|\sum_{i=1}^{n} \alpha_i e_i\right\| = \left\|\sum_{s=1}^{k} \sum_{i \in \tau_s} \alpha_i T e_i\right\| \geq \max_{1 \leq s \leq k} \left\|\sum_{i \in \tau_s} |\alpha_i| T e_i\right\|$$
$$= \max_{1 \leq s \leq k} \left\|\sum_{i \in \tau_s} |\alpha_i| \chi_{\eta_i}\right\|_\infty \geq \max_{1 \leq s \leq k} y^*_{i(s)}\left(\sum_{i \in \tau_s} |\alpha_i| \chi_{\eta_i}\right)$$
$$\geq \max_{1 \leq s \leq k} \sum_{i \in \tau_s} |\alpha_i|/8\lambda.$$

In order to prove the other inequality we put, for every $i \in \tau_s$, $1 \leq s \leq k$,

$$y_i = \sum_{j \in \eta_i \cap \eta_{i(s)}} u_{i,j}.$$

Notice that

$$y_i^*(y_i) = y_i^*(\chi_{\eta_{i(s)}}) \geq 1/8\lambda; \qquad i \in \tau_s, \quad 1 \leq s \leq k,$$

and hence,

$$8\lambda^2 \left\|\sum_{i=1}^{n} \alpha_i y_i\right\| \geq 8\lambda \left\|Q \sum_{i=1}^{n} \alpha_i y_i\right\| = 8\lambda \left\|\sum_{i=1}^{n} \alpha_i y_i^*(y_i) T e_i\right\| \geq \left\|\sum_{i=1}^{n} \alpha_i T e_i\right\|.$$

Consequently,

$$\left\|\sum_{i=1}^{n} \alpha_i e_i\right\| \leq 8\lambda^2 \left\|\sum_{s=1}^{k} \sum_{i \in \tau_s} \alpha_i y_i\right\|$$
$$= 8\lambda^2 \left\|\sum_{s=1}^{k} \sum_{i \in \tau_s} |\alpha_i| \chi_{\eta_i \cap \eta_{i(s)}}\right\|_\infty \leq 8\lambda^2 \left\|\sum_{s=1}^{k} \sum_{i \in \tau_s} |\alpha_i| \chi_{\eta_{i(s)}}\right\|_\infty$$
$$\leq 8\lambda^2 \max_{1 \leq s \leq k} \sum_{i \in \tau_s} |\alpha_i| \left\|\sum_{s=1}^{k} \chi_{\eta_{i(s)}}\right\|_\infty.$$

Thus, we have only to show that

$$\left\|\sum_{s=1}^{k} \chi_{\eta_{i(s)}}\right\|_\infty \leq 4\lambda.$$

Assume this is false. Then there would exist $[4\lambda]$ integers, say $1, 2, \ldots, [4\lambda]$, so that $\|\sum_{s=1}^{[4\lambda]} \chi_{\eta_{i(s)}}\|_\infty = [4\lambda]$, i.e.,

$$\eta_{i(1)} \cap \eta_{i(2)} \cap \cdots \cap \eta_{i([4\lambda])} \neq \varnothing.$$

Put now, for $1 \leq s \leq [4\lambda]$,

$$\tilde{\eta}_s = \eta_{i(s)} - \bigcup \{\eta_{i(l)},\ 1 \leq l \leq [4\lambda],\ l \neq s \text{ and } y^*_{i(s)}(\chi_{\eta_{i(l)}}) < 1/8\lambda\}.$$

Since $i(s_1)$ is not a friend of $i(s_2)$ for $s_1 \neq s_2$ it follows that the sets $\tilde{\eta}_s$ are mutually disjoint and hence,

$$z = \sum_{s=1}^{[4\lambda]} \sum_{j \in \tilde{\eta}_s} u_{i(s),j}$$

is an element of norm 1 in Z. Note also that, for every $1 \leq s \leq [4\lambda]$,

$$y_{i(s)}^*(\chi_{\tilde{\eta}_s}) \geq y_{i(s)}^*(\chi_{\eta_{i(s)}}) - [4\lambda]/8\lambda \geq \tfrac{1}{2}.$$

Hence,

$$\begin{aligned} \|Qz\| &= \left\| \sum_{s=1}^{[4\lambda]} y_{i(s)}^*(\chi_{\tilde{\eta}_s}) T e_{i(s)} \right\| \geq \frac{1}{2} \left\| \sum_{s=1}^{[4\lambda]} T e_{i(s)} \right\| \\ &= \frac{1}{2} \left\| \sum_{s=1}^{[4\lambda]} \chi_{\eta_{i(s)}} \right\| \quad = [4\lambda]/2 > \lambda, \end{aligned}$$

and this contradicts the fact that $\|Q\| \leq \lambda$. ■

THEOREM 1. *Let X be a λ-injective lattice with* $\dim X < \infty$. *Then X is order isomorphic to a* 1*-injective lattice by an order isomorphism U so that* $\|U\| \, \|U^{-1}\| \leq 2^{18} \lambda^{14}$.

Proof: By Lemma 1, X is order isomorphic to a lattice Y of the form appearing in the statement of Lemma 2 and the norm of this isomorphism is $\leq 4\lambda^2$. Consequently, Y is $4\lambda^3$-injective. By Lemma 2, Y is isomorphic to a 1-injective lattice by an isomorphism of norm $\leq 8(4\lambda^3)32(4\lambda^3)^3$. ■

Remark: By combining the two steps of the proof of Theorem 1 we can get a better estimate for the bound $f(\lambda)$ for the norm of the isomorphism. Since anyhow we get an estimate which seems far from the best possible we chose to give the presentation above for the sake of clarity.

We pass now to the characterization of injective discrete lattices. Recall that a lattice X is called *discrete* if it coincides with the band generated by its atoms.

THEOREM 2. *A discrete injective lattice is order isomorphic to a* 1*-injective lattice. In particular, a discrete injective lattice with countably many atoms is order isomorphic to one of the following lattices*

$$l_\infty,\ l_1,\ l_\infty \oplus l_1,\ \left(\sum_{n=1}^{\infty} \oplus l_1^n \right)_\infty,\ l_1 \oplus \left(\sum_{n=1}^{\infty} \oplus l_1^n \right)_\infty,\ \text{and}\ \left(\sum_{n=1}^{\infty} \oplus l_1 \right)_\infty.$$

Proof: Let X be a discrete λ-injective lattice. We shall prove the theorem only in the case where X has a countable number of atoms. The proof in the general case is similar but the notation is somewhat more complicated.

Before we proceed we recall the trivial fact that X is order complete. Indeed, let $\{x_\alpha\}_\alpha$ be an upward directed order bounded net in X and let x^{**} be its supremum in X^{**}. Let P be a positive projection from X^{**} onto X. Then Px^{**} is the supremum of $\{x_\alpha\}_\alpha$ in X.

Let $\{x_i\}_{i=1}^\infty$ be the atoms of X normalized so that $\|x_i\| = 1$ for every i. For every integer n the lattice spanned by $\{x_i\}_{i=1}^n$ is λ-injective. Hence, for every n, there is a partition of $\{1, 2, \ldots, n\}$ into m_n pairwise disjoint sets $\{\sigma_{j,n}\}_{j=1}^{m_n}$ so that, for all $\{\alpha_i\}_{i=1}^n$ and a suitable function $g(\lambda)$,

$$\left\|\sum_{i=1}^n \alpha_i x_i\right\| \Big/ g(\lambda) \le \sup_{1 \le j \le m_n} \sum_{i \in \sigma_{j,n}} |\alpha_i| \le g(\lambda) \left\|\sum_{i=1}^n \alpha_i x_i\right\|.$$

We may clearly assume that the sets $\sigma_{j,n}$ are numbered so that the smallest element of $\sigma_{j,n}$ is less than the smallest element of $\sigma_{j+1,n}$ and thus, for every $i \le n$,

$$i \in \sigma_{j_n(i),n} \qquad \text{with} \quad j_n(i) \le i.$$

By a diagonal procedure we can find a subsequence $\{n_k\}_{k=1}^\infty$ of the integers so that

$$j(i) = \lim_{k \to \infty} j_{n_k}(i)$$

exists for every i. For every j, put $\sigma_j = \{i; j(i) = j\}$. Clearly, $\{\sigma_j\}_{j=1}^\infty$ forms a partition of the integers into disjoint subsets (of course, some of the σ_j may be empty). It is clear from the construction that, for every choice of scalars $\{\alpha_i\}_{i=1}^\infty$, which are eventually zero,

$$\left\|\sum_{i=1}^\infty \alpha_i x_i\right\| \Big/ g(\lambda) \le \sup_j \sum_{i \in \sigma_j} |\alpha_i| \le g(\lambda) \left\|\sum_{i=1}^\infty \alpha_i x_i\right\|.$$

Consequently, the closed linear span of $\{x_i\}_{i=1}^\infty$ is order isomorphic to the lattice $U = (\sum_j \oplus l_1^{n_j})_0$, where $n_j = \bar{\bar{\sigma}}_j$. Denote by T the order isomorphism from U into X and let $V = (\sum_j \oplus l_1^{n_j})_\infty \supset U$. Since X is λ-injective there is a positive extension $\hat{T}$ of T from V into X with $\|\hat{T}\| \le \lambda \|T\| \le \lambda g(\lambda)$. For every $0 \le v \in V$ put

$$T_0 v = \sup\{Tu; u \in U, u \le v\}.$$

This supremum exists since X is order complete and $Tu \le \hat{T}v$ for every $u \le v$. We have

$$\|v\|/g(\lambda) = \sup_{0 \le u \le v} \|Tu\| \le \|T_0 v\| \le \|\hat{T}v\| \le \lambda g(\lambda)\|v\|.$$

It is easily checked that T_0 is additive on the positive cone of V and that $v_1 \wedge v_2 = 0 \Rightarrow T_0 v_1 \wedge T_0 v_2 = 0$. Hence, T_0 extends uniquely to an order isomorphism from V into X. Since V and thus $T_0 V$ is order complete, since

X is discrete and T_0V contains all its atoms, it follows that $X = T_0V$. Thus, X is order isomorphic to the 1-injective lattice V.

In order to prove that V and thus also X are order isomorphic to one of the six spaces appearing in the statement of the theorem, we have just to use the following obvious observation. Assume that $\{\sigma_j\}_{j=1}^{\infty}$ is a partition of the integers N into disjoint finite sets with $\sup_j \overline{\overline{\sigma}}_j = \infty$. Then there exists a partition of N into disjoint sets $\{\eta_k\}_{k=1}^{\infty}$ with $\overline{\overline{\eta}}_k = k$ for every k and a permutation π of N so that, for every j, $\pi\sigma_j$ is contained in the union of two of the η_ks and, for every k, $\pi^{-1}\eta_k$ is contained in the union of two of the σ_js. ■

Remark. The main difficulty encountered in trying to extend Theorem 2 to general injective lattices stems from the fact that we do not know how to guarantee the existence of nicely complemented finite-dimensional sublattices in an injective lattice. From the representation theorem of 1-injective lattices it follows easily that these lattices are spanned by finite-dimensional sublattices which are complemented by contractive projections. For general injective lattices we know only that they have the bounded approximation property and this information does not seem to suffice.

We pass now to the characterization of block injective unconditional basic sequences defined in the introduction.

THEOREM 3. *The only normalized block injective unconditional basic sequences (up to an equivalence and permutation) are the unit vector bases of the spaces*

$$c_0,\ l_1,\ c_0 \oplus l_1,\ \left(\sum_{n=1}^{\infty} \oplus l_1^n\right)_0,\ l_1 \oplus \left(\sum_{n=1}^{\infty} \oplus l_1^n\right)_0,\ \textit{and}\ \left(\sum_{n=1}^{\infty} \oplus l_1\right)_0.$$

It is obvious that all the basic sequences listed above are block injective. The proof of the converse requires besides Theorem 1 also a uniformity argument which is based on the following lemma.

LEMMA 3. *Let $\{e_j\}_{j=1}^{\infty}$ be a normalized unconditional basis of a Banach space X whose unconditional constant is 1. Assume that, for every k there is a finite normalized basic sequence $\{w_i\}_{i=n_k}^{n_{k+1}-1}$ with an unconditional constant equal to 1 and a block basis $\{v_j\}_{j=m_k}^{m_{k+1}-1}$ of $\{w_i\}_{i=n_k}^{n_{k+1}-1}$ which is isometrically equivalent to $\{e_j\}_{j=m_k}^{m_{k+1}-1}$ (where $n_1 = 1 < n_2 < \cdots$; $m_1 = 1 < m_2 < \cdots$). Then there exists a norm $|||\cdot|||$ on $\operatorname{span}\{w_i\}_{i=1}^{\infty}$ so that*

(i) *The unconditional constant of $\{w_i\}_{i=1}^{\infty}$ is 1.*

(ii) *The norm $|||\cdot|||$ agrees with the original norm on $\operatorname{span}\{w_i\}_{i=n_k}^{n_{k+1}-1}$ for every k.*

(iii) *With the norm $|||\cdot|||$, the block basis $(v_j\}_{j=1}^{\infty}$ of $\{w_i\}_{i=1}^{\infty}$ is isometrically equivalent to $\{e_j\}_{j=1}^{\infty}$.*

Proof. Denote all the given norms (on $\operatorname{span}\{e_j\}_{j=1}^{\infty}$ and on $\operatorname{span}\{w_i\}_{i=n_k}^{n_{k+1}-1}$, $k = 1, 2, \ldots$) by $\|\cdot\|$. We may assume that, for every j,

$$v_j = \sum_{i=p_j}^{p_{j+1}-1} a_i w_i,$$

where $a_i \geq 0$ for all i and $p_1 = 1 < p_2 < \cdots$ (note that, in particular, $p_{m_k} = n_k$ for every k). We define the norm $|||\cdot|||$ on the algebraic span of $\{w_i\}_{i=1}^{\infty}$ as that which is generated by the convex hull C of all the vectors of the form

$$\sum_{j=1}^{l} \sum_{i=p_j}^{p_{j+1}-1} b_j \theta_i a_i w_i,$$

whenever $\|\sum_{j=1}^{l} b_j e_j\| \leq 1$ and $\theta_i = \pm 1, i = 1, 2, \ldots$, as well as all the vectors of norm ≤ 1 in $\operatorname{span}\{w_i\}_{i=n_k}^{n_{k+1}-1}$, $k = 1, 2, \ldots$.

From the definition of C it is clear that (i) holds. It is also clear that

(*) $|||w||| \leq \|w\|$, $w \in \operatorname{span}\{w_i\}_{i=n_k}^{n_{k+1}-1}$, $k = 1, 2, \ldots$,

(**) $|||\sum_{j=1}^{l} b_j v_j||| \leq \|\sum_{j=1}^{l} b_j e_j\|$, for every choice of scalars $\{b_j\}_{j=1}^{\infty}$.

Actually, we have equalities in (*) and (**). We shall prove this for (**), the proof of (*) being similar and simpler.

Assume that $\|\sum_{j=1}^{l} \hat{b}_j e_j\| = 1$ with $\hat{b}_j \geq 0$ for every j. By the Hahn–Banach theorem and the fact that $\{e_j\}_{j=1}^{\infty}$ has unconditionality constant one there exist nonnegative scalars $\{c_j\}_{j=1}^{l}$ so that

$$\sum_{j=1}^{l} c_j \hat{b}_j = 1 \quad \text{and} \quad \left\|\sum_{j=1}^{\infty} b_j e_j\right\| \leq 1 \Rightarrow \sum_{j=1}^{l} |b_j c_j| \leq 1.$$

By the Hahn–Banach theorem, again, there exists, for every k, a positive functional φ_k on span $\{w_i\}_{i=n_k}^{n_{k+1}-1}$ so that

$$\|\varphi_k\| \leq 1 \qquad \text{and} \quad \varphi_k(v_j) = c_j, \qquad m_k \leq j \leq m_{k+1} - 1.$$

(For $j > l$ we put $c_j = 0$.) Define now a linear functional φ on $\operatorname{span}\{w_i\}_{i=1}^{\infty}$ by putting

$$\varphi(w_i) = \varphi_k(w_i); \qquad n_k \leq i < n_{k+1}, \quad k = 1, 2, \ldots.$$

Clearly, $\varphi(\sum_{j=1}^{l} \hat{b}_j v_j) = \sum_{j=1}^{l} \hat{b}_j c_j = 1$. It is also easily verified that, for every $w \in C$, $|\varphi(w)| \leq 1$. Hence, $|||\varphi||| \leq 1$ and thus $|||\sum_{j=1}^{l} \hat{b}_j v_j||| \geq 1$. ■

Proof of Theorem 3. Let $\{e_j\}_{j=1}^{\infty}$ be a block injective unconditional basis. We shall prove that there is a $\lambda < \infty$ so that, for every n, the lattice generated by $\{e_j\}_{j=1}^{n}$ is λ-injective. Once this is established, the proof proceeds in complete analogy to the proof of Theorem 2.

Assume that no such λ exists. Then we can find integers $m_1 = 1 < m_2 < \cdots$ so that, for each k, the lattice generated by $\{e_j\}_{j=m_k}^{m_{k+1}-1}$ is not k-injective.

Hence, one can construct a sequence of integers $n_1 = 1 < n_2 < \cdots$ and, for each k, a basic sequence $\{w_i\}_{i=n_k}^{n_{k+1}-1}$, whose unconditional constant is 1, and a block basis $\{v_j\}_{j=m_k}^{m_{k+1}-1}$ of $\{w_i\}_{i=n_k}^{n_{k+1}-1}$ which is isometrically equivalent to $\{e_j\}_{j=m_k}^{m_{k+1}-1}$ and so that every positive projection from $\operatorname{span}\{w_i\}_{i=n_k}^{n_{k+1}-1}$ onto $\operatorname{span}\{v_j\}_{j=m_k}^{m_{k+1}-1}$ is of norm at least k. Notice that there is no loss of generality in assuming that $\{v_j\}_{j=1}^{\infty}$ are blocks of the form $v_j = \sum_{i=p_j}^{p_{j+1}-1} a_i w_i$ with $a_i \geq 0$ for all i.

We construct now a norm on $\operatorname{span}\{w_i\}_{i=1}^{\infty}$ as in Lemma 3 and conclude that, with respect to this norm, $\{w_i\}_{i=1}^{\infty}$ is an unconditional basic sequence which contains a block basis $\{v_j\}_{j=1}^{\infty}$ equivalent to $\{e_j\}_{j=1}^{\infty}$. Since $\{e_j\}_{j=1}^{\infty}$ is block injective it follows that there exists a bounded projection P from $[w_i]_{i=1}^{\infty}$ onto $[v_j]_{j=1}^{\infty}$ and, by using its diagonal matrix (with respect to the supports of $\{v_j\}_{j=1}^{\infty}$) we get a projection Q from $[w_i]_{i=1}^{\infty}$ onto $[v_j]_{j=1}^{\infty}$ so that $\|Q\| \leq \|P\|$ and $Qw_i = \lambda_i v_j$, $p_j \leq i \leq p_{j+1} - 1$, $j = 1, 2, \ldots$. We can further replace Q by a positive projection R with $\|R\| \leq \|Q\|$ by putting

$$Rw_i = \frac{|\lambda_i| v_j}{\sum_{h=p_j}^{p_{j+1}-1} a_h |\lambda_h|}, \qquad p_j \leq i \leq p_{j+1} - 1, \quad j = 1, 2, \ldots.$$

This of course, contradicts the fact that every positive projection from $\operatorname{span}\{w_i\}_{i=n_k}^{n_{k+1}-1}$ onto $\operatorname{span}\{v_j\}_{j=m_k}^{m_{k+1}-1}$ is of norm $\geq k$ for $k > \|R\|$. ■

References

1. D. I. Cartwright, Extensions of positive operators between Banach lattices, *Mem. Amer. Math. Soc.* **164** (1975).
2. R. Haydon, Injective Banach lattices, *Math. Z.* **156** (1977), 19–47.
3. J. Lindenstrauss and L. Tzafriri, "Classical Banach Spaces, Vol. 1, Sequence Spaces," Springer-Verlag, Berlin and New York, 1977.
4. J. Lindenstrauss and L. Tzafriri, "Classical Banach Spaces, Vol. 2, Function Spaces," Springer-Verlag, Berlin and New York, 1979.
5. H. P. Lotz, Extensions and liftings of positive linear mappings on Banach lattices, *Trans. Amer. Math. Soc.* **211** (1975), 85–100.

MATHEMATICAL ANALYSIS AND APPLICATIONS, PART B
ADVANCES IN MATHEMATICS SUPPLEMENTARY STUDIES, VOL. 7B

Remarks on New Systems of Partial Differential Equations Related to Optimal Control

J. L. LIONS

Collège de France
Paris, France

DEDICATED TO L. SCHWARTZ

Contents

INTRODUCTION

We consider a system the state of which is given by the *linearized* Navier–Stokes equations. We assume that the control is *pointwise*, i.e., in mathematical terms of the form $v(t)\delta(x - b)$ where $\delta(x - b)$ denotes the Dirac mass at a point b given in the domain where the flow is considered. The state of the system is then defined as a "very weak" solution, in a suitable space of distributions, by the transposition method as in Magenes and this author (see Lions and Magenes [4]). The cost function involves the "terminal state," i.e., the value of the state at time T where T is given, and in order for this cost function to make sense one has to introduce (as it has been done in a systematic manner in Lions [2]), a new space, denoted here by $\mathscr{U}$.

This space is briefly studied here and it is used to obtain the *optimality system* (given in Section 1.6). This optimality system is in general a *nonlinear problem*; it becomes linear if there are *no constraints*, i.e., if v spans the whole space $\mathscr{U}$. In that case one can *uncouple* the optimality system. By using the Schwartz's kernel theorem (see Schwartz [5]) one can show in general that the *kernel* of the "uncoupling mapping" is the *solution of a nonlinear partial differential equation* (with a quadratic nonlinearity); this equation is referred

ISBN 0-12-512802-9

to as a nonlinear equation of the Riccati's type (see Lions [1]). When the state equation is given—as it is the case here—by the linearized Navier–Stokes equations, this equation has an interesting structure, which is indicated in Section 2.3.

1. Pointwise Control of the Linearized Navier–Stokes System

1.1. *State Equation*

Let Ω be a bounded open set in $\mathbb{R}^n$, with smooth boundary Γ. Let b be given in Ω. The control functions $v(t)$ satisfy

$$v(t) \in (L^2(0, T))^n. \tag{1.1}$$

If $\delta(x - b)$ denotes the Dirac mass at point b, we consider the system the state of which is given by the solution of

$$\frac{\partial y}{\partial t} - \Delta y = -\nabla q + v(t)\delta(x - b), \qquad \text{div } y = 0 \qquad \text{in} \quad Q = \Omega \times]0, T[\tag{1.2}$$

subject to

$$y = 0 \qquad \text{on} \quad \Gamma \times]0, T[= \Sigma, \tag{1.3}$$

$$y(x, 0) = 0 \quad \text{in } \Omega. \tag{1.4}$$

In (1.2)

$$y = \{y_1, \ldots, y_n\}, \tag{1.5}$$

and

$$v(t)\delta(x - b) = \{v_1(t)\delta(x - b), \ldots, v_n(t)\delta(x - b)\}. \tag{1.6}$$

We have first to make precise what we mean by a solution of (1.2), (1.3), (1.4); since the right-hand side $v(t)\delta(x - b)$ is "irregular," we have to define *weak solutions.*

Remark 1.1. The system (1.2) is the *linearized* Navier–Stokes system.

1.2. *Weak Solutions*

We use the transposition method, as in Lions and Magenes [4]. Let Ψ be given in $(L^2(Q))^n$. We define φ as the solution of

$$\begin{gathered} -\frac{\partial \varphi}{\partial t} - \Delta \varphi = \Psi - \nabla \pi, \qquad \text{div } \varphi = 0 \quad \text{in } Q, \\ \varphi = 0 \quad \text{on } \Sigma, \qquad \varphi = 0 \quad \text{for} \quad t = T \end{gathered} \tag{1.7}$$

This problem admits a unique solution φ (π is defined up to an additive constant) which satisfies

$$\varphi_j \in L^2(0, T; H^2(\Omega)) \quad \forall j, \tag{1.8}$$

where $H^2(\Omega) = \{f \,|\, D^\alpha f \in L^2(\Omega) \forall \alpha, |\alpha| \le 2\}$.

If we multiply (formally) (1.2) by φ and use integration by parts, we find

$$\int_Q (-\nabla q)\varphi \, dx\, dt + \int_0^T \varphi(b, t)v(t)\, dt = \int_Q y(\Psi - \nabla \pi)\, dx\, dt. \tag{1.9}$$

But

$$\int_Q (-\nabla q)\varphi \, dx\, dt = \int_Q q(\operatorname{div} \varphi)\, dx\, dt = 0$$

and similarly $\int_Q (-\nabla \pi) y \, dx\, dt = 0$, so that (1.9) reduces to

$$\int_Q y\Psi \, dx\, dt = \int_0^T \varphi(b, t)v(t)\, dt. \tag{1.10}$$

We assume from now on that

$$n = 2 \text{ or } 3. \tag{1.11}$$

Remark 1.2. If (1.11) is *not* satisfied, one will use *weaker* solutions.

It classically follows from (1.11) that

$$H^2(\Omega) \subset C^0(\Omega) \qquad \text{(space of continuous functions in } \Omega) \tag{1.12}$$

so that (1.8) implies

$$\text{“}t \to \varphi_j(b, t)\text{''} \in L^2(0, T); \tag{1.13}$$

moreover

$$\Psi \to \varphi_j(b, \cdot) \text{ is continuous from } (L^2(Q))^n \to L^2(0, T). \tag{1.14}$$

Therefore

$$\Psi \to \int_0^T \varphi(b, t)v(t)\, dt = \sum_{j=1}^{n} \int_0^T \varphi_j(b, t)v_j(t)\, dt \\ \text{is a continuous linear form on } (L^2(Q))^n, \tag{1.15}$$

and consequently *it uniquely defines* y *in* $(L^2(Q))^n$ *satisfying* (1.10) $\forall \Psi \in (L^2(Q))^n$. We now *define* y by (1.10).

Let us verify that there exists a distribution q in Q such that (1.2) holds true in the sense of distributions in Q. Indeed, let φ be given arbitrarily in $(\mathscr{D}(Q))^n$ † satisfying

$$\operatorname{div} \varphi = 0. \tag{1.16}$$

† $\mathscr{D}(Q)$ = space of C^∞ functions with compact support in Q, provided with the Schwartz topology.

We then *define* $\Psi = -\partial\varphi/\partial t - \Delta\varphi$ and $\pi = 0$; we have (1.7) so that (1.10) gives

$$\int_Q y\left(-\frac{\partial\varphi}{\partial t} - \Delta\varphi\right) dx\, dt = \int_0^T \varphi(b, t)v(t)\, dt$$

$$\forall\varphi \in (\mathscr{D}(Q))^n \text{ satisfying (1.16).} \qquad (1.17)$$

This implies the existence of $q \in \mathscr{D}'(Q)$† such that the first equation (1.2) holds true.

Let now π be given in $\mathscr{D}(Q)$ and let Ψ be defined by

$$\Psi = \nabla\pi. \qquad (1.18)$$

For this choice of Ψ, the solution of (1.7) is $\varphi = 0$ and (1.10) reduces to

$$\int_Q y(\nabla\pi)\, dx\, dt = 0 \quad \forall\pi \in \mathscr{D}(Q),$$

i.e., the second equation (1.2).

One can then show, by techniques similar to those used in Lions and Magenes [4], that (1.3), (1.4) are satisfied in a generalized sense. But since $v(t)\delta(x - b)$ has its support on the line $\{x = b, 0 < t < T\}$ in Q, it follows that *y is smooth outside this line*, the regularity being up to the boundary $\Gamma \times]0, T[$ and on $t = 0$, $x \neq b$. If Γ is C^∞, y will be C^∞ up to the boundary of Q, except in the neighborhood of $x = b$, $t = 0$ and $t = T$. Consequently $y(x, T; v)$ is defined in $\Omega\backslash\{b\}$.

Of course, the solution y as defined by (1.10) depends on v; it is denoted by $y(v)$ and one easily verifies that

$$v \to y(v) \textit{ is continuous from } (L^2(0, T))^n \to (L^2(Q))^n. \qquad (1.19)$$

1.3. *Cost Function*

The cost function is formally given by

$$J(v) = |y(T; v) - z_d|^2 + N \int_0^T |v(t)|^2\, dt, \qquad (1.20)$$

where z_d is given in $(L^2(\Omega))^n$, N is given >0 and where the notations are as follows:

$$|\varphi|^2 = \sum_{j=1}^n \int_\Omega \varphi_j^2\, dx, \qquad \int_0^T |v(t)|^2\, dt = \sum_{j=1}^n \int_0^T v_j^2(t)\, dt.$$

† $\mathscr{D}'(Q)$ = dual space of $\mathscr{D}(Q)$ = space of distributions over Q.

In (1.21), $y(T; v)$ denotes the function $x \rightarrow y(x, T; v)$, which is defined in $\Omega \backslash \{b\}$; but—as we shall see below—this function *does not belong in general* (i.e., $\forall v$) *to* $(L^2(\Omega))^n$ so that (1.20) is *not* defined for $v \in (L^2(0, T))^n$.

We introduce now the best space of v's such that (1.20) makes sense.

1.4. *Space* $\mathscr{U}$

We define

$$\mathscr{U} = \{v \mid v \in (L^2(0, T))^n,\ y(T; v) \in (L^2(\Omega))^n\}. \tag{1.21}$$

This definition makes sense, since $y(T; v)$ is defined as a smooth function in $\bar{\Omega} \backslash \{b\}$.

We observe that

$$\text{if}\quad v \in (L^2(0, T))^n \quad \text{and if}\quad v \equiv 0 \text{ in a neighborhood of } T, \qquad \text{then}\quad v \in \mathscr{U}. \tag{1.22}$$

Indeed by the regularizing property of the linearized Navier–Stokes system, if $v = 0$ in $]T - \alpha, T[$, then y is C^∞ in $\bar{\Omega} \times]T - \alpha, T]$ (if Γ is C^∞), so that *in particular* $y(T; v)$ belongs to $(L^2(\Omega))^n$.

We equip $\mathscr{U}$ with the norm given by

$$\|v\|_{\mathscr{U}} = \left[\int_0^T |v(t)|^2\, dt + |y(T; v)|^2\right]^{\frac{1}{2}}; \tag{1.23}$$

one has

$$\textit{provided with the norm } (1.23),\ \mathscr{U} \textit{ becomes a Hilbert space.} \tag{1.24}$$

One has to show that $\mathscr{U}$ is complete. If v_j is a Cauchy sequence for the norm (1.23), then $v_j \rightarrow v$ in $(L^2(0, T))^n$ and $y(T; v_j) \rightarrow g$ in $(L^2(\Omega))^n$. But $y(T; v_j) \rightarrow y(T; v)$ in, say, $\mathscr{D}'(\Omega \backslash \{b\})$, so that $y(T; v) = g$, and the result follows [because $y(T; v)$ can be proven to be an element of V' † and therefore cannot have a Dirac mass at b].

One can also show, by techniques entirely similar to those used in Lions [2], that

$$(\mathscr{D}(]0, T[))^n \textit{ is dense in } \mathscr{U}. \tag{1.25}$$

Therefore we can identify $\mathscr{U}'$, the dual of $\mathscr{U}$, to a space of vector distributions on $]0, T[$:

$$\mathscr{D}(]0, T[)^n \subset \mathscr{U} \subset (L^2(0, T))^n \subset \mathscr{U}' \subset (\mathscr{D}'(]0, T[))^n. \tag{1.26}$$

We shall return on the space $\mathscr{U}$ in Section 1.7.

† $V = \{\varphi \mid \varphi \in (H^1(\Omega))^n,\ \varphi = 0 \text{ on } \Gamma,\ \operatorname{div} \varphi = 0\}$; V' is the dual of V.

1.5. *Problem of Optimal Control*

Let now $\mathcal{U}_{\text{ad}}$ be given satisfying

$$\mathcal{U}_{\text{ad}} = \text{closed convex subset of } \mathcal{U}, \text{ nonempty.} \tag{1.27}$$

The problem of optimal control is now to find

$$\inf J(v), \qquad v \in \mathcal{U}_{\text{ad}}. \tag{1.28}$$

The function $v \to J(v)$ is continuous, strictly convex, infinite at infinity on $\mathcal{U}$; therefore

problem (1.28) *admits a unique solution* $u \in \mathcal{U}_{\text{ad}}$ *such that* $J(u) = \inf J(v), \quad v \in \mathcal{U}_{\text{ad}}$; (1.29)

the control u is called the *optimal control*, and we set

$$y(u) = y \qquad \text{(optimal state).} \tag{1.30}$$

We want now to characterize u.

1.6. *Optimality System*

The control u is characterized by

$$(y(T) - z_{\text{d}}, y(T; v) - y(T)) + N \int_0^T u(v - u)\, dt \geq 0 \quad \forall v \in \mathcal{U}_{\text{ad}}, \ u \in \mathcal{U}_{\text{ad}}, \tag{1.31}$$

(where $(\varphi, \psi) = \sum_{j=1}^{n} \int_\Omega \varphi_j(x)\psi_j(x)\, dx$).

We transform (1.31) by using the *adjoint state* p, defined by

$$-\frac{\partial p}{\partial t} - \Delta p = -\nabla \pi, \qquad \operatorname{div} p = 0 \quad \text{in } Q, \tag{1.32}$$

$$p(x, T) = y(T) - z_{\text{d}} \quad \text{in } \Omega, \tag{1.33}$$

$$p = 0 \quad \text{on } \Sigma. \tag{1.34}$$

We multiply the first equation (1.32) by $y(v) - y$ and we *formally* apply Green's formula; we obtain:

$$0 = -(y(T) - z_{\text{d}}, y(T; v) - y(T)) + \int_0^T \left(p, \left(\frac{\partial}{\partial t} - \Delta\right)(y(v) - y)\right) dt.$$

Therefore

$$\begin{aligned}(y(T) - z_{\text{d}}, y(T; v) - y(T)) &= \int_0^T (p, (v - u)\delta(x - b) - \nabla(q(v) - q))\, dt \\ &= \int_0^T p(b, t)(v - u)\, dt. \end{aligned} \tag{1.35}$$

But one can show (by techniques similar to those of Lions [2]) that

$$"t \to p(b, t)" \in \mathscr{U}' \tag{1.36}$$

and that (1.35) is justified, provided that in the right-hand side we understand the integral as giving the scalar product between $\mathscr{U}'$ and $\mathscr{U}$.

Using (1.35) in (1.31) gives

$$\int_0^T (p(b,t) + Nu(t))(v(t) - u(t))\, dt \geq 0 \quad \forall v \in \mathscr{U}_{\text{ad}}. \tag{1.37}$$

We have proved

THEOREM 1.1. *The optimal control u solving* (1.28) *is unique. It is given through the unique solution* $\{y, p, u\}$ *of the optimality system*:

$$\begin{aligned} &\frac{\partial y}{\partial t} - \Delta y = -\nabla q + u\delta(x - b), \qquad \operatorname{div} y = 0, \\ &-\frac{\partial p}{\partial t} - \Delta p = -\nabla \pi, \qquad \operatorname{div} p = 0 \quad \text{in } Q, \end{aligned} \tag{1.38}$$

$$y(x, 0) = 0, \qquad p(x, T) = y(x, T) - z_{\text{d}}(x) \quad \text{on } \Omega, \tag{1.39}$$

$$y = 0, \qquad p = 0 \quad \text{on } \Sigma = \Gamma \times]0, T[, \tag{1.40}$$

and

$$u \in \mathscr{U}_{\text{ad}}, \qquad \int_0^T (p(b,t) + Nu(t))(v(t) - u(t))\, dt \geq 0 \quad \forall v \in \mathscr{U}_{\text{ad}}. \tag{1.41}$$

We recall once again that the integral

$$\int_0^T p(b, t)v(t)\, dt$$

expresses the scalar product between $\mathscr{U}'$ and $\mathscr{U}$.

1.7. *Characterization of* $\mathscr{U}$

One shows first, by techniques similar to those of Lions [2], that

$$\mathscr{U} \text{ does not depend on } \Omega, \tag{1.42}$$

i.e., in particular, we obtain the same space $\mathscr{U}$ by considering the problem in $\Omega = \mathbb{R}^n$.

There is no restriction in assuming then that $b = 0$ and therefore $y(v) = y$ is defined by

$$\frac{\partial y}{\partial t} - \Delta y = -\nabla q + v\delta(x), \qquad \operatorname{div} y = 0 \quad \text{in } \mathbb{R}^n \times]0, T[, \tag{1.43}$$

$$y(x, 0) = 0. \tag{1.44}$$

By Fourier transform in x we obtain:

$$\frac{\partial \hat{y}_i}{\partial t} + |\xi|^2 \hat{y}_i = -\xi_i \hat{q} + v_i, \tag{1.45}$$

$$\xi_i \hat{y}_i = 0^\dagger. \tag{1.46}$$

It follows from (1.45), (1.46) that

$$-|\xi|^2 \hat{q} + \xi_i v_i = 0,$$

so that (1.45) becomes

$$\frac{\partial \hat{y}_i}{\partial t} + |\xi|^2 \hat{y}_i = v_i - \frac{\xi_i \xi_j v_j}{|\xi|^2}. \tag{1.47}$$

Since $\hat{y}_i(\xi, 0) = 0$, (1.47) gives

$$\hat{y}_i(\xi, T) = \int_0^T e^{-(T-s)|\xi|^2} \left[v_i(s) - \frac{\xi_i \xi_j}{|\xi|^2} v_j(s) \right] ds. \tag{1.48}$$

We want to find functions v such that $y_i(T; v) \in L^2(\mathbb{R}^n)\, \forall i$, i.e.,

$$\sum_i M_i < \infty, \tag{1.49}$$

where

$$M_i = \int (\hat{y}_i(\xi, T))^2 \, d\xi$$
$$= \int d\xi \int_0^T e^{-(T-s)|\xi|^2 - (T-t)|\xi|^2} \left[v_i(s) - \frac{\xi_i \xi_j}{|\xi|^2} v_j(s) \right] \left[v_i(t) - \frac{\xi_i \xi_j}{|\xi|^2} v_k(t) \right] ds\, dt. \tag{1.50}$$

We make the change of variable

$$(2T - (s + t))^{\frac{1}{2}} \xi_i = \pi^{\frac{1}{2}} \eta_i.$$

We observe that

$$\int e^{-\pi|\eta|^2} \frac{\eta_i \eta_j}{|\eta|^2} \, d\eta = 0 \qquad \text{if} \quad i \neq j,$$

$$\int e^{-\pi|\eta|^2} \frac{\eta_i^2 \eta_j \eta_k}{|\eta|^4} \, d\eta = 0 \qquad \text{if}\, j \neq k.$$

We observe that $\int e^{-\pi|\eta|^2} \, d\eta = 1$,

$$\int e^{-\pi|\eta|^2} \eta_i^2 \, d\eta = \frac{1}{n};$$

† We adopt the summation convention.

we set

$$\int e^{-\pi|\eta|^2} \frac{\eta_i^2 \eta_j^2}{|\eta|^4} \, d\eta = A_{ij},$$

and we observe that $\sum_i A_{ij} = 1/n$.

Therefore

$$M_i = \pi^{n/2} \int_0^T \int_0^T (2T - (s + t))^{-n/2} \times \left[\left(1 - \frac{2}{n}\right) v_i(s) v_i(t) + \sum_j A_{ij} v_j(s) v_j(t)\right] ds \, dt,$$

so that

$$\sum M_i = \pi^{n/2} \int_0^T \int_0^T (2T - (s + t))^{-n/2} \sum_i \left(1 - \frac{2}{n} + \frac{1}{n}\right) (v_i(s) v_i(t)) \, ds \, dt,$$

so that finally (1.49) is equivalent to

$$\int_0^T \int_0^T (2T - (s + t))^{-n/2} v_i(s) v_i(t) \, ds \, dt < \infty \quad \forall i. \tag{1.51}$$

Let us consider now a similar problem for the *heat equation*: Let z be the solution of

$$\frac{\partial z}{\partial t} - \Delta z = w(t) \delta(x) \quad \text{in } \mathbb{R}^n \times]0, T[, \qquad z(x, 0) = 0, \tag{1.52}$$

where w is a scalar function and let us define

$$\mathscr{U}_0 = \{w \mid w \in L^2(0, T), \ z(T; w) \in L^2(\mathbb{R}^n)\}. \tag{1.53}$$

One easily checks that $\mathscr{U}_0$ consists of those functions w such that

$$\int_0^T \int_0^T (2T - (s + t))^{-n/2} w(s) w(t) \, ds \, dt < \infty, \tag{1.54}$$

and therefore one has

$$\mathscr{U} = (\mathscr{U}_0)^n. \tag{1.55}$$

2. The Case without Constraints.

2.1. *Optimality System*

One says that *there are no constraints* if

$$\mathscr{U}_{\text{ad}} = \mathscr{U}. \tag{2.1}$$

Then (1.41) reduces to

$$p(b,t) + Nu(t) = 0. \tag{2.2}$$

The optimality system is now

$$\frac{\partial y}{\partial t} - \Delta y + \frac{1}{N} p(b,t)\delta(x-b) = -\nabla q, \qquad \operatorname{div} y = 0, \tag{2.3}$$

$$-\frac{\partial p}{\partial t} - \Delta p = -\nabla \pi, \qquad \operatorname{div} p = 0 \quad \text{in} \quad Q = \Omega \times]0,T[,$$

$$y(x,0) = 0, \qquad p(x,T) = y(x,T) - z_{\mathrm{d}}(x) \quad \text{in } \Omega, \tag{2.4}$$

$$y = 0, \qquad p = 0 \quad \text{on } \Sigma. \tag{2.5}$$

2.2. *Uncoupling*

We are going now to decouple system (2.3), by adaptation to the present situation of the method given in Lions [1].

For the linear Stokes system, built without Dirac mass (and without having to introduce the functional space $\mathscr{U}$) see Lions [3].

We introduce $s \in]0,T[$ and we consider the following system in the time interval $t \in]s,T[$:

$$\begin{aligned} &\frac{\partial \alpha}{\partial t} - \Delta\alpha + \frac{1}{N}\beta\delta(x-b) = -\nabla\rho, \qquad \operatorname{div}\alpha = 0, \\ &-\frac{\partial \beta}{\partial t} - \Delta\beta = -\nabla\sigma, \qquad \operatorname{div}\beta = 0, \end{aligned} \tag{2.6}$$

with

$$\alpha(s) = h, \qquad h \text{ given in } (L^2(\Omega))^n, \tag{2.7}$$

and

$$\beta(T) = \alpha(T) - z_{\mathrm{d}} \quad \text{on } \Omega. \tag{2.8}$$

This problem admits a unique solution since it is the optimality system of a problem of optimal control similar to the previous one, with $(0,T)$ replaced by (s,T) and with initial state h instead of 0.

Therefore $\beta(s)$ is uniquely defined and it depends of h in a way which is linear plus constant. Therefore

$$\beta(s) = P(s)h + r(s), \tag{2.9}$$

where $P(s) \in \mathscr{L}((L^2(\Omega))^n; (L^2(\Omega))^n)$, and $r(s)$ denotes the rest.

If we take $h = y(s)$ where now $\{y, p\}$ is the solution of (2.3), (2.4), and (2.5), then $\{\alpha, \beta\}$ is simply the restriction of $\{y, p\}$ to $]s, T[$; therefore

$$p(s) = P(s)y(s) + r(s). \tag{2.10}$$

The problem is now to give a direct characterization of $P(s)$ (the characterization of r will easily follow).

2.3. *The Kernel of* $P(s)$. *Riccati's Equation*

According to the Schwartz's kernel theorem, $P(s)$ is uniquely defined by its kernel $\|P_{ij}(x, \xi, s)\|$, where the P'_{ij} are distributions on $\Omega_x \times \Omega_\xi$, for every fixed s. One has

$$(P(s)h)_i = \int_\Omega P_{ij}(x, \xi, s)h_j(\xi)\, d\xi, \tag{2.11}$$

and we want to characterize the $P_{ij}(x, \xi, s)'$ as solutions of a system of partial differential equations.

Symmetry of $P(s)$. Let us denote by $\hat{\alpha}$, $\hat{\beta}$, $\hat{\rho}$, $\hat{\sigma}$ the solution of (2.6) when $\hat{\alpha}(s) = \hat{h}$ *and assuming that* $z_{\mathrm{d}} = 0$. Therefore

$$\beta(s) = P(s)h, \qquad \hat{\beta}(s) = P(s)\hat{h}. \tag{2.12}$$

We multiply the first equation (2.6) by $\hat{\beta}$ and we apply Green's formula—which is valid. We obtain

$$(h, P(s)\hat{h}) = (\alpha(T), \hat{\alpha}(T)) + \frac{1}{N}\int_0^T \beta(b, t)\hat{\beta}(b, t)\, dt, \tag{2.13}$$

which shows that

$$(P(s)h, \hat{h}) = (h, P(s)\hat{h}) \quad \forall h, \hat{h},$$

i.e.,

$$P_{ij}(x, \xi, s) = P_{ij}(\xi, x, s) \quad \forall i, j. \tag{2.14}$$

We use now the *identity* (2.10) in (2.3). It becomes[†]

$$-\frac{\partial P}{\partial t} y - P\frac{\partial y}{\partial t} - \Delta P y - \frac{\partial r}{\partial t} - \Delta r = -\nabla\pi; \tag{2.15}$$

using the first equation (2.3), it gives

$$-\frac{\partial P}{\partial t} y - P\left[\Delta y - \frac{1}{N} p(b, t)\delta(x - b) - \nabla q\right] - \Delta P y - \frac{\partial r}{\partial r} - \Delta r = -\nabla\pi. \tag{2.16}$$

[†] One proves, by methods analogous to those of Lions [1] that the following calculations are indeed justified. We replace s by t.

We observe that since $\operatorname{div} p(t) = 0 \quad \forall h$, we should have

$$\operatorname{div} P(t)h = 0 \quad \forall h, \tag{2.17}$$

$$\operatorname{div} r(t) = 0. \tag{2.18}$$

We can express (2.17) under the equivalent form:

$$\frac{\partial}{\partial x_i} P_{ij}(x, \xi, t) = 0 \quad \forall j; \tag{2.19}$$

using (2.14), (2.19) implies

$$\frac{\partial}{\partial \xi_j} P_{ij}(x, \xi, t) = 0 \quad \forall i. \tag{2.20}$$

But (2.20) implies that $P(\nabla q) = 0$ so that (2.16) reduces to

$$-\frac{\partial P}{\partial t} y - P\,\Delta y - \Delta P y + \frac{1}{N} P[(Py + r)\delta(x - b)] - \frac{\partial r}{\partial t} - \Delta r = -\nabla \pi. \tag{2.21}$$

But σ depends linearly on h [in (2.6)] so that applying again the Schwartz kernel theorem, we know the existence of distributions $R_j(x, \xi, t)$ such that

$$(\nabla \pi)_i = \frac{\partial}{\partial x_i}\left[\int_\Omega R_j(x, \xi, t) y_j(\xi, t)\, d\xi + \lambda(x, t)\right]. \tag{2.22}$$

Therefore the ith component of (2.21) can be written:

$$\begin{aligned}
&-\int_\Omega \left[\frac{\partial P_{ij}}{\partial t} - (\Delta_x + \Delta_\xi) P_{ij}(x, \xi, t)\right] y_j(\xi, t)\, d\xi \\
&\quad + \frac{1}{N} P_{ik}(x, b, t) \int_\Omega P_{kj}(\xi, b, t) y_j(\xi, t)\, d\xi \\
&\quad - \frac{\partial r_i}{\partial t} - \Delta r_i + \frac{1}{N} P_{ij}(x, b, t) r_j(b, t) \\
&= -\frac{\partial}{\partial x_i} \int_\Omega R_j(x, \xi, t) y_j(\xi, t)\, d\xi - \frac{\partial \lambda}{\partial x_i}.
\end{aligned} \tag{2.23}$$

This relation is an identity for all vector functions $\{y_j(\xi, t)\}$ such that $y_j = 0$ on $\partial\Omega = \Gamma$ and such that

$$\frac{\partial}{\partial \xi_j} y_j(\xi, t) = 0.$$

Therefore there exists $S_j(x, \xi, t)$, distributions on $\Omega_x \times \Omega_\xi$ such that

$$-\left[\frac{\partial P_{ij}}{\partial t} - (\Delta_x + \Delta_\xi) P_{ij}\right] + \frac{1}{N} P_{ik}(x, b, t) P_{kj}(\xi, b, t) = -\frac{\partial R_j}{\partial x_i} - \frac{\partial S_j}{\partial \xi_i}, \tag{2.24}$$

and the rest in (2.23) gives:

$$-\frac{\partial r_i}{\partial t} - \Delta r_i + \frac{1}{N} P_{ij}(x, b, t) r_j(b, t) = -\frac{\partial \lambda}{\partial x_i}. \tag{2.25}$$

We have also *boundary conditions*: Since $\beta(s) = 0$ on Γ in (2.9), we see that $P(s)$ maps $(L^2(\Omega))^n$ into the space of functions which are smooth outside b and which are zero on Γ; therefore

$$P_{ij}(x, \xi, t) = 0 \qquad \text{if} \quad x \in \Gamma \quad \forall i, j \tag{2.26}$$

and by virtue of (2.14) *the same is true* if $\xi \in \Gamma$. We have also

$$r_i(x, t) = 0 \qquad \text{if} \quad x \in \Gamma \quad \forall i. \tag{2.27}$$

Final condition. We apply (2.10) for $s = T$; it gives

$$p(T) = P(T)y(T) + r(T)$$

and comparing to the second condition (2.4), we obtain

$$P(T) = \text{identity operator}, \tag{2.28}$$

$$r(T) = -z_{\mathrm{d}}. \tag{2.29}$$

We can now summarize the informations obtained in the

THEOREM 2.1. *The kernels $P_{ij}(x, \xi, t)$ satisfy the following equations*:

$$-\left[\frac{\partial P_{ij}}{\partial t} - (\Delta_x + \Delta_\xi) P_{ij}(x, \xi, t)\right] + \frac{1}{N} P_{ik}(x, b, t) P_{kj}(\xi, b, t) = -\frac{\partial R_j}{\partial x_i} - \frac{\partial S_j}{\partial \xi_i}, \tag{2.30}$$

$$\frac{\partial}{\partial x_i} P_{ij}(x, \xi, t) = 0, \qquad \frac{\partial}{\partial \xi_j} P_{ij}(x, \xi, t) = 0;$$

$$P_{ij}(x, \xi, t) = 0 \qquad \textit{if} \quad x \in \Gamma, \xi \in \Omega \quad \textit{and} \quad \textit{if} \quad \xi \in \Gamma, \quad x \in \Omega; \tag{2.31}$$

$$P_{ij}(x, \xi, T) = \delta(x - \xi)\delta_i^j (\delta_i^j = \textit{Kronecker index}), \tag{2.32}$$

$$P_{ij}(x, \xi, t) = P_{ji}(\xi, x, t) \quad \forall i, j. \tag{2.33}$$

The functions $r_j(x, t)$ satisfy

$$-\frac{\partial r_i}{\partial t} - \Delta r_i + \frac{1}{N} P_{ij}(x, b, t) r_j(b, t) = -\frac{\partial \lambda}{\partial x_i}, \qquad \operatorname{div} r = 0, \tag{2.34}$$

$$r = 0 \quad \textit{on} \quad \Gamma \times]0, T[, \tag{2.35}$$

$$r(x, T) = -z_{\mathrm{d}}(x) \quad \textit{on} \quad \Omega. \tag{2.36}$$

Remark 2.1. The system (2.30) is a *nonlinear system of the type of Navier–Stokes* with a *quadratic nonlinearity*; it is an extension to the infinite-dimensional case of Riccati's equations appearing in the optimal control of system governed by *ordinary* differential equations.

Remark 2.2. A system of the type (2.30) was introduced in Lions [3] but *without* nonlinearity; a *direct* solution is then indicated.

Remark 2.3. By methods analogous to those of Lions [1] one can *reconstruct* the optimal control starting from P and r, which can be used to show uniqueness in the class of *kernels* which map $(L^2(\Omega))^n$ into the space of vectors which are in $(H^1(\Omega))^n$, null on Γ and divergence free.

References

1. J. L. Lions, "Contrôle optimal de systèmes gouvernés par des équations aux dérivées partielles," Dunod, Gauthier-Villars, Paris, 1968; English transl. by S. K. Mitter, Springer-Verlag, Berlin and New York, 1970.
2. J. L. Lions, Functions spaces and optimal control of distributed systems, Lecture Notes UFRJ, 1980 (see an introduction in C.R.A.S., Paris, 1979).
3. J. L. Lions, Optimal control and new systems of the Navier Stokes type, *Proc. 1st French–South East Asia Math. Conf., Singapore, 1979.*
4. J. L. Lions and E. Magenes, "Problèmes aux limites non homogènes et applications," Vols. 1, 2, Dunod, Paris, 1968; English transl. by D. Kenneth, Springer-Verlag, Berlin and New York, 1970, 1971.
5. L. Schwartz, Théorie des noyaux, *Proc. Internat. Congr. Mathematicians, 1950*, Vol. 1, pp. 220–230.

MATHEMATICAL ANALYSIS AND APPLICATIONS, PART B
ADVANCES IN MATHEMATICS SUPPLEMENTARY STUDIES, VOL. 7B

Modules microdifférentiels et classes de Gevrey

BERNARD MALGRANGE

Institut Fourier
Laboratoire de Mathématiques Pures associé au CNRS
Saint-Martin-d'Héres, France

A LAURENT SCHWARTZ, EN TÉMOIGNGE D'AMITIÉ ET DE RECONNAISSANCE

1. POSITION DU PROBLÈME

Dans la suite, s désigne un réel ≥ 0, ou $+\infty$. Pour $s \neq +\infty$, on note $\mathbb{C}\{x\}_s$ l'ensemble des séries formelles à une indéterminée $\sum_{n\geq 0} a_n x^n$ qui vérifient la condition suivante, dite "condition de Gevrey d'ordre s": la série $\sum [a_n/(n!)^s]\rho^n$ a un rayon de convergence >0. On a évidemment $\mathbb{C}\{x\}_0 = \mathbb{C}\{x\}$, l'anneau des séries entières convergentes; pour $s > 0$, la formule de Stirling montre que la condition de Gevrey d'ordre s équivaut à la suivante: la série $\sum_{n\geq 1} [a_n/\Gamma(ns)]\rho^n$ a un rayon de convergence >0. Enfin, on pose $\mathbb{C}\{x\}_\infty = \mathbb{C}[[x]]$, l'anneau des séries formelles.

Il est facile de voir que, pour tout s, $\mathbb{C}\{x\}_s$ est un anneau pour l'addition et la multiplication usuelle des séries entières, et que son corps des fractions est égal à $\mathbb{C}\{x\}_s[x^{-1}]$, corps que nous noterons K_s.

Soit E un espace vectoriel de dimension finie μ sur K_s; rappelons qu'une *connexion* sur E est définie par la donnée d'une application $\mathbb{C}$-linéaire $\partial: E \to E$ qui vérifie, pour $a \in K_s$, et $e \in E$ la formule $\partial(ae) = a\,\partial e + (da/dx)e$.

Soit $\mathbf{e} = (e_1, \ldots, e_\mu)$ une base de E; la *matrice de la connexion* $M = (m_{ij}) \in \mathfrak{Gl}(n, K_s)$ dans cette base est définie par la formule $\partial e_j = \sum_i m_{ij} e_i$ ou en abrégé $\partial\mathbf{e} = \mathbf{e}M$ (nous écrivons l'action de K_s indifféremment à droite ou à gauche).

Soit $f = \sum a_i e_i$ un élément de E; on a alors $\partial f = \sum b_i e_i$, avec $b_i = da_i/dx + \sum m_{ij} a_j$, ou encore, en abrégé $B = (dA/dx) + MA$ avec

$$A = \begin{pmatrix} a_1 \\ \vdots \\ a_\mu \end{pmatrix}, \qquad B = \begin{pmatrix} b_1 \\ \vdots \\ b_\mu \end{pmatrix};$$

ainsi, la notion de connexion n'est-elle qu'une version débarrassée des coordonnées de celle de système différentiel du premier ordre. De tels systèmes ont été étudiés depuis longtemps dans les cas $s = 0, \infty$. Dans le cas formel ($s = \infty$) une classification complète est connue, voir notamment [6]. Dans le cas analytique ($s = 0$) des travaux de Birkhoff généralisés et

ISBN 0-12-512802-9

complétés récemment par Jurkat et d'autres auteurs permettent de décrire les connexions à partir de leurs formalisées, et des "multiplicateurs de Stokes" déduits des développements asymptotiques des solutions dans des secteurs angulaires suffisamment petits et de leurs raccordements d'un secteur à l'autre (voir [2, 5]; voir aussi dans [1] l'esquisse d'une présentation, due à Deligne, de ces questions).

Pour s quelconque, l'étude de ces systèmes a été entreprise récemment par Ramis, dans une série d'articles en cours de publication (voir notamment [10, 11]). Ramis montre entre autres que l'étude du cas général permet de mieux comprendre les théorèmes d'indice et de comparaison formel-analytique de [7], et aussi qu'elle est indispensable à l'étude de la resommation des solutions formelles d'une connexion sur K_0; c'est dire qu'il ne s'agit nullement d'une généralisation gratuite. A ces motivations, nous en ajouterons une autre à la Section 3, à savoir le lien avec les systèmes micro-différentiels.

Les résultats que nous avons en vue ici montrent, en gros, que, du point de vue de leur structure, les K_s-connexions se comportent

(a) comme les connexions formelles en ce qui concerne les parties très polaires de la matrice de connexion,

(b) comme les connexions analytiques en ce qui concerne les parties peu polaires de ladite matrice.

Pour énoncer ces résultats de façon précise, rappelons rapidement la notion de "rang" ou "irrégularité de Katz" d'une connexion, en renvoyant à Deligne [4] pour les détails.

(1.1) Soit e un vecteur cyclique de E, i.e. un vecteur tel que les $\partial^k e$ ($k \in \mathbb{N}$) engendrent E sur K_s (pour l'existence d'un tel vecteur, voir par exemple [4]). Soit $\delta^\mu e + \sum_{i \geq 1} a_i \delta^{\mu - i} e = 0$ l'équation minimale de e, avec $a_1, \ldots, a_\mu \in K_s$, et $\delta = x\partial$. Posons d'autre part, pour $a \in K_s$: $v(a) = \sup\{l \in \mathbb{Z} \mid a \in x^l \mathbb{C}\{x\}_s\}$. Alors le rationnel $r(E) = \sup(0, \sup_i [-v(a_i)]/i)$ est indépendant du vecteur cyclique choisi; par définition, c'est le "rang de Katz" de E. Une autre manière de dire la même chose est la suivante: $r(E)$ est la plus grande pente du polygone de Newton, au sens de [10], de l'opérateur $\delta^\mu + \sum a_i \delta^{\mu - i}$.

(1.2) Rappelons qu'un réseau Λ de E est un sous-$\mathbb{C}\{x\}_s$-module de type fini qui engendre E sur K_s; un tel Λ est nécessairement libre de rang μ sur $\mathbb{C}\{x\}_s$ (ceci résulte de la structure des idéaux de $\mathbb{C}\{x\}_s$, analogue à celle de $\mathbb{C}\{x\}_0$; nous omettrons ces points, de vérification facile).

Cela étant, soit r_0 un entier ≥ 0. On a alors les propriétés suivantes:

Pour qu'on ait $r(E) \leq r_0$, il faut et il suffit qu'il existe un réseau Λ de E tel qu'on ait $x^{r_0 + 1} \partial \Lambda \subset \Lambda$. Sous cette hypothèse, prenons n'importe quel réseau Λ vérifiant cette condition, prenons une base $\mathbf{e}$ de Λ sur $\mathbb{C}\{x\}_s$, soit M la matrice de la connexion dans cette base, et soit $\bar{M}$ le coefficient de x^{-r_0-1} dans M. Alors pour qu'on ait $r(E) = r_0$ (resp., $r(E) < r_0$) il faut et il suffit que $\bar{M}$ soit non nilpotente (resp., nilpotente).

(1.3) Soit q un entier >0, et soit t une racine de l'équation

$$t^q = x; \qquad \text{on a alors un isomorphisme évident} \quad \mathbb{C}[[x]][t] \simeq \mathbb{C}[[t]];$$

par restriction aux classes de Gevrey, on obtient un isomorphisme $\mathbb{C}\{x\}_s[t] \simeq \mathbb{C}\{t\}_\sigma$, avec $\sigma = s/q$, ou encore un isomorphisme $K_s[t] \simeq \mathbb{C}\{t\}_\sigma[t^{-1}] =_{\text{déf}} L_\sigma$. Ceci étant, soit E un K_s-vectoriel muni d'une connexion; alors la connexion s'étend de façon immédiate au L_σ-vectoriel $E' = E \otimes_{K_s} L_\sigma$, et l'on a $r(E') = qr(E)$. Ceci permet notamment, après ramification, de caractériser $r(E)$ dans le cas général par une propriété analogue à (1.2).

Cela posé, le but de cet article est la démonstration des théorèmes suivants, qui donnent un sens précis, respectivement aux assertions (b) et (a) ci-dessus.

THÉORÈME (1.4). *Soient $s \in]0, +\infty[$, et E un K_s-vectoriel à connexion; on suppose qu'on a $r(E) < 1/s$. Alors il existe un K_0-vectoriel E' à connexion, tel qu'on ait un isomorphisme (de K_s-vectoriels à connexion): $E \simeq E' \otimes_{K_0} K_s$. De plus, E' est unique au sens suivant: si un autre K_0-vectoriel à connexion E'' vérifie la même condition, l'isomorphisme $E' \otimes_{K_0} K_s \simeq E'' \otimes_{K_0} K_s$ provient d'un isomorphisme $E' \simeq E''$.*

Cet énoncé peut se traduire ainsi en terme de bases: soit E vérifiant les hypothèses du théorème (1.4); alors (a) il existe une base de E sur K_s dans laquelle la matrice de connexion soit à coefficients convergents. (b) Si l'on a deux telles bases de E, disons $\mathbf{e}$ et $\mathbf{f}$, la matrice $S \in \mathrm{Gl}(\mu, K_s)$ définie par $\mathbf{f} = \mathbf{e}S$ est à coefficients convergents, i.e. est dans $\mathrm{Gl}(n, K_0)$.

Le théorème qui suit est une extension aux classes de Gevrey d'un résultat classique de Turrittin–Sibuya relatif au cas formel (voir, p.ex. [6] ou [13]). Pour ce théorème, voir aussi [11].

THÉORÈME (1.5). *Soit $s \in]0, +\infty[$, soit E un K_s-vectoriel à connexion, et soit r un entier $\geq 1/s$. On suppose que E possède une base $\mathbf{e} = (e_1, \ldots, e_\mu)$ possédant les propriétés suivantes:*

(i) *la matrice M de la connexion dans la base $\mathbf{e}$ a un pôle d'ordre $r + 1$;*
(ii) *le coefficient $\bar{M}$ de x^{-r-1} dans M se décompose en deux blocs $\bar{M} = \begin{pmatrix} \bar{M}' & 0 \\ 0 & \bar{M}'' \end{pmatrix}$; de plus, $\bar{M}'$ et $\bar{M}''$ n'ont pas de valeur propre commune.*

Designons par Λ le réseau engendré sur $\mathbb{C}\{x\}_s$ par $e_1, \ldots, e_\mu$; dans ces conditions, il existe une autre base $\mathbf{f} = (f_1, \ldots, f_\mu)$ de Λ, avec $f_i - e_i \in x\Lambda$, telle que, dans la base $\mathbf{f}$, la matrice N de la connexion ait la forme $N = \begin{pmatrix} N' & 0 \\ 0 & N'' \end{pmatrix}$.

On remarquera que, dans la situation considérée ici, on a nécessairement $r(E) = r$ à cause de (1.2); on a donc $r(E) \geq 1/s$, ce qui est la condition opposée à celle du théorème (1.4).

La démonstration de (1.4) occupera les paragraphes 2–5; celle du théorème (1.5), beaucoup plus facile, sera faite au paragraphe 6.

2. Reduction au cas $s = 1$

(2.1) Montrons d'abord que, dans l'énoncé (1.4), l'unicité est une conséquence immédiate des théorèmes de comparaison de Ramis: soit $F = \mathrm{Hom}_{K_0}(E', E'')$; on munit F, de la manière usuelle d'une connexion en posant, pour $\varphi \in F$: $(\partial\varphi)(e') = \partial(\varphi(e')) - \varphi(\partial e')$; alors, un $\varphi \in F$ qui commute à ∂ est une *section horizontale* de F, i.e., une solution de $\partial\varphi = 0$. Par hypothèse, on dispose d'une flèche $\varphi: E' \otimes_{K_0} K_s \to E'' \otimes_{K_0} K_s$, i.e., d'une section horizontale de $F \otimes_{K_0} K_s$; les résultats de [10] montrent alors qu'on a $\varphi \in F$, ce qui est le résultat cherché, pourvu qu'on démontre qu'on a $r(F) < 1$.

Pour établir ce dernier point, prenons une base $\mathbf{e}' = (e'_1, \ldots, e'_\mu)$ de E' telle que la matrice M' de ∂ dans $\mathbf{e}'$ ait un pôle d'ordre ≤ 2; alors, d'après (1.2), le coefficient $\bar{M}'$ de x^{-2} dans M' est nilpotent; choisissons aussi une base $\mathbf{e}''$ de E'', possédant la même propriété, et définissons M'' et $\bar{M}''$ de la même manière. Un calcul immédiat montre alors que, dans la base $e'^*_i \otimes e''_j$ de $F \simeq E'^* \otimes E''$, la matrice de la connexion est égale à $-{}^t M' \otimes \mathrm{id} + \mathrm{id} \otimes M''$; sa partie polaire d'ordre 2, vaut $-{}^t\bar{M}' \otimes \mathrm{id} + \mathrm{id} \otimes \bar{M}''$, et elle est donc nilpotente; donc, d'après (1.2) on a bien $r(F) < 1$.

Remarque. Supposant acquise la partie "existence" du théorème (1.4), le même raisonnement permet en fait de démontrer que le foncteur $E' \to E' \otimes_{K_0} K_s$ est une équivalence de catégorie entre les K_0 et les K_s-vectoriels à connexion, de rang de Katz $< 1/s$.

(2.2) Pour démontrer l'existence de E', on peut utiliser la théorie des "développements asymptotiques Gevrey": voir une esquisse dans [11, Section 7], et les détails dans un article ultérieur de Ramis. Nous emploierons ici une autre méthode, qui le double intérêt de se relier à la théorie des opérateurs microdifférentiels, et de mieux se prêter aux extensions à plusieurs variables; ce dernier point sera explicité dans une autre publication [9].

Dans les paragraphes qui suivent, nous démontrerons le théorème (1.4) dans le cas $s = 1$. Le reste du paragraphe 2 est consacré à montrer comment, en supposant ce résultat acquis, on peut en déduire le cas général. Ceci se fait en 3 étapes: $s = 1/q$, s rationnel, et enfin s quelconque.

(A) *Le cas* $s = 1/q$ (q *entier* >1). Posons $x^q = t$, $L = \mathbb{C}\{t\}[t^{-1}]$, et $L_1 = \mathbb{C}\{t\}_1[t^{-1}]$; alors K_s est un surcorps de degré q de L_1. Désignons par F l'ensemble E, considéré comme vectoriel à connexion sur L_1; on vérifie immédiatement qu'on a $r(F) < 1$, donc il existe F', vectoriel à connexion sur L, muni d'un isomorphisme $F' \otimes_L L_1 \simeq F$; on peut considérer F' comme plongé dans F, et tout revient donc à démontrer que l'action de x sur F laisse stable F'; désignons par $u \in \mathrm{Hom}_{L_1}(F, F) = \mathrm{Hom}_L(F', F') \otimes_L L_1$ cette action; elle vérifie l'équation différentielle $(t\partial_t u) = pu$; d'autre part, on voit comme en (2.1) que le "rang" de $\mathrm{Hom}_L(F', F')$ est <1. Alors, par [10], u provient d'un élément de $\mathrm{Hom}_L(F', F')$. D'où le résultat cherché.

(B) *Le cas* $s = p/q$ (p, q *entiers* ≥ 1). Soit ici t une racine de $t^p = x$; posons $\sigma = 1/q$, et $L_\sigma = \mathbb{C}\{t\}_\sigma[t^{-1}]$; L_σ est un surcorps de degré p de K_s. Soit $F = E \otimes_{K_s} L_\sigma$, muni de la connexion sur L_σ qui étend celle de E; d'après (1.3), on a $r(F) = \mathrm{pr}(E)$, donc $r(F) < q = 1/\sigma$. D'après (A), il existe F', vectoriel à connexion sur $L = \mathbb{C}\{t\}[t^{-1}]$ muni d'un isomorphisme $F' \otimes_L L_\sigma \simeq F$, et on peut encore considérer F' comme plongé dans F.

Le groupe de Galois $\Gamma = \mathrm{Gal}\, L_\sigma/K_s$ agit sur F, et E est le sous-ensemble des éléments de F invariants par ce groupe; montrons que Γ laisse stable F'. Pour cela, considérons F (resp., F') comme vectoriel à connexion sur K_s (resp., K_0); il est immédiat que le "rang" de ces vectoriels à connexion est égal à $r(E)$, et donc $<1/s$; d'autre part, tout $\gamma \in \Gamma$ commute à ∂_x, et peut donc être considéré comme une section horizontale de $\mathrm{Hom}_{K_s}(F, F) = \mathrm{Hom}_{K_0}(F', F') \otimes_{K_0} K_s$; par [10], on en déduit que γ provient d'un endomorphisme de F', ce qui démontre l'assertion cherchée. Il est alors immédiat de vérifier que l'on peut prendre pour E' l'ensemble des éléments de F' invariants par Γ.

(C) *Le cas général.* On choisit un σ rationnel qui vérifie $s < \sigma < 1/r(E)$, et l'on pose $E_\sigma = E_s \otimes_{K_s} K_\sigma$; d'après (B) il existe E', vectoriel à connexion sur K_0 muni d'un isomorphisme $E' \otimes_{K_0} K_\sigma \simeq E_\sigma$; cela s'écrit aussi $E'_s \otimes_{K_s} K_\sigma \simeq E \otimes_{K_s} K_\sigma$, avec $E'_s = E' \otimes_{K_0} K_s$; par [10], on en déduit que E'_s et E sont isomorphes, ce qui est le résultat cherché.

3. Le cas $s = 1$: Interpretation microdifférentielle

Nous aurons à utiliser, dans le cas d'une variable, la définition et quelques propriétés des opérateurs microdifférentiels analytiques; pour les démonstrations, et le cas général, voir [12] ou [3]. On pourra aussi voir un résumé dans [8].

Soit $\mathcal{O}$ (resp., $\mathcal{D}$) le faisceau des fonctions holomorphes sur $\mathbb{C}$ (resp., le faisceau des opérateurs différentiels linéaires à coefficients holomorphes sur $\mathbb{C}$); et soit $\mathscr{E}$ le faisceau des opérateurs microdifférentials sur $T^*\mathbb{C}\backslash\{0\}$. Rappelons que $\mathscr{E}$ est la réunion pour $k \in \mathbb{Z}$ des $\mathscr{E}(k)$, appelés "faisceau des opérateurs microdifférentiels d'ordre $\leq k$", et que la fibre de $\mathscr{E}(k)$ au point $\lambda = (y_0, \eta_0), \eta_0 \neq 0$ est par définition l'ensemble des séries $p = \sum_{l \geq -k} q_l(y)\eta^{-l}$ avec a_l holomorphe au voisinage de $y_0 \in \mathbb{C}$ qui vérifient la condition suivante:

(3.1) la série

$$\sum_{l \geq \sup(-k,0)} \frac{a_l(y)}{l!} x^l$$

est uniformément convergente au voisinage de $(y_0, 0) \in \mathbb{C}^2$.

On plonge $\mathcal{D}_{y_0}$ dans $\mathscr{E}_\lambda$ par la formule suivante: à l'opérateur différentiel $\sum a_m(y)\partial_y^m$, on associe le "symbole" $\sum a_m(y)\eta^m$; on prolonge alors la composition des opérateurs différentiels par la formule

$$p \circ q \underset{\text{déf}}{=} \sum_{m \geq 0} \frac{1}{m!} (\partial_\eta^m p)(\partial_y^m q)$$

(le produit figurant au second membre étant le produit usuel). On montre que l'on a $\mathscr{E}(k) \circ \mathscr{E}(l) \subset \mathscr{E}(k + l)$; muni de ce produit, $\mathscr{E}$ est donc un faisceau d'anneaux gradués non commutatif, dont on montre qu'il est cohérent, et à fibres noethériennes à droite et à gauche.

On montre aussi ceci: pour que $p = \sum_{l \geq -k} a_l(y)\eta^{-l} \in \mathscr{E}_\lambda(k)$ admette un inverse dans $\mathscr{E}_\lambda(-k)$, il faut et il suffit qu'on ait $a_{-k}(y_0) \neq 0$. En particulier, $\mathscr{E}_\lambda(0)$ est un anneau local.

Soit π la projection $T^*\mathbb{C}\backslash\{0\} \to \mathbb{C}$, et U un ouvert de $\mathbb{C}$; un $\mathscr{E}|\pi^{-1}(U)$-Module cohérent F s'appelle par définition un "système microdifférentiel sur $\pi^{-1}(U)$"; il est constant sur les fibres de π; une filtration de F est une suite de sous-faisceaux $F(k), k \in \mathbb{Z}$ telle qu'on ait, pour tout (k, l): $\mathscr{E}(k)F(l) \subset F(k+l)$; à cause des propriétés d'inversibilité dans $\mathscr{E}$, ceci équivaut à $\mathscr{E}(k)F(l) = F(k+l)$. La filtration est dite "bonne" si c'est la filtration quotient d'un morphisme surjectif $\mathscr{E}^p \to F$ (par définition des faisceaux cohérents il en existe toujours localement).

Si l'on a une bonne filtration de F, son symbole $\sigma(F) =_{\text{déf}} F(0)/F(-1)$ est constant sur les fibres de π et cohérent sur $\mathscr{E}(0)/\mathscr{E}(-1) = \pi^{-1}(\mathcal{O})$; on montre que le support de F est égal au support de $\sigma(F)$; donc, si U est connexe on a, soit $\operatorname{supp}(F) = \pi^{-1}(U)$, soit $\operatorname{supp}(F) = \pi^{-1}(D)$, D un sous-ensemble discret de U; dans le second cas, on dira que F est *holonome*.

Ces définitions étant posées, soit F un germe de module microdifférentiel holonome au voisinage de $\pi^{-1}(0)$; vu la constance le long des fibres, on peut considérer F comme $\mathscr{E}_\lambda$-module, $\lambda = (0, 1)$; prenons une bonne filtration de

F, et soit $\sigma(F)$ son symbole; alors $\sigma(F)$ est fini sur $\mathcal{O}_0$, et de support réduit à l'origine; donc il est fini sur $\mathbb{C}$. Soit μ son rang sur $\mathbb{C}$, et soient $(\bar{f}_1, \ldots, \bar{f}_\mu)$ une base de $\sigma(F)$ sur $\mathbb{C}$; le *théorème de préparation pour les opérateurs microdifférentiels nous affirme alors ceci*:

Soit $\tilde{\mathscr{E}}_\lambda \subset \mathscr{E}_\lambda$ l'ensemble des p qui ne dépendent pas de y, i.e. sont de la forme $\sum_{l \geq -k} a_l \eta^{-l}$, $a_l \in \mathbb{C}$; alors $F(0)$ est libre de rang μ sur $\tilde{\mathscr{E}}_\lambda(0)$, et la base $(\bar{f}_1, \ldots, \bar{f}_\mu)$ se relève en une base $(f_1, \ldots, f_\mu) = \mathbf{f}$ de $F(0)$ sur $\tilde{\mathscr{E}}_\lambda(0)$; il en résulte aussitôt que F est libre de base $\mathbf{f}$ sur $\tilde{\mathscr{E}}_\lambda$.

Maintenant, la remarque essentielle et d'ailleurs triviale est la suivante: la correspondance $\sum a_l \eta^{-l} \mapsto \sum a_l x^l$ établit, à cause de (3.1) un isomorphisme entre $\tilde{\mathscr{E}}_\lambda(0)$ et $\mathbb{C}\{x\}_1$, et aussi entre $\tilde{\mathscr{E}}_\lambda$ et K_1; par suite, F est un espace vectoriel de dimension μ sur K_1, et $F(0)$ en est un réseau.

Montrons que F est muni naturellement d'une connexion: soit $a = \sum a_l x^l \in K_1$, et soit $a(\eta^{-1})$ son image dans $\tilde{\mathscr{E}}_\lambda$; pour $f \in F$, il résulte de la formule de composition des opérateurs microdifférentiels qu'on a $ya(\eta^{-1})f = a(\eta^{-1})(yf) + b(\eta^{-1})f$, avec $b = x^2\, da/dx$; par suite $\partial_x = \eta^2 \circ y$ possède les propriétés voulues. (Dans la suite, on écrira simplement $\eta^{-1} = x$ ou encore $\partial_y^{-1} = x$, et $y = x^2\partial_x$). De plus, on a $yF(0) \subset F(0)$ et l'action de y sur $F(0)/F(-1) = \sigma(F)$ est nilpotente, puisque $\sigma(F)$ est de support l'origine. Donc, on a $r(F) < 1$. Enfin, si G est un autre $\mathscr{E}_\lambda$-module holonome, et u un morphisme $F \to G$, il est clair que u est un morphisme de K_1-vectoriels à connexion. On a ainsi défini un foncteur Φ:$\mathscr{E}_\lambda$-modules holonomes, avec $\lambda = (0, 1) \in T^*\mathbb{C} \mapsto K_1$-vectoriels à connexion, de rang de Katz <1. On a alors le résultat suivant:

Dictionnaire (3.2). *Φ est une équivalence de catégories.*

Il suffit de trouver un quasi-inverse de Φ; pour cela, donnonsnous un K_1-vectoriel à connexion F; par les identifications faites ci-dessus, ceci nous donne une action sur F du sous-anneau $\tilde{\mathscr{E}}_\lambda[y]$ de $\mathscr{E}_\lambda$; pour obtenir le quasi-inverse cherché, il suffit d'établir que l'application $f \mapsto 1 \otimes f$ de F dans $\mathscr{E}_\lambda \otimes_{\tilde{\mathscr{E}}_\lambda[y]} F$ est bijective, pourvu qu'on ait $r(F) < 1$.

Pour cela, prenons une base $\mathbf{f}$ de F dans laquelle la matrice M de la connexion ait un pôle d'ordre 2, la partie polaire d'ordre 2 étant nilpotente. Soit P la matrice carrée d'ordre μ, à coefficients dans $\tilde{\mathscr{E}}_\lambda(0)$, définie par $P = {}^t(\eta^{-2}M(\eta^{-1}))$, et soit u la multiplication à droite dans $\mathscr{E}^\mu$ par $y - P$. Désignant par i l'injection naturelle $\tilde{\mathscr{E}} \to \mathscr{E}$, l'assertion à démontrer équivaut à la suivante: l'application (i, u):$\tilde{\mathscr{E}}^\mu_\lambda \otimes \mathscr{E}^\mu_\lambda \to \mathscr{E}^\mu_\lambda$ est bijective. Or l'assertion analogue est vraie pour les gradués associés, parce que la partie d'ordre 0 de P est nilpotente; l'assertion résulte alors du théorème de préparation pour les opérateurs microdifférentiels, p. ex. sous la forme où il est énoncé dans [3].

4. Un théorème de finitude

Nous continuons avec les notations de la Section 3: $\lambda = (0, 1) \in T^*\mathbb{C}$, F est un module holonome sur $\mathscr{E}_\lambda$ muni d'une bonne filtration $\{F(k)\}$; enfin $\mathbf{f} = (f_1, \ldots, f_\mu)$ est une base de $F(0)$ sur $\tilde{\mathscr{E}}_\lambda(0)$ provenant d'une base $\bar{\mathbf{f}} = (\bar{f}_1, \ldots, \bar{f}_\mu)$ de $\sigma(F)$ sur $\mathbb{C}$. On désigne par $P = (p_{ij})$ la matrice à coefficients dans $\tilde{\mathscr{E}}_\lambda(0)$ définie par $yf_j = \sum p_{ij} f_i$, ou encore $y\mathbf{f} = \mathbf{f}P$. Le résultat essentiel de ce paragraphe est le théorème suivant, qui sera la clef de la démonstration de (1.4).

Théorème (4.1). *$F(0)$ est fini de rang μ sur $\mathbb{C}\{y\} = \mathscr{O}_0$.*

Utilisons le dictionnaire précédent: alors, via la base $\mathbf{f}$, $F(0)$ s'identifie à $\mathbb{C}\{x\}_1^\mu$ et l'action de y à celle de l'opérateur différentiel x^2D sur $\mathbb{C}\{x\}_1^\mu$ défini par $(x^2D)\Phi = x^2\, d\Phi/dx + P\Phi$ avec $P = \sum_{k \geq 0} P_k \eta^{-k} = \sum_{k \geq 0} P_k x^k$.

Nous allons d'abord rappeler les raisonnements de Ramis qui donnent l'indice de x^2D. Pour cela soit $\rho > 0$ et soit $H(\rho)$ l'espace des séries entières sans terme constant $\varphi = \sum_{n \geq 1} a_n x^n$ qui vérifient $|\varphi|_\rho = \sum (|a_n|/(n-1)!)\rho^n < +\infty$.

Lemme (4.2). *L'application $\varphi \mapsto (x^2/\rho)(d\varphi/dx)$ est une application isométrique de $H(\rho)$ sur le sous-espace de $H(\rho)$ des séries sans terme de degré 1.*

Evident.

Lemme (4.3). *L'application $\varphi \mapsto x\varphi$ est compacte de $H(\rho)$ dans lui-même.*

En effet, sur le sous-espace de $H(\rho)$ formée des séries d'ordre $\geq n_0$, la norme de la multiplication par x est égale à $1/n_0$. Donc, sur $H(\rho)$, la multiplication par x est limite uniforme d'opérateurs de rang fini, donc compacte.

Lemme (4.4). *La multiplication est bilinéaire continue de $H(\rho) \times H(\rho)$ dans $H(\rho)$, et de norme ≤ 1.*

Soient $\varphi, \psi \in H(\rho)$, avec $\varphi = \sum a_n x^n$, $\psi = \sum b_n x^n$; on a $\varphi\psi = \sum a_m b_n x^{m+n}$, d'où $|\varphi\psi|_\rho \leq \sum (|a_m b_n|/(m+n-1)!)\rho^{m+n} \leq |\varphi|_\rho |\psi|_\rho$.

Lemme (4.5). *Pour $\rho \leq \rho_0$, l'application $\Phi \mapsto x^2(d\Phi/dx) + P\Phi$ $(= x^2D\Phi)$ envoie $H(\rho)^\mu$ continuement dans luimême, et est d'indice $-\mu$.*

Posons $P = P_0 + xQ$; il résulte des lemmes précédents que xQ est compact de $H(\rho)$ dans lui-même, pourvu que ρ soit assez petit, disons $\leq \rho_0$. Il suffit alors d'établir que $x^2\, d/dx + P_0$ est d'indice $-\mu$; comme P_0 est nilpotente, on peut la supposer réduite à la forme de Jordan. Le résultat est alors immédiat par récurrence sur μ, à partir de (4.2).

Notons $(x^2D)_\rho$ l'application x^2D restreinte à $H(\rho)^\mu$; pour $\rho \le \rho_0$, on a des applications $\mathrm{Ker}(x^2D)_{\rho_0} \to \mathrm{Ker}(x^2D)_\rho$ et $\mathrm{Coker}(x^2D)_{\rho_0} \to \mathrm{Coker}(x^2D)_\rho$.

LEMME (4.6). *Ces applications sont bijectives.*

En effet, l'application $H(\rho_0) \to H(\rho)$ est injective donc $\mathrm{Ker}(x^2D)_{\rho_0} \to \mathrm{Ker}(x^2D)_\rho$ est injective. D'autre part, $H(\rho_0)$ est dense dans $H(\rho)$, et l'image de $(x^2D)_\rho$ est fermée, puisque de codimension finie, d'après un lemme classique de L. Schwartz. Donc l'application $\mathrm{Coker}(x^2D)_{\rho_0} \to \mathrm{Coker}(x^2D)_\rho$ est d'image dense, donc surjective (puisque ces espaces sont de dimension finie). L'égalité des deux indices entraîne alors le résultat cherché.

Maintenant, il est clair que, avec les identifications qu'on a faites, on a $F(-1) = \lim_{\rho>0} H(\rho)^\mu$. Des résultats précédents, on déduit donc que $x^2\partial$ est d'indice $-\mu$ dans $F(-1)$; comme $F(-1)$ est de codimension finie dans $F(0)$, le même résultat vaut pour $F(0)$; cette assertion n'est autre, dans le cas particulier qui nous occupe que le théorème de l'indice de Ramis.

Démontrons maintenant (4.1). D'après les résultats précédents, le noyau de $x^2\partial_x$ dans $F(-1)$ est de dimension finie; pour k assez grand, ce noyau ne recontre pas $F(-k)$. Alors, quitte à décaler la filtration, on peut déjà supposer $k = 1$. Alors, le conoyau de $x^2\partial_x$ dans $F(-1)$ est de codimension μ, et il est égal au conoyau de $(x^2D)_\rho$, pour $\rho \le \rho_0$. Choisissons $g_1, \ldots, g_\mu \in H(\rho_0)\mu$ qui forment une base d'un supplémentaire de $\mathrm{Im}(x^2D)_{\rho_0}$, et prenons un $\rho < \rho_0$. Nous allons voir que tout élément $f \in H(\rho)\mu$ s'écrit d'une manière et d'une seule $f = \sum a_i(x^2D)g_i$, $a_i \in \mathbb{C}\{y\}$. Cela établira le résultat cherché.

Le théorème du graphe fermé montre qu'il existe $C > 0$ (dépendant de ρ) qui possède la propriété suivante: tout $f \in H(\rho)\mu$ s'écrit d'une manière unique

$$f = \sum a_i^0 g_i + (x^2D)f^1, \qquad \text{avec} \quad |a_i^0| \le C|f|_\rho, \quad |f^1|_\rho \le C|f|_\rho;$$

en itérant,

$$f^p = \sum a_i^p g_i + (x^2D)f^{p+1} \qquad \text{avec} \quad |a_i^p| \le C|f^p|_\rho, \quad |f^{p+1}|_\rho \le C|f^p|_\rho;$$

d'où en regroupant

$$f = \left[\sum_0^p a_i^k (x^2D)^k\right] g_i + (x^2D)^{p+1} f^{p+1},$$

avec

$$|a_i^k| \le C^{k+1}|f|_\rho, \quad |f^{p+1}|_\rho \le C^{k+1}|f|_\rho. \tag{4.7}$$

En particulier, les séries $a_i = \sum a_i^k y^k$ sont convergentes. Pour établir l'égalité $f = \sum a_i g_i$, on peut procéder de deux manières:

(a) La théorie des opérateurs micro-différentiels définit a priori le second membre de cette formule; il suffit alors de voir que, pour tout k, on a $f - \sum a_i g_i \in F(-k)$; or, du fait que P_0 est nilpotente, on a $P_0^\mu = 0$. On en déduit facilement qu'on a $(x^2\partial_x)^{\mu l} F(-1) \subset F(-1-l)$ et le résultat s'ensuit.

(b) On peut établir aussi directement la convergence des séries figurant au second membre, et la convergence du reste vers 0 dans $H(\rho')^\mu$, $\rho' < \rho$ convenable.

En effet, posons encore $x^2 D = x^2\, d/dx + P_0 + xQ$; les lemmes (4.2) à (4.5) montrent que pour $\rho \leq \rho_0$, la norme de $x^2\, d/dx + xQ$ dans $H(\rho)^\mu$ est $\leq C'\rho$, $C' > 0$ indépendant de ρ. Montrons que $(x^2\, d/dx + P)^\mu$ possède la même propriété. On a $(x^2\, d/dx + P)^\mu = \sum \Phi(1) \ldots \Phi(\mu)$, la somme étant étendue aux applications $\Phi: \{1, \ldots, \mu\} \to \{x^2\, d/dx + xQ, P_0\}$. Tous les termes figurant au second membre contiennent $x^2\, d/dx + xQ$ en facteur, puisque $P_0^\mu = 0$; ceci établit la propriété

Les convergences cherchées se déduisent immédiatement de là, et des majorations (4.7). Enfin, l'unicité de l'expression de f est évidente, par récurrence sur l'unicité des formules (4.7). Ceci établit le théorème.

Notons H le $\mathscr{D}_0$-module obtenu en restreignant les scalaires de $\mathscr{E}_\lambda$ à $\mathscr{D}_0$ dans F. On a les propriétés suivantes:

H est fini sur $\mathscr{D}_0$; en effet on a: pour $k \geq 0$, $F(k) = \partial_y^k F(0)$ donc H est engendré sur $\mathscr{D}_0$ par $F(0)$, (4.8)

H est holonome sur $\mathscr{D}_0$, c'est-à-dire que sa variété caractéristique est contenue dans $\pi^{-1}(0) \cup$ (la section nulle). (4.9)

En effet, considérons la filtration de H définie par $H(k) = F(k)$, $k \geq 0$. Cette filtration est bonne (e.g., au sens de [8]), et la variété caractéristique est donc le support de $\operatorname{gr} H$ en tant que $\operatorname{gr} \mathscr{D}_0 = \mathbb{C}\{y\}[\eta]$-module; or on a $y^\mu H(k) \subset H(k-1)$ pour $k \geq 1$ donc $y^\mu(\operatorname{gr} H) \subset H(0)$ et le résultat suit aussitôt.

∂_y est bijectif sur H. (4.10)

L'application "restriction des scalaires de $\mathscr{E}_\lambda$ à $\mathscr{D}_0''$ définit donc un foncteur ψ de la catégorie des $\mathscr{E}_\lambda$ modules holonomes dans la catégories des $\mathscr{D}_0$-modules holonomes sur lesquelles ∂_y est bijectif. On a le résultat suivant:

Dictionnaire (4.11). *ψ est une équivalence de catégories.*

On construit un quasi-inverse de ψ en établissant le résultat suivant: soit G un $\mathscr{D}_0$-module holonome sur lequel ∂_y soit bijectif. Alors l'application $g \to 1 \otimes g: G \to \mathscr{E}_\lambda \otimes_{\mathscr{D}_0} G$ est bijective; quoique cela ne soit probablement pas indispensable, nous allons utiliser les résultats qui précèdent, et opérer en deux temps.

(a) On suppose que G est déjà de la forme $\psi(F)$; on dispose alors de deux applications

$$u: f \mapsto 1 \otimes f, \qquad F \to \mathscr{E}_\lambda \otimes_{\mathscr{D}_0} F$$
$$v: a \otimes f \mapsto af, \qquad \mathscr{E}_\lambda \otimes_{\mathscr{D}_0} F \to F.$$

On a évidemment $v \circ u = \text{id}$. Montrons qu'on a aussi $u \circ v = \text{id}$; il suffit d'établir la formule $a \otimes f = 1 \otimes af$; cette formule est vraie pour η^{-1}, car on a $1 \otimes \eta^{-1}f = \eta^{-1}\eta \otimes \eta^{-1}f = \eta^{-1} \otimes f$; donc elle est vraie pour les $a \in \mathscr{D}_0[\eta^{-1}]$. Pour passer à la limite, remarquons que les formules précédentes entraînent qu'on a, pour tout $l: \mathscr{E}_\lambda(l) \otimes F(0) = \mathscr{E}_\lambda(0) \otimes F(l)$ et que les $\mathscr{E}_\lambda(l) \otimes F(0)$ forment une filtration de $\mathscr{E}_\lambda \otimes_{\mathscr{D}_0} F$; en prenant des générateurs de F sur $\mathscr{D}_0$ dans $F(0)$, on écrit F comme un quotient $\mathscr{D}_0^\nu \to F$, donc $\mathscr{E}_\lambda \otimes_{\mathscr{D}_0} F$ comme une quotient de $\mathscr{E}_\lambda^\nu$, et la filtration précédete est la filtration quotient de celle de $\mathscr{E}_\lambda^\nu$, donc elle est bonne. Le passage à la limite cherché s'obtient alors en remarquant qu'une bonne filtration est séparée [3].

(b) Dans le cas général, on écrit une suite exacte de $\mathscr{D}_0$-Modules

$$0 \to G' \to G \to \mathscr{E}_\lambda \otimes_{\mathscr{D}_0} G \to G'' \to 0$$

en appliquant le résultat précédent à $\mathscr{E}_\lambda \otimes_{\mathscr{D}_0} G$ qui est holonome sur $\mathscr{E}_\lambda$, et en utilisant la platitude de $\mathscr{E}_\lambda$ sur $\mathscr{D}_0$, on voit qu'on a $\mathscr{E}_\lambda \otimes_{\mathscr{D}_0} G' = 0$, $\mathscr{E}_\lambda \otimes_{\mathscr{D}_0} G'' = 0$; par suite, G' et G'' ont leur variété caractéristique réduite à la section nulle, donc sont isomorphes à des $\mathscr{O}_0^\nu$ en tant que $\mathscr{D}_0$-modules; mais, comme ∂_y est inversible sur G' et G'', ceci implique $G' = G'' = 0$.

Exemple (4.12). Prenons $E = \mathscr{D}_0/\mathscr{D}_0 y$, module noté habituellement $\mathscr{D}_0\delta$. Alors, $F = \mathscr{E}_\lambda \otimes_{\mathscr{D}_0} E$ est engendré sur $\mathscr{D}_0$ par $\eta^{-1}\delta$, la fonction de Heaviside.

Exemple (4.13). Prenons $E = \mathscr{D}_0 \exp(1/y)$ et $F = \mathscr{E}_\lambda \otimes_{\mathscr{D}_0} E$, muni de la filtration définie par les $\mathscr{E}_\lambda(l) \otimes \exp(1/y)$. On vérifie que $F(0)$ est libre de rang 2 sur $\mathscr{O}_0$, de générateurs $\exp(1/y)$ et $\eta^{-1}\exp(1/y)$.

Remarque (4.14). Soit $\hat{\mathscr{E}}_\lambda$ l'anneau des opérateurs microdifférentiels formels, i.e., des expressions $\sum_{l \geq -k} a_l(y)\eta^{-l}$, les a_l convergeant dans un voisinage fixe de 0, mais sans imposer (3.1). Une théorie entièrement analogue peut être faite pour les $\hat{\mathscr{E}}_\lambda$-modules holonomes F. En particulier, le dictionnaire (3.2) sera vrai, avec K_1 remplacé par K_∞; les démonstrations sont d'ailleurs plus simples, puisqu'il n'y a aucune convergence à vérifier. La seule différence est qu'ici, $F(0)$ sera fini sur $\mathbb{C}[[y]]$ de rang $\nu \leq \mu$, avec $\mu - \nu$ égal à l'irrégularité de $\Phi(F)$, au sens de [7]; ceci à cause de la différence entre l'indice formel et l'indice "Gevrey 1". Par exemple, on voit que $\hat{\mathscr{E}}_\lambda(0)\exp(1/y)$ est libre de rang 1 sur $\mathbb{C}[[y]]$.

Notons encore $\hat{\hat{\mathscr{E}}}_\lambda$ l'anneau des opérateurs micro-différentiels formels en η et y [i.e., les $a_\lambda(y)$ sont des séries formelles]. On voit ici, par contre, que le passage de $\hat{\mathscr{E}}_\lambda$ à $\hat{\hat{\mathscr{E}}}_\lambda$ n'introduit aucune différence dans la structure des modules micro-différentiels holonomes; nous laissons le lecteur préciser ce point.

5. Transformation de Laplace et microlocalisation

Soit G un module holonome sur $\mathscr{D}_0$; l'idée de la construction qui va être faite consiste à montrer qu'on peut récupérer son microlocalisé $\mathscr{E}_\lambda \otimes_{\mathscr{D}_0} G$ par la succession des opérations suivantes: algébrisation, transformation de Laplace, localisation à l'infini.

(5.1) *Algébrisation.* Les résultats de ce n° sont essentiellement classiques; dans leur principe, ils remontent à Birkhoff. Prenons une présentation $\mathscr{D}_0^q \to \mathscr{D}_0^p \to G \to 0$ de G, et munissons G de la filtration quotient; comme G est holonome, les $G(k)/G(k-1)$ ont leurs supports à l'origine pour $k \gg 0$; quitte à décaler la filtration, et à changer de présentation, on peut supposer que ceci se produit dès qu'on a $k \geq 1$. La présentation de G se prolonge alors en une présentation d'un $\mathscr{D}$-Module $\mathscr{G}$ sur une disque ouvert $D \subset \mathbb{C}$ de centre 0, muni de la filtration quotient $\mathscr{G}(k)$; quitte à rétrécir D, on peut supposer que, sur D^*, on a $\mathscr{G}(0) = \cdots = \mathscr{G}(k) = \cdots$. Donc sur D^*, $\mathscr{G}$ est simplement un fibré vectoriel muni d'une connexion, donc est isomorphe à $\mathscr{O} \otimes_{\mathbb{C}} \mathrm{sol}(\mathscr{G})$, $\mathrm{sol}(\mathscr{G})$ étant le faisceau des solutions de $\partial_y g = 0$.

Comme $\mathrm{sol}(\mathscr{G})$ est un système local sur D^*, donc déterminé par sa monodromie, il s'étend en un système local Γ sur $\mathbb{C}^*$; on prolonge alors $\mathscr{G}$ sur $\mathbb{C}^*$ par $\mathscr{G} = \mathscr{O} \otimes_{\mathbb{C}} \Gamma$; on pose aussi, sur $\mathbb{C}^*: \mathscr{G} = \mathscr{G}(0) = \cdots = \mathscr{G}(k) = \cdots$.

On prolonge $\mathscr{O}$ à $Y = P_1(\mathbb{C})$ par le faisceau $\mathscr{O}[y]$ des fonctions méromorphes ayant au plus un pôle à l'∞, et on prolonge $\mathscr{D}$ à Y par le faisceau $\mathscr{D}[y]$ défini de la même manière. Enfin, on prolonge $\mathscr{G}$ en un $\mathscr{D}[y]$-Module en lui imposant d'être libre sur $\mathscr{O}[y]$ au voisinage de l'infini, et à singularité régulière (voir, p.ex. [4]); au voisinage de l'infini, on pose encore $\mathscr{G} = \mathscr{G}(0) = \cdots = \mathscr{G}(k) = \cdots$.

L'unicité à isomorphisme près de $\mathscr{G}$ se voit ainsi: si l'on a un autre $\mathscr{G}'$ vérifiant les mêmes conditions, on a sur D un isomorphisme de $\mathscr{D}$-Modules $i: \mathscr{G} \to \mathscr{G}'$; montrons que i se prolonge de manière unique en un isomorphisme $\mathscr{G} \to \mathscr{G}'$ sur Y. Sur D^*, i s'interprète comme une section horizontale du fibré vectoriel à connexion $Hom_{\mathscr{O}}(\mathscr{G}, \mathscr{G}')$; la théorie usuelle des équations différentielles donne alors un prolongement unique de i à $\mathbb{C}^*$; enfin, la régularité à l'infini de $\mathscr{G}$ et $\mathscr{G}'$ montre que cette section est à croissance modérée à l'infini, d'où le prolongement cherché à l'infini. Pour montrer que ce prolongement est un isomorphisme, il suffit de faire de même avec i^{-1} au lieu de i.

Plus généralement, le même procédé montre que le foncteur $\mathscr{G} \mapsto \mathscr{G}_0$ établit une équivalence avec la catégorie des $\mathscr{D}_0$-modules holonomes de la catégorie des $\mathscr{D}[y]$-Modules cohérents $\mathscr{G}$ qui possèdent les propriétés suivantes:

(5.1.1a) $\mathscr{G}_0$ est holonome;

(5.1.1b) Sur $Y\backslash\{0\}$, $\mathscr{G}$ est localement libre de rang fini (donc libre) sur $\mathscr{O}[y]$, et à singularité régulière à l'infini.

Posons maintenant $\tilde{G} = \Gamma(Y, \mathscr{G})$ et $\tilde{G}(k) = \Gamma(Y, \mathscr{G}(k))$. Comme $\mathscr{G}(k)$ est cohérent sur $\mathscr{O}[y]$, il vérifie les "théorèmes A et B" (classique). On en déduit les propriétés suivantes:

(5.1.2) On a $\mathscr{G} = \mathscr{O}[y] \otimes_{\mathbb{C}[y]} \tilde{G}$; en particulier $\tilde{G}$ détermine $\mathscr{G}$.

(5.1.3) L'application naturelle $\tilde{G}(k)/\tilde{G}(0) \to G(k)/G(0)$ est un isomorphisme.

Ceci se voit en prenant les sections de la suite exacte $0 \to \mathscr{G}(0) \to \mathscr{G}(k) \to \mathscr{G}(k)/\mathscr{G}(0) \to 0$, dont le dernier terme est de support l'origine, et égal en ce point à $G(k)/G(0)$.

(5.1.4) $\tilde{G}$ est fini sur l'algèbre $\mathbb{C}[y, \partial_y]$ des opéreteurs différentiels à coefficients polynomiaux ("l'algèbre de Weyl").

Tout d'abord, comme $\mathscr{G}(0)$ est cohérent sur $\mathscr{O}[y]$, $\tilde{G}(0)$ est fini sur $\mathbb{C}[y]$. D'autre part, les propriétés de la filtration $G(k)$ impliquent que l'application $\partial_y: G(k)/G(k-1) \to G(k+1)/G(k)$ est surjective pour $k \geq 0$; donc la même formule est vraie avec G remplacé par $\tilde{G}$; ceci s'écrit encore $\tilde{G}(k+1) = \tilde{G}(k) + \partial_y \tilde{G}(k)$; par récurrence sur k, on en déduit que G est engendré sur $\mathbb{C}[y, \partial_y]$ par $\tilde{G}(0)$. D'où le résultat cherché.

(5.2) *Transformation de Laplace.* L'algèbre de Weyl $\mathbb{C}[y, \partial_y]$ peut s'interpréter comme l'algèbre des polynômes en deux variables non-commutatives, y et η, avec $[y, \eta] = -1$. On peut donc aussi bien écrire $y = -\partial_n$, et considérer $\mathbb{C}[y, \partial_y]$ comme l'algèbre $\mathbb{C}[\eta, \partial_\eta]$ des opérateurs differentiels en η à coefficients polynomiaux. Nous noterons $\hat{G}$ le $\mathbb{C}[\eta, \partial_\eta]$-module $\tilde{G}$, avec ce changement de point de vue; soit Z une autre copie de $\mathbb{P}_1(\mathbb{C})$, de coordonnée η; on définit $\mathscr{O}[\eta]$ et $\mathscr{D}[\eta]$ comme ci-dessus, et l'on considère le $\mathscr{D}[\eta]$-Module $\hat{\mathscr{G}} = \mathscr{D}[\eta] \otimes_{\mathbb{C}[\eta, \partial_\eta]} \hat{G}$, qui est aussi égal à $\mathscr{O}[\eta] \otimes_{\mathbb{C}[\eta]} \hat{G}$.

Notre but est d'établir une relation entre $\hat{\mathscr{G}}_\infty$ et le microlocalisé $\mathscr{E}_\lambda \otimes_{\mathscr{D}_0} G$ de G. Posons $K = \mathbb{C}\{\eta^{-1}\}[\eta]$, et $K_1 = \mathbb{C}\{\eta^{-1}\}_1[\eta]$; on a $K = \mathscr{O}_\infty[\eta]$, d'où $\hat{\mathscr{G}}_\infty = K \otimes_{\mathbb{C}[\eta]} \hat{G}$, et $K_1 \otimes_K \hat{\mathscr{G}}_\infty = K_1 \otimes_{\mathbb{C}[\eta]} \hat{G}$; d'autre part, on a un isomorphisme $K_1 \xrightarrow{\sim} \tilde{\mathscr{E}}_\lambda$ (notations du Section 3), et une application $\hat{G} = \tilde{G} \to G$

à savoir la restriction des sections à l'origine. D'où une application $K_1 \otimes_K \hat{\mathscr{G}}_\infty \to \tilde{\mathscr{E}}_\lambda \otimes_{\mathbb{C}[\eta]} G$; on en déduit de façon évidente une application $\mathscr{L}: K_1 \otimes_K \hat{\mathscr{G}}_\infty \to \mathscr{E}_\lambda \otimes_{\mathscr{D}_0} G$, et $\mathscr{L}$ commute à l'action de $K_1 = \tilde{\mathscr{E}}_\lambda$ et à celle de $y = -\partial_\eta$; le résultat est alors le suivant:

Théorème (5.2.1). *$\mathscr{L}$ est bijective.*

Supposons le résultat acquis pour G de la forme $\psi(F)$, avec les notations de (4.11), et montrons comment le cas général en résulte. Soit G un $\mathscr{D}_0$-module holonome; posons $H = \mathscr{E}_\lambda \otimes_{\mathscr{D}_0} G$, et considérons comme en (4.11.b) la suite exacte $0 \to G' \to G \to H \to G'' \to 0$; on a $\mathscr{E}_\lambda \otimes_{\mathscr{D}_0} G' = 0$ et de même avec G''. Pour établir le résultat, il suffit donc de montrer qu'on a $\hat{\mathscr{G}}'_\infty = \hat{\mathscr{G}}''_\infty = 0$; or G' et G'' sont de la forme $\mathscr{O}_0^p$; et, pour $G = \mathscr{O}_0$, on a $\mathscr{G} = \mathscr{O}[y]$, $\tilde{G} = \mathbb{C}[y]$, et $\hat{G} = \mathbb{C}[\eta, \partial_\eta]\delta$, donc $\hat{\mathscr{G}}$ est de support l'origine et $\hat{\mathscr{G}}_\infty = 0$.

Supposons donc à partir de maintenant que G provient d'un $\mathscr{E}_\lambda$-module F par restriction des scalaires à $\mathscr{D}_0$; soit $F(k)$, $k \in \mathbb{Z}$ une bonne filtration de F, et soit $\mu = \dim_{\mathbb{C}} \sigma(F)$. La filtration de G définie par $G(k) = F(k)$, $(k \geq 0)$ vérifie les conditions énoncées en (5.1). D'autre part, ici, $\mathscr{L}$ est simplement l'application $K_1 \otimes_K \hat{\mathscr{G}}_\infty = K_1 \otimes_{\mathbb{C}[\eta]} \hat{G} \to G$ définie par $a(\eta) \otimes \hat{g} \mapsto a(\eta)g$, g la restriction de $\hat{g}$ à l'origine dans Y [cf. démonstration de (4.11)].

Montrons d'abord que $\mathscr{L}$ est surjective; soit $G'(1) \subset G(1)$ l'image par $\mathscr{L}$ de $\mathbb{C}\{\eta^{-1}\}_1 \otimes \hat{G}(1)$; c'est un sous-$\mathbb{C}\{\eta^{-1}\}_1$-module de $G(1)$; comme, par (5.13), l'application $\hat{G}(1)/\hat{G}(0) \to G(1)/G(0)$ est bijective, on a $G(1) = G'(1) + G(0) = G'(1) + \eta^{-1}G(1)$; par Nakayama, on a donc $G'(1) = G(1)$ et le résultat s'ensuit:

Reste à montrer l'injectivité de $\mathscr{L}$; comme $\mathscr{L}$ est une application de K_1-vectoriels, et que le second membre est de dimension μ sur K_1, il suffit de montrer que le premier membre est de même dimension, ou encore que $\hat{\mathscr{G}}_\infty$ est de dimension μ sur K. Ceci va encore exiger un peu de travail.

Lemme (5.2.2). *La multiplication par η est bijective sur $\hat{G}$.*

Il s'agit de démontrer que $\partial_y: \tilde{G} \to \tilde{G}$ est bijective; soit $b \in \tilde{G}$; il existe $a_0 \in G$ unique tel qu'on ait $\partial_y a_0 = b_0$, la restriction de b à $0 \in Y$; par le théorème d'existence et d'unicité des équations différentielles, a_0 se prolonge de manière unique en un $a \in \Gamma(\mathbb{C}, \mathscr{G})$ vérifiant $\partial_y a = b$; enfin, comme $\mathscr{G}$ est régulier à l'infini, a est à croissance modérée et l'on a $a \in \tilde{G}$; d'où le résultat.

Le résultat précédent munit $\hat{G}$ d'une structure de $\mathbb{C}[\eta, \eta^{-1}]$-module.

Lemme (5.2.3). *$\hat{G}$ est fini sur $\mathbb{C}[\eta, \eta^{-1}]$.*

Soit $\mathscr{O}[y, y^{-1}]$ le faisceau sur Y des fonctions méromorphes n'ayant de pôles au plus qu'aux points 0 et ∞, et posons $\mathscr{G}[y^{-1}] = \mathscr{G} \otimes_{\mathscr{O}[y]} \mathscr{O}[y, y^{-1}]$. On a $\mathscr{G}[y^{-1}] = \mathscr{G}(0)[y^{-1}]$, donc, par (4.1) $\mathscr{G}[y^{-1}]$ est localement libre de rang μ sur $\mathcal{O}[y, y^{-1}]$; en fait il est même libre puisque tout faisceau localement libre sur $\mathcal{O}[y]$ est déjà libre (ceci résulte immédiatement de la classification des fibrés vectoriels sur $\mathbb{P}_1$).

Comme la singularité à l'infini est régulière, on peut trouver un réseau $\Lambda \subset \mathcal{G}_\infty = \mathcal{G}_\infty[y^{-1}]$ stable par $y\partial_y$; le sous-faisceau de $\mathcal{G}[y^{-1}]$ formé des sections qui, à l'infini, appartiennent à Λ est libre sur le faisceau des fonctions méromorphes sur Y avec pôle en 0 (même raisonnement que ci-dessus). En prenant une base $e_1, \ldots, e_\mu$ des sections de ce sous-faisceau, on trouve une base de $\mathcal{G}[y^{-1}]$ vérifiant $y\partial_y e_j = \sum a_{ij} e_i$, $a_{ij} \in \mathbb{C}[y^{-1}]$.

En revenant à $\mathcal{G}$, on en déduit l'existence de μ sections $(g_1, \ldots, g_\mu)$ de $\mathcal{G}$ qui possèdent les propriétés suivantes:

(a) elles engendrent un sous-$\mathcal{D}[y]$-Module cohérent $\mathcal{G}'$ de $\mathcal{G}$, qui coincide avec $\mathcal{G}$ en dehors de l'origine;

(b) elles vérifient des équations $y^{k+1}\partial_y g_j = \sum b_{ij} g_j, b_{ij} \in \mathbb{C}[y], \deg(b_{ij}) \leq k$.

Ces équations s'écrivent aussi, en faisant $y = -\partial_\eta$, $\partial_y = \eta$ et en appliquant Leibniz: $\eta\partial_\eta^{k+1} g_j = \sum c_{ij} g_i$, $c_{ij} \in \mathbb{C}[\partial_\eta]$, $\deg(c_{ij}) \leq k_\zeta$

Par suite, le sous-$\mathbb{C}[\eta, 1/\eta, \partial_\eta]$-module de $\hat{G}$ engendré par les g_i est fini sur $\mathbb{C}[\eta, 1/\eta]$, et engendré par les $\partial_\eta^l g_i$, $l \leq k$, et il est clair que ce sous-module contient $\hat{G}'$. On aura donc terminé si l'on montre que $\hat{G}/\hat{G}'$ est fini sur $\mathbb{C}[\eta]$. Or $\mathcal{G}$ et $\mathcal{G}'$ sont des limites inductives de $\mathcal{O}[y]$-Modules cohérents, donc sont Y-acycliques; en prenant les sections de la suite exacte $0 \to \mathcal{G}' \to \mathcal{G} \to \mathcal{G}/\mathcal{G}' \to 0$, on trouve un isomorphisme $\tilde{G}/\tilde{G}' \simeq \mathcal{G}_0/\mathcal{G}'_0$. Ce dernier espace est un $\mathcal{D}_0$-Module cohérent de support l'origine, donc isomorphe par un résultat classique à $(\mathcal{D}_0\delta)^p$; alors $\hat{G}/\hat{G}' \simeq \mathbb{C}[\eta]^p$, et le lemme s'ensuit.

Le faisceautisé $\hat{\mathcal{G}}$ de $\hat{G}$ est donc fini sur $\mathcal{O}$ en dehors de 0, ∞; comme il est muni d'une connexion, il est localement libre, et finalement $\hat{\mathcal{G}}$ sera localement libre (donc libre) sur $\mathcal{O}[\eta, \eta^{-1}]$. (En fait, la démonstration précédente montre aussi que $\hat{\mathcal{G}}_0$ est à singularité régulière, mais ce fait ne nous servira pas ici).

Pour montrer que $\hat{\mathcal{G}}_\infty$ est de rang μ, il suffit d'établir que $\hat{\mathcal{G}}_a$ est de rang μ sur $\mathcal{O}_a$ pour $a \in \mathbb{C}^*$. Considérant la suite exacte $0 \to \hat{\mathcal{G}} \xrightarrow{\eta - a} \hat{\mathcal{G}} \to \hat{\mathcal{G}}/(\eta - a)\hat{\mathcal{G}} \to 0$, et prenant les sections, il suffit finalement d'établir le lemme suivant:

LEMME (5.2.4). La multiplication par $(\eta - a)$ dans $\hat{G}$ est d'indice $-\mu$.

Ceci revient à démontrer que, dans $\tilde{G}$, l'application $(\partial_y - a)$ est d'indice $-\mu$. Pour cela, on va appliquer un théorème de comparaison de [7]. Soit i l'injection $\mathbb{C} \to \mathbb{P}_1(\mathbb{C}) = Y$, et posons $\mathcal{H} = i_* i^* \mathcal{G}$ ($\mathcal{H}$ est le faisceau des "sections de $\mathcal{G}$ avec éventuellement une singularité essentielle à l'infini"). En prenant les sections de la suite exacte $0 \to \mathcal{G} \to \mathcal{H} \to \mathcal{H}/\mathcal{G} \to 0$, on trouve une suite exacte $0 \to \tilde{G} \to \Gamma(Y, \mathcal{H}) \to \mathcal{H}_\infty/\mathcal{G}_\infty \to 0$.

D'une part, on a $\Gamma(Y, \mathscr{H}) = \Gamma(\mathbb{C}, \mathscr{G}|\mathbb{C})$; l'opérateur $\partial_y - a$ est visiblement bijectif sur $\mathscr{G}_0 = G$; le théorème d'existence et d'unicité des équations différentielles montre alors, comme en (5.2.2) que $\partial_y - a$ est bijectif sur $\Gamma(\mathbb{C}, \mathscr{G}|\mathbb{C})$.

D'autre part, d'après [7], l'indice de $(\partial_y - a)$ dans $\mathscr{H}_\infty/\mathscr{G}_\infty$ est égal à l'irrégularité de la connexion $\partial_y - a = -z^2\partial_z - a$ sur $\mathscr{G}_\infty$, avec $z = y^{-1}$; comme ∂_z est régulière, il existe une base $(g_1, \ldots, g_\mu)$ de $\mathscr{G}_\infty$ dans laquelle la matrice de $\partial_z + a/z^2$ ait un pôle d'ordre 2, de partie polaire d'ordre 2 égale à $a\,\mathrm{id}$. Donc, par [7], l'irrégularité de cette connexion est égale à μ et le lemme en résulte. Ainsi, le théorème (5.2.1) est démontré.

Le théorème (1.4), dans le cas $s = 1$, en résulte immédiatement, via le dictionnaire (3.2): à E, $\mathbb{C}\{x\}_1[x^{-1}]$-vectoriel à connexion, avec $r(E) < 1$, on associe le $\mathscr{E}_\lambda$-module $\Phi(E)$; on pose $G = \psi(F)$ [dictionnaire (4.12)]; d'après (5.2.1), on a un isomorphisme $K_1 \otimes_K \hat{\mathscr{G}}_\infty \simeq G$; alors on peut prendre $E' = \hat{\mathscr{G}}_\infty$.

6. Demonstration du théorème (1.5)

Nous allons employer une variante-Gevrey de la méthode que Sibuya utilise dans le cas formel. Soit p (resp., q) l'ordre de $\bar{M}'$ (resp., $\bar{M}''$); écrivons

$$M = \begin{pmatrix} M_{11} & M_{12} \\ M_{21} & M_{22} \end{pmatrix},$$

et cherchons le changement de base $\mathbf{f} = \mathbf{e}S$ sous la forme

$$S = \begin{pmatrix} \mathrm{id} & S' \\ S'' & \mathrm{id} \end{pmatrix}$$

avec S' (resp., S'') de type (p, q) [resp., (q, p)], et sans terme constant. Soit N la matrice de la connexion dans la base $\mathbf{f}$; on a $dS/dx = SN - MS$; un calcul immédiat montre alors que N aura la forme voulue $\left(\begin{smallmatrix} N' & 0 \\ 0 & N'' \end{smallmatrix}\right)$ si et seulement si S' et S'' vérifient les équations suivantes

$$x\,dS'/dx + M_{12} + M_{11}S' - S'M_{22} - S'M_{21}S' = 0, \tag{6.1}$$

$$x\,dS''/dx + M_{21} + M_{22}S'' - S''M_{11} - S''M_{12}S'' = 0. \tag{6.2}$$

Les deux équations sont analogues; regardons par exemple (6.1). En changeant les notations, cette équation est de la forme suivante:

$$x^{r+1}\,dF/dx + P + QF + (RF, F) = 0, \tag{6.3}$$

avec P, Q, R à coefficients dans $\mathbb{C}\{x\}_s$, et P et R sans terme constant; de plus, l'hypothèse "$\bar{M}'$ et $\bar{M}''$ sans valeur propre commune" signifie que l'application $\sum' \mapsto \bar{M}'\sum' - \sum'\bar{M}''$ de $\mathrm{Hom}(\mathbb{C}^q, \mathbb{C}^p)$ dans lui-même est *bijec-*

tive, par un lemme classique d'algèbre linéaire. Avec les notations de (6.3), cela signifie que $Q(0)$ est inversible. Comme d'après les hypothèses, on a $r \geq 1$, il est alors immédiat de vérifier que l'équation (6.3) a une solution formelle et une seule, qui est sans terme constant. Il s'agit donc de démontrer que, sous la condition $r \geq 1/s$, cette solution est à coefficients dans $\mathbb{C}\{x\}_s$. Pour cela, nous allons utiliser le théorème des fonctions implicites.

Soit $\rho > 0$; désignons par $H(\rho)$ l'espace des séries formelles $\varphi = \sum_{n \geq 1} a_n x^n$ sans terme constant qui vérifient $|\varphi|_\rho = \sum [|a_n|/(n!)^s]\rho^n < +\infty$ (pour $s = 1$, ceci diffère légèrement des notations du Section 4, mais peu importe). Enonçons d'abord deux lemmes.

LEMME (6.4). *La multiplication est continue de $H(\rho) \times H(\rho)$ dans $H(\rho)$.*

Evident, par un calcul analogue à (4.4).

LEMME (6.5) *Soit r un entier avec, $r \geq 1/s$; l'application $\varphi \mapsto x^{r+1}\, d\varphi/dx$ est continue de $H(\rho)$ dans lui-même.*

En effet, si $\varphi = \sum a_n x^n$, on a $x^{r+1}\, d\varphi/dx = \sum n a_n x^{n+r}$, d'où

$$\left| x^{r+1} \frac{d\varphi}{dx} \right|_\rho = \sum \frac{n|a_n|}{[(n+r)!]^s} \rho^{n+r} \leq \rho^r \sum \frac{|a_n|}{(n!)^s} \rho^n = \rho^r |\varphi|_\rho .$$

Démontrons maintenant le théorème. On procède par dilatation, suivant une méthode connue dans la théorie des équations elliptiques. Soit $\lambda \in \mathbb{C}$; posons $F_\lambda(x) = F(\lambda x)$, et définissons de même P_λ, Q_λ, R_λ; au lieu de résoudre (6.3) dans $H(\rho)^m$ $(m = pq)$, pour ρ assez petit, il revient au même de résoudre dans $H(1)$, pour λ petit, l'équation

$$\lambda^r x^{r+1}\, dF_\lambda/dx + P_\lambda + Q_\lambda F_\lambda + (R_\lambda F_\lambda, F_\lambda) = 0. \tag{6.6}$$

Pour cela, considérons l'application $(\lambda, G) \mapsto \psi(\lambda, G)$ définie au voisinage de 0 dans $\mathbb{C} \times H(1)^m$, à valeurs dans $H(1)^m$, qui est définie par

$$\psi(\lambda, G) = \lambda^r x^{r+1}\, dG/dx + P_\lambda + Q_\lambda G + (R_\lambda G, G).$$

En utilisant (6.4) et (6.5), il est immédiat de vérifier que cette application est analytique; on a $\psi(0, G) = Q(0)G$, donc $\partial/\partial G\, \psi(0, 0)$ est la multiplication par $Q(0)$, qui est inversible. On conclut alors par le théorème des fonctions implicites.

Remarque (6.7). Dans le cas $s = 1$, $r = 1$, l'interprétation microdifférentielle donnée à la Section 3 permet facilement de redémontrer le théorème (1.5) comme une conséquence du théorème de préparation pour les opérateurs microdifférentiels. Peut-être le cas général peut-il se traiter de façon analogue (je n'ai pas examiné cette question en détail). Pour cela, il faudrait étudier systématiquement le "calcul différentiel d'ordre infini dans les classes de Gevrey" qui généralise celui qu'on obtient pour $s = 1$ en faisant subir au calcul microdifférentiel le changement de variables $x = \eta^{-1}$, $x^2\partial_x = y$. Je reviendrai peut-être ultérieurement sur ce sujet.

Références

1. D. Bertrand, Travaux récents sur les points singuliers des équations différentielles linéaires, *Sém. Bourbaki, 1978–1979*, N°. 538.
2. G. D. Birkhoff, The generalized Rieman problem for linear differential equations. . . , *Proc. Amer. Acad. Arts Sci.* **49** (1913), 521–568.
3. L. Boutet de Monvel, Opérateurs pseudo-différentiels analytiques, *Sém. Grenoble, 1975–1976*, Institut Fourier.
4. P. Deligne, "Equations différentielles à points singuliers réguliers," Springer Lecture Notes, N°. 163, Springer-Verlag, Berlin and New York, 1970.
5. W. Jurkat, "Meromorphe Differentialgleichungen," Springer Lecture Notes, N°. 637, Springer-Verlag, Berlin and New York, 1978.
6. A. Levelt, Jordan decomposition for a class of singular differential operators, *Ark. Mat.* **13** (1975), 1–27.
7. B. Malgrange, Sur les points singuliers des équations différentielles linéaires, *Enseign Math.* **20** (1974), 147–176.
8. B. Malgrange, L'involutivité des caractéristiques des systèmes différentiels et microdifférentiels, *Sém. Bourbaki, 1977–1978*, N°. 522.
9. B. Malgrange, Déformation de modules différentiels et microdifférentiels, to appear.
10. J.-P. Ramis, Dévissage Gevrey, *Astérisque* **59/60** (1978), 173–204.
11. J.-P. Ramis, Les séries k-sommables et leurs applications, Springer Lecture Notes in Physics, N°. 126, Springer-Verlag, Berlin and New York, 1980.
12. M. Sato, T. Kawai, and M. Kashiwara, "Microfunctions and Pseudodifferential Equations, "Springer Lecture Notes, N°. 287, pp. 264–529, Springer-Verlag, Berlin and New York, 1973.
13. W. Wasow, "Asymptotic Expansions for Ordinary Differential Equations," Wiley, New York, 1963.

MATHEMATICAL ANALYSIS AND APPLICATIONS, PART B
ADVANCES IN MATHEMATICS SUPPLEMENTARY STUDIES, VOL. 7B

Invariant Subsets for Area Preserving Homeomorphisms of Surfaces

JOHN N. MATHER†

Department of Mathematics
Princeton University
Princeton, New Jersey

TO LAURENT SCHWARTZ ON HIS 65TH BIRTHDAY

Throughout this paper, S will be a connected surface without boundary, provided with a Borel measure μ, for which open nonvoid sets have positive measure, and any compact subset has finite measure. The surface will be said to be *smooth* if it is provided with the structure of a C^∞ manifold, and the measure μ will be said to be *smooth* if for any open subset U of S and any C^∞ local coordinate system x, y defined in U, we have $\mu | U = u(x, y)\, dx\, dy$, where u is C^∞ and positive.

Let $f: S \to S$ be a homeomorphism. A subset A of S will be said to be (f-) *invariant* if $fA = A$. The mapping f will be said to be *area preserving* if $f_* \mu = \mu$. If A is a μ-measurable subset of S, then $\mu(A)$ will be called the *area* of A. In this paper, we will obtain some results about f-invariant sets, which are valid for certain area preserving f. Our results will be valid for sufficiently differentiable "generic" area preserving f, in a sense which we now describe.

We let $\mathscr{A}$ denote the set of all area preserving homeomorphisms of S. In Section 5, we will define a subset $\mathscr{X}$ of $\mathscr{A}$. Our main results will be valid for $f \in \mathscr{X}$. In the case S and μ are smooth, we let $\mathscr{A}^r$ denote the space of C^r area preserving diffeomorphisms of S provided with the C^r topology. We let $\mathscr{X}^r = \mathscr{X} \cap \mathscr{A}^r$. It follows from known results (explained in Section 6) that $\mathscr{X}^r$ is a residual subset of $\mathscr{A}^r$, if $r \geq 4$, in the sense of Baire category. This is the sense in which our results are valid "generically."

If U is a surface, we will denote its "ideal boundary" (Section 1) by $b_I U$. Points of $b_I U$ are what are usually called "ends" of U by topologists. However, we need also another notion of "ends," due to Carathéodory. So, to avoid confusion, we will use a terminology which is sometimes used by analysts: we will call points of $b_I U$ "ideal boundary points" of U. The

† Supported by an NSF contract MCS 79-02017.

ISBN 0-12-512802-9

disjoint union $c_I U = U \coprod b_I U$, together with a natural topology, defined in Section 1, will be called the "ideal completion of U."

By a *domain*, we will always mean a nonempty connected open subset of S. If U is a domain, and p is an ideal boundary point of U, we let $Z(p)$ denote the set of limit points in $c_I S$ of sequences in U converging in $c_I U$ to p (Section 2). If $Z(p) \subset S$ and $Z(p)$ contains more than one point, we will say p is *regular*.

If f is a homeomorphism of S onto itself, and U is an f-invariant domain, then $f|U$ has a unique continuous extension $f_U: c_I U \to c_I U$, and f_U is a homeomorphism. If p is a regular ideal boundary point of U, $f_U(p) = p$, and f_U is locally orientation preserving at p, then it is possible to associate a rotation number to the triple (f, U, p). The definition of this rotation number, which depends on a deep theorem of Carathéodory, will be given in Section 4. We will call this number the *Carathéodory rotation number* associated to (f, U, p).

Theorem 5.1 is the first main theorem of this paper. If implies that if $f \in \mathscr{X}$, then all the Carathéodory rotation numbers associated to f are irrational.

Suppose S is smooth, and f is a C^1 diffeomorphism of S. Let $\mathbf{O}$ be a periodic point of f. We define

$$W_f^s(\mathbf{O}) = \{x \in S : f^{nk}(x) \to \mathbf{O} \text{ as } k \to +\infty\}$$
$$W_f^u(\mathbf{O}) = \{x \in S : f^{nk}(x) \to \mathbf{O} \text{ as } k \to -\infty\}$$

where n is the period of $\mathbf{O}$. If $\mathbf{O}$ is hyperbolic, i.e., one eigenvalue of $df^n(\mathbf{O})$ has absolute value < 1, and the other has absolute value > 1, then these are immersed 1-manifolds in S, by the Hadamard–Perron theory.

We will call the two components (with respect to the manifold topology) of $W_f^s(\mathbf{O}) - \mathbf{O}$ (resp., $W_f^u(\mathbf{O}) - \mathbf{O}$), the two *branches* of the stable (resp., unstable) manifold of $\mathbf{O}$. A *branch associated to* $\mathbf{O}$ will mean a branch of the stable or unstable manifold of $\mathbf{O}$. The second main theorem (Theorem 5.2) of this paper implies that if S is compact, $f \in \mathscr{X}^1$, $\mathbf{O}$ is a hyperbolic periodic point of f, and b_1, b_2 are two branches associated to $\mathbf{O}$, then b_1 and b_2 have the same closure.

1. Ideal Boundary Points

Let $H = \{(x, y) \in \mathbb{R}^2 : x \geq 0\}$. Throughout this paper, *surface* will mean a Hausdorff topological space, with a countable basis for its topology, each point of which has an open neighborhood homeomorphic to $\mathbb{R}^2$ or to H. Points having arbitrarily small open neighborhood of the latter type constitute the *boundary* of the surface. Other points constitute the *interior* of the surface. If X is a surface, we let ∂X denote its boundary. If the boundary is empty, the surface is said to be *without boundary*.

In this section, we recall the definition and fundamental properties of what analysts sometimes call *ideal boundary points* of a surface.

Let U be a surface, possibly with boundary. For any compact set K in U, let $C(U - K)$ denote the set of connected components of $U - K$. Let

$$b_I U = \varprojlim_{K} C(U - K), \tag{1.1}$$

where K ranges over all compact subsets of U. An element of $b_I U$ is called an *ideal boundary point* of U, and $b_I U$ is called the ideal boundary of U. The disjoint union

$$c_I U = U \coprod b_I U$$

will be called the *ideal completion* of U.

Let $\mathscr{T}(U)$ denote the topology of U. Let $\mathscr{T}_0(U)$ denote the set of all connected open subsets V of U whose frontier $\mathscr{F}(V)$ is compact. Clearly,

$$\mathscr{T}_0(U) = \bigcup \{C(U - K): K \text{ is a compact subset of } U\},$$

since if $V \in \mathscr{T}_0(U)$, then $V \in C(U - \mathscr{F}(V))$. If K is a compact subset of U, let

$$\pi_K : b_I U \to C(U - K)$$

denote the projection corresponding to the representation (1.1) of $b_I U$. If $V \in \mathscr{T}_0 U$, define

$$V^* = V \coprod \pi^{-1}_{\mathscr{F}(V)}\{V\}.$$

Let $\mathscr{T}_0(c_I U) = \{V^* : V \in \mathscr{T}_0(U)\}$. It is easily checked that $\mathscr{T}_0(c_I U)$ is the basis of a topology $\mathscr{T}(c_I U)$ on $c_I U$. From now on, we provide $c_I U$ with the topology $\mathscr{T}(c_I U)$.

A thorough account of the theory of ideal boundary points of surfaces without boundary is given in Richards [14]. Here, we summarize some of the main facts which we will need.

Let $\{U_1, U_2, \ldots\}$ be the set of connected components of a surface U. It is easily seen that $c_I U$ is the disjoint union $c_I U_1 \coprod c_I U_2 \coprod \cdots$.

It is easy to determine the topologies on U and $b_I U$ induced by the inclusions $U \subset c_I U$ and $b_I U \subset c_I U$. The topology on U is the original topology of U. The topology of $b_I U$ is the inverse limit topology for the representation (1.1), where $C(U - K)$ is given the discrete topology.

A classical theorem says that every surface can be triangulated. Assume U is triangulated. The set of subpolyhedra K of U which are also surfaces forms a cofinal system for the representation (1.1). If U has only finitely many components, and $K \subset U$ is a subpolyhedron, then $C(U - K)$ is finite. It follows that in the case U has only finitely many components, the representation (1.1) has a cofinal subsystem such that each $C(U - K)$ is finite. Hence $b_I U$ is compact. It follows easily from this that $c_I U$ is compact. Of course, $c_I U$ is not compact when U has infinitely many components.

Let $p \in b_I U$. A set $V \in \mathcal{T}_0 U$ will be said to be a *fundamental set* of p if $p \in V^*$. A family $\{V_\alpha\}$ of members of $\mathcal{T}_0 U$ will be said to be a *fundamental system* of p, if $\{V_\alpha^*\}$ is a basis of the family of neighborhoods of p in $c_I U$.

Suppose $V_1 \supset V_2 \supset V_3 \supset \cdots$ is a system of members of $\mathcal{T}_0 U$, and $\bar{V}_1 \cap \bar{V}_2 \cap \bar{V}_3 \cap \cdots = \emptyset$, where $\bar{V}_I$ denotes the closure of V_i in U. In this case, it is easily seen that $\{V_i\}_{i=1,2,\ldots}$ is a fundamental system for a (necessarily unique) ideal boundary point of U.

For, let W_i be the closure of V_i in $c_I U$. Clearly W_i is connected, and since each component of $c_I U$ is compact, it follows that W_i is compact. Hence $W = W_1 \cap W_2 \cap W_3 \cap \cdots$ is compact, connected, and nonempty, and $W \subset b_I U$. Since $b_I U$ is totally disconnected, it follows that W is a single point p, and clearly $W_1 \supset W_2 \supset \cdots$ is a basis of the family of neighborhoods of p in $c_I U$.

Suppose U is provided with a triangulation. If $p \in b_I U$, it is easily seen that there is a fundamental system $V_1 \supset V_2 \supset \cdots$ for p, where each V_i is connected, and each $\mathcal{F} V_i$ is a polyhedral circle or arc, and $\mathcal{F} V_i \cap \mathcal{F} V_j = \emptyset$, for $i \neq j$.

By the *genus* of a surface U, one means half the rank of the intersection form on $H_1(U, \mathbb{Z}/2)$. An ideal boundary point of U is said to be of *finite genus* if there exists a fundamental set of p which has finite genus. In this case, there is a fundamental set of p of genus 0. We need the following result, which is a consequence of the main result in Richards [14]:

PROPOSITION 1.1. *Let $p \in b_I U$ be of finite genus, and suppose p is not adherent to the boundary of U. Then $c_I U$ is locally homeomorphic to $\mathbb{R}^2$.*

2. IDEAL BOUNDARY POINTS OF OPEN SUBSETS OF S

Throughout this section, we let U be an open subset of S, and p an ideal boundary point of U. We let $V_1 \supset V_2 \supset \cdots$ be a fundamental system for p.

If some V_i is relatively compact in S (i.e., has compact closure in S), we will say p is *compact* (*rel* S). Obviously this notion is independent of the particular system chosen.

If A is a subset of a topological space B, we denote the closure of A in B by $\mathrm{cl}_B A$ or $\mathrm{cl}(A; B)$, the frontier of A in B by $\mathcal{F}_B A$ or $\mathcal{F}(A; B)$, and the interior of A in B by $\mathrm{int}_B A$ or $\mathrm{int}(A; B)$. We will always use "boundary" as in "manifold with boundary." We will use "frontier" for the point-set notion of "boundary." The word "interior" will mean "complement of the boundary" in a manifold with boundary, except when we say interior of A "in B," when it will have the point-set meaning.

We let $Z(p) = \bigcap_i \mathrm{cl}(V_i; c_I S)$. Obviously, $Z(p)$ is independent of the particular fundamental system chosen. Obviously, $Z(p)$ is a compact, nonempty subset of $c_I S$, contained in the frontier of U. Since each V_i is connected, $Z(p)$ is connected. Obviously, p is compact (rel. S) if and only if $Z(p) \subset S$.

From Proposition 1.1., we easily obtain:

PROPOSITION 2.1. *Suppose p is a compact (rel. S) ideal boundary point of U. Then p has an open neighborhood in $c_I U$ homeomorphic to $\mathbb{R}^2$.*

Proof. Some V_i is relatively compact, and hence of finite genus. ■

Associated isolated ideal boundary point. We shall construct a pair $(\hat{U}_p, \hat{p})$ where $\hat{U}_p$ is an open, connected subset of S and $\hat{p}$ is an ideal boundary point of $\hat{U}_p$. In the case p is compact (rel. S) we will show that $\hat{p}$ is compact (rel. S) and isolated in $b_I \hat{U}_p$. In this case, we will call $\hat{p}$ [or $(\hat{U}_p, \hat{p})$] the *isolated ideal boundary point associated to p* [or (U, p)].

We let U_p be the component of U to which p is adherent. We let $\hat{U}_p$ be the component of $S - Z(p)$ which contains U_p.

By assumption, each V_i is connected. Let K be a compact set in $\hat{U}_p$. Since $\bigcap_i \mathrm{cl}(V_i; c_I S) = Z(p)$, we have that there exists i_0 such that $K \cap \mathrm{cl}(V_i; c_I S) = \varnothing$ for all $i \geq i_0$. Since V_i is connected, and it is contained in $\hat{U}_p - K$, it lies in one component W_K of $\hat{U}_p - K$ for sufficiently large i, and this component is independent of i. Clearly W_K is independent of the choice of fundamental system $V_1 \supset V_2 \supset \cdots$ of p.

LEMMA 2.2. *$\{W_K : K \subset \hat{U}_p$ is compact$\}$ is a fundamental system of an ideal boundary point $\hat{p}$ of $\hat{U}_p$. If p is compact (rel. S), then so is $\hat{p}$.*

Proof. If $K \subset K'$, then $W_K \supset W_{K'}$, so $\{W_K : K \subset \hat{U}_p$ is compact$\}$ is an element $\hat{p}$ of $\varprojlim C(\hat{U}_p - K) = b_I \hat{U}_p$. Clearly $\{W_K\}$ is a fundamental system of $\hat{p}$.

If p is compact (rel. S), let N be a compact neighborhood of $Z(p)$ in S, and let $K = U_p \cap \mathscr{F} N$. Clearly, K is compact, and $W_K \subset N$, so W_K is relatively compact. Hence, $\hat{p}$ is compact (rel. S). ■

It remains to show that $\hat{p}$ is isolated in $b_I \hat{U}_p$. This is an immediate consequence of the following lemma (with $A = Z(p)$ and $V = \hat{U}_p$).

DEFINITION. If A is a closed subset of S, a *residual domain of A* will mean a connected component of $S - A$.

Note that whenever we use the term "domain" or "residual domain", the ambient space is *always* S.

LEMMA 2.3. *Let A be a compact, connected subset of S. Let V be a residual domain of A. Then V has only finitely many ideal boundary points q such that $Z(q) \cap S \neq \emptyset$. Each such ideal boundary point q is compact (rel. S) and isolated in $b_I V$, and $Z(q) \subset A$.*

Proof. From the fact that S can be triangulated, it follows easily that A has a compact neighborhood N which is a surface. Let $K = V \cap \partial N$ and $\pi: b_I V = \varprojlim_K C(V - K) \to C(V - K)$ be the projection. Let $q \in b_I V$. Clearly $\pi(q) \subset N$ or $\pi(q) \cap N = \emptyset$, since $\pi(q)$ is connected and does not meet ∂N. If $Z(q) \cap S \neq \emptyset$, it follows that $\pi(q) \subset N$. Hence q is compact (rel. S), and $Z(q) \subset A$.

By Lefschetz–Alexander duality,

$$H_2(N, N - A) \simeq \check{H}^0(A),$$

where the right-hand side denotes Čech cohomology, and the coefficient group is $\mathbb{Z}/2$. Hence

$$H_2(N, N - A) \simeq \mathbb{Z}/2.$$

By the exact sequence for homology, we obtain that $H_1(N - A)$ is finite.

Let W be a component of $N - A$. By Proposition 1.1, $c_I W$ is a compact surface (usually with boundary). Hence $H_2(c_I W) = 0$ or $\mathbb{Z}/2$. Since $H_1(N - A)$ is finite, so is $H_1(W)$. From the exact sequence

$$H_2(c_I W) \to H_2(c_I W, W) \to H_1(W),$$

we obtain that $H_2(c_I W, W)$ is finite. By Lefschetz–Alexander duality,

$$\check{H}_0(b_I W) \simeq H_2(c_I W, W),$$

so $\check{H}_0(b_I W)$ is finite. However, $b_I W$ is compact and totally disconnected, since it is the inverse limit of finite sets. Hence $b_I W$ is finite.

If V is in the interior of N, then V is a connected component of $N - A$. Otherwise, $V \cap N$ is a union of various connected components of $N - A$, each of which contains a boundary component of N. Since N has only finitely many boundary components, $V \cap N$ is the union of at most finitely many components of $N - A$. If q is an ideal boundary point of V such that $Z(q) \neq \emptyset$, then $q \in b_I W$ for one of the components W of $N - A$ in $V \cap N$. It follows that there are only finitely many such q and each is isolated in $b_I V$. ∎

LEMMA 2.4. $Z(\hat{p}) = Z(p)$.

3. Carathéodory's Theory of Prime Ends

In this section, we state results from Carathéodory's theory of prime ends which we shall need. The results we state are somewhat more general than the results Carathéodory proved in his paper [3]. Carathéodory considered only open, connected, simply connected, bounded sets in the complex plane. We shall need to consider relatively compact, connected surfaces in S with compact boundary, and only finitely many ideal boundary points. However, it is possible to generalize Carathéodory's proofs to obtain the results which we state.

In a recent exposition of Carathéodory's theory [5], Epstein proves several of the generalizations of Carathéodory's theorems which we need.

Throughout this section we let $U \subset S$ be a relatively compact, connected surface. We suppose ∂U is compact and U has only finitely many ideal boundary points.

Definition. (a) A *chain* means a sequence $V_1 \supset V_2 \supset \cdots$ of connected open nonempty subsets of U such that:

(1) For each i, $\mathscr{F}_U V_i$ is connected;
(2) For $i \neq j$, $\text{cl}_S \mathscr{F}_U V_i \cap \text{cl}_S \mathscr{F}_U V_j = \varnothing$.

(b) If $\sigma = \{V_1 \supset V_2 \supset \cdots\}$ and $\tau = \{W_1 \supset W_2 \supset \cdots\}$ are two chains, we say τ divides σ if for each i, there is a j such that $W_j \subset V_i$. We say τ and σ are *equivalent* if each divides the other. We say σ is *prime* if any chain which divides it is equivalent to it. A *prime end* of U is an equivalence class of prime chains. We let $E(U)$ denote the set of all prime ends of U.

Example. Let $x \in U$. We may choose a sequence $V_1 \supset V_2 \supset \cdots$ of open subsets of U such that: $\text{cl}_S V_i \subset U$, and it is homeomorphic to a closed disk; $\text{cl}_S V_{i+1} \subset V_i$; and $\bigcap_i \text{cl}_S V_i = x$. It is easily seen that $V_1 \supset V_2 \supset \cdots$ is a prime chain. We denote the prime end which it represents by $\omega(x)$. Clearly, $\omega(x)$ is independent of the choice of chain $V_1 \supset \cdots$, as long as it satisfies the conditions we listed above.

Let e be a prime end of U, and let $V_1 \supset V_2 \supset \cdots$ be a chain which represents it. By the *impression* of e, we will mean the set

$$Y(e) = \bigcap_i \text{cl}_S(V_i).$$

This is clearly independent of the chain which represents e, and is compact, connected and nonempty.

If $x \in U \cap Y(e)$, it is easy to check that $\omega(x)$ divides e [i.e., a chain representing $\omega(x)$ divides a chain representing e]. Since e is prime, it follows that $e = \omega(x)$, and hence $x = Y(e)$.

The set

$$\alpha(e) = \bigcap_i \text{cl}(V_i; c_I U)$$

is compact, connected and nonempty. If $\alpha(e)$ contains a point x of U, then $x \in Y(e)$, so $e = \omega(x)$, and $\alpha(e) = x$. Otherwise $\alpha(e) \subset c_I U - U$, which is a finite set. Since $\alpha(e)$ is connected and nonempty, it follows that $\alpha(e)$ is a single point in this case, too. Thus α is a mapping of $E(U)$ into $c_I U$. Clearly, $\alpha \circ \omega = \text{id}_U$ and if $\alpha(e) = p \in b_I U$, then $Y(e) \subset Z(p)$.

Let x be a point of $\text{cl}_S(U)$ and e a prime end. If there is a chain $V_1 \supset V_2 \supset \cdots$ which represents e, such that $\mathscr{F}_U V_1, \mathscr{F}_U V_2, \ldots$ converges in S to x, then x is said to be a *principal point* of e. The set of principal points of e is called the *principal set* of e, and will be denoted by $X(e)$.

Our definitions differ from Carathéodory's definitions, even in the case he considered. Our definitions also differ from Epstein's, and Epstein's differ from Carathéodory's. In order to show the equivalence of these definitions, we need the following result.

LEMMA 3.1. *Suppose S is provided with a triangulation. Let e be a prime end of U.*

(a) *There is a chain $V_1 \supset V_2 \supset \cdots$ representing e, such that for each i, $\text{cl}_S(\mathscr{F}_U V_i)$ is a PL-circle or a PL-arc. In the case it is a PL-circle, it has at most one point not in U. In the case it is a PL-arc, all points of the arc except possibly the endpoints are in U.*

(b) *If x is a principal point of e, then we may choose the chain $V_1 \supset V_2 \supset \cdots$ so that $\mathscr{F}_U V_i \to x$, in addition to the conditions listed above.*

Proof. (a) In the case $\alpha(e) \in U$, the conclusion of the lemma is obvious. So, we suppose $\alpha(e)$ is a point p of the ideal boundary of U. In the case $Z(p)$ is a single point x, we have that e is the only prime end of U such that $\alpha(e) = p$, and we may choose a chain $V_1' \supset V_2' \supset \cdots$ defining e, consisting of a family of concentric disks (in some coordinate system) centered at x. The conclusion of the lemma is clear in this case, too. So, we may suppose $Z(p)$ contains more than one point.

Let $V_1' \supset V_2' \supset \cdots$ be a chain defining e. Since p is isolated in $b_I U$, there is a closed disk D in $c_I U$ with p in its interior, such that $D \cap b_I U = p$. For sufficiently large i, $V_i' \subset D - p$. Thus, we might as well suppose that for all i, $V_i' \subset D - p$, and $\text{cl}_D V_i'$ does not meet ∂D.

Step 1. (For sufficiently large i, p is not an interior point relative to D, of $\text{cl}_D V_i'$.) Let Δ be a PL-disk in S whose interior meets $Z(p)$, and which does not meet ∂D. Since $Z(p) \subset \mathscr{F}U$, we may choose $x \in U \cap \Delta$. Since $Z(p)$ is connected and has at least two points, $Z(p) \cap \Delta$ has at least two points, so we may connect x to points of $Z(p) \cap \Delta$ by PL-arcs $\alpha', \beta' \subset \Delta$ whose only common point is x. Let y (resp., z) be the first point of $Z(p)$ on α' (resp., β') from x. Let α (resp., β) be the subarc xy (resp., xz) of α' (resp., β').

Let $\mu_1 = \alpha \cup \beta$. Let D'_1 be the component of $U - \mu_1$ which is contained in $D - p$.

By induction, we will construct a sequence $\mu_1, \mu_2, \ldots$ of PL-arcs in S, and a sequence $D'_1, D'_2, \ldots$, where D'_i is a component of $U - \mu_i$. We have already constructed μ_1 and D'_1. Assume μ_i and D'_i have been constructed, μ_i is a PL-arc in S whose endpoints lie in $Z(p)$, but which is otherwise in U, and D'_i is a component of $U - \mu_i$ which is contained in $D - p$.

Then $\mathscr{F}D'_i \subset \mu_i \cup Z(p)$. Since μ_i cannot separate in S, there is at least one point in $\mathscr{F}D'_i \cap Z(p)$ not in μ_j for $j \leq i$. Hence we may choose a PL-disk Δ_{i+1} in S whose interior meets $\mathscr{F}D'_i \cap Z(p)$, and which does not meet μ_j for $j \leq i$ or ∂D. Let $x_{i+1} \in D'_i \cap \Delta_{i+1}$. We repeat the construction of μ_i with x_{i+1} in place of x and Δ_{i+1} in place of Δ, to get a PL-arc μ_{i+1} in Δ_{i+1}, whose endpoints are in $Z(p)$, but which is otherwise in D'_i. Let D'_{i+1} be the component of $U - \mu'_{i+1}$ which lies in $D - p$. This completes the inductive step.

If the interior μ_i^0 of μ_i lies in D'_{i+1}, then $D'_{i+1} \cup D'_i = U$, contrary to the assumption that $D'_i \cup D'_{i+1} \subset D - p$. Hence $\mu_i \cap D'_{i+1} = \varnothing$. Since $\mu_{i+1}^0 \subset D'_i$, it follows that $D'_{i+1} \subset D'_i$. Since $\mu_i \cap \mu_{i+1} = \varnothing$, it follows that $\{D'_1 \supset D'_2 \supset \cdots\}$ is a chain.

In choosing the disk Δ_{i+1}, we could have supposed it had arbitrarily small diameter (with respect to some metric). We will suppose that the diameters of these disks tend to 0 as $i \to \infty$.

Then, if p is an interior point, relative to D, of $\mathrm{cl}_D V'_i$, there exists j such that $D'_j \subset V'_i$. Hence the chain $\{D'_1 \supset D'_2 \supset \cdots\}$ divides the chain $\{V'_1 \supset V'_2 \supset \cdots\}$, if p is an interior point of $\mathrm{cl}_D V'_i$, for all i. It is clear, however, that under this condition, $\{V'_1 \supset V'_2 \supset \cdots\}$ cannot divide $\{D'_1 \supset D'_2 \supset \cdots\}$. Hence $\{V'_1 \supset V'_2 \supset \cdots\}$ is not prime. This contradiction proves our assertion.

Step 2. (We may assume $\mathrm{cl}_S \mathscr{F}_U V'_i$ meets $Z(p)$ for all i.) If p is not an interior point of $\mathrm{cl}_D V'_i$, then $\mathrm{cl}_S \mathscr{F}_U V'_i$ meets $Z(p)$. Since this is true for all sufficiently large i, we may assume it is true for all i, by replacing $V'_1 \supset V'_2 \supset \cdots$ with a subsequence, if necessary.

Step 3. (Construction of V_i). Let N_i be a neighborhood of $\mathrm{cl}_S(\mathscr{F}_U V'_{i+1})$ in S which is a compact subpolyhedron, a surface, and does not meet $\partial D \cup \mathrm{cl}_S(\mathscr{F}_U V'_i)$. Such a neighborhood exists since $\mathrm{cl}_S(\mathscr{F}_U V'_{i+1})$ does not meet $\partial D \cup \mathrm{cl}_S(\mathscr{F}_U V'_i)$.

Let γ_i be a PL-arc in $D - p \subset U \subset S$ joining $\mathscr{F}_U V'_i$ to $\mathscr{F}_U V'_{i+1}$, and meeting ∂N_i in only finitely many points. By replacing γ_i with a subarc if necessary, we may assume that only its endpoints meet $\mathscr{F}_U V_i' \cup \mathscr{F}_U V_{i+1}'$. Since ∂N_i separates $\mathrm{cl}_S(\mathscr{F}_U V_i')$ from $\mathrm{cl}_S(\mathscr{F}_U V_{i+1}')$, it is clear that γ_i crosses ∂N_i an odd number of times. Hence, there is one component μ_i of $\mathscr{F}_U(U \cap N_i)$ which γ_i crosses an odd number of times. Since γ_i is in $D - p$ and $N_i \cap \partial D = \varnothing$, it follows that $\mu_i \subset D - p$.

Since $\mathscr{F}_U V_i'$, $\mathscr{F}_U V_{i+1}'$, μ_i and γ_i are all in $D - p$, μ_i does not meet ∂D, and γ_i crosses μ_i an odd number of times, it follows that μ_i separates $\mathscr{F}_U V_i'$ from $\mathscr{F}_U V_{i+1}'$ in $D - p$.

By Step 2, μ_i cannot be homeomorphic to a circle. It follows that $\mu_i \cup p$ is homeomorphic to a circle.

Let V_i be the component of $D - \mu_i - p$ which contains $\mathscr{F}_U V_{i+1}'$.

Step 4. ($V_{i+1}' \subset V_i \subset V_i'$). First, $\mu_i \not\subset V_{i+1}'$. Clearly, γ_i contains a subarc γ_i' whose endpoints lie in μ_i and $\mathscr{F}_U V_i'$, respectively, and γ_i' does not meet $\mathscr{F}_U V_{i+1}'$. Therefore, if we had $\mu_i \subset V_{i+1}'$, it would follow that $\gamma_i' \subset V_{i+1}'$ and hence $\mathscr{F}_U V_i' \subset V_{i+1}'$, which is impossible, since $V_{i+1}' \subset V_i'$, and $\mathscr{F}_U V_{i+1}' \cap \mathscr{F}_U V_i' = \varnothing$.

Next $\mu_i \cap V_{i+1}' = \varnothing$, since μ_i is connected, $\mu_i \subset U$, $\mu_i \cap \mathscr{F}_U V_{i+1}' = \varnothing$, and $\mu_i \not\subset V_{i+1}'$.

Since $V_i \supset \mathscr{F}_U V_{i+1}'$, V_{i+1}' is connected, and $\mathscr{F}_U V_i = \mu_i$ does not meet V_{i+1}', it follows that $V_{i+1}' \subset V_i$.

Since μ_i separates $\mathscr{F}_U V_i'$ from $\mathscr{F}_U V_{i+1}'$ and $V_i \supset \mathscr{F}_U V_{i+1}'$, it follows that $V_i \cap \mathscr{F}_U V_i' = \varnothing$. Since V_i is connected, $V_i \cap \mathscr{F}_U V_i' = \varnothing$ and $V_i \cap V_i' \supset V_{i+1}' \neq \varnothing$, it follows that $V_i \subset V_i'$.

Step 5. ($\mathrm{cl}_S(\mathscr{F}_U V_i) \cap \mathrm{cl}_S(\mathscr{F}_U V_j) = \varnothing$ if $i \neq j$.) To obtain this result, we will have to modify the construction of N_i. First, however, notice that $\mathscr{F}_U V_i = \mu_i \subset \partial N_i$ does not meet $\mathscr{F}_U V_i'$, by construction of N_i. For $j < i$, we have $V_i \subset V_i' \subset V_j$, and hence $\mathscr{F}_U V_i \subset V_i' \subset V_j$, so $\mathscr{F}_U V_i$ does not meet $\mathscr{F}_U V_j$. Therefore, all we have to show is that $\mathrm{cl}_S \mu_i - \mu_i$ does not meet $\mathrm{cl}_S \mu_j - \mu_j$ for $i \neq j$.

We modify the construction of N_i in the following way. We construct N_1 as before. Assuming N_j has been constructed, we construct μ_j as before. Assuming $N_1, \ldots, N_{i-1}$ and $\mu_1, \ldots, \mu_{i-1}$ have been constructed, we construct N_i as before, but with the additional condition that ∂N_i does not meet $\bigcup_{j=1}^{i-1} (\mathrm{cl}_S \mu_j - \mu_j)$. Since $\mathrm{cl}_S \mu_j - \mu_j$ has at most two points, this is possible. Since $\mathrm{cl}_S \mu_i \subset \partial N_i$, we obtain the desired conclusion.

Step 6. From the fact that $\mathscr{F}_U V_i = \mu_i$ is connected and Step 5, we obtain that $V_1 \supset V_2 \supset \cdots$ is a chain. From Step 4, we obtain that it defines e.

This completes the proof of (a).

(b) We may assume that $\mathscr{F}_U V_i' \to x$. By taking N_i to be a sufficiently small neighborhood of $\mathrm{cl}_S(\mathscr{F}_U V_i')$, we may arrange that $\partial N_i \to x$. Then $\mathscr{F}_U V_i = \mu_i \to x$. ■

From Lemma 3.1, it follows that our notion of prime ends is the same as Carathéodory's, in the case he considers.

To show that Epstein's definition is the same as ours, we will need the following straightforward generalization of the result in [3, Section 18].

LEMMA 3.2. *Let* $\sigma = \{V_1 \supset V_2 \supset \cdots\}$ *and* $\tau = \{W_1 \supset W_2 \supset \cdots\}$ *be two chains. Suppose* $\mathscr{F}_U V_1 \neq \varnothing$ *and* $\mathscr{F}_U V_i$ *converges to a point* $x \in S$ *(i.e., for every neighborhood* N *of* x*, we have that* $\mathscr{F}_U V_i \subset N$ *for all sufficiently large* i*). Suppose that* $V_i \cap W_j \neq \varnothing$ *for all* i, j*. Then* σ *divides* τ.

Proof. Let $y \in U - V_1$. For any $n \geq 1$, choose a point $y_n \in \mathscr{F}_U W_n$ and a point $z_n \in W_n$. Let γ_n and μ_n be curves in U joining y to y_n and z_n, respectively. By the definition of chain, there can be at most one n_0 such that $x \in \mathrm{cl}_S(\mathscr{F}_U W_{n(0)})$. If $n \neq n_0$, then $x \notin \gamma_n \cup \mu_n \cup \mathrm{cl}_S(\mathscr{F}_U W_n)$, and hence for all sufficiently large m, we have that $\mathrm{cl}_S(\mathscr{F}_U V_m)$ does not intersect $\gamma_n \cup \mu_n \cup \mathrm{cl}_S(\mathscr{F}_U W_n)$.

Let $w_{nm} \in W_n \cap V_m$. Let τ_{nm} be a curve in W_n joining z_n and w_{nm}. Since $y \notin V_1$, we have $y \notin V_m$. Since μ_n joins y and z_n, and does not meet $\mathscr{F}_U V_m$, it follows that $z_n \notin V_m$. Since z_n and w_{nm} are on opposite sides of $\mathscr{F}_U V_m$, it follows that τ_{nm} crosses $\mathscr{F}_U V_m$. Since $\tau_{nm} \subset W_n$ and $\mathscr{F}_U V_m$ does not meet $\mathscr{F}_U W_n$, it follows that $\mathscr{F}_U V_m \subset W_n$.

Since γ_n joins y and y_n and does not meet $\mathscr{F}_U V_m$, it follows that $y_n \notin V_m$. Since $\mathscr{F}_U W_n$ is connected and does not meet $\mathscr{F}_U V_m$, it follows that $\mathscr{F}_U W_n$ does not meet V_m.

From the facts that $\mathscr{F}_U W_n$ does not meet V_m and $\mathscr{F}_U V_m \subset W_n$, it follows that $V_m \subset W_n$. Thus σ divides τ. ■

COROLLARY 3.3. *If* $\sigma = \{V_1 \supset V_2 \supset \cdots\}$ *is a chain such that* $\mathscr{F}_U V_1 \neq \varnothing$ *and* $\mathscr{F}_U V_i$ *converges to a point in* S*, then* σ *is prime.*

This corollary shows that a prime end in Epstein's sense is a prime end in our sense.

The next lemma generalizes another result of Carathéodory [3, Theorem VII] and is proved by Carathéodory's method.

LEMMA 3.4. *Every prime end of* U *has at least one principal point.*

Proof. Let e be a prime end of U and let $\sigma = \{V_1 \supset V_2 \supset \cdots\}$ be a chain representing it. Let $p = \alpha(e)$. If $p \in U$, the lemma is obvious. So, we will assume $p \in b_I U$. If $Z(p)$ consists of one point, the lemma is obvious, so we will assume $Z(p)$ has more than one point. We let D be a closed disk in $c_I U$ having p in its interior, such that $D \cap b_I U = p$.

For the rest of the proof, we provide S with a PL-structure and a metric d. By the *diameter* of a subset β of S, we mean

$$\sup_{x,\, y \in \beta} d(x, y).$$

We let $x_1, x_2, \ldots$ be a sequence of points in S such that every i, there exists $j(i)$ such that $x \in V_i$ for $j \geq j(i)$.

Step 1. (For any compact set K in U, there exists a closed PL-arc β in S, of arbitrarily small diameter, such that $\beta - \partial\beta \subset D - p$, $\partial\beta \subset Z(p)$, and $\beta - \partial\beta$ separates U into two components, one of which contains K, and the other of which contains an infinite number of elements of the sequence $x_1, x_2, \ldots$.) We let K' be the compact connected set in U which contains $K \cup \partial D$. We provide S with a triangulation t, compatible with PL-structure, such that any simplex of the triangulation which meets K' is contained in U, and at least one limit point of the sequence $x_1, x_2, \ldots$ is in the interior of a 2-simplex. We let L' be the union of all closed simplices of the triangulation which lie in U, and we let L be the connected component of L' which contains K'.

For any 2-simplex Δ of the triangulation t, which is not in L, but which meets $L \cap (D - p)$, we let $\varepsilon\Delta$ denote union of the connected components of $\partial\Delta \cap (\mathrm{D} - \mathrm{p})$ which meet L. Thus, $\varepsilon\Delta$ is the union of at most three PL open arcs. We let $\mathscr{C}$ denote the collection of all arcs obtained in this way.

Let U_0 be the component of $D - L - p$ which is a punctured neighborhood of p in D. The arcs of $\mathscr{C}$ which meet U_0 divide U_0 into a finite number of components $U_1, \ldots, U_k$. Each $\mathscr{F}_U U_i$ is a member of $\mathscr{C}$. It is obvious that $\mathscr{F}_U U_i$ is a disjoint union of members of $\mathscr{C}$. If $\mathscr{F}_U U_i$ contained two distinct members β and β' of $\mathscr{C}$, then we could connect two points q and q' in $\beta \cap L$ and $\beta' \cap L$ by an arc γ, which lies in U except for its endpoints, and also by an arc γ' in $L \cap (D - p)$, since the latter set is connected. The circle $\gamma \cap \gamma'$ divides $D - p$ into two components. Since β crosses $\gamma \cup \gamma'$ exactly once, the two ends of β lie on different sides of $\gamma \cup \gamma'$. But both ends approach p in D, which shows that they lie on the same side after all. This contradiction shows that $\mathscr{F}_U U_i$ is a single arc of $\mathscr{C}$.

One U_i must contain an infinite number of the x_j, since at least one limit point of the sequence $x_1, x_2, \ldots$ is in the interior of a 2-simplex. Let $\beta = \mathrm{cl}_S \mathscr{F}_U \mathrm{U}_i$. By choosing a sufficiently small triangulation t to begin with, we may arrange for β to have as small diameter as we want. Since U_i is one component of $U - \beta$, and K lies in the other component, β has the required properties.

Step 2. We take $K = \partial D$. We may construct β_1 with $\operatorname{diam} \beta_1 \leq 1$, having the properties listed in Step 1. We let $x_{11} = x_1$, and let $x_{12}, x_{13}, \ldots$ be a subsequence of $x_2, x_3, \ldots$ such that each x_{1i}, $i \geq 2$, is separated by

β_1 from K. We then may construct β_2 with diam $\beta_2 \leq \frac{1}{2}$, having the properties listed in Step 1, but in relation to the sequence $x_{11}, x_{12}, \ldots$ in place of the sequence $x_1, x_2, \ldots$. We take $x_{21} = x_{11}, x_{22} = x_{12}$, and let $x_{23}, x_{24}, \ldots$ be a subsequence of $x_{13}, x_{14}, \ldots$ such that each $x_{2i}, i \geq 3$ is separated by β_2 from K.

Assuming β_i and $x_{i1}, x_{i2}, \ldots$ have been constructed, we construct β_{i+1} with diam $\beta_{i+1} \leq 1/(i+1)$, having the properties listed in Step 1, but in relation to the sequence $x_{i1}, x_{i2}, \ldots$ in place of the sequence $x_1, x_2, \ldots$. We take $x_{i+1,1} = x_{i1}, \ldots, x_{i+1,i+1} = x_{i,i+1}$, and let $x_{i+1,i+2}, \ldots$ be a subsequence of $x_{i,i+2}, \ldots$ such that each $x_{i+1,j}, j \geq i+2$, is separated by β_{i+1} from K.

Thus, we get a sequence of closed PL-arcs $\beta_1, \beta_2, \ldots$ and a sequence $x_{11}, x_{22}, \ldots$ of points such that $\beta_i - \partial\beta_i \subset D - p$, $\partial\beta_i \subset Z(p)$, $\beta_i - \partial\beta_i$ separates U into two components, one of which contains $K = \partial D$, and the other of which contains $\{x_{i,i+1}, x_{i,i+2}, \ldots\}$. Moreover, since $x_{11}, x_{22}, \ldots$ is a subsequence of $x_1, x_2, \ldots$, there exists for each i a positive integer $j(i)$, such that $x_{jj} \in V_i$ for all $j \geq j(i)$.

Step 3. Let $\beta_1, \beta_2, \ldots$, and $x_{11}, x_{22}, \ldots$ be as in Step 2. Since diam $\beta_i \leq 1/i$, $\partial\beta_i \subset Z(p)$, and $Z(p)$ is compact, the β_i have a limit point x in $Z(p)$. Let $D_1, D_2, \ldots$ be a sequence of disks in S, such that $D_{i+1} \subset \operatorname{int}_S D_i$, $D_i \cap \partial D = \varnothing$, and $x = \bigcap_i D_i$.

For each i, choose $k = k(i)$ such that $\beta_k \subset \operatorname{int}_S D_i$. Such a k exists, since x is a limit point of the β_k.

We can connect ∂D to β_k by an arc μ_i in U which crosses ∂D_i a finite number of times. Since ∂D_i separates ∂D from β_k, the arc μ_i must cross ∂D_i an odd number of times. Thus, μ_i must cross one component of $\partial D_i \cap U$ an odd number of times. Let γ_i be the closure of this component in S. Thus, γ_i is a closed arc on ∂D_i, $\gamma_i - \partial\gamma_i \subset D - p$, and $\partial\gamma_i \subset Z(p)$. Let W_i be the component of $U - \gamma_i$ which contains β_k.

We will show that the sequence $W_1, W_2, \ldots$ has a subsequence $W_{i(1)}$, $W_{i(2)}, \ldots$, such that $W_{i(1)} \supset W_{i(2)} \supset \cdots$, and that this chain represents e.

Since μ_i crosses γ_i an odd number of times, γ_i separates ∂D from β_k in U. Thus $\partial D \cap W_i = \varnothing$.

Let B_k be the component of $U - \beta_k$ which does not meet ∂D. Since there is a subarc μ_i' of μ_i, connecting ∂D to γ_i, and not meeting β_k, we have $B_k \cap \gamma_i = \varnothing$. From $\beta_k \subset W_i$ and $\gamma_i \cap B_k = \varnothing$, we obtain $B_k \subset W_i$.

Hence $x_{ll} \in W_i$ for $l > k = k(i)$, and we obtain $W_i \cap W_{i'} \neq \varnothing$ for all i, i'.

Since μ_i' is a compact subset of U, $x \in Z(p)$, and $x = \bigcap_i D_i$, we may choose j so large that μ_i' does not meet D_j. Since $\partial D \cap W_j = \varnothing$, and μ_i' connects ∂D and γ_i, and is disjoint from γ_j, it follows that $\gamma_i \cap W_j = \varnothing$.

From $\gamma_i \cap W_j = \varnothing$ and $W_i \cap W_j \neq \varnothing$, it follows that $W_j \subset W_i$. Thus, we may choose a subsequence $W_{i(1)}, W_{i(2)}$ such that $W_{i(1)} \supset W_{i(2)} \supset \cdots$.

Clearly $\mathscr{F}_U W_i = \gamma_i \subset \partial D_i$ converges to x, as $i \to \infty$. Moreover, $W_i \cap V_j$ contains x_{ll} for all sufficiently large l, and hence is not empty. It then follows from Lemma 3.2 that $\{W_{i(1)} \supset W_{i(2)} \supset \cdots\}$ divides $\{V_1 \supset V_2 \supset \cdots\}$. Since the latter chain is prime, it follows that the two chains represent the same prime end e. Since $\mathscr{F}_U W_i \to x$, as $i \to \infty$, it follows that x is a principal point of e. ■

Lemma 3.4 shows that a prime end in our sense is a prime end in Epstein's sense.

Topology on the set of prime ends. Let V be an open set in U. A chain $\{V_1 \supset V_2 \supset \cdots\}$ will be said to *divide* V if for i sufficiently large, $V_i \subset V$. Clearly, this is independent of the chain chosen within an equivalence class, so we have a notion of a prime end *dividing* V. We let $\tilde{V}$ denote the set of all prime ends dividing V. Clearly, if V and W are open sets in U, then $\widetilde{V \cap W} = \tilde{V} \cap \tilde{W}$, so the collection of sets

$$\{\tilde{V}: V \text{ is open in } U\}$$

is the basis of a topology on the set $E(U)$ of all prime ends of U.

It is easily seen that ω (see Example in beginning of this section) is an embedding of U onto an open subset of $E(U)$.

We state a generalization of the purely topological corollary of Carathéodory main theorem [3, Theorem XIII] on prime ends.

PROPOSITION 3.5. *$E(U)$ is a compact surface. If $p \in b_I U$ and $Z(p)$ is not a single point, then $\alpha^{-1}(p)$ is a boundary component of $E(U)$.*

If $Z(p)$ is a point, then $\alpha^{-1}(p)$ is obviously a single point, in the interior of $E(U)$.

Proposition 3.5 may be proved by Carathéodory's method. In his exposition, Epstein has proved a generalization of Carathéodory's theorem [5, Theorem 6] which obviously implies Proposition 3.5.

If $p \in b_I U$ and $Z(p)$ has more than one point, we will write $C(p)$ for $\alpha^{-1}(p)$. By Proposition 3.5, $C(p)$ is homemorphic to a circle. We will call it the *Carathéodory circle* associated to p.

We will need the following result, which generalizes [3, Theorem XX] and may be proved by Carathéodory's method.

DEFINITION. A prime end e of U is said to be *accessible* if there is a curve $\gamma:[0,1] \to S$ such that $\gamma(0,1] \subset U$ and $\gamma(t) \to e$ in $E(U)$, as $t \to 0$, $t \in (0,1]$.

Note that if $V_1 \supset V_2 \supset \cdots$ is a chain representing e, then γ passes through $\mathscr{F}_U V_i$ for all sufficiently large i. Thus if x is a principal point of e, then $\gamma(0) = x$.

So, if ϵ is accessible, it has only one principal point. The result we need is:

PROPOSITION 3.6. *The principal set $X(e)$ is compact and connected. If it has only one point, then e is accessible.*

A proof in this generality is given in Epstein [5, Theorems 7.1 and 7.4].

The following easy result generalizes [3, Theorem XVII]. We follow Carathéodory's proof.

PROPOSITION 3.7. *Let γ be a closed arc in S and let x be one of its endpoints. Suppose $x \in \mathscr{F}_S U$ and $\gamma - x \subset U$. Let $\gamma^* = \mathrm{cl}_{E(U)}(\gamma - x)$. Then $\gamma^* - (\gamma - x)$ is one point, so γ^* is a closed arc.*

Proof. Since $E(U)$ is compact, $\gamma^* - (\gamma - x)$ contains at least one point e. Let $V_1 \supset V_2 \supset \cdots$ be a chain defining e. Since $\mathrm{cl}_S(\mathscr{F}_U V_i) \cap \mathrm{cl}_S(\mathscr{F}_U V_j) = \varnothing$ for $i \neq j$, there is at most one i such that $x \in \mathrm{cl}_S(\mathscr{F}_U V_i)$. If there is such an i, we may delete the corresponding V_i from the chain. So, we may assume $x \notin \mathrm{cl}_S(\mathscr{F}_U V_i)$ for all i.

There is a sequence $y_1, y_2, \ldots$ in $\gamma - x$ which converges to e. This sequence also converges to x. For any i, $y_n \in V_i$, for sufficiently large n, since $y_n \to e$. The subarc $y_n x$ of γ does not meet $\mathscr{F}_U V_i$, for sufficiently large n, since $y_n \to x$ and $x \notin \mathrm{cl}_S(\mathscr{F}_U V_i)$. If n is so large that both these conditions are satisfied then $y_n x \subset V_i \cup x$.

It follows immediately that any sequence in $\gamma - x$ which converges to x also converges to e. Hence $e = \gamma^* - (\gamma - x)$. ■

4. GENERALIZATION TO ARBITRARY OPEN SETS

Suppose U is open in S and $p \in b_I U$. We will say p is a *regular* ideal boundary point of U, if p is compact (rel. S) and $Z(p)$ contains more than one point.

Let $(\hat{U}_p, \hat{p})$ be the isolated ideal boundary point associated to p. Clearly, $\hat{p}$ is regular. Let D be a disk in $c_I \hat{U}_p$ such that D is a neighborhood of $\hat{p}$ in $c_I \hat{U}$ and $D \cap b_I \hat{U}_p = \hat{p}$. Let $V = D - \hat{p} \subset \hat{U}_p \subset S$. Clearly V is homeomorphic to $S^1 \times [0, 1)$ and the unique ideal boundary point $\hat{p}'$ of V satisfies $Z(\hat{p}') = Z(p') = Z(p)$.

We define the *Carathéodory circle $C(p)$ associated to p* to be $C(\hat{p}')$. This is independent of the choice of D.

In some cases, both definitions we have given of $C(p)$ apply. However, in these cases, it can easily be seen that there is a natural identification between the two versions of $C(p)$, so no ambiguity results.

The Carathéodory rotation number. Suppose f is a homeomorphism of S onto itself, U is an open f-invariant subset of S, and p is an f_U-fixed regular ideal boundary point of U. Obviously f induces a homeomorphism $f_*: C(p) \to C(p)$. Moreover, f_* is orientation preserving or reversing according to whether f_U is locally orientation preserving or reversing at p.

In the case f_U is locally orientation preserving at p, we define the *Carathéodory rotation number* $\rho(f, U, p)$ to be the usual rotation number of $f_*: C(p) \to C(p)$. This rotation number was first studied by Birkhoff [1].

5. The Main Results

Throughout this section we let f be an area preserving homeomorphism of S. We will state the main results under purely topological hypotheses on f. However, the main interest in our results lies in the fact that they are satisfied for a residual set in $\mathscr{A}^r$, if $r \geq 4$.

One of our hypotheses concerns periodic points.

Moser stable periodic points. Let $\mathbf{O}$ be a fixed point of f. We will say $\mathbf{O}$ is Moser stable if every neighborhood of $\mathbf{O}$ contains an f-invariant disk D, with $\mathbf{O}$ in its interior, such that $f \mid \partial D$ is transitive (i.e., has a dense orbit).

If $\mathbf{O}$ is a periodic point of period n, we say that $\mathbf{O}$ is *Moser stable* if it is a Moser stable fixed point for f^n in the sense just defined.

Sectorial periodic points. Let $\mathbb{R}_+$ denote the set of nonnegative numbers. Let c be a contracting homeomorphism of $\mathbb{R}_+$, i.e., a homeomorphism such that $c^n(t) \to 0$, for all $t \in \mathbb{R}^+$. Let $\mathbf{O}$ be a fixed point of f, and let $U \subset S$ be a closed surface such that $\mathbf{O} \in \partial U$ and germ of U at $\mathbf{O}$ is mapped onto itself, i.e., there is a neighborhood N of $\mathbf{O}$ in U such that $fN \subset U$ and $f^{-1}N \subset U$. We will say U is an *elementary sector for* (f, Q) if the germ $(f \mid U)_{\mathbf{O}}$ of $f \mid U$ at $\mathbf{O}$ is topologically conjugate to the germ $(c \times c^{-1})_0$. Here, $(c \times c^{-1})(x, y) = (c(c), c^{-1}(y))$. This definition is clearly independent of the choice of c.

We will say that a closed surface $U \subset S$ is a *sector* for (f, Q), if the germ of U at $\mathbf{O}$ is a finite union of germs at $\mathbf{O}$ of elementary sectors. This definition applies both in the case $\mathbf{O} \in \partial U$ and the case $\mathbf{O} \notin \partial U$. We will say $\mathbf{O}$ is a *sectorial fixed point* of f if S is a sector for $(f, \mathbf{O})$.

Clearly, if U is a sector for $(f, \mathbf{O})$, its germ at $\mathbf{O}$ is expressible as the union of elementary sectors in exactly one way.

If $\mathbf{O}$ is a periodic point of f, we will say that $\mathbf{O}$ is a *sectorial periodic point* of f, if it is a sectorial fixed point of f^n, for some n.

Examples. Suppose $0 < \lambda < 1 < \mu$ and let $L = L(\lambda, \mu): \mathbb{R}^2 \to \mathbb{R}^2$ be defined by $L(\lambda, \mu)(x, y) = (\lambda x, \mu y)$. Then 0 is a sectorial fixed point of $L(\lambda, \mu)$: the elementary sectors for $(L, 0)$ are the quadrants of $\mathbb{R}^2$.

In this example, the union of the upper left and lower right quadrants is not a sector, because it is not a surface.

Now suppose f is C^1, $\mathbf{O}$ is a fixed point of f, and λ and μ are the eigenvalues of $df(\mathbf{O})$ and $0 < \lambda < 1 < \mu$. A theorem of Hartman–Grobman [6–9, 12, 13] implies that the germ of f at $\mathbf{O}$ is topologically conjugate to the germ of $L(\lambda, \mu)$ at 0. It follows immediately that $\mathbf{O}$ is a sectorial fixed point of f.

A *fixed* point may be a sectorial *periodic* point, without being a sectorial *fixed* point. For example, suppose $\mathbf{O}$ is a fixed point of f, which is C^1, and the eigenvalues of $df(\mathbf{O})$ satisfy $0 < |\lambda| < 1 < |\mu|$, but not both λ and μ are positive. In this case $\mathbf{O}$ is a sectorial fixed point of f^2, but not of f.

Our other hypothesis depends on the following notion.

Connections. By a *fixed connection* for f, we mean an f-invariant arc α in S, both of whose endpoints are f-fixed, or an f-invariant circle α, such that $f|\alpha$ is orientation preserving and α contains an f-fixed point. By a *periodic connection*, we mean a fixed connection for f^n for some n.

Now we state our first main theorem:

THEOREM 5.1. *Suppose every fixed point of f is sectorial periodic or Moser stable, and f has no fixed connections. Then $\rho(f, U, p) \neq 0$, whenever it is defined, i.e., whenever U is an open invariant set in S and p is a regular f_U-fixed ideal boundary point of f, such that f_U is locally orientation preserving in a neighborhood of p.*

Let $\mathscr{X}^r$ be the set of all $f \in \mathscr{A}^r$ such that all periodic points of f are sectorial or Moser stable and there are no periodic connections for f. Robinson's results [15] imply that $\mathscr{X}^r$ is residual (in the sense of Baire category) in $\mathscr{A}^r$, for $r \geq 4$. See Section 6. We let $\mathscr{X} = \mathscr{X}^0$.

COROLLARY. *Let $f \in \mathscr{X}$. Let U be an open subset of S such that $f^n U = U$ for some positive number n. Let p be a regular ideal boundary point of U, and suppose $f_U^{nm}(p) = p$ for some positive number m. Then f_U^{nm} is locally orientation preserving at p, and $\rho(f^{nm}, U, p)$ is irrational.*

Proof. Obviously if $f \in \mathscr{X}$, then $f^l \in \mathscr{X}$ for all l. Assuming that f_U^{mn} is locally orientation preserving at p, it follows immediately from Theorem 5.1 that $\rho(f^{kmn}, U, p) = k\rho(f^{mn}, U, p) \not\equiv 0 \pmod 1$, for all $k \in \mathbb{Z}$. Hence $\rho(f^{mn}, U, p)$ is irrational.

If f^{mn} is locally orientation reversing at p, then f^{2mn} is locally orientation preserving at p, and $\rho(f^{2mn}, U, p) = 0$, a contradiction. ∎

Let $\mathbf{O}$ be a periodic point of f of period n. By the *stable* (resp., *unstable*) set of $\mathbf{O}$, we mean $\{x \in S: f^{nk}(x) \to \mathbf{O} \text{ as } k \to +\infty\}$ (resp., $\{x \in S: f^{nk}(x) \to \mathbf{O}$ as $k \to -\infty\}$).

If $\mathbf{O}$ is sectorial periodic, and U is an elementary sector for $(f^n, \mathbf{O})$, then there is a neighborhood N of $\mathbf{O}$ in ∂U, such that N is an arc, $N = N_+ \cup N_-$, where N_+ and N_- are arcs, $N_+ \cap N_- = \mathbf{O}$, $f^n(N_+) \subset N_+$, $f^{-n}(N_-) \subset N_-$, $f^{nk}(x) \to \mathbf{O}$ as $k \to +\infty$ if $x \in N_+$, and $f^{nk}(x) \to \mathbf{O}$ as $k \to -\infty$ if $x \in N_-$. Let $b_+ = \{x \in S : x \neq \mathbf{O}$ and $\exists k > 0$ such that $f^{nk}(x) \in N_+\}$ and $b_- = \{x \in S : x \neq \mathbf{O}$ and $\exists k < 0$ such that $f^{nk}(x) \in N_-\}$.

We shall call b_+ and b_- stable and unstable *branches* of $\mathbf{O}$. Each branch of $\mathbf{O}$ is an injectively immersed, connected, noncompact 1-manifold. The stable (resp., unstable) set of $\mathbf{O}$ is the union of $\mathbf{O}$ and a finite collection of mutually disjoint stable (resp., unstable) branches of $\mathbf{O}$.

Our second main theorem states:

THEOREM 5.2. *Suppose S is compact, every fixed point of f is sectorial periodic or Moser stable, and f has no fixed connections. Let $\mathbf{O}$ be any sectorial fixed point of f and let b_1 and b_2 be two branches (either stable or unstable) of $\mathbf{O}$. Then $\mathrm{cl}_S\, b_1 = \mathrm{cl}_S\, b_2$.*

EXAMPLE. One can construct an area preserving homeomorphism of the 2-sphere having two hyperbolic fixed points, such that the stable set of either is the unstable set of the other, and such that the branches of both the stable or unstable set of either point are interchanged by the homeomorphism. One can construct such an example satisfying all the hypotheses of the theorem. But the two hyperbolic fixed points do not satisfy the conclusion. This does not contradict the theorem: the fixed points are sectorial periodic, not sectorial fixed.

COROLLARY. *Let $f \in \mathscr{X}$. Let $\mathbf{O}$ be any sectorial periodic point of f. If b_1 and b_2 are any two branches of $\mathbf{O}$, then $\mathrm{cl}_S\, b_1 = \mathrm{cl}_S\, b_2$.*

Proof. $\mathbf{O}$ is sectorial fixed for some f^n. Thus, Theorem 5.2 applies. ∎

6. GENERICITY CONDITIONS

This section contains a brief exposition of the known facts which imply that $\mathscr{X}^r$ is residual in $\mathscr{A}^r$ for $r \geq 4$.

Throughout this section, we suppose S and μ are smooth. We let f be a C^1 area preserving diffeomorphism of S.

Let $\mathbf{O}$ be a fixed point of f. Let λ and μ be the characteristic values of $df(\mathbf{O})$. Since f is area preserving, $|\lambda\mu| = 1$. The fixed point $\mathbf{O}$ is said to be *elliptic* if λ is imaginary, *parabolic* if $\lambda = \pm 1$, and *hyperbolic* if λ is real and not equal to ± 1.

In the hyperbolic case, the Hartman–Grobman theorem [6–9, 12, 13] implies that the germ of f at $\mathbf{O}$ is topologically conjugate to the germ of $df(\mathbf{O})$ at 0. Since $|\mu| = |\lambda|^{-1}$, one of the eigenvalues must have absolute value >1, and the other must have absolute value <1. Thus, $\mathbf{O}$ is sectorial periodic in this case.

In the elliptic case, $\mu = \bar{\lambda}$ and $\lambda\mu = |\lambda|^2 = 1$, so f is locally orientation preserving at f. Suppose f is C^4. The methods of Birkhoff in [2] show that if $\lambda^3 \neq 1$, $\lambda^4 \neq 1$, then there exist C^∞ local coordinates x, y, centered at $\mathbf{O}$, such that the measure on S equals $dx\,dy$ in a neighborhood of $\mathbf{O}$, and f is given, locally near $\mathbf{O}$, by

$$f(z) = \lambda z e^{ib|z|^2} + \mathbf{o}(|z|^4),$$

in terms of the complex coordinate $z = x + iy$. In other words, a suitable representative of the 4-jet of f at $\mathbf{O}$ preserves the circles $x^2 + y^2 = r^2$ and rotates each such circle with the rotation number $(1/2\pi)(br^2 - i\log\lambda)$. The number b is called the (first) Birkhoff invariant of f at $\mathbf{O}$. We will not be concerned with the higher Birkhoff invariants.

Moser and Rüssman proved that if f is C^4, $\mathbf{O}$ is an elliptic fixed point of f, an eigenvalue λ of $df(\mathbf{O})$ is not a third or fourth root of unity, and the first Birkhoff invariant does not vanish, then $\mathbf{O}$ is Moser stable in the sense we have defined (see [11]). This is one of the many important results in the Kolmogorov–Arnol'd–Moser theory.

Obviously, these comments extend to periodic points. Let $\mathbf{O}$ be a periodic point of f of period n. We say $\mathbf{O}$ is elliptic, hyperbolic, or parabolic if it satisfies the corresponding condition when thought of as a fixed point of f^n. A hyperbolic periodic point is sectorial. If $\mathbf{O}$ is elliptic periodic, and the Moser conditions discussed above are satisfied for f^n, then $\mathbf{O}$ is Moser stable.

Robinson [15] proved that there is a residual set $\mathscr{R}^r$ in $\mathscr{A}^r$ $(r \geq 4)$ such that if $f \in \mathscr{R}^r$, then every periodic point $\mathbf{O}$ is either elliptic or hyperbolic, and in the case it is elliptic it satisfies the Moser conditions we discussed above. Thus every periodic point of such an f is sectorial or Moser stable.

Robinson showed, moreover, that for $f \in \mathscr{R}^r$ all intersections of stable and unstable manifolds of hyperbolic points are transversal. Thus, $f \in \mathscr{R}^r$ has no periodic connection. Hence $\mathscr{R}^r \subset \mathscr{X}^r$, and we obtain that $\mathscr{X}^r$ is residual in $\mathscr{A}^r$ for $r \geq 4$.

7. Beginning of the Proof of Theorem 5.1

We will prove several lemmas. Throughout this section, f will be an area preserving homeomorphism of S, U an f-invariant open subset of S, and p an f_U-fixed regular ideal boundary point of U. We let $(\hat{U}_p, \hat{p})$ be the isolated ideal boundary point associated to p. Since $C(p) = C(\hat{p})$, by definition, it makes sense to speak of a principal point of an element e of $C(p)$.

Our first lemma and its proof are due to Cartwright and Littlewood [4].

LEMMA 7.1. *Let $e \in C(p)$ be such that $f_*(e) = e$. Let $\mathbf{O}$ be a principal point of e. Then $f(\mathbf{O}) = \mathbf{O}$.*

Proof. By Lemma 3.1b, we may choose a chain $V_1 \supset V_2 \supset \cdots$ in $\hat{U}_p$ defining e, such that $\mathrm{cl}_S(\mathscr{F}_{U(p)} V_i)$ converges to $\mathbf{O}$.

Let D be a small disk about $\hat{p}$ in $c_I \hat{U}_p$ such that $D \cap b_I \hat{U}_p = p$. For sufficiently large i, we have that V_i and fV_i are in $D - \hat{p}$ and do not meet ∂D. Since the sequences $V_1 \supset V_2 \supset \cdots$ and $fV_1 \supset fV_2 \supset \cdots$ define the same prime end, we have $V_i \cap fV_i \neq \varnothing$. Since f is area preserving, neither V_i nor fV_i can be properly contained in the other. It follows that for sufficiently large i the frontiers in $D - \hat{p}$ of V_i and fV_i must have at least one common point.

Thus, there exists $x_i \in \mathscr{F}_{\hat{U}(p)} V_i$ such that $f(x_i) \in \mathscr{F}_{\hat{U}(p)} V_i$. Since $\mathscr{F}_{\hat{U}(p)} V_i \to \mathbf{O}$ it follows that $x_i \to \mathbf{O}$ and $f(x_i) \to \mathbf{O}$. Hence $f(\mathbf{O}) = \mathbf{O}$. ∎

LEMMA 7.2. *$Z(p)$ contains no Moser stable fixed point.*

Proof. Suppose $\mathbf{O} \in Z(p)$ is a Moser stable fixed point. Let D be a small invariant disk about $\mathbf{O}$ such that $f \mid \partial D$ is transitive. It is a well-known property of homeomorphisms of the circle that this implies that $f \mid \partial D$ is topologically conjugate to an irrational rotation, and hence every orbit is dense.

We can take D to be an arbitrarily small neighborhood of $\mathbf{O}$, so we can suppose that $Z(p) \not\subset D$ and $\hat{U}_p \not\subset D$. Since $\mathbf{O} \in Z(p)$, $Z(p) \not\subset D$, and $Z(p)$ is connected, it follows that $Z(p)$ meets ∂D. Since $Z(p)$ is f-invariant and closed, and every orbit of $f \mid \partial D$ is dense, it follows that $\partial D \subset Z(p)$. Hence $\partial D \cap \hat{U}_p = \varnothing$. Since $\hat{U}_p$ is connected, and $\hat{U}_p \not\subset D$, it follows that $\hat{U}_p \cap D = \varnothing$. Thus, $\mathbf{O} \notin \mathrm{cl}_S \hat{U}_p$, contrary to the hypothesis that $\mathbf{O} \in Z(p)$. ∎

To prove Theorem 5.1, we begin by supposing that it is false, i.e., $\rho(f, U, p) = 0$. Then $f_*: C(p) \to C(p)$ has fixed point e by a theorem of Poincaré. By Lemma 7.1, every point of the principal set $X(e)$ is fixed. However, by Lemma 7.2, no such point can be Moser stable, since $X(e) \subset Z(p)$. Moreover, since $X(e)$ is connected (by Proposition 3.6), it must be a single point under the hypotheses of Theorem 5.1, and this point must be sectorial periodic.

Thus, the proof of Theorem 5.1 is reduced to the study of sectorial periodic points in $Z(p)$.

8. Invariant Continua

A *continuum* is a compact, connected space with more than one point. Throughout this section, we let S_0 be an open subset of S, and $f: S_0 \to S$ an injective area preserving homeomorphism of S_0 onto an open subset fS_0 of S. We let A be a continuum in S_0 such that $fA = A$.

There is a canonical inclusion of $b_I S$ in $b_I(S - A)$. For, let $p \in b_I S$ and let $V_1 \supset V_2 \supset \cdots$ be a fundamental system of p. Then there exists i_0 such that $\mathscr{F}_S V_i \cap A = \varnothing$ for $i \geq i_0$. Then $V_{i(0)} \supset V_{i(0)+1} \supset \cdots$ is a fundamental system of an element q of $b_I(S - A)$ and q is independent of the original fundamental system chosen. Moreover, it is easily checked that the mapping $b_I S \to b_I(S - A)$ is a homeomorphism onto a subset of $b_I(S - A)$ which is both open and closed. Let $p \in b_I(S - A)$. If $p \in b_I S$ then $Z(p) = p$; otherwise $Z(p) \subset S$. From Lemma 2.3, it follows that $b_I(S - A) - b_I S$ is a discrete topological space. Since $S - A$ may have infinitely many components, $b_I(S - A) - b_I S$ may be infinite, however.

The mapping f induces a mapping $f_*: b_I(S - A) - b_I S \to b_I(S - A) - b_I S$ defined in the following obvious way. If $p \in b_I(S - A) - b_I S$, and $V_1 \supset V_2 \supset \cdots$ is a fundamental system for p, then there exists i_0 such that $V_i \subset S_0$ for $i \geq i_0$, and $fV_{i(0)} \supset fV_{i(0)+1} \supset \cdots$ is a fundamental system for some element $f_* p$ of $b_I(S - A) - b_I S$. Clearly, $f_* p$ is independent of the choice of fundamental system for p.

Lemma 8.1. *For any $p \in b_I(S - A) - b_I S$, there exists n such that $f_*^n(p) = p$.*

Proof. Let N be a compact neighborhood of A in S_0 which is a surface. Any residual domain of A which is not contained in N contains a boundary component of N. Since N has only finitely many boundary components, all but finitely many residual domains of A lie in N.

Let ε be the smallest area of a residual domain of A which does not lie in N. (We may have $\varepsilon = \infty$, but in any case, $\varepsilon > 0$.) The sum of the areas of the residual domains of A which lie in N is finite; hence there are only finitely many residual domains of A of area $\geq \varepsilon$.

Let U be a residual domain of A of area $< \varepsilon$. Then $U \subset N \subset S_0$ and U is a connected component of $S_0 - A$, which is relatively compact in S_0. Therefore, f is defined on U, and fU is a connected component of $fS_0 - A$, which is relatively compact in fS_0, so fU is a residual domain of A of area $< \varepsilon$.

Thus, f is defined on every residual domain U of A of area $< \varepsilon$, and fU is a residual domain of A of area $< \varepsilon$. However, for any $a > 0$, there are only finitely many residual domains of A whose area equals a, since all but finitely many residual domains of A lie in N, and the sum of the areas of those which lie in N is finite. Thus f permutes the set of residual domains of fixed area

$a < \varepsilon$ among themselves, so we must have $f^n U = U$ for some n, if U is such a domain.

By Lemma 2.3, such a U has only finitely many ideal boundary points, so any ideal boundary point of U is periodic for f_*.

All but finitely many members of $b_I(S - A) - b_I S$ are in $b_I U$ for some residual domain U of A of area $<\varepsilon$, by Lemma 2.3 and the fact that all but finitely many residual domains of A have area $<\varepsilon$.

Thus, we have shown that all but at most finitely many members of $b_I(S - A) - b_I S$ are periodic points of f_*. Since f_* is a bijection, it follows that all members of $b_I(S - A) - b_I S$ are periodic points of f_*. ■

For the next lemma, we provide S with a metric d which is compatible with its topology. For any subset $\alpha \subset S$, we define its diameter to be $\sup\{d(x, y): x, y \in \alpha\}$. If $x \in S_0$, and $n > 0$, we say $f^n x$ is defined if $f^{n-1}x$ is defined and $f^{n-1}x \in S_0$. In this case, we set $f^n x = f(f^{n-1}x)$.

LEMMA 8.2. *Suppose α is an arc in S_0 such that $f^n x$ is defined for all $x \in \alpha$ and all $n > 0$. Suppose diam $f^n\alpha \to 0$ as $n \to +\infty$. Suppose the endpoints of α are in A. Then $\alpha \subset A$.*

Proof. If $\alpha \not\subset A$, there is a subarc β of α whose endpoints are in A, but none of whose other points are in A. Let β^0 denote the interior of β, and let

$$\beta^* = \mathrm{cl}(\beta^0; c_I(S - A)).$$

Then β^* is either an arc, whose endpoints q and q' lie in $b_I(S - A) - b_I S$, or a circle such that $\beta^* - \beta^0$ is one point $q \in b_I(S - A) - b_I S$.

Let D be a closed disk in $c_I(S - A)$ such that q is in the interior of D and $D \cap b_I(S - A) = q$. Since ∂D is a compact subset of $S - A$, diam $f^n\beta \to 0$ as $n \to \infty$, and the endpoints of β lie in A, it follows that $f^n\beta \cap \partial D = \varnothing$ for n sufficiently large.

Let k be the period of q for f_*. Take n so large that $f^{kn}\beta \cap \partial D = \varnothing$. Then $f^{kn}\beta^* \subset D$, since one endpoint of β^* is q, and $f^{kn}\beta^* \cap \partial D = \varnothing$.

Then $f^{kn}\beta^*$ is a circle, and it separates D into two regions, one of which, W_n, does not meet ∂D. Since we could have taken D to be an arbitrarily small neighborhood of q in $c_I(S - A)$, we have that $f^{kn}\beta^* \to q$ as $n \to +\infty$, and hence $W_n \to q$ as $n \to +\infty$. Hence area $W_n \to 0$ as $n \to +\infty$.

On the other hand, it is clear that $f^k W_n = W_{n+1}$ for k sufficiently large. This contradicts the area preserving property of f^k. ■

COROLLARY 8.3. *Suppose $S = S_0 = fS_0$. If $\mathbf{O}$ is a sectorial periodic point of f and b is a branch of $\mathbf{O}$, then $b \cap A \neq \varnothing$ implies $b \subset A$.*

Remark. By definition $\mathbf{O} \notin b$, so this corollary does not imply that $b \subset A$ if $\mathbf{O} \in A$.

Proof. Let $x \in b \cap A$. Since A is closed, and $\mathbf{O}$ is a limit point of $\{f^n x\}$, we have $\mathbf{O} \in A$. Hence the arc $x\mathbf{O}$ on $b \cup \mathbf{O}$ satisfies the hypothesis of Lemma 8.2 (possibly with f^{-1} in place of f), so $x\mathbf{O} \subset A$. But, clearly $b \cup \mathbf{O} \subset \bigcup_{n=-\infty}^{\infty} f^n(x\mathbf{O})$, so $b \cup \mathbf{O} \subset A$. ■

9. Sectorial Fixed Points

Throughout this section, we let f be an area preserving homeomorphism of an open subset S_0 of S onto an open subset fS_0 of S. We let A be a continuum in S_0 such that $fA = A$. We let $\mathbf{O}$ be a sectorial fixed point in A.

By a *stable* (resp., *unstable*) *semibranch* β of $\mathbf{O}$, we will mean $\gamma - \mathbf{O}$, where γ is a closed arc in S_0 (resp., fS_0), such that $\mathbf{O}$ is one endpoint of γ, and $f^n\gamma \to \mathbf{O}$ as $n \to +\infty$ (resp., as $n \to -\infty$).

Let W be an (elementary) sector for $(f, \mathbf{O})$, which is not a neighborhood of $\mathbf{O}$. We will say that $\operatorname{int}_S W$ is an (*elementary*) *open sector* for $(f, \mathbf{O})$. We will also say that a subset V of S is an *open sector* for $(f, \mathbf{O})$, if its germ at $\mathbf{O}$ is the complement of a semibranch of $\mathbf{O}$.

Example. Let $S = \mathbb{R}^2$ and $f(x, y) = (2x, 2^{-1}y)$. The sets $\{x > 0, y > 0\}$, $\{x > 0\}$, $\{x > 0\} \cup \{y > 0\}$, $\{x > 0\} \cup \{y \neq 0\}$ are open sectors for $(f, \mathbf{O})$. But $\{x > 0, y > 0\} \cup \{x < 0, y < 0\}$ is not an open sector, because it is not the interior of a closed surface in $\mathbb{R}^2$.

Our definition of $E(U)$ in Section 3 does not apply to $U = S - A$, since $S - A$ need not be either connected, or relatively compact. We will define $E(S - A)$ in the following way.

Let N be a compact neighborhood of A in S which is a surface. We let $E(N - A) = \coprod E(U)$ as a topological space, where the disjoint union is taken over all connected components U of $N - A$. Obviously, ∂U is compact, and by Lemma 2.3, U has only finitely many ideal boundary points, so the conditions we imposed on U in the beginning of Section 3 are fulfilled. We will identify U with its image in $E(U)$ under ω; recall that ω is a homeomorphism onto an open subset of $E(U)$. This gives an identification of $N - A$ as an open subset of $E(N - A)$.

We let $E(S - A) = (S - A)\coprod_{N-A} E(N - A)$. [That is, we identify a point in $N - A \subset S - A$ with its image in $E(N - A)$ under ω.]

This is independent of N: If N' is a second compact neighborhood of A which is a surface, the inclusion $N' - A \subset N - A$ extends to an inclusion $E(N' - A) \subset E(N - A)$, and the resulting mapping

$$(S - A) \coprod_{N'-A} E(N' - A) \to (S - A) \coprod_{N-A} E(N - A)$$

is a homeomorphism, which permits us to identify the two spaces.

We will call an element of $E(S - A)$ a *prime end* of $S - A$. Let $e \in \partial E(S - A)$. A sequence $V_1 \supset V_2 \supset \cdots$ will be said to be a *chain representing* e if it is such in one component U of $N - A$ (for any admissible choice of N).

DEFINITION. Let $e \in \partial E(S - A)$ and let V be an open sector for $(f, \mathbf{O})$. Let (x, y) be some continuous local coordinate system for S centered at $\mathbf{O}$, such that the germ of V at $\mathbf{O}$ is the set germ defined by $y > 0$. Let $V_i = \{v \in V : x(v)^2 + y(v)^2 \leq i^{-1}\}$. We will say e is a *sector end associated* to V if $V_{i(0)} \supset V_{i(0)+1} \supset \cdots$ is a chain defining e, for $i(0)$ sufficiently large.

Obviously, if there is a sector end $e \in E(S - A)$ associated to V, it is unique, the germ of V at $\mathbf{O}$ is contained in $S - A$, and the two (or one) semibranches of $\mathbf{O}$ which bound V are in A. Furthermore $X(e) = Y(e) = \mathbf{O}$. Thus, sector ends are very special.

PROPOSITION 9.1. *Let β be a semibranch of $\mathbf{O}$, such that $\beta \not\subset A$. Then $\beta \subset S - A$ and there is a sector end $e \in E(S - A)$ such that the germ of β at $\mathbf{O}$ is in the sector to which e is associated.*

Proof. Suppose $\beta \cap A \neq \varnothing$, and let $x \in \beta \cap A$. By Lemma 8.2 (applied possibly to f^{-1} in place of f), the arc $x\mathbf{O}$ on $\beta \cup \mathbf{O}$ is in A. Since $fA = A$ and $f^n\beta \subset x\mathbf{O}$ for some integer n, it follows that $\beta \subset A$. This contradiction shows that $\beta \subset S - A$.

Let $\sum_1, \sum_2$ be the two elementary open sectors which β borders. We will next show that the germs of $\sum_1$ and $\sum_2$ at $\mathbf{O}$ lie in $S - A$. Consider first the case when β is a stable semi-branch of $\mathbf{O}$. Choose any $x \in \beta$ and let $xf(x)$ be the arc on β with endpoints x and $f(x)$. Let N be a neighborhood of $xf(x)$ in S_0. Clearly, the union of the forward iterates of N under f contains the germs of $\sum_1$ and $\sum_2$ at $\mathbf{O}$. On the other hand, by taking N to be a sufficiently small neighborhood of $xf(x)$ in S_0, we may guarantee that $N \cap A = \varnothing$. It follows that the germs of $\sum_1$ and $\sum_2$ at $\mathbf{O}$ lie in $S - A$.

We may obtain this result when β is an unstable semibranch of $\mathbf{O}$, by using f^{-1} in place of f.

Let β_i $(i = 1, 2)$ be a semibranch of $\mathbf{O}$ which borders $\sum_i$ and does not meet β. If $\beta_1, \beta_2 \subset A$, then $V = \sum_1 \cup \beta \cup \sum_2$ is the required sector, to which e is associated, and the proof is finished. Otherwise, one $\beta_i \not\subset A$, say β_1. Then $\beta_1 \cap A = \varnothing$, and the argument which we have just given shows that the germ at $\mathbf{O}$ of the remaining elementary sector which borders it is in $S - A$. Adjoining this elementary sector and β_1 to V, we obtain a larger open sector whose germ at $\mathbf{O}$ is in $S - A$. If its borders are in A, the same argument as above shows that it has the properties required for the conclusion of the proposition. If not, we can construct a still larger open sector in $S - A$.

Continuing until we can go no further, we obtain an open sector having the required properties, and the proposition is proved. ■

From Proposition 3.5, it follows that $E(S - A)$ is a surface, and its interior is precisely $S - A$. We extend the definition of $\alpha_U: E(U) \to c_I U$ (Section 3) to obtain a mapping $\alpha: E(S - A) \to c_I(S - A)$, in the following way. Let N be a compact neighborhood of A which is a surface. We define

$$\alpha_N: E(N - A) = \coprod E(U) \to c_I(N - A) = \coprod c_I U$$

to be α_U on $E(U)$. We let $\alpha | E(N - A) = \alpha_N$ and $\alpha | S - A =$ inclusion. Then

$$\partial E(S - A) = \coprod \{\alpha^{-1}(p): p \in b_I(S - A) - b_I S\},$$

by Proposition 3.5, since $Z(p)$ is never reduced to a single point, for $p \in b_I(S - A) - b_I S$.

There are obvious inclusions:

$$E(S_0 - A) \subset E(S - A) \qquad \text{and} \qquad E(fS_0 - A) \subset E(S - A).$$

The mapping $f: S_0 - A \to fS_0 - A$ extends in an obvious way to a mapping $E(S_0 - A) \to E(fS_0 - A)$. We let f_* be the restriction of this mapping to $\partial E(S - A)$, so f_* is a mapping of the latter set into itself.

PROPOSITION 9.2. *Suppose $e \in \partial E(S - A)$ is a fixed point of f_* and $\mathbf{O}$ is a principal point of e. Then e is a sector end.*

Proof. Let $V_1 \supset V_2 \supset \cdots$ be a chain defining e, with $V_1 \subset S_0 \cap fS_0$. We consider two cases:

Case 1. For some semibranch β of $\mathbf{O}$, we have that $\mathbf{O}$ is adherent to $\beta \cap V_i$ for every i.

By Proposition 9.1, $\beta \subset S - A$, and there is a sector end e', such that the germ of β at $\mathbf{O}$ is in the sector to which e' is associated. There is a sequence $x_1, x_2, \ldots$ of points in β which converge to $\mathbf{O}$ (in S) and to e [in $E(S - A)$], by our hypothesis on β. But also $x_i \to e'$ in $E(S - A)$, so $e' = e$.

Case 2. There is no semibranch as in Case 1.

By Lemma 7.1, every point of $X(e)$ is fixed. By Proposition 3.6, $X(e)$ is connected. However, $\mathbf{O}$ is isolated in the fixed point set of f, so $X(e) = \mathbf{O}$. By Proposition 3.6, e is accessible. This means that there is a closed arc γ in S, such that one endpoint of γ is $\mathbf{O}$, $\gamma - \mathbf{O} \subset S - A$, and $\gamma^* = (\gamma - \mathbf{O}) \cup e$, where $\gamma^* = \mathrm{cl}_{E(S-A)}(\gamma - \mathbf{O})$, so γ^* is a closed arc in $E(S - A)$. We will call such an arc γ an *endcut* of e.

Since $\mathbf{O}$ is sectorial, we may choose a finite set $b_1, \ldots, b_k$ of semibranches of $\mathbf{O}$ such that every semibranch of $\mathbf{O}$ intersects one of the semibranches $b_1, \ldots, b_k$. From our assumptions for Case 2, it follows that by replacing γ

with a possibly smaller endcut, if necessary, we may assume that $\gamma \cap b_i = \varnothing$, $i = 1, \ldots, k$. Furthermore, we may assume that $\gamma - \mathbf{O}$ is contained in some elementary open sector W. By replacing γ with a smaller endcut, if necessary, we may assume that $f\gamma - \mathbf{O} \subset W$, as well.

Let II be the mapping of $\mathbb{R}^2_+$ into itself defined by $\mathrm{II}(x, y) = (2x, \frac{1}{2}y)$. (Recall that $\mathbb{R}_+$ is the set of nonnegative real numbers.)

We may assume that $W \subset S_0 \cap fS_0$ and that there is a homeomorphism h of $\mathrm{cl}_S(W)$ onto $\{(x, y) \in \mathbb{R}^2_+ : x^2 + y^2 \leq 1\}$ such that $\mathrm{II}h(w) = hf(w)$, whenever w and $f(w)$ are in W. For, if this is not possible for the original W, we may arrange for it to hold by replacing W with $W \cap N$, where N is a suitable neighborhood of $\mathbf{O}$ in S. By shrinking γ, we may continue to suppose $\gamma - \mathbf{O} \subset W$ and $f\gamma - \mathbf{O} \subset W$.

Obviously, we may choose γ so that $h(\gamma - \mathbf{O})$ is PL, although in general we must permit this half-open arc to have infinitely many vertices, converging to 0.

Let $\beta_1 = h^{-1}\{x = 0, 0 < y \leq 1\}$, $\beta_2 = h^{-1}\{0 < x \leq 1, y = 0\}$. Then β_1 and β_2 are stable and unstable semibranches of $\mathbf{O}$ which border W.

Let U be the connected component of $S - A$ which contains $\gamma - \mathbf{O}$. We will suppose that the chain $V_1 \supset V_2 \supset \cdots$ is chosen so that $\mathscr{F}_U V_i \to \mathbf{O}$, $V_1 \cap \beta_j = \varnothing$ for $j = 1, 2$, and $\gamma - \mathbf{O} \subset V_i$. It is possible to choose $V_1 \supset V_2 \supset \cdots$ to have the first property, since $\mathbf{O}$ is a principal point of e. Then, since $\mathscr{F}_U V_i \to \mathbf{O}$ and since there exists a small neighborhood N of $\mathbf{O}$ such that $\beta_j \cap N \cap V_i = \varnothing$ for $j = 1, 2$ and i sufficiently large, it follows that $\mathscr{F}_U V_i \cap \beta_j = \varnothing$ for $j = 1, 2$ and i sufficiently large. But $\beta_j \cap N \cap V_i = \varnothing$ and $\beta_j \cap \mathscr{F}_U V_i = \varnothing$ imply $\beta_j \cap V_i = \varnothing$, since $\beta_j \subset U$ or $\beta_j \cap U = \varnothing$. Therefore, we may arrange for the second and third conditions to hold, by replacing $V_1, V_2, \ldots$ with a subsequence.

Then $\gamma - \mathbf{O}$ meets $\mathscr{F}_U V_i$, for every i, since $\gamma - \mathbf{O} \not\subset V_i$ and $V_i \cap (\gamma - \mathbf{O}) \neq \varnothing$. From the facts that $\gamma - \mathbf{O} \subset W$, $\mathscr{F}_U V_i$ is connected, $\mathscr{F}_U \mathrm{V}_i \cap \beta_j = \varnothing$ for $j = 1, 2$, $\mathscr{F}_U V_i \to \mathbf{O}$ as $i \to \infty$, and $\gamma - \mathbf{O}$ meets $\mathscr{F}_U V_i$, it follows that $\mathscr{F}_U V_i \subset W$, for i sufficiently large.

Therefore, we may assume that $\mathscr{F}_U V_i \subset W$, for all i. By the proof of Lemma 3.1, we may assume in addition that $h(\mathrm{cl}_S \mathscr{F}_U V_i)$ is a PL-circle or PL-arc for each i.

Since $f_* e = e$, $f^{-1}V_1 \supset f^{-1}V_2 \supset \cdots$ is a chain defining e, and $(f^{-1}V_i \cap \gamma) \cup \mathbf{O}$ is a neighborhood of $\mathbf{O}$ in γ, for all i. For sufficiently large i, $\gamma - \mathbf{O} \not\subset f^{-1}V_i$ [since $(\gamma - \mathbf{O}) \cap Y(e) = \varnothing$]; hence γ meets $\mathscr{F}_U(f^{-1}V_i)$. Hence, for sufficiently large i, $\mathscr{F}_U V_i$ meets $f\gamma$, as well as γ. We will suppose this holds for all i.

Let y_i (resp., z_i) denote the first point (from $\mathbf{O}$) on γ (resp., $f\gamma$), where γ (resp., $f\gamma$) meets $\mathscr{F}_U V_i$. Let γ_i (resp., γ_i') denote the subarc $\mathbf{O}y_i$ (resp., $\mathbf{O}z_i$) of γ (resp., $f\gamma$). Let μ_i denote the subarc $y_i z_i$ of $\mathscr{F}_U V_i$.

Let $\eta_i = \gamma_i \cup \mu_i \cup \gamma'_i$. Then η_i is a closed curve in S, $\eta_i - \mathbf{O}$ is PL, and $\eta_i - \mathbf{O} \subset W$. Therefore, there is exactly one residual domain R_i of η_i in S which is not a subset of W. Let $R'_i = S - R_i$. Then R'_i is a compact subset of $W \cup \mathbf{O}$.

Let $W'_a = \{x^2 + y^2 < 1, xy < a, x > 0, y > 0\} \subset \mathbb{R}^2_+$, where $0 < a < 1$. Let $W_a = h^{-1}(W'_a) \subset W$.

LEMMA 9.3. *If a is sufficiently small (depending on i), then*

$$W_a \subset \bigcup_{n=-\infty}^{\infty} f^n(R'_i).$$

Here, we define $f^n(R'_i)$ to be $f^n(\operatorname{dom} f^n \cap R'_i)$, where $\operatorname{dom} f^n$ (the domain of f^n) is the set of x for which $f^n(x)$ is defined. Recall, that $f(x)$ is defined if $x \in S_0$. For $n > 1$, $f^n(x)$ is defined if $f^{n-1}(x)$ is defined and $f^{n-1}(x) \in S_0$. In this case $f^n(x) = f(f^{n-1}(x))$. If $x \in fS_0$, then $f^{-1}x_0$ is defined. For $n < -1$, $f^n(x)$ is defined if $f^{n+1}(x)$ is defined and $f^{n+1}(x) \in fS_0$. In this case $f^n(x) = f^{-1}(f^{n+1}(x))$.

Proof. (See Fig. 1) Let

$$X_a = W_a \cap h^{-1}\{(x/2)^2 + (2y)^2 \geq 1\},$$
$$Y_a = W_a \cap h^{-1}\{(2x)^2 + (y/2)^2 \geq 1\}.$$

For each $x \in W_a$, there is a largest integer $n(x) \leq 0$ such that $f^{n(x)}(x) \in X_a$ and a smallest integer $m(x) \geq 0$ such that $f^{m(x)}(x) \in Y_a$. Then $f^i(x) \in W_a$ for $n(x) \leq i \leq m(x)$.

We shall suppose that a is chosen so small that the following conditions are satisfied:

$$(\mathrm{cl}_S(X_a) \cap \mathrm{cl}_S(Y_a)) \cap (\gamma \cup f\gamma) = \varnothing,$$

$$\mathrm{cl}_S(W_a) \cap (\mu_i \cup (\gamma - \gamma_i) \cup (f\gamma - \gamma'_i)) = \varnothing.$$

For each $x \in W_a$ which is not in γ_i, we define $\pi(x) \in \mathbb{Z}/2$, as follows. Join x to a point in X_a by a PL-curve in W_a, and let $\pi(x)$ be the number of times (mod 2) that this curve crosses γ_i. It is easily checked that $\pi(x)$ is well defined, $\pi(x) = 0$ if $x \in X_a$, and $\pi(x) = 1$ if $x \in Y_a$.

Thus, for any $x \in W_a$, we have either $f^j(x) \in \gamma_i$ for some j satisfying $n(x) \leq j \leq m(x)$, or we have that $\pi(f^j(x)) = 1$ and $\pi(f^{j-1}(x)) = 0$, for some j satisfying $n(x) < j \leq m(x)$. In the first case, $f^j(x) \in R'_i$, since $\gamma_i \subset R'_i$. In the second case, $f^j(x) \in R'_i$, because a PL-curve connecting $f^j(x)$ to a point of X_a crosses η_i an odd number of times, since it crosses γ_i an odd number of times, γ'_i an even number of times, and does not cross μ_i.

Thus in either case, $x \in f^{-j}(R'_i)$, and Lemma 9.3 is proved. ■

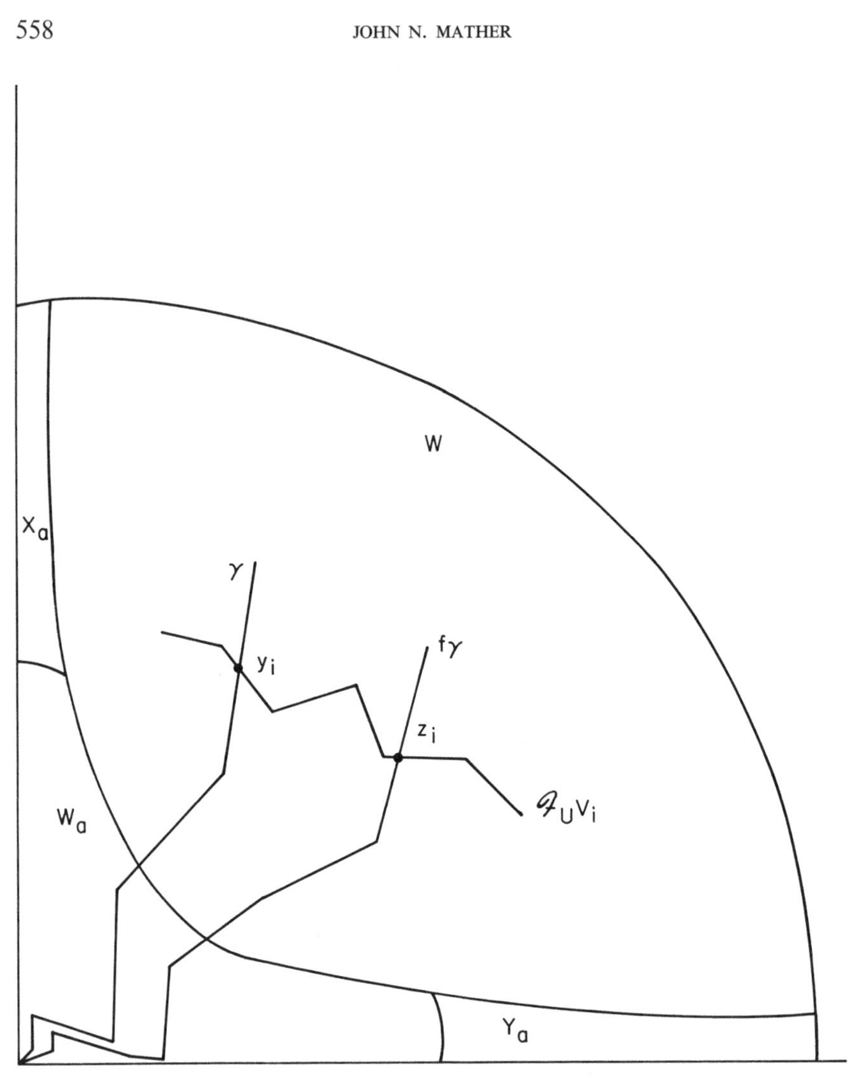

FIGURE 1

Before returning to Case 2 in Proposition 9.2, we prove:

LEMMA 9.4. *A meets only one residual domain of* η_i.

Proof. Suppose A meets two residual domains R_i'' and R_i''' of η_i. Let N be a compact neighborhood of $A \cup \eta_i$ in S which is a connected surface. Let U be the component of $N - A$ which contains $\eta_i - \mathbf{O}$. We may connect a point in $R_i'' \cap A$ (resp., $R_i''' \cap A$) to a point in η_i by a closed arc in N not meeting $\mathbf{O}$. By replacing this arc by a subarc, we may assume that only the endpoints are in A or η_i. Then the rest is in U, since it is in $N - A$ and one endpoint is in $\eta_i - \mathbf{O}$. Thus, we get a closed arc ζ_i'' (resp., ζ_i''') with one endpoint z_i'' (resp., z_i''') in $A \cap R_i''$ (resp., $A \cap R_i'''$) and otherwise in U. Let ζ_i^2 (resp., ζ_i^3) be a subarc of ζ_i'' (resp., ζ_i'''), one of whose endpoints is z_i'' (resp., z_i''') but which does not meet η_i.

By Proposition 3.7, ζ_i^2 and ζ_i^3 are endcuts of prime ends e_i^2 and e_i^3 in $E(U)$. Since γ is an endcut of e and $f_*e = e$, $f\gamma$ is an endcut of e, also. Hence γ_i and γ_i' are endcuts of e. Since $\mathbf{O}$, z_i'', and z_i''' are distinct, e, e_i^2, and e_i^3 are distinct.

Let $\eta_i^* = \mathrm{cl}_{E(U)}(\eta_i - \mathbf{O}) = (\eta_i - \mathbf{O}) \cup e$, $\zeta_i^{2*} = \mathrm{cl}_{E(U)}(\zeta_i^2 - z_i'') = (\zeta_i^2 - z_i'') \cup e_i^2$, $\zeta_i^{3*} = \mathrm{cl}_{E(U)}(\zeta_i^3 - z_i''') = (\zeta_i^3 - z_i''') \cup e_i^3$. Since ζ_i^2 and ζ_i^3 lie in different residual domains of η_i, ζ_i^{2*} and ζ_i^{3*} lie in different components of $E(U) - \eta_i^*$. But this is impossible, since the closed curve η_i^* meets $\partial E(U)$ in only one point, e, while ζ_i^{2*} and ζ_i^{3*} each meets $\partial E(U)$ in a point (e_i^2 or e_i^3) which is distinct from e.

This contradiction proves the Lemma. ∎

Now, we return to Case 2 of Proposition 9.2. By taking i sufficiently large, we may suppose η_i is in an arbitrarily small disk about $\mathbf{O}$. Then R_i' is in the same disk. Thus for i sufficiently large $A \not\subset R_i'$, and it follows from Lemma 9.4 and the fact that $\eta_i \cap A = \mathbf{O}$ that $A \subset R_i \cup \mathbf{O}$.

Since A is f-invariant, it then follows from Lemma 9.3 that $A \cap W_a = \varnothing$.

Since $\mathscr{F}_U V_i \to \mathbf{O}$ and $\mathscr{F}_U V_i \subset W$, it follows that $\mathscr{F}_U V_i$ is in W_a, for i sufficiently large. However, the endpoints of $\mathscr{F}_U V_i$ are in A. Since $A \cap W_a = \varnothing$, each of these endpoints is in β_1 or β_2. From the fact that the V_i form a chain, and $\mathscr{F}_U V_i \to \mathbf{O}$, it follows that $\mathscr{F}_U V_i$ must have one endpoint in each of β_1 and β_2, for i sufficiently large.

Consider an i so large that $\mathscr{F}_U V_i \subset W_a$ and $\mathscr{F}_U V_i$ has one endpoint P_{ij} in each of β_j, for $j = 1, 2$, and set $P_{i1}P_{i2} = \mathscr{F}_U V_i$. Let V_i' be the residual domain of the simple closed curve $\mathbf{O}P_{i1}P_{i2}\mathbf{O}$ which lies in W_a. Since $\mathscr{F}_U V_j \to \mathbf{O}$ as $j \to \infty$, and $\mathscr{F}_U V_j \subset W_a$ for j large, it follows that $\mathscr{F}_U V_j \subset V_i'$, for j large.

Since the endpoints of $\mathscr{F}_U V_i$ are in A, and $\mathscr{F}_U V_i$ has one endpoint in each of β_1 and β_2, for i sufficiently large, it follows that $\beta_1 \cap A \neq \varnothing$ and $\beta_2 \cap A \neq \varnothing$. By Lemma 8.2 (applied possibly to f^{-1} in place of f), $\beta_1 \subset A$ and $\beta_2 \subset A$. Hence $\mathbf{O}P_{ij}$, which is in β_j for $j = 1, 2$, does not intersect V_i. Hence the boundary $\mathbf{O}P_{i1}P_{i2}\mathbf{O}$ of V_i' does not intersect V_i. On the other hand, $V_i \cap V_i' \neq \varnothing$, since both V_i and V_i' contain $\mathscr{F}_U V_j$, for j large. Since V_i is connected, it follows that $V_i \subset V_i'$, for sufficiently large i.

Clearly, $\mathscr{F}_S V_i - \mathscr{F}_U V_i \subset A$. Since $\mathscr{F}_U V_i \subset \mathscr{F}_U V_i'$, $A \cap W_a = \varnothing$, and $V_i' \subset W_a$, it follows that $\mathscr{F}_S V_i \cap V_i' = \varnothing$. Since V_i' is connected, we then obtain $V_i = V_i'$, for all sufficiently large i.

Hence e is a sector end associated to W. ∎

10. Proof of Theorem 5.1.

Suppose Theorem 5.1 were false. At the end of Section 7, we have observed that there exists $e \in C(p)$ such that $f_*(e) = e$. By Lemma 3.4, e has a principal point $\mathbf{O}$. By Lemma 7.1, $f(\mathbf{O}) = \mathbf{O}$. By Lemma 7.2, $\mathbf{O}$ is not Moser stable. From the hypothesis of Theorem 5.1, it therefore follows that $\mathbf{O}$ is sectorial periodic. Hence there exists n such that $\mathbf{O}$ is sectorial fixed for f^n.

Let $A = Z(p)$. By Proposition 9.2, applied to f^n in place of f, e is a sector end of $E(S - A)$. Let Σ be the sector to which it is associated. Since $f_*(e) = e$, we get $f(\Sigma)_{\mathbf{O}} = \Sigma_{\mathbf{O}}$ (equality of germs at $\mathbf{O}$). Let b_1 and b_2 be the two branches of $\mathbf{O}$ which border Σ. Then $f(b_i) = b_i$, $i = 1, 2$; for otherwise, f would be locally orientation reversing at p, and we assumed this was not the case.

Obviously, there exists a unique mapping $E: b_1 \cup \mathbf{O} \cup b_2 \to C(p)$, with the following properties: $E(\mathbf{O}) = e$; if $x \in b_1 \cup \mathbf{O} \cup b_2$, then $X(E(x)) = Y(E(x)) = x$. If we provide $b_1 \cup \mathbf{O} \cup b_2$ with the manifold topology, then E is a continuous immersion.

There are two possibilities: either $E(b_1) = E(b_2) = C(p) - e$, or E is an embedding onto an open circular arc $e_1 e e_2$, possibly with $e_1 = e_2$. In the first case, we set $e_1 = e_2 = e$.

Then $f_*(e_i) = e_i$, $i = 1, 2$. By Lemma 3.4, e_i has a principal point $\mathbf{O}_i$. By the argument just given, e_i is a sector end, for $i = 1, 2$. Let Σ_i be the sector to which it is associated. Let b_{i1}, b_{i2} be the two branches of $\mathbf{O}$ which border Σ_i. There is a mapping $E_i: b_{i1} \cup \mathbf{O} \cup b_{i2} \to C(p)$ having properties anlogous to E.

Obviously, $E(b_1) = E_1(b_{11})$ or $E_1(b_{12})$. For $x \in b_1$, we have $x = X(E(x)) = X(E_1(E_1^{-1}E(x))) = E_1^{-1}E(x) \in b_{1j}$ where $j = 1$ or 2, whichever is appropriate.

Then $b_1 = b_{1j}$ and we have obtained a fixed connection. This contradicts our hypothesis and proves Theorem 5.1. ■

11. Proof of Theorem 5.2

Throughout this section, we suppose f is an area preserving homeomorphism of an open subset S_0 of S onto an open subset fS_0 of S. We let A be a continuum in S_0 such that $fA = A$. We suppose that every fixed point of f in A is sectorial periodic or Moser stable, and f has no fixed connections in A. We shall prove the following generalization of Theorem 5.2:

Proposition 11.1. *Let $\mathbf{O}$ be a sectorial fixed point of f in A. Then every semibranch of $\mathbf{O}$ is a subset of A.*

Theorem 5.2 follows: Let the hypotheses of Theorem 5.2 be satisfied and let b_1, b_2 be two branches of $\mathbf{O}$. Then $\mathrm{cl}_S b_2$ is an invariant continuum, so, by Proposition 11.1, it contains b_1.

Proof. Let β be a semibranch of **O**. Suppose $\beta \not\subset A$. By Proposition 9.1, $\beta \subset S - A$ and there is a sector end $e \in E(S - A)$ such that the germ of β at **O** is in the sector to which e is associated.

Since **O** is sectorial fixed, $f(\beta) \subset \beta$ or $f^{-1}(\beta) \subset \beta$, and we obtain $f_*(e) = e$. Let Σ be the sector to which e is associated. Then $f(\Sigma)_{\mathbf{O}} = \Sigma_{\mathbf{O}}$ (equality of germs). Let β_1 and β_2 be two semibranches of **O** which border Σ. Then $\beta_1, \beta_2 \subset A$, so they are contained in branches b_1 and b_2. Since **O** is sectorial fixed, $f(b_i) = b_i$, $i = 1, 2$.

Let $p = \alpha(e) \in b_I(S - A)$. We construct the mapping $E: b_1 \cup \mathbf{O} \cup b_2 \to C(p)$, the points e_1, e_2, and the branches b_{i1}, b_{i2} just as in the last section. This is possible, because e_i is a sector end by Proposition 9.2.

The reasoning of the last section shows that $b_1 = b_{1j}$ for $j = 1$ or 2. This gives a fixed connection in A, and a contradiction. Thus, $\beta \subset A$. ■

Acknowledgments

I would like to thank M. Handel and J. Guckenheimer for very helpful comments concerning the material in this paper. When I first worked on these problems, I suspected that there should be a fixed point theorem relevant to this type of problem. M. Handel pointed out to me that Cartwright and Littlewood had already proved a fixed point theorem of the type I wanted in [4]. In a previous paper [10] I have used their theorem. While I do not use their theorem in this paper, I do depend heavily on their ideas; particularly on their observation that Carathéodory's theory has many consequences for dynamical systems.

References

1. G. D. Birkhoff, Sur quelques courbes fermées remarqables, *Bull. Soc. Math. France* **60** (1932), 1–26; *Collect. Math. Papers* **2** (1950), 418–443.
2. G. D. Birkhoff, Surface transformations and their dynamical applications, *Acta Math.* **43** (1920), 1–119; *Collect. Math. Papers* **2** (1950), 111–229.
3. C. Carathéodory, Über die Begrenzung einfach zusammenhangender Gebiete, *Math. Ann.* **73** (1913), 323–370.
4. M. L. Cartwright and J. E. Littlewood, Some fixed point theorems, *Ann. of Math.* **54** (1951), 1–37.
5. D. B. A. Epstein, Prime ends, preprint, Warwick Univ., 1978.
6. D. M. Grobman, Homeomorphism of systems of differential equations *Dokl. Akad. Nauk SSSR* **128** (1959), 880–881 [in Russian].
7. D. M. Grobman, Topological classification of neighborhoods of a singularity in n-space, *Mat. Sb.* (*N.S.*) **56**, **(98)** (1962), 77–94 [in Russian].
8. P. Hartman, On the local linearization of differential equations, *Proc. Amer. Math. Soc.* **14** (1963), 568–573.
9. P. Hartman, "Ordinary Differential Equations," Wiley, New York, 1964.
10. J. Mather, Area preserving twist homeomorphisms of the annulus, *Comment. Math. Helv.* **54** (1979), 397–404.
11. J. Moser, "Stable and Random Motions in Dynamical Systems," Annals of Mathematics Studies, No. 77, Princeton Univ. Press, Princeton, New Jersey, 1973.
12. J. Palis, Local structure of hyperbolic fixed points in Banach spaces, *An. Acad. Brasil. Ciênc.* **40** (1968), 263–266.

13. C. Pugh, On a theorem of P. Hartman, *Amer. J. Math.* **91** (1969), 363–367.
14. I. Richards, On the classification of noncompact surfaces, *Trans. Amer. Math. Soc.* **106** (1963), 259–269.
15. R. C. Robinson, Generic properties of conservative systems I, II, *Amer. J. Math.* **92** (1970), 562–603, 897–907.

MATHEMATICAL ANALYSIS AND APPLICATIONS, PART B
ADVANCES IN MATHEMATICS SUPPLEMENTARY STUDIES, VOL. 7B

A New Class of Symmetric Functions

N. METROPOLIS†

Los Alamos Scientific Laboratory
University of California
Los Alamos, New Mexico

G. NICOLETTI

Istituto di Geometria
Università di Bologna
Bologna, Italia

AND

GIAN-CARLO ROTA‡

Department of Mathematics
Massachusetts Institute of Technology
Cambridge, Massachusetts

and

Los Alamos Scientific Laboratory
University of California
Los Alamos, New Mexico

DEDICATED TO LAURENT SCHWARTZ ON THE OCCASION OF HIS SIXTY-FIFTH BIRTHDAY

··· MAESTRO, IL MIO VEDER S'AVVIVA
SI' NEL TUO LUME, CH'IO DISCERNO CHIARO
QUANTO LA TUA RAGION PORTI O DESCRIVA . . .
PURGATORIO, XVIII, 10–12.

Contents

† This work was supported in part by U.S. DOE.
‡ This work was supported in part by NSF: MCS 7308445.

ISBN 0-12-512802-9

1. Introduction

Let A be an abelian group; a *symmetric* function f is a function

$$f(x_1, x_2, \ldots)$$

of infinitely many variables x_i $(i = 1, 2, \ldots)$ defined when almost all variables are set equal to zero, ranging over A, and which is invariant under all the permutations σ of the variables:

$$f(x_{\sigma(1)}, x_{\sigma(2)}, \ldots) = f(x_1, x_2, \ldots).$$

Classically, A is a commutative ring with identity, and only polynomial symmetric functions are studied, such as, for example the *elementary symmetric functions*

$$a_n(x_1, x_2, \ldots) = \sum_{i_1 < i_2 < \cdots < i_n} x_{i_1} x_{i_2} \cdots x_{i_n}$$

and the *complete homogeneous functions*

$$h_n(x_1, x_2, \ldots) = \sum_{i_1 \leq i_2 \leq \cdots \leq i_n} x_{i_1} x_{i_2} \cdots x_{i_n}.$$

We introduce here two classes of symmetric functions which we call *sampling functions* and *bisymmetric functions*. We believe both of these concepts to be new and introduced here for the first time.

Our motivation is partly combinatorial, in that the algebra of these functions gives a better insight into the algebra of symmetric functions, and partly the idea (not developed here) that these functions may lead to a new class of representations of the infinite symmetric group and the general linear group.

Sampling functions are not always polynomials, but it will be seen that—in a natural way—the algebra of sampling functions is the closure of the algebra of polynomial symmetric functions.

In fact, we are able to give a simple harmonic analysis of the algebra of sampling functions by their Fourier–Gelfand transforms on the space of maximal ideals. Those functions on such a space which are Fourier–Gelfand transforms of sampling functions can be given a simple characterization. Such a characterization is not available for the smaller algebra of polynomial symmetric functions.

Sampling functions have several applications. Here we give only one. We show how they can be used to investigate algebraic dependencies (syzygies) of symmetric functions over finite fields. It is known [1] that over a finite field the elementary symmetric functions are no longer algebraically independent. We explain this phenomenon by establishing a completeness result of Stone–Weierstrass type over the space of maximal ideals of the algebra of sampling functions, giving a simple criterion ensuring that the algebra generated by a given set of functions is the whole algebra of polynomial symmetric functions.

Another application of sampling functions, which we hope to develop elsewhere, is to the study of inequalities among symmetric functions, even polynomial symmetric functions.

The notion of bisymmetric function is again motivated by viewing symmetric functions as functions on the space of maximal ideals. The concept has a difference-theoretic origin which is made clear below, but there is also a representation-theoretic interpretation which we hope to develop in a later note.

The main idea of this note is extremely simple. It harks back to the two notations for a multiset, alternatively used in combinatorics for a long time. A multiset can be described either by listing its elements in any order, each element repeated as often as its multiplicity. Or else, by first giving a set—the support of the multiset—together with a multiplicity assigned to each element of the support. Our Fourier–Gelfand transform is the recognition that this double notation is an instance, perhaps the simplest instance, of harmonic analysis ("spec" analysis) on a commutative ring. We believe that the duality between support and multiplicity, i.e., between set and multiset (elsewhere thought of as the duality between the commutative and the anticommutative), far from being just a notational quirk, is instead by its very simplicity, the primordial case which may throw light on the foundations of harmonic analysis.

This note is the first version of a more extensive memoir now in preparation.

2. Sampling Functions

Let $\Sigma(A)$ be the algebra of all symmetric functions over the commutative ring A with identity. If b is an element of A, we define the *mock-translation operator* E_b over $\Sigma(A)$ as follows: For each $f \in \Sigma(A)$, set

$$E_b f(x_1, x_2, \ldots) = f(b, x_1, x_2, \ldots).$$

Obviously, mock-translations are endomorphisms of $\Sigma(A)$, and they commute.

The *mock-difference operator* Δ_b is defined as

$$\Delta_b := E_b - I,$$

where b is a nonzero element of A, and I is the identity operator; thus, for example, one computes $\Delta_b a_n = b a_{n-1}$.

We say that a symmetric function f is a *sampling function* when for every element $b \in A$, there exists a nonnegative integer k depending on b such that

$$\Delta_b^k f \equiv 0.$$

Sampling functions form a subalgebra of $\Sigma(A)$ which we will write $\Sigma_s(A)$; we further denote by $\Sigma_p(A)$ the subalgebra of polynomial symmetric functions. It is easily shown that every polynomial symmetric function is a sampling function, but not vice versa. In fact, for a nonnegative integer n and an element $b \in A$, $b \neq 0$, set

$$\beta_{b,n}(x_1, x_2, \ldots)$$

to be the function taking the value $\binom{m}{n}$ (the binomial coefficient $\binom{m}{n}$ is here understood as $1 + 1 + \cdots + 1$ as many times as $\binom{m}{n}$, where 1 is the identity of A). Here m is the number of variables x_i set equal to b.

These *beta functions* are sampling functions; indeed, one computes $\Delta_b \beta_{b,n} = \beta_{b,\,n-1}$ and $\Delta_c \beta_{b,n} = 0$ if $c \neq b$. But beta functions are not polynomial functions, except when A is a finite field, in which case—as will be seen later—$\Sigma_p(A) = \Sigma_s(A)$ [but $\Sigma_p(A) \neq \Sigma(A)$ even when A is a finite field].

Let $\Phi^+(A)$ be the set of all functions ϕ defined on the set of all nonzero elements of A, taking nonnegative integer values, and equal to zero almost everywhere; namely, *finite multisets*. Finite multisets can be added and be multiplied by constants. For each such finite multiset ϕ, set

$$\Delta_\phi := \prod_{b \in A,\, b \neq 0} \Delta_b^{\phi(b)} \qquad \text{and} \qquad \beta_\phi := \prod_{b \in A,\, b \neq 0} \beta_{b,\phi(b)}.$$

We have then the following fundamental expansion of Newtonian type:

THEOREM 1. *Every sampling function f can be uniquely expanded in the series*

$$f(x_1, x_2, \ldots) = \sum_\phi a_\phi \beta_\phi(x_1, x_2, \ldots), \qquad a_\phi \in A, \quad \phi \in \Phi^+,$$

where $a_\phi := \varepsilon(\Delta_\phi f)$ and $\varepsilon(f(x_1, x_2, \ldots)) := f(0, 0, \ldots)$.

Here, the convergence of the series on the right is to be understood in the weak topology of $\Sigma_s(A)$, defined as follows: for every assignment of values (almost all zero) to the x_i, the series on the right is finite and equals the value on the left.

Under this topology, $\Sigma_s(A)$ becomes a topological algebra, and we shall understand it as such henceforth.

We note that if A is a commutative group, $\Sigma_s(A)$ is a $\Sigma_s(Z)$-module. Sampling functions are defined even in this more general case, and an analog of Theorem 1 still holds, but the beta functions take values in Z, and the products $a_\phi \beta_\phi$ have to be understood as multiples of $a_\phi \in A$. We shall occasionally and tacitly make use of this generalization.

3. Harmonic Analysis

For simplicity, let A now be a field. In this case, every finite multiset $\phi \in \Phi^+$ defines a maximal ideal G_ϕ of $\Sigma_s(A)$, as follows: G_ϕ is the set of all sampling functions which vanish when for each $b \in A$, exactly $\phi(b)$ of the variables x_i are set equal to b.

We set $\hat{f}$ to be the Fourier–Gelfand transform of an arbitrary sampling function f relative to this set of maximal ideals. Specifically, $\hat{f}(G_\phi)$ equals the value of $f(x_1, x_2, \ldots)$ when for each $b \in A$, $b \neq 0$, exactly $\phi(b)$ of the variables are set equal to b. Let us write $\hat{f}(\phi)$ in place of $\hat{f}(G_\phi)$, and thus identify the set of finite multisets with the space of the maximal ideals defined above.

We shall now explicitly characterize the algebra of those Φ^+-valued functions which are the Fourier–Gelfand transforms of sampling functions. To this end, we introduce *factorial functions*, a notion we believe to be of independent interest.

Let S be a nonempty set with a distinguished element which we denote by 0; let $\Phi^+(S)$ be the set of all functions defined over the nonzero elements of S, taking nonnegative integer values, and such that only a finite number of elements of S take a nonzero value; such functions will be called *finite multisets* of S. Denote by δ_b ($b \in S$, $b \neq 0$) the *Dirac multiset*, i.e., the function defined as follows:

$$\delta_b(c) := \begin{cases} 1 & \text{if } \ b = c, \\ 0 & \text{otherwise.} \end{cases}$$

We define the *translation operator* $\hat{E}_b$ as follows: If g is a function from $\Phi^+(S)$ to A, and if $\phi \in \Phi^+(S)$, set

$$\hat{E}_b g(\phi) := g(\phi + \delta_b).$$

Let the *difference operator* $\hat{\Delta}_b$ equal $\hat{E}_b - \hat{I}$, where $\hat{I}$ is the identity operator. Finally, if $\phi \in \Phi^+$, set

$$\hat{\Delta}_\phi := \prod_{b \in A,\, b \neq 0} \Delta_b^{\phi(b)}.$$

We say that an A-valued function g is a *factorial function* when for every $b \in S$, there is a nonnegative integer k such that $\hat{\Delta}_b^k f = 0$. The set of factorial functions will be denoted by $(\Phi^+(S), A)$.

THEOREM 2. *The Fourier–Gelfand transform establishes an isomorphism between the algebra $\Sigma_s(A)$ of sampling functions and the algebra $(\Phi^+(A), A)$ of factorial functions. Under this isomorphism, the mock-translation and the mock-difference operator E_b and Δ_b are carried onto the translation and difference operators $\hat{E}_b$ and $\hat{\Delta}_b$.*

As an example, the Fourier–Gelfand transforms of the beta functions are:

$$\hat{\beta}_{b,n}(\phi) = \binom{\phi(b)}{n},$$

where $\phi \in \Phi^+$, and if $\psi \in \Phi^+$,

$$\hat{\beta}_{\psi}(\phi) = \binom{\phi}{\psi}$$

where

$$\binom{\phi}{\psi} = \prod_{b \in A,\, b \neq 0} \binom{\phi(b)}{\psi(b)}.$$

Not unexpectedly, we obtain an expansion of Newtonian type for the factorial functions.

THEOREM 3. *Every factorial function g can be uniquely expanded in the form*

$$g(\phi) = \sum_{\psi} a_{\psi} \binom{\phi}{\psi}, \qquad a_{\psi} \in A, \qquad \psi \in \Phi^+,$$

where $a_{\psi} := \hat{\varepsilon}(\hat{\Delta}_{\psi} g)$, and $\hat{\varepsilon} g$ is the evaluation of g at the zero multiset.

As an application of the theorem above, we give the expansion of the Fourier–Gelfand transforms of the elementary symmetric functions a_n:

$$\hat{a}_n(\phi) = \sum_{\psi} b^{\psi} \binom{\phi}{\psi}, \qquad \psi \in \Phi^+, \qquad |\psi| = n,$$

where

$$b^{\psi} := \prod_{b \in A,\, b \neq 0} b^{\psi(b)}.$$

Here $|\psi|$ is the cardinality of the multiset ψ, namely,

$$|\psi| = \sum_{b} \psi(b).$$

The same expansion formula allows us to compute the product of the Fourier–Gelfand transforms of two beta functions:

$$\beta_{\psi}(\phi)\beta_{\chi}(\phi) = \sum_{\rho} \binom{\phi}{\psi} \binom{\psi}{\rho - \chi} \beta_{\rho}(\phi), \qquad \rho, \phi, \chi, \psi \in \Phi^+.$$

Using this formula, one can compute the expansion of the product of any two factorial functions whose expansions are given. The classical identities holding among polynomial symmetric functions are easily derived by use of Theorems 1 and 2.

This motivates us to investigate the structure of the algebra $(\Phi^+(A), A) = \hat{\Sigma}_s(A)$.

4. Structure Theorems

We can give a description by generators and relations of the topological ring $\Sigma_s(A)$ of sampling functions. We shall do this here when A is the ring of integers $\mathbf{Z}$ or when A is a field, the general case not being essentially different.

We consider a set of generators x_ϕ indexed by finite multisets $\phi \in \Phi^+(S)$; these generators will be subject to the following relations:

$$\text{(i)} \quad x_\phi = \prod_{b \in S,\, b \neq 0} x_{\phi(b)\delta_b},$$

where $\phi(b)\delta_b$ is the constant $\phi(b)$ multiplying the multiset δ_b,

$$\text{(ii)} \quad x_{m\delta_b} x_{n\delta_b} = \sum_k \binom{k}{m}\binom{m}{k-n} x_{k\delta_b},$$

where $m\delta_b$ equals the constant m times the multiset δ_b, etc.

An infinite sum (or series)

$$\sum_\phi a_\phi x_\phi, \qquad a_\phi \in \mathbf{Z}, \quad \phi \in \Phi^+(S)$$

is said to be *locally finite* if for every finite multiset ϕ, only a finite number of coefficients $a_{m\phi}$ (with integer $m \geq 0$) are different from zero. A locally finite series

$$\sum_\phi a_\phi x_\phi, \qquad \phi \in \Phi^+(S)$$

is said to converge if for every finite multiset $\psi \in \Phi^+(S)$, the following sum is finite:

$$\sum_\phi a_\phi x_\phi,$$

where ϕ ranges over all multisets such that $\phi \leq n\psi$ for some nonnegative integer n.

Every locally finite series is convergent.

The topological ring (algebra over $\mathbf{Z}$) thus generated is isomorphic to the algebra of factorial functions $(\Phi^+(S), \mathbf{Z})$.

If $b \in S$, we define the operator E_b as follows: If ϕ is a finite multiset

$$E_b x_\phi := \begin{cases} x_\phi + x_{\phi - \delta_b} & \text{if} \quad \phi(b) \geq 1, \\ x_\phi & \text{if} \quad \phi(b) = 0. \end{cases}$$

Using identities (i) and (ii), it can be shown that E_b turns out to be an endomorphism of $(\Phi^+(S), \mathbf{Z})$, that is, that

$$E_b(x_\phi + x_\psi) = E_b x_\phi + E_b x_\psi, \qquad E_b(x_\phi x_\psi) = E_b x_\phi E_b x_\psi.$$

From Theorem 3 and the preceding remark we infer

THEOREM 4. *The topological algebra of sampling functions* $\Sigma_s(\mathbf{Z})$ *is isomorphic to the algebra over* $\mathbf{Z}$ *generated by the variables* x_ϕ *as above, with* $S = \mathbf{Z}$, *under the identities* (i) *and* (ii).

When A is a field of characteristic 0, set $S = A$, define as above the variables x_ϕ, with $\phi \in \Phi^+(A)$, and take the (topological) ring of all locally finite series

$$\sum_\phi a_\phi x_\phi, \qquad a_\phi \in A, \qquad \phi \in \Phi^+(A)$$

under the identities (i) and (ii).

This topological algebra over A is isomorphic to the algebra $\Sigma_s(A)$. Hence, we have

THEOREM 5. *The topological algebra of sampling functions over a field* A *of characteristic* 0 *is isomorphic to the tensor product*

$$A \otimes (\Phi^+(A), \mathbf{Z}).$$

If we set now, for each $b \in A$, $b \neq 0$,

$$y_b^n := \sum_i i! S(n, i) x_{i\delta_b},$$

where the $S(n, i)$ are the Stirling numbers of the second kind, it is verified that

$$y_b^m \cdot y_b^n = y_b^{m+n}.$$

This new set of variables is a set of generators for the same algebra of sampling functions, because $y_b^1 = x_{\delta_b}$. Thus, we have

COROLLARY 6. *Over a field of characteristic* 0, *the algebra of sampling functions is isomorphic to the topological algebra of polynomials in the* (*in general infinitely many*) *variables* y_b, *with coefficients in* A.

We now determine a set of generators and relations of the algebra $(\Phi^+(S), \mathbf{Z}_p)$ of all factorial functions ranging over the finite field $\mathbf{Z}_p$ with p

elements. To this end, we introduce, as before, the ϕ-indexed generators x_ϕ, $\phi \in \Phi^+(S)$. For ease of notation, we write $x(\phi)$ in place of x_ϕ. We assume identities (i) and (ii) of the preceding paragraph, and in addition the following identities:

(iii) $x(m\delta_b) = \prod_i x(m_i p^i \delta_b), \qquad m \geq 0,$

where

$$m = \sum_i m_i p^i, \qquad 0 \leq m_i \leq p - 1.$$

In other words, the m_i are the digits in the p-ary expansion of the integer m. We have now:

THEOREM 7. *The topological algebra of sampling functions $\Sigma_s(\mathbf{Z}_p)$ over the finite field $\mathbf{Z}_p$ is isomorphic to the algebra over $\mathbf{Z}_p$ generated by the variables x_ϕ as above, with $S = \mathbf{Z}_p$, under identities* (i), (ii), *and* (iii).

When A is a field of characteristic p, set $S = A$; define as above the variables x_ϕ, with $\phi \in \Phi^+(A)$, and take the (topological) ring of all locally finite series

$$\sum_\phi a_\phi x_\phi, \qquad a_\phi \in A, \qquad \phi \in \Phi^+(A)$$

under the identities (i), (ii), and (iii).

This topological algebra over A is isomorphic to the algebra $\Sigma_s(A)$. Hence, we have

THEOREM 8. *The topological algebra of sampling functions over the field A of characteristic p is isomorphic to the tensor product*

$$A \otimes (\phi^+(A), \mathbf{Z}_p).$$

Let now, for each $b \in A$, $b \neq 0$, for each nonnegative integer i and for each m, $0 \leq m \leq p - 1$,

$$y((i, m), b) := x(mp^i \delta_b).$$

We define a new set of variables, called the *digit variables*, as follows

$$z_{i,b}^m = \sum_{j=0}^{p-1} j!S(m, j)y((i, j), b),$$

where $b \in A$, $b \neq 0$, $i \geq 0$, $m \geq 0$.

One shows that these new variables satisfy the following identities

$$z_{i,b}^m z_{i,b}^n = z_{i,b}^{m+n}$$

and Fermat's identity

$$z_{i,b}^{p} = z_{i,b}.$$

Consequently, we have:

COROLLARY 9. *Over a field A of characteristic p, the algebra of sampling functions is isomorphic to the topological algebra of polynomials in the infinitely many variables $z_{i,b}$, subject to Fermat's identity, $z_{i,b}^{p} = z_{i,b}$, and with coefficients in A.*

The results in this paragraph show that the structure of the algebra of sampling functions over the field A depends only in a trivial way on A, while it depends critically on the characteristic of A.

5. BISYMMETRIC FUNCTIONS

The mock translations E_b and the mock differences Δ_b suggest the possibility of defining (as in the calculus of finite differences) the operators E_b^{-1} and ∇_b. However, since E_b increases by 1 the multiplicity of the element $b \in A$ (that is, it sets equal to b a variable x_i which was zero), the operator E_b^{-1} is not well defined, because it should *decrease* the multiplicity of b. (That is, it should set to zero a variable x_i which was set equal to b).

This leads us to generalize the concept of symmetric function in such a way that it is possible that a "negative" number of variables are set equal to an element $b \in A$, $b \neq 0$.

To this end, we define a *bisymmetric function* f to be a function

$$f(x_1, x_2, \ldots | y_1, y_2, \ldots)$$

in two sequences of infinitely many variables x_i, y_i, symmetric in the x_i and the y_i separately, and such that, for every $b \in A$

$$f(b, x_1, x_2, \ldots | b, y_1, y_2, \ldots) = f(x_1, x_2, \ldots | y_1, y_2, \ldots).$$

The value of such a function depends only on the *difference* of the multiplicities of each $b \in A$, $b \neq 0$, with respect to the x_i and the y_i.

In the algebra of all bisymmetric functions

$$\Sigma(A | A)$$

the operators E_b^{-1} and $\nabla_b := I - E_b^{-1}$ can now be defined, by setting

$$E_b^{-1} f(x_1, x_2, \ldots | y_1, y_2, \ldots) = f(x_1, x_2, \ldots | b, y_1, y_2, \ldots).$$

We note that if in a bisymmetric function $f(x_i | y_i)$ we set to zero all the variables y_i, we obtain a symmetric function in the variables x_i. Conversely, every symmetric function f can be extended, in general, in many ways to a bisymmetric function.

The analog of the concept of sampling function is here that of bisampling function: a bisymmetric function is said to be a *bisampling function* if for every $b \in A$, $b \neq 0$, there exists a nonnegative integer m such that $\Delta_b^m f = 0$. This is equivalent to assuming that $\nabla_b^m f = 0$.

Bisampling functions form a subalgebra of $\Sigma(A|A)$ which we shall write as $\Sigma_s(A|A)$. Let $\Phi(A)$ be the set of all functions ϕ defined on the set of all nonzero elements of A, taking integer values (negative values are now allowed) and equal to zero almost everywhere, called *finite multiplicities.*

We can define bisymmetric analogs of the beta functions. In fact, set

$$\beta_{b,n}(x_1, x_2, \ldots | y_1, y_2, \ldots), \qquad b \in A, \quad b \neq 0, \quad n \geq 0,$$

to be the function taking the value $\binom{m}{n}$, where m is now the *difference* between the number of variables x_i and the number of variables y_i set equal to b. Note that $\binom{m}{n}$ make sense also if m is negative, upon setting $\binom{m}{n} = (-1)^n \left\langle {-m \atop n} \right\rangle$. The beta functions here defined turn out to be bisampling functions. We have now

THEOREM 10. *Every bisampling function f can be uniquely expanded in the series*

$$f(x_i | y_i) = \sum_{\phi} a_\phi \beta_\phi(x_i | y_i), \qquad a_\phi \in A, \quad \phi \in \Phi^+,$$

where $a_\phi := \varepsilon(\Delta_\phi f)$ and $\varepsilon(f(x_1, x_2, \ldots | y_1, y_2, \ldots)) = f(0, 0, \ldots | 0, 0, \ldots)$, and with the same meaning of the symbols Δ_ϕ and β_ϕ as in the preceding paragraphs.

If f is a sampling function, we denote by $\tilde{f}$ the unique bisampling function having, in its expansion, the same coefficients as the function f. We have:

THEOREM 11. *The map $f \to \tilde{f}$ is an isomorphism of the algebra $\Sigma_s(A)$ into the algebra $\Sigma_s(A|A)$. The inverse isomorphism sends $\tilde{f}(\mathbf{x}|\mathbf{y})$ to $f(\mathbf{x}|\mathbf{0})$.*

COROLLARY 12. *Every sampling function f can be extended in an unique way to a bisampling function $\tilde{f}$, such that $\tilde{f}(\mathbf{x}|\mathbf{0}) = f(\mathbf{x})$.*

In the algebra of bisymmetric functions we define a duality, that is an involutory automorphism, as follows:

$$f^*(x_i | y_i) = f(y_i | x_i).$$

Let us now define the functions $c_n(x_i | y_i)$ by the following generating function:

$$\sum_{n \geq 0} c_n(x_i | y_i) t^n = \prod_{i \geq 1} \frac{(1 - x_i t)}{(1 - y_i t)}.$$

Then the functions c_n are bisymmetric and bisampling functions, and

$$c_n(x_1, x_2, \ldots | \mathbf{0}) = (-1)^n a_n(x_1, x_2, \ldots),$$
$$c_n(\mathbf{0} | y_1, y_2, \ldots) = h_n(y_1, y_2, \ldots).$$

These identities give a generalization of the classic Newton identities between the a_n and the h_n in the theory of polynomial symmetric functions.

We thus define a *bipolynomial function* to be a bisymmetric function which is a polynomial in the variables x_i and y_i. The algebra $\Sigma_p(A|A)$ of bipolynomial functions is a subalgebra of $\Sigma_s(A|A)$. Obviously the morphism ~ defined above carries the a_n into the c_n:

$$\tilde{a}_n = c_n.$$

Hence:

THEOREM 13. *The map $f \to \tilde{f}$ is an isomorphism of the algebra $\Sigma_p(A)$ of polynomial symmetric functions into the algebra $\Sigma_p(A|A)$. Furthermore, the involution $f(x_1, x_2, \ldots | y_1, y_2, \ldots) \to f(y_1, y_2, \ldots | x_1, x_2, \ldots)$ of the algebra of bisymmetric functions when restricted to bipolynomial functions, gives the involution of the algebra of polynomial symmetric functions, sending $\tilde{a}_n$ to $(-1)^n \tilde{h}_n$.*

The structure of the algebra of bisampling functions is obviously—because of Theorem 11—the same as the structure of the algebra of sampling functions, and it can be described by generators and relations in the same way.

6. APPLICATIONS TO FINITE FIELDS

If A is a finite field, the functions $\beta_{b,n}$ are polynomial functions. For example, if $A = \mathrm{GF}(p)$ then we have

$$\beta_{b,n}(x_1, x_2, \ldots,) = \sum_{i_1 < i_2 < \cdots < i_n} (1 - (x_{i_1} - b)^{p-1})(1 - (x_{i_2} - b)^{p-1}) \cdots,$$

whence:

THEOREM 14. *If A is a finite field, then every sampling function is a polynomial symmetric function.*

In this case, we are able to prove a result of Stone–Weierstrass type:

THEOREM 15. *If A is a finite field, and if F is a family of sampling functions which distinguishes the ideals G_ϕ and contains the unit function 1, then the subalgebra spanned by F is precisely the algebra $\Sigma_s(A)$ of all sampling functions.*

As applications, we can prove that the subset a_n for $n = up^t, 0 \leq u < p - 1$, $t = 0, 1, 2, \ldots$ generates $\Sigma_s(\mathrm{GF}(p^n))$, and the same happens for h_n for $n = up^t$.

We conclude with yet another notion of symmetric function which we hope to develop elsewhere.

A *completely symmetric function* is a symmetric function which is invariant also under every permutation of the elements of the field A. In order to study the structure of the algebra $\Sigma\Sigma(A)$ of all completely symmetric functions, we define the *Heaviside functions* H_i *and* H_{ij} as follows:

$$H_i(x_1, x_2, \ldots) = \begin{cases} 1 & \text{if } x_i \neq 0, \\ 0 & \text{otherwise,} \end{cases}$$

$$H_{ij}(x_1, x_2, \ldots) = \begin{cases} 1 & \text{if } x_i \neq x_j, \\ 0 & \text{otherwise.} \end{cases}$$

Then, we have:

THEOREM 16. *Every completely symmetric function f over a finite field A is a polynomial in the Heaviside functions.*

If $A = \mathbf{Z}_p$, then

$$H_i(x_1, x_2, \ldots) = x_i^{p-1},$$

$$H_{ij}(x_1, x_2, \ldots) = (x_i - x_j)^{p-1},$$

so that the Heaviside functions are polynomials. For arbitrary A, the Heaviside functions are no longer polynomials, nonetheless a generalization of Theorem 16 can be proved.

In closing we mention that $\Sigma_s(A)$ has a Hopf algebra structure which we intend to study elsewhere.

REFERENCES

1. O. ABERTH, The elementary symmetric functions in a finite field of prime order, *Illinois J. Math.* **8**, (1964), 132–138.
2. D. FOATA, ED., "Combinatoire et représentation du groupe symétrique, Strasbourg, 1976," Springer-Verlag, Berlin and New York, 1977.
3. L. GEISSINGER, Hopf algebras of symmetric functions and class functions, *in* "Combinatoire et représentation du groupe symétrique, Strasbourg, 1976" (D. Foata, ed.), pp. 168–181, Springer-Verlag, Berlin and New York, 1977.
4. D. E. LITTLEWOOD, "The Theory of Group Characters," Oxford Univ. Press (Clarendon), London and New York, 1958.
5. D. E. LITTLEWOOD, "A University Algebra," Heinemann, London, 1961.
6. P. A. MACMAHON, "Collected Papers," MIT Press, Cambridge, Massachusetts, 1978.
7. G.-C. ROTA, "Finite Operator Calculus," Academic Press, New York, 1975.

MATHEMATICAL ANALYSIS AND APPLICATIONS, PART B
ADVANCES IN MATHEMATICS SUPPLEMENTARY STUDIES, VOL. 7B

Sur les semimartingales au sens de L. Schwartz

P. A. MEYER ET C. STRICKER

Département de Mathématiques
Université de Strasbourg
Strasbourg, France

Tout le monde connait l'oeuvre de L. Schwartz en analyse, c'est pourquoi nous tenons à lui dédier plus particulièrement cet article, en témoignage d'admiration pour son oeuvre en probabilités.

Dans le travail récent [8], où il étudie les semimartingales à valeurs dans les variétés, Schwartz a eu besoin d'étendre la notion de semimartingale à des processus définis seulement dans un ouvert aléatoire. Il ne s'agit cependant pas d'une notion locale au sens usuel, car un même processus peut être une semimartingale dans deux ouverts aléatoires sans l'être dans leur réunion. Cependant, l'un des théorèmes de Schwartz affirme que, si un processus X est une semimartingale dans chacun des ouverts d'une suite (A_n), et si $\bigcup_n A_n = \bar{\mathbb{R}}_+ \times \Omega$, alors X est une semimartingale au sens usuel (sur $\bar{\mathbb{R}}_+ \times \Omega$). Nous nous proposons ici de mieux expliquer ce résultat, en introduisant une définition des semimartingales dans un ouvert qui soit véritablement locale, et en donnant alors une forme plus générale du résultat de Schwartz. Après quoi nous étudions les semimartingales dans un intervalle de la forme $[0, T[$ ou $]0, T[$, et nous abordons la théorie de l'intégrale stochastique pour les semimartingales au sens de Schwartz. Nous terminons par une application des résultats obtenus au cas particulier des martingales locales.

SEMIMARTINGALES DANS UN OUVERT ALEATOIRE

$(\Omega, \mathscr{F}, P)$ est un espace probabilisé complet, muni d'une filtration $(\mathscr{F}_t)$ satisfaisant aux conditions habituelles de la théorie des processus. Nous ne rappelons pas les définitions les plus courantes: temps d'arrêt, ensembles prévisibles ou optionnels, semimartingales ... pour lesquelles on renvoie à [4, 5]. Une partie A de $\mathbb{R}_+ \times \Omega$ est dite *évanescente* si sa coupe $A(\omega)$ est vide pour presque tout $\omega \in \Omega$; nous identifions toujours deux parties ou processus *indistinguables*, i.e., ne différant que sur une partie évanescente de $\mathbb{R}_+ \times \Omega$. On dit ainsi, par abus de langage, qu'une partie de $\mathbb{R}_+ \times \Omega$ est

ISBN 0-12-512802-9

mesurable si elle est indistinguable d'un élément de la tribu $\mathscr{B}(\mathbb{R}_+) \times \mathscr{F}$; les parties mesurables de $\mathbb{R}_+ \times \Omega$ sont aussi appelées *ensembles aléatoires.*

Un ensemble aléatoire A est dit *ouvert* (ouvert droit, gauche†) si presque toutes ses coupes $A(\omega)$ sont ouvertes pour la topologie de $\mathbb{R}_+$ (la topologie droite, gauche)—on prendra garde que $[0, t[$ est un voisinage de 0 dans $\mathbb{R}_+$. Etant donnée une famille $(A_i)_{i \in I}$ d'ouverts (d'ouverts droits, gauches) aléatoires,‡ il existe une partie dénombrable J de I telle que $\bigcup_{i \in J} A_i$ contienne, à un ensemble évanescent près, tout ouvert de la famille. Ce résultat est démontré dans [3] pour les ouverts droits ou gauches; pour les ouverts ordinaires, il a été remarqué par de nombreux auteurs. On dit que $\bigcup_{i \in J} A_i$, qui est défini à un ensemble évanescent près, est la *réunion essentielle* de la famille (A_i). Pour alléger langage et notations, nous l'appellerons simplement la *réunion* de la famille, et la noterons $\bigcup_{i \in I} A_i$.

DEFINITION 1. Soient X un processus, A un ouvert aléatoire. On dit que X est une *semimartingale dans* A s'il existe une famille $(A_i)_{i \in I}$ d'ouverts aléatoires, une famille $(Y_i)_{i \in I}$ de semimartingales, telles que $A = \bigcup_{i \in I} A_i$ et que pour tout $i \in I$ on ait $X = Y_i$ dans A_i.

Remarques. (a) On aurait pu remplacer partout *ouvert* par *ouvert droit, ouvert gauche.*

(b) Il est commode de supposer que X est un *processus*, i.e., indexé par $[0, \infty[$, mais dans les cas concrets où l'on rencontre des semimartingales dans A, celles-ci ne sont souvent définies que dans A ou dans $\bar{A}$. La définition 1 garde son sens dans ce cas, et nous nous permettrons de temps en temps de l'utiliser ainsi.

(c) D'après les rappels précédant la définition 1, on peut aussi bien remplacer les familles $(A_i)_{i \in I}$ par des suites $(A_n)_{n \in \mathbb{N}}$, et c'est toujours ainsi que l'on travaille dans les démonstrations. Mais la définition telle que nous l'avons donnée montre immédiatement, par exemple, que si X est une semimartingale dans chacun des ouverts d'une famille $(B^j)_{j \in J}$, X est une semimartingale dans leur réunion. Voir aussi plus bas la remarque (e).

(d) Supposons que X soit un processus optionnel. Alors l'intérieur B_i de l'ensemble aléatoire $\{X = Y_i\}$ est un ouvert optionnel contenant A_i, et X est une semimartingale dans l'ouvert optionnel $\bigcup_{i \in I} B_i$ qui contient A. On aurait donc pu supposer A et les A_i optionnels dans la définition 1. Les mêmes considérations valent pour les ouverts gauches. Mais pour les ouverts droits, on peut seulement affirmer que l'intérieur droit de $\{X = Y_i\}$ est *progressif*. Pour plus de détails, voir [4, Chapitre IV, No 90].

† Par convention, $\{0\}$ est un ouvert gauche.

‡ L'adjectif *aléatoire* sera fréquemment omis.

(e) De même, en considérant la réunion des intérieurs des ensembles $\{X = Y\}$, Y parcourant l'ensemble de toutes les semimartingales, on voit qu'il existe *un plus grand ouvert* (ouvert droit, gauche) *dans lequel un processus mesurable X est une semimartingale.* Si X est optionnel, cet ouvert est optionnel (toutefois, le plus grand ouvert droit dans lequel X est une semimartingale est seulement progressif).

(f) Voici un exemple simple de semimartingale dans un ouvert, non prolongeable au delà—nous en verrons plus loin un autre, plus intéressant, mais exigeant un peu plus de théorie. Soit (B_t) un mouvement brownien réel issu de 0, et soit A l'ouvert prévisible $\{B \neq 0\}$. Montrons que le processus $X = 1/B$, défini seulement dans A, y est une semimartingale. En effet, pour tout rationnel r et tout entier n, soit $A_{n,r} =]r, T_{n,r}[$ où $T_{n,r} = \inf\{t > r : |B_t| < 1/n\}$, et soit $Y_t^{n,r} = 0$ pour $t \leqq r$, $1/B_t$ pour $r \leqq t < T_{n,r}$, n pour $t \geqq T_{n,r}$. On vérifie sans peine que $Y^{n,r}$ est une semimartingale, que $Y^{n,r} = X$ dans $A_{n,r}$, et que A est la réunion des $A_{n,r}$. Donc X satisfait à la définition 1. Comme X n'admet pas de prolongement càdlàg. au delà de A, X n'est pas non plus prolongeable en une semimartingale.

Plus généralement, si Z est une semimartingale, $X = 1/Z$ est une semimartingale dans l'ouvert aléatoire $\{Z \neq 0 \text{ et } Z_- \neq 0\}$.

Fermes optionnels et semimartingales

Le théorème suivant est une généralisation de Schwartz [8, Proposition (2.4)]. Notre démonstration est inspirée de celle de Schwartz; une démonstration antérieure, publiée dans la note [7], était plus compliquée pour un résultat moins précis.

Théorème 1. *Soit X une semimartingale dans un ouvert optionnel A, et soit F un fermé optionnel contenu dans A. Il existe alors une semimartingale Z telle que $Z = X$ dans un ouvert droit optionnel contenant F.*

Démonstration. A étant supposé optionnel, nous pouvons supposer que X l'est aussi, en le remplaçant par XI_A si nécessaire [inversement, d'ailleurs, si l'on sait à l'avance que X est optionnel, il est inutile de supposer A optionnel: remplacer A si nécessaire par le plus grand ouvert dans lequel X est une semimartingale (cf. plus haut (e)) qui est optionnel et contient A].

Par définition, il existe une suite (Y_n) de semimartingales, une suite (A_n) d'ouverts—que nous pouvons supposer optionnels—telles que $X = Y_n$ dans A_n et que $A = \bigcup_n A_n$. Soit D un ensemble dénombrable dense dans $\mathbb{R}_+$; rangeons en une suite (r_k, n_k) l'ensemble dénombrable $D \times \mathbb{N}$, et désignons par I_k l'intervalle stochastique $[r_k, S_k[$, où

$$S_k = \inf\{t > r_k, t \notin A_{n_k}\}.$$

On a $X = Y_{n_k}$ dans I_k, et les intérieurs des I_k recouvrent A, donc aussi F. Posons $B_p = I_p \backslash (\bigcup_{k<p} I_k)$, $C_p = \bigcup_{k \leqq p} I_k = \bigcup_{k \leqq p} B_k$; B_k est réunion finie d'intervalles stochastiques semi-ouverts, donc I_{B_k} est une semimartingale, et X est égal dans C_p à la semimartingale

$$Z_p = \sum_{k \leqq p} Y_{n_k} I_{B_k}.$$

D'autre part, les intérieurs des C_p croissent et recouvrent F. D'après le théorème de Borel–Lebesgue, le début T_p de $F \backslash C_p^\circ$ tend donc en croissant vers $+\infty$, et le processus

$$Z = Z_0 I_{[0, T_0[} + \sum_{n \geqq 1} Z_n I_{[T_{n-1}, T_n[}$$

est une semimartingale, et on a $X = Z$ sur F. Plus précisément, sur $[T_{n-1}, T_n[$ on a $F \subset C_n^\circ$, et l'ensemble $\{X = Z\} \cap [T_{n-1}, T_n[= \{X = Z_n\} \cap [T_{n-1}, T_n[$ contient $C_n^\circ \cap [T_{n-1}, T_n[$. L'ouvert droit optionnel $\bigcup_n (C_n^\circ \cap [T_{n-1}, T_n[)$ est donc contenu dans $\{X = Z\}$, et contient F. On peut même affirmer que c'est un voisinage de tout point de F distinct des T_n (qui sont en nombre fini sur tout intervalle fini).

Complements au théorème 1

Cette section peut être omise sans inconvénient pour la suite: elle contient en effet une amélioration du théorème 1, dont l'intérêt n'est pas certain, et quelques résultats de théorie des processus qui nous semblent utiles, mais qui sortent de notre sujet principal.

Voici le résultat que nous avons en vue. Avec les notations du théorème 1, *si le fermé F et l'ouvert A sont prévisibles, il existe une semimartingale X telle que $X = Z$ sur un voisinage de F* (et non plus seulement un voisinage droit). Il suffit évidemment pour établir cela de trouver un fermé optionnel G contenu dans A, qui soit un voisinage de F, et d'appliquer le théorème 1 à G. Un tel fermé existe, et l'on a en toute généralité l'énoncé suivant, qui permet de traduire beaucoup de résultats de topologie en résultats de théorie des processus. Citons par exemple le "théorème d'Urysohn": si U et V sont deux fermés prévisibles disjoints, il existe un processus adapté continu qui vaut 1 sur U, 0 sur V.

Théorème 2. *Soient A un ouvert prévisible, F un fermé prévisible contenu dans A. Il existe un fermé prévisible G contenu dans A, qui est un voisinage de F.*

Démonstration. Nous allons construire séparément un fermé aléatoire $G_1 \subset A$ qui est un voisinage droit de F, un fermé $G_2 \subset A$ qui est un voisinage de $F \backslash G_1^\circ$, et il restera à les réunir.

Pour le voisinage droit, nous procédons comme pour le théorème 1. A tout rationnel r nous associons le temps prévisible $S_r = \inf\{t > r, t \notin A\}$. Nous considérons pour tout r une suite $S_{r,n}$ de temps prévisibles[†] annonçant S_r, et rangeons en une suite I_k les intervalles stochastiques prévisibles $[r, S_{r,n} \vee r[$. Puis les B_p, C_p et les temps d'arrêt T_p (T_p est le début de $F \backslash C_p^\circ$) comme dans le théorème 1. Ici, B_p et C_p sont prévisibles, et T_p, début d'un fermé prévisible, est un temps prévisible. Le fermé G_1 sera alors l'adhérence de l'ensemble prévisible

$$J = (C_0 \cap [0, T_0[) \cup \bigcup_{n \geqq 1} (C_n \cap [T_{n-1}, T_n[)$$

Comme J est un fermé droit, G_1 est aussi l'adhérence de J pour la topologie gauche, qui est prévisible.[‡] D'autre part, J est contenu dans A puisque les C_n le sont; lorsqu'on le ferme, on ne fait que lui ajouter certains points appartenant aux graphes $[T_k]$, donc à $F \subset A$. Donc G_1 est bien un fermé prévisible contenu dans A, voisinage droit de F, et $F \backslash G_1^\circ$ est contenu dans la réunion des graphes $[T_k]$.

Il nous reste donc à construire un fermé prévisible G_2, contenu dans A, qui soit un voisinage gauche de la réunion des graphes $[T_k]$. Pour cela, il nous suffit de construire pour chaque k un fermé prévisible $G^k \subset A$, contenu dans $[0, T_k]$, qui soit un voisinage gauche de $[T_k]$, et tel en outre que

$$P\{G^k \subset [T_k - 1, T_k]\} \geqq 1 - 2^{-k}$$

($T_k - 1$ n'est pas un temps d'arrêt). En effet, d'après le lemme de Borel–Cantelli, cette condition entraîne que $G^k \subset [T_k - 1, T_k]$ presque surement (p.s.) pour n grand, donc G^k s'éloigne à l'infini, et la réunion des G_u^k est alors le fermé G_2 cherché.

Nous supprimons maintenant l'indice k de la notation: nous écrivons T pour T_k, G pour G^k; T est un temps prévisible strictement positif dont le graphe passe dans A. Choisissons une suite (R_n) de temps prévisibles annonçant T. L'ensemble $A \cap (\bigcup_n [R_n])$ est prévisible, discret, et admet T comme seul point d'accumulation; quitte à remplacer R_1 par le premier point de cet ensemble, R_2 par le second, etc., nous pouvons supposer que les R_n prennent leurs valeurs dans A. Quitte à remplacer enfin R_n par R_{p+n}, où l'entier p est suffisamment grand, nous pouvons supposer que $P\{R_n < T - 1\} < 2^{-k}$.

[†] Voir Dellacherie [4, Chapitre IV, No 77].

[‡] Voir Dellacherie [4, Chapitre IV, No 91, 1].

Pour chaque n, soit $S_n = \inf\{t > R_n, t \notin A\} \wedge R_{n+1}$; comme A^c est un fermé prévisible, R_{n+1} un temps prévisible, S_n est prévisible. Pour n assez grand, R_n est dans la composante connexe de A contenant T, donc $S_n = R_{n+1}$, de sorte que la réunion $J = [T] \cup (\bigcup_n [R_n, S_n])$ est un fermé prévisible, qui est un voisinage de T pour la topologie gauche. Mais J n'est pas contenu dans A, car sur $\{S_n < T\}$ la valeur S_n n'appartient pas à A. Pour lever cette difficulté, nous poserons $G = [T] \cup (\bigcup_n [R_n, S'_n])$, où les temps prévisibles S'_n seront assujettis aux conditions

$$R_n \leqq S'_n \leqq S_n ; \qquad S'_n < S_n \quad \text{sur } \{S_n < T\}, \quad P\{S'_n < S_n, S_n = T\} < 2^{-n},$$

la dernière condition assurant, d'après le lemme de Borel–Cantelli, que G reste un voisinage gauche de $[T]$. L'existence de tels temps d'arrêt est une conséquence du lemme suivant, que la tradition orale attribue à C. Herz (pour appliquer le lemme, prendre $S = S_n$, $S'_n = R_n \vee U_k$ avec k assez grand).

LEMME 1. *Soient S et T deux temps prévisibles, avec $S \leqq T$. Il existe alors une suite croissante (U_k) de temps prévisibles, telle que*

(1) $\lim_k U_k = S$, $U_k < S$ *pour tout k sur* $\{0 < S < T\}$
(2) *sur* $\{S = T\}$, *on a p.s.* $U_k = S$ *pour k assez grand.*

Demonstration. Il suffit de démontrer ce lemme lorsque $T = +\infty$. En effet, en appliquant ce cas particulier à $S^* = S_{\{S<T\}}$ et $T^* = +\infty$, on construit une suite $(U_k^*) \uparrow S^*$, avec $U_k^* < S^*$ sur $\{S^* < \infty\}$ et $U_k^* = +\infty$ pour k grand sur $\{S^* = \infty\}$. Il ne reste plus qu'à poser $U_k = U_k^* \wedge T$.

Traitons donc le cas où $T = +\infty$. Soit (W_n) une suite de temps prévisibles annonçant S, et soit (M_t) la martingale $P\{S < \infty | \mathscr{F}_t\}$. Posons

$$V_n = W_n \qquad \text{si } M_{W_n -} > \tfrac{1}{2}, \quad V_n = +\infty \quad \text{sinon.}$$

Sur $\{S = \infty\}$ la martingale tend p. s. vers 0, donc $V_n = +\infty$ pour n grand. Sur $\{S < \infty\}$, la martingale tend vers 1, donc $V_n = W_n$ pour n grand. Pour obtenir la suite cherchée, il ne reste plus qu'à poser

$$U_k = \inf_{n \geqq k} V_n.$$

SEMIMARTINGALES DANS UN INTERVALLE STOCHASTIQUE (I)

Dans ce paragraphe, et dans le suivant, nous étudions les semimartingales dans un intervalle stochastique. Il est clair que l'on peut se borner à étudier les intervalles de la forme $[0, T[$ ou $]0, T[$. Dans le premier paragraphe, nous étudions ce qui se passe à l'extrémité droite de l'intervalle, et nous en déduisons des conséquences importantes pour les semimartingales dans des ouverts aléatoires quelconques. Dans le paragraphe suivant, nous examinons ce qui se passe à l'extrémité gauche.

L'énoncé suivant suppose que X est défini sur tout $\mathbb{R}_+$ (il suffirait de connaître X sur $[0, T]$, mais en tout cas la conaissance de X sur $[0, T[$ est insuffisante; si X n'est donné que sur $[0, T[$, on commencera par le prolonger de manière appropriée).

THÉORÈME 3. *Soient T un temps d'arrêt, X un processus optionnel défini sur $\mathbb{R}_+$, qui est une semimartingale dans $[0, T[$. Il existe alors une suite croissante (S_n) de temps d'arrêt, tendant vers T p.s., telle que pour tout n le processus arrêté X^{S_n} soit une semimartingale. On peut supposer en outre que pour tout n, on a $S_n < S_{n+1}$ sur $\{S_n < T\}$.*

Démonstration. Soit $\mathscr{S}$ l'ensemble des temps d'arrêt $S \leqq T$ tels que X^S soit une semimartingale, et soit Σ la borne supérieure essentielle de $\mathscr{S}$. Comme $\mathscr{S}$ est évidemment stable pour l'opération $\vee$, il existe une suite croissante (S_n) d'éléments de $\mathscr{S}$ qui converge vers Σ p.s. Il suffit donc de montrer que $\Sigma = T$ p.s. Montrons que la supposition $P\{\Sigma < T\} > 0$ est absurde.

Par définition des semimartingales dans $[0, T[$, il existe des semimartingales Y^n, des ouverts (optionnels) A^n recouvrant $[0, T[$, tels que $X = Y^n$ dans A^n. Notre hypothése entraîne l'existence d'un n tel que $P\{\Sigma < T, \Sigma \in A^n\} > 0$, puis d'un rationnel r tel que $P\{r < \Sigma, [r, \Sigma] \subset A^n, \Sigma < T\} > 0$, enfin d'un m tel que $P\{r < S_m < \Sigma, [r, \Sigma] \subset A^n, \Sigma < T\} > 0$. Soit

$$D = T \wedge \inf\{t \geqq S_m : t \notin A^n\}.$$

Nous avons $S_m \leqq D \leqq T$, et les conditions précédentes entraînent que $P\{D > \Sigma\} > 0$. D'autre part, on a $X = Y^n$ sur $[S_m, D[$ donc X coïncide sur $[S_m, D]$ avec la semimartingale

$$Z = Y^n I_{[S_m, D[} + X_D I_{[D, \infty[}$$

et alors $X^D = X^{S_m} I_{[0, S_m[} + Z I_{[S_m, \infty[}$ est une semimartingale, ce qui entraîne la conclusion absurde $D \leqq \Sigma$.

Il reste à établir la dernière phrase de l'énoncé. A cet effet, nous remarquons que, si U est un temps d'arrêt tel que, pour tout ω, la valeur $U(\omega)$ figure parmi les valeurs $S_n(\omega)$, alors X^U est une semimartingale. Pour voir cela, posons $A_0 = \{U = S_0\}$, $A_n = \{U = S_n\} \setminus \bigcup_{m<n} \{U = S_m\}$ pour $n > 0$; les A_n forment une partition de Ω. Soit $\lambda_n = P(A_n)$, et soit P_n la loi conditionnelle $P\{\cdot | A_n\}$ si $\lambda_n \neq 0$. Comme $X^U = X^{S_n}$ sur A_n, on a $X^U = X^{S_n} P_n$—p.s., donc X^U est une semimartingale sous P_n (théorème de Girsanov) et, d'après le théorème de convexité de Jacod [5, p. 235], X^U est une semimartingale sous la loi $P = \Sigma_{\lambda_n \neq 0} \lambda_n P_n$.

Ce résultat auxiliaire étant établi, nous énumérons par ordre croissant les points de l'ensemble $H = \bigcup_n [S_n]$:

$$T_0 = \inf[t, t \in H\}, \ldots, T_{n+1} = \inf\{t > T_n, t \in H\}$$

et posons $S'_n = T \wedge T_n$; pour tout ω, la valeur $S'_n(\omega)$ figure parmi les valeurs $S_k(\omega)$, donc $X^{S'_n}$ est une semimartingale, et la suite (S'_n) satisfait à toutes les conditions exigées.

Remarques. (a) On notera que $X_{S_n -}$ existe sur l'ensemble $\{S_n < \infty\}$, donc X_{T-} existe sur l'ensemble $\{S_n = T < \infty\}$: les S_n annoncent toujours T sur l'ensemble où X_{T-} n'existe pas.

(b) Si X n'est donné que sur $[0, T[$, le prolongement le plus naturel pour l'application du théorème 3 consiste à définir, pour $t \geqq T$

$$X_t = X_{T-} \quad \text{si cette limite existe,} \qquad X_t = 0 \quad \text{sinon.}$$

(c) Supposons que T soit totalement inaccessible. Alors le théorème 3 permet d'affirmer que X^T est une semimartingale. En effet, posons $R_n = S_n$ si $S_n < T$, $R_n = +\infty$ sinon. Alors $(X^T)^{R_n} = X^{S_n}$ est une semimartingale. D'autre part, T étant totalement inaccessible la suite (S_n) converge stationnairement vers T, donc $R_n \uparrow +\infty$, et X^T est une semimartingale.

(d) Dans la démonstration du résultat auxiliaire, à la fin de la preuve du théorème 3, la phrase "la valeur $U(\omega)$ figure parmi les valeurs $S_n(\omega)$" peut évidemment être remplacée par "la valeur $U(\omega)$ est majorée par l'une des valeurs $S_n(\omega)$", à condition de remplacer ensuite $\{U = S_n\}$ par $\{U \leqq S_n\}$. En conséquence, si K est l'ensemble prévisible $\bigcup_n [0, S_n]$, X^U est une semimartingale pour tout temps d'arrêt U tel que $[0, U] \subset K$.

Avant de poursuivre, donnons la version du théorème 3, relative à un intervalle $]0, T[$.

THÉORÈME 3'. *Supposons que X soit une semimartingale dans* $]0, T[$. *Alors il existe une suite croissante (S_n) de temps d'arrêt, tendant vers T p.s., et telle que pour tout n X^{S_n} soit une semimartingale dans* $]0, \infty[$. *On peut en outre supposer que $S_n > 0$ sur $\{T > 0\}$, et que $S_n < S_{n+1}$ sur $\{S_n < T\}$.*

Démonstration. Pour donner tout son sens à l'énoncé, remarquons d'abord qu'un processus est une semimartingale sur $]0, \infty[$ si et seulement si c'est une semimartingale sur $[1/n, \infty[$, au sens usuel, pour tout $n > 0$. D'autre part, on ne peut décalquer la démonstration du théorème 3, car rien ne permet d'affirmer que Σ, la borne supérieure des temps S tels que X^S soit une semimartingale sur $]0, \infty[$, n'est pas nulle.

Pour tout k, appliquons le théorème 3 à l'intervalle $[1/k, T \vee 1/k[$: nous en déduisons une suite croissante de temps d'arrêt $S_{kn} \geqq 1/k$, tendant vers $T \vee 1/k$, et telle que pour tout n et tout k $X^{S_{kn}}$ soit une semimartingale sur

$[1/k, \infty[$. Choisissons $n = n(k)$ assez grand pour que $P\{S_{kn} < T - 1/k\} < 2^{-k}$; alors, d'après le lemme de Borel–Cantelli, la suite $S_{kn(k)}$ tend p.s. vers T, avec pour k assez grand $1/k < T$, $T - 1/k < S_{kn(k)} \leqq T$. Si l'on pose

$$S_m = \inf_{k \geqq m} S_{kn(k)}$$

la suite (S_m) tend vers T en croissant, et pour tout m le processus X^{S_m} est une semimartingale dans $[1/k, \infty[$ pour tout k, donc une semimartingale dans $]0, \infty[$. Nous laisserons au lecteur la dernière phrase de l'énoncé.

Maintenant, nous allons déduire du théorème 1 une importante propriété. Comme d'habitude, X est défini dans $\mathbb{R}_+$ tout entier, mais voir les commentaires suivant la démonstration.

THÉORÈME 4. *Soient A un ouvert optionnel, X un processus optionnel, qui est une semimartingale dans A. Il existe une suite croissante (K_n) de fermés* prévisibles, *une suite (Y^n) de semimartingales, telles que*

(1) $X = Y^n$ *sur* K^n,
(2) $A \subset \bigcup_n K_n \subset \bar{A}$, $A \cap K_n \subset \mathring{K}_{n+1}$.

De plus, si X est continu sur $\bar{A}$, on peut supposer chaque Y^n continu dans un voisinage de K_n.

Démonstration. Pour tout u rationnel, posons $T_u = \inf\{t \geqq u, t \in A^c\}$. D'après le théorème 3, il existe une suite croissante de temps d'arrêt $(S_n(u))$, tendant vers T_u, telle que $S_n(u) < S_{n+1}(u)$ sur $\{S_n(u) < T_u\}$, et que pour tout n le processus $X^{S_n(u)} I_{[u, \infty[}$ soit une semimartingale. Nous rangeons alors les rationnels en une suite u_n, et construisons les fermés K_n:

$$K_1 = [u_1 + 1, S_1(u_1)],$$
$$K_2 = [u_1 + \tfrac{1}{2}, S_2(u_1)] \cup [u_2 + \tfrac{1}{2}, S_2(u_2)],$$
$$\vdots$$
$$K_n = \bigcup_{i \leqq n} [u_i + 1/n, S_n(u_i)],$$

où l'on convient qu'un intervalle $[a, b]$ tel que $a > b$ est vide. La suite (K_n) vérifie bien les conditions (2). Il reste à définir convenablement les semimartingales Y^n. Nous posons

$$U_1^n = \inf\{t, t \in K_n\}, \qquad V_1^n = \inf\{t > U_1^n, t \notin K_n\},$$
$$U_2^n = \inf\{t > V_1^n, t \in K_n\} \qquad \ldots$$

et ainsi de suite jusqu'à l'indice n (après lequel la construction ne donnerait plus que des temps d'arrêt infinis, car K_n est réunion de n intervalles fermés).

On a alors

$$K_n = [U_1^n, V_1^n] \cup \cdots \cup [U_n^n, V_n^n], \qquad V_i^n < U_{i+1}^n \quad \text{sur } \{V_i^n < \infty\}$$

La semimartingale cherchée est alors

$$Y^n = \sum_{i \leq n} I_{[U_i^n, U_{i+1}^n[} X^{V_i^n}.$$

La vérification que Y^n est une semimartingale se fait au moyen du théorème de convexité de Jacod, comme dans la démonstration du théorème 3. Comme on a $[U_i^n, V_i^n] \subset [U_i^n, U_{i+1}^n[$, on a $X = Y^n$ sur $[U_i^n, V_i^n]$ pour tout i, donc $X = Y^n$ sur K_n. On peut même dire un peu mieux: si l'on remplace Y^n par $Y'^n = Y^{n+1}$, on a $X = Y'^n$ sur K_{n+1}, donc sur un voisinage de K_n pour la topologie *gauche* (comparer avec le théorème 1, en notant qu'ici K_n est prévisible, et n'est pas contenu dans A).

Supposons enfin que X soit une semimartingale continue; alors Y^n, telle qu'elle est définie plus haut, est continue dans $[U_i^n, V_i^n[$ où elle est égale à X, donc (par arrêt) dans $[U_i^n, U_{i+1}^n[$. Elle est donc continue dans un voisinage *droit* de K_n. Mais alors Y'^n est continue dans un voisinage droit de K_{n+1}, donc dans un voisinage ordinaire de K_n.

COROLLAIRE. *Si X (défini dans $\mathbb{R}_+$) est une semimartingale dans un ouvert optionnel A, X est aussi une semimartingale dans un ouvert gauche prévisible contenant A.*

Demonstration. Soit K'_n l'intérieur de K_n pour la topologie gauche; on a $K_n \subset K'_{n+1} \subset K_{n+1}$, donc $U_n K_n = U_n K'_n$ est un ouvert gauche prévisible, et le théorème 4 montre que X y est une semimartingale.

Remarque. Supposons que X soit défini seulement dans l'ouvert optionnel A. Soit A' la fermeture de A pour la topologie gauche; comme on a

$$I_{A'}(t) = \limsup_{s \uparrow\uparrow t} I_A(s).$$

A' est un ensemble prévisible. En un point t de $A' \backslash A$, prolongeons X en posant

$$X_t = \lim_{s \uparrow\uparrow t} X_s \text{ si cette limite existe et est finie}$$
$$= 0 \qquad \text{sinon}$$

puis prolongeons encore X par 0 hors de A'. Nous prolongeons ainsi X en un processus optionnel sur $\mathbb{R}_+$, auquel on peut appliquer le théorème 4. L'ouvert gauche prévisible $\bigcup_n K_n = \alpha$ est tel que $A \subset \alpha \subset A'$, et en tout point t de $\alpha \backslash A$ on peut affirmer que la limite X_{t-} existe, et que $X_t = X_{t-}$. Il est parfois plus agréable de travailler dans un ouvert gauche prévisible que dans un ouvert optionnel.

Semimartingales dans un intervalle stochastique (II)

Nous nous proposons ici d'étudier le prolongement en 0 d'une semimartingale dans un ouvert de la forme $]0, T[$. Compte tenu du théorème 3′, on ne restreint pas la généralité de manière essentielle en se bornant à l'étude des semimartingales dans $]0, \infty[$. Si X est une telle semimartingale (contrairement à notre habitude, nous ne supposons pas X définie dans $[0, \infty[$ entier), les processus

$$X_t^n = (X_t - X_{1/n}) I_{\{t \geqq 1/n\}}$$

sont des semimartingales ordinaires. Nous pouvons donc associer à X les mesures aléatoires suivantes sur $]0, \infty[$:

— dans tous les cas, par recollement des mesures $d[X^n, X^n]_t$, la mesure $d[X, X]_t$;
— Si les X^n sont spéciales, admettant des décompositions canoniques $M^n + A^n$, les mesures aléatoires dA_t^n se recollent en une mesure aléatoire prévisible dA_t.

Si X est la restriction *à* $]0, \infty[$ d'une semimartingale ordinaire, les conditions suivantes sont évidemment satisfaites.

(a) $\lim_{t \downarrow 0} X_t$ existe p.s. (nous noterons X_0 cette limite).
(b) $\lim_{s \downarrow 0} \int_s^t d[X, X]_u < \infty$ (nous noterons $[X, X]_t$ cette limite, mais avec un abus de notation: normalement $[X, X]_t$ inclut le terme X_0^2).
(c) Pour tout processus prévisible borné H sur $]0, \infty[$, $\lim_{s \downarrow 0} \int_s^t H_u \, dX_u$ existe p.s. (cette intégrale stochastique se définit par un recollement immédiat).

Si X est restriction d'une semimartingale spéciale, on a les deux propriétés supplémentaires:

(d) $[X, X]_t^{1/2}$ (existe et) est localement intégrable.
(e) $\lim_{s \downarrow 0} \int_s^t |dA_u| < \infty$ p.s. (nous noterons alors $A_t = \int_{]0,t]} dA_s$).

Voici d'abord un résultat simple.

Théorème 5. *Pour que X soit la restriction à $]0, \infty[$ d'une semimartingale spéciale, il faut et il suffit que les conditions (d) et (e) soient satisfaites.*

Demonstration. Nous avons déjà dit plus haut que ces conditions sont nécessaires. En sens inverse, on peut se ramener par arrêt au cas où la v.a. $[X, X]_\infty^{1/2}$ est intégrable. Il en est alors de même des v.a. plus petites $[X^n, X^n]_\infty^{1/2}$, donc chaque X^n est spéciale. Soit $X^n = M^n + A^n$ la décomposition canonique de X^n; l'inégalité (due à Yor [13])

$$E[[M^n - M^p, M^n - M^p]_\infty^{1/2}] \leqq 2E[[X^n - X^p, X^n - X^p]_\infty^{1/2}]$$

montre alors que les M^n forment une suite de Cauchy dans $\mathscr{H}^1$, qui converge vers une martingale M nulle en 0. D'après la condition (e), nous pouvons définir le processus à variation finie prévisible $A_t = \int_{0+}^t dA_s$, nul en 0, et poser $Y = M + A$, semimartingale spéciale nulle en 0. Pour tout n, nous avons sur $[1/n, \infty[$

$$X_t - Y_t = X_{1/n} - Y_{1/n}$$

v.a. $\mathscr{F}_{1/n}$-mesurable. Cette v.a. ne pouvant dépendre de n, nous voyons qu'elle est en fait $\mathscr{F}_0$-mesurable, et que c'est la limite X_{0+} de X en 0. Ainsi nous avons $X_t = X_{0+} + M + A$ pour tout t, et le théorème est établi.

Pour les semimartingales quelconques, on obtient alors sans peine un critère simple. On notera que la condition (c′) est plus faible que (c): la convergence p.s. est remplacée par la convergence en probabilité, et on n'y considère que des indicatrices.

THÉORÈME 6. *Soit X une semimartingale dans $]0, \infty[$. Pour que X soit la restriction à $]0, \infty[$ d'une semimartingale, il faut et il suffit que soient satisfaites la condition* (b) (*existence de $[X, X]_t$*) *et la condition*

(c′) *Pour tout ensemble prévisible K, $\lim_{s \downarrow 0} \int_s^t I_K(u)\, dX_u$ existe au sens de la convergence en probabilité.*

Démonstration. Nous savons que ces conditions sont nécessaires. Pour montrer qu'elles sont suffisantes, nous allons nous ramener au théorème 5 par changement de loi. Sans changer de notation, nous remplaçons P par une loi équivalente, de telle sorte que les v.a. $[X, X]_t$ soient intégrables. Alors la condition (d) est satisfaite, les semimartingales X^n sont spéciales, et nous pouvons considérer les processus à variation finie A^n, la mesure prévisible dA, les martingales M^n qui (d'après la démonstration précédente) convergent dans $\mathscr{H}^1$ vers une martingale M nulle en 0. Soit H une densité prévisible de la mesure $|dA|$ par rapport à dA; H étant la différence de deux indicatrices, la condition (c′) nous dit que

$$\int_s^t |dA_u| = \int_s^t H_u\, dA_u = \int_s^t H_u\, dX_u - \int_s^t H_u\, dM_u$$

a une limite finie lorsque $s \downarrow 0$. Donc la condition (e) est satisfaite.

Remarques. (a) On peut montrer (Stricker [10]) que *la condition* (c′) *entraîne la condition* (b), mais ce résultat exige une assez longue démonstration, et nous nous bornons à le mentionner ici.

(b) Voir à la fin de cet article un exemple de semimartingale sur $]0, \infty[$, admettant une limite en 0, telle que $[X, X]_t$ et A_t existent, mais non prolongeable en une semimartingale sur $[0, \infty[$.

MESURES-SEMIMARTINGALES ET INTEGRALES STOCHASTIQUES

Soit X une semimartingale dans un ouvert A, et soit H un processus prévisible défini sur $\mathbb{R}_+$, que nous supposerons borné pour simplifier. Le problème de l'intégrale stochastique consiste à trouver une semimartingale Y dans l'ouvert A, telle que $dY = H\,dX$ (en un sens qui sera précisé plus bas). Nous ne savons pas résoudre ce problème, dont on apercevra aussitôt les difficultés en considérant le cas où $A =]0, \infty[$. Mais nous savons au moins définir quel être mathématique est ce "$H\,dX$", donner un sens à la relation précédente, et utiliser les notions ainsi introduites pour étudier la structure des semimartingales dans un ouvert aléatoire.

Commençons par le cas déterministe. Soit A un sous-ensemble de $\mathbb{R}_+$, qui est une réunion d'intervalles A_i (ouverts, fermés ou semi-ouverts) d'intérieur non vide. On peut alors représenter A comme réunion dénombrable d'intervalles maximaux, qui sont les composantes connexes de A. Nous dirons qu'une fonction définie dans A est *localement constante* si elle est constante dans chaque composante connexe de A. Les êtres que nous allons considérer sont des classes de fonctions définies dans A, modulo les fonctions localement constantes, et il est commode de les considérer comme des mesures simplement additives (m.s.a) de la manière suivante: soit $\mathfrak{I}(A)$ la plus petite classe d'ensembles, stable pour les opérations $\cup$, $\cap$, $\setminus$, contenant les intervalles $]u, v]$ tels que $[u, v] \subset A^{\dagger}$; si f est une fonction réelle[‡] sur A, il existe une mesure simplement additive unique sur $\mathfrak{I}(A)$, notée μ_f ou df, telle que $\mu_f(]u, v]) = f(v) - f(u)$ pour tout intervalle $[u, v] \subset A$, et on a $df = dg$ si et seulement si $f - g$ est localement constante. Inversement, toute m.s.a. sur $\mathfrak{I}(A)$ est de la forme précédente (avec une fonction f qui s'annule, par exemple, au milieu de chaque composante connexe). Les m.s.a. fournissent donc une bonne représentation des classes qui nous intéressent.

Considérons maintenant un ensemble aléatoire A dont les coupes $A(\omega)$ sont du type précédent, et pour tout $\omega \in \Omega$ une m.s.a. $\mu(\omega)$ sur $\mathfrak{I}(A(\omega))$—nous dirons, sans chercher à préciser le langage, que μ est une m.s.a. aléatoire. Nous identifions deux m.s.a. aléatoires μ et μ' telles que $\mu(\omega) = \mu'(\omega)$ pour presque tout ω. Il conviendrait en fait d'ajouter à la définition des m.s.a. aléatoires une condition de mesurabilité, assurant que $\mu = dX$, où X est un processus mesurable défini dans A,[§] mais nous nous bornerons en fait aux

† Si $0 \in A$, adjoindre $\{0\}$ à $\mathfrak{I}(A)$ avec $\mu_f(\{0\}) = 0$.

‡ Si f est càdlàg. dans A, on peut remplacer $\mathfrak{I}(A)$ par la classe des réunions finies d'intervalles (u, v) (de types quelconques) tels que $[u, v] \subset A$. Nous le ferons sans mention spéciale lorsque ce sera utile.

§ Cependant, si l'on définit de manière naturelle la notion de m.s.a. aléatoire μ *adaptée* (par exemple, si $A =]0, \infty[$, cela signifiera que $\mu(]s, t])$ est $\mathscr{F}_t$-measurable pour $s < t$), on ne pourra représenter une telle mesure au moyen d'un processus X *adapté*.

m.s.a. du type suivant, pour lesquelles aucune théorie générale n'est nécessaire:

DEFINITION 2. Soit A un ouvert optionnel. On dit qu'une m.s.a. aléatoire μ est une *mesure-semimartingale* s'il existe des ouverts optionnels A^n recouvrant A à un ensemble évanescent près, des semimartingales Y^n, tels que l'on ait $\mu = dY^n$ dans A^n.

Nous utiliserons cette définition, avec les modifications évidentes, pour des ouverts droits ou gauches. Comme dans la définition 1, on aurait pu considérer des familles $(A^i)_{i \in I}$ et $(Y^i)_{i \in I}$ au lieu de suites.

Se donner une mesure-semimartingale dans A revient à se donner une suite d'ouverts optionnels A^n recouvrant A (à un ensemble évanescent près), une suite de semimartingales Y^n satisfaisant à la condition de compatibilité

(1) pour tout (n, m), pour presque tout ω, on a $dY^n(\omega) = dY^m(\omega)$ dans $A^n(\omega) \cap A^m(\omega)$.

En effet, en remplaçant chaque $Y^n(\omega)$ par 0 sur l'ensemble négligeable

$$N = \{\omega : \exists n, \exists m, dY^n(\omega) \neq dY^m(\omega) \text{ sur } A^n(\omega) \cap A^m(\omega)\},$$

on peut supposer que la condition de compatibilité a lieu sans ensemble exceptionnel, et la construction de la m.s.a. μ est alors ramenée à un problème banal de recollement déterministe. On pourrait même supposer que chaque Y^n est seulement une semimartingale dans A^n (détails laissés au lecteur).

Les mesures semimartingales constituent une généralisation des semimartingales dans les ouverts:

LEMME 2. *Soit A un ouvert optionnel.*

(a) *Soit X un processus optionnel. Alors la m.s.a. aléatoire dX est une mesure-semimartingale si et seulement si X est une semimartingale dans A.*

(b) *Si A est de la forme $[0, T[$, toute mesure-semimartingale μ dans A est de la forme dX, où X est une semimartingale dans A.*

Démonstration. Si X est une semimartingale dans A, il existe des ouverts A^n recouvrant A, des semimartingales Y^n telles que $X = Y^n$ dans A^n. Mais alors $dX = dY^n$ dans A^n, et dX est une mesure-semimartingale. Inversement, s'il existe des A^n recouvrant A et des semimartingales Y^n, tels que $dX = dY^n$ dans A^n, X est égal dans A^n à $Y^n + (X - Y^n)$, où $X - Y^n$ est un processus optionnel localement constant dans A^n. Il suffit donc d'établir le résultat suivant:

Si B est un ouvert optionnel, Z un processus optionnel localement constant dans B, Z est une semimartingale dans B.

La démonstration est très simple: pour tout r rationnel, soit $S_r = \inf\{t > r, t \notin B\}$, et soit B_r l'ouvert $B \cap]r, S_r]$; les B_r recouvrent B, et sur B_r le processus Z est égal à la semimartingale $Z_t^r = Z_r I_{\{t \geqq r\}}$.

Enfin, pour la dernière phrase de l'énoncé il suffit de poser $X_t = \mu(\{0\}) + \mu(]0, t])I_{\{t < T\}}$. Nous laissons le lecteur vérifier que X est optionnel; la relation $dX = \mu$ dans $[0, T[$ est évidente, et (a) entraîne que X est une semimartingale dans A.

Les théorèmes 1 et 4 s'étendent aux mesures-semimartingales, avec les mêmes démonstrations.

THÉORÈME 7. *Soit μ une mesure-semimartingale dans un ouvert optionnel A.*

(a) *Si F est un fermé optionnel contenu dans A, il existe une semimartingale Y telle que $\mu = dY$ dans un ouvert droit contenant F; si F et A sont prévisibles, le mot "ouvert droit" peut être remplacé par "ouvert".*

(b) *Il existe une suite croissante (K_n) de fermés prévisibles contenus dans $\bar{A}$, dont chacun est une réunion finie d'intervalles stochastiques fermés, et une suite (Y^n) de semimartingales, de sorte que*

(1) $\mu = dY^n$ *dans* K_n°,

(2) $A \subset \bigcup_n K_n$, $A \cap K_n \subset K_{n+1}^\circ$.

Il est clair que nous pouvons modifier les semimartingales Y^n, de telle sorte que $Y^n = Y_-^n$ en tout point de $K_n \backslash A$. Alors nous avons $dY^n = dY^m$ dans $K_n \cap K_m$, et nous pouvons prolonger la mesure-semimartingale μ à l'ouvert gauche prévisible $\bigcup_n K_n = \alpha$ (ce prolongement ne comporte pas de masses aux points de $\alpha \backslash A$).

Passons à la définition de l'intégrale stochastique. Soit μ une mesure semimartingale dans l'ouvert optionnel A. Donnons nous des ouverts A^n recouvrant A, des semimartingales Y^n telles que $\mu = dY^n$ dans A^n. Si H est un processus prévisible borné, le caractère local de l'i.s. prévisible entraîne que l'on a $d(H \cdot Y^n) = d(H \cdot Y^m)$ dans $A^n \cap A^m$, d'où par recollement une nouvelle mesure-semimartingale, que nous noterons $H \cdot \mu$.

Cette définition étant posée, nous pouvons établir le théorème suivant, qui élucide bien la structure des mesures-semimartingales.

THÉORÈME 8. *Soit μ une mesure-semimartingale dans un ouvert optionnel A. Il existe un processus prévisible borné H, partout >0 dans A, et une semimartingale Y, tels que l'on ait $dY = H \cdot \mu$ dans A. On peut supposer en outre que Y est portée par A au sens suivant: il existe un ensemble prévisible α tel que $A \subset \alpha \subset \bar{A}$, que $\alpha \backslash A$ soit à coupes dénombrables, que $I_\alpha \cdot Y = Y$, et que Y ne saute pas sur $\alpha \backslash A$.*[†]

[†] Pour le lecteur qui connaît les intégrales stochastiques optionnelles non compensées de Yor [14], cela signifie simplement que $I_A \cdot Y$ existe et vaut Y.

Démonstration. Nous reprenons les ensembles K_n du théorème 7b. Quitte à remplacer K_n par $K_n \cap [0, n]$, Y^n par $I_{K_n} \cdot Y^n$, nous pouvons supposer à la fois que $Y^n = I_{K_n} \cdot Y^n$, et qu'elle est arrêtée a l'instant n. Il existe d'après [11] une loi Q équivalente à P telle que toutes les semimartingales Y^n appartiennent à l'espace de semimartingales $\mathscr{H}^1(Q)$. Posons alors

$$Y = I_{K_0} \cdot Y_0 + \sum_{n \geqq 1} a_n I_{K_n \setminus K_{n-1}} \cdot Y^n,$$

où les constantes a_n ($0 < a_n \leqq 1$) assurent que la série converge normalement dans $\mathscr{H}^1(Q)$. Alors Y est aussi une semimartingale (jusqu'à l'infini) pour la loi P, et l'on a $Y = I_\alpha \cdot Y$, α désignant l'ouvert gauche prévisible $\bigcup_n K_n$ (rappelons que $A \subset \alpha \subset \bar{A}$). Si les Y^n ont été choisies de manière à ne pas sauter sur $K_n \setminus A$, Y ne saute pas sur $\alpha \setminus A$. Enfin, on a $dY = H \cdot \mu$ dans A, en prenant pour H le processus prévisible borné

$$H = I_{K_0} + \sum_{n \geqq 1} a_n I_{K_n \setminus K_{n-1}}$$

qui est partout >0 dans A, et même borné inférieurement par $c_n = \inf_{m \leqq n} a_m$ dans l'ensemble K_n.

Remarque. Soit $J = (1/H) I_{\{H > 0\}}$, et soit $[U, V]$ un intervalle stochastique contenu dans $\alpha = \bigcup_n K_n$. Alors pour tout ω il existe un n tel que $[U(\omega), V(\omega)] \subset K_n(\omega)$, donc $J(\omega)$ est borné sur $[U(\omega), V(\omega)]$. D'après une remarque de Lenglart, le processus prévisible $J I_{]U, V]}$ est localement borne, et on a alors $\mu = J \cdot dY$ dans $]U, V] \cap A$†.

Nous pouvons maintenant donner un sens global à l'intégrale stochastique (i.s.) $G \cdot \mu$, où G est un processus prévisible (que nous supposerons borné pour simplifier).

DÉFINITION 3. Soit μ une mesure-semimartingale dans un ouvert optionnel A, et soit G un processus prévisible borné. On dit que $\int G\mu$ existe s'il existe une semimartingale jusqu'à l'infini (Z_t), un ensemble prévisible α contenant A, tels que $\alpha \setminus A$ soit à coupes dénombrables, que $I_\alpha \cdot Z = Z$, que Z ne saute pas sur $\alpha \setminus A$, et que $dZ = G \cdot \mu$ dans A. On pose alors $\int G \cdot \mu = Z_\infty$.‡

Ces conditions déterminent uniquement Z, et donc $\int G \cdot \mu = Z_\infty$. En effet, soit U la différence de deux semimartingales satisfaisant aux conditions précédentes; comme $dU = 0$ dans A, U n'a pas de saut dans A; elle n'en a pas non plus dans $\alpha \setminus A$, ni dans α^c puisque $I_\alpha \cdot U = U$. Donc U est continue,

† On peut aussi dire quelque chose lorsque $V = D_U = \inf\{t \geqq U, t \notin A\}$. Posant $U_n = \inf\{t \geqq U, t \notin K_n\}$ on a $U_n \uparrow V$, $H^{-1} I_{]U, U_n]}$ est localement borné, et $\mu = H^{-1} I_{]U, U_n]} dY$ dans $]U, U_n]$.

‡ Ici encore, on a une interprétation au moyen des intégrales stochastiques de Yor: $\int G \cdot \mu$ existe si et seulement s'il existe une semimartingale jusqu'à l'infini Z, telle que $G \cdot \mu = dZ$ dans A, et que l'i.s. de Yor $I_A \cdot Z$ existe et soit égale à Z. Alors $\int G \cdot \mu = Z_\infty$.

nulle en 0 puisque U n'a pas de saut en 0. Soit β la fermeture à gauche de A, qui est prévisible; la différence symétrique de α et β est un ensemble à coupes dénombrables sur lequel U ne saute pas, donc $I_{\alpha \Delta \beta} \cdot U = 0$, et $I_\beta \cdot U = U$. Comme $dU = 0$ dans A, U est constant dans toute composante connexe de A, donc de β. On en déduit sans peine que $d[U, U]$ et $d\langle U, U \rangle$ sont nuls dans β, puis partout puisque $I_\beta \cdot U = U$. Donc U n'a pas de partie martingale, c'est un processus à variation finie continu. A nouveau on a $dU = 0$ sur β, $I_\beta \cdot U = U$, donc finalement $U = 0$.

Le théorème suivant (dont la démonstration est facile et ennuyeuse) permet alors de ramener cette intégrale stochastique généralisée à l'intégrale stochastique ordinaire (des processus prévisibles non localement bornés; voir [2,5,12]), de sorte qu'il n'est pas nécessaire de développer une théorie nouvelle.

THÉORÈME 9. *Les notations étant celles des énoncés précédents, pour que l'intégrale* $\int G \cdot \mu$ *existe, il faut et il suffit que l'intégrale stochastique ordinaire* $\int_{[0,\infty]} (G_s/H_s) I_{\{H_s > 0\}}\, dY_s$ *existe,*[†] *et ces deux intégrales sont égales.*

LE THÉORÈME DE CONVEXITÉ DE JACOD

Nous commençons par un lemme, d'un caractère assez intuitif. Soit K un élément de $\mathscr{F}$ de probabilité non nulle, et soit Q la loi P conditionnée par K. Nous désignons par A l'ouvert aléatoire $[0, \infty[\times K$, par $(\mathscr{F}'_t)$ la filtration obtenue en augmentant $(\mathscr{F}_t)$ de tous les ensembles Q-négligeables. Soit enfin X un processus optionnel par rapport à $(\mathscr{F}_t)$.

THÉORÈME 10. *Pour que X soit une semimartingale dans A, il faut et il suffit que X soit une $((\mathscr{F}'_t), Q)$-semimartingale.*

Démonstration. Si X est une semimartingale dans A pour la loi P, X est aussi une semimartingale dans A pour la loi Q, mais A est-Q-indistinguable de $\mathbb{R}_+ \times \Omega$, donc X est une semimartingale ordinaire pour la loi Q.

Inversement, supposons que X soit une semimartingale pour la loi Q. Quitte à arrêter X à un temps fixe, nous pouvons supposer que X est une semimartingale jusqu'à l'infini. Quitte à remplacer P par une loi équivalente, nous pouvons supposer que X est une quasimartingale pour la loi Q, et par la décomposition de Rao, que X est une surmartingale positive pour la loi Q. Alors, si M est la P-martingale $M_t = dQ/dP|\mathscr{F}_t$, on sait que MX est une surmartingale pour la loi P. Il suffit donc de vérifier que $1/M$ est

[†] Nous exigeons que cette i.s. représente une semimartingale sur $[0, \infty]$.

une semimartingale dans A. Or soit $T_n = \inf\{t: M_t \leqq 1/n\}$; on a $A \subset U_n[0, T_n[$, et il suffit devérifier que $1/M$ est une semimartingale dans $[0, T_n[$ pour tout n. Or soit $N_t = M_t$ sur $[0, T_n[$, $N_t = 1$ sur $[T_n, \infty[$; N est une semimartingale bornée inférieurement par $1/n$, donc $1/N$ est une semimartingale par la formule d'Ito, et on a $1/M = 1/N$ dans $[0, T_n[$.

Comme première application de ce lemme, nous étendons aux semimartingales dans un ouvert aléatoire le *théorème de convexité de Jacod*, que nous avons utilisé plusieurs fois dans cet exposé pour les semimartingales ordinaires. Voici nos notations: P est une combinaison convexe $\sum_n \lambda_n P_n$ de lois de probabilité; pour tout n, $(\mathscr{F}_t^n)$ est la filtration $(\mathscr{F}_t)$ augmentée de tous les ensembles P_n-négligeables. Enfin, X est un processus optionnel par rapport à $(\mathscr{F}_t)$, et A est un ouvert aléatoire non nécessairement optionnel.

THÉORÈME 11. *Si X est, pour tout n, une $((\mathscr{F}_t^n), P_n)$-semimartingale dans A, X est aussi une $((\mathscr{F}_t), P)$-semimartingale dans A.*

Démonstration. Par un raisonnement familier, on se ramène au cas où A est un ouvert de la forme $[0, T[$, où T est une v.a. positive (non nécessairement un temps d'arrêt). On peut représenter T comme une réunion dénombrable d'intervalles de la forme $[0, S_i[$, où S_i ne prend que deux valeurs, 0 et $s_i > 0$, et il suffit de démontrer le théorème dans chacun d'eux. Nous omettons l'indice i et posons $K = \{S = s\}$. Soit Q (resp., Q_n) la loi conditionnée de P (P_n) par K; X est une semimartingale pour P_n dans $\mathbb{R}_+ \times K$, donc, d'après le lemme 10, pour tout $r < s$ X^r est une semimartingale pour Q_n. D'après le théorème de convexité de Jacod ordinaire, Q étant une combinaison convexe des lois Q_n, X^r est une semimartingale pour Q. Appliquant à nouveau le lemme 10, on voit que X est une semimartingale pour P dans $[0, S[$.

Une démonstration tout à fait semblable permet d'étendre aux semimartingales dans un ouvert le résultat principal de [11].

THÉORÈME 12. *Soient X un processus optionnel par rapport à $(\mathscr{F}_t)$, A un ouvert aléatoire (non nécessairement optionnel), $(\mathscr{H}_t)$ une filtration, satisfaisant aux conditions habituelles et contenant $(\mathscr{F}_t)$. Si X est une $((\mathscr{H}_t), P)$-semimartingale dans A, X est aussi une $((\mathscr{F}_t), P)$ semimartingale dans A—et donc aussi dans un ouvert $(\mathscr{F}_t)$—optionnel contenant A.*

Démonstration. On se ramène comme ci-dessus au cas où $A = [0, T[$, puis $[0, S[$ où S ne prend qu'une valeur $\neq 0$, et enfin au théorème analogue pour les vraies semimartingales grâce au théorème 10.

Remarque. Il n'y a pas de difficulté à étendre ces deux théorèmes aux mesures-semimartingales.

Deux exemples

Nous donnons ici des exemples de semimartingales dans des ouverts, qui ne sont pas prolongeables. Le premier concerne $]0, \infty[$, le second un ouvert beaucoup plus complexe.

Exemple 1. Considérons une suite (D_k) de v.a. indépendantes, de lois données par

$$\begin{aligned} D_k &= (-1)^k/k && \text{avec probabilité } 1 - 1/k^2 \\ &= (-1)^{k+1}(k - 1/k) && \text{avec probabilité } 1/k^2 \end{aligned}$$

La série $\sum_k D_k$ converge p.s. d'après le lemme de Borel–Cantelli, de même que la série $\sum_k D_k^2$. Nous pouvons donc poser $X_t = \sum_{k>1/t} D_k$, processus càdlàg. sur $]0, \infty[$, adapté à la famille $\mathscr{F}_t = \sigma(D_k, k > 1/t)$. Pour tout n, le processus $X_t^n = (X_t - X_{1/n})I_{(t \geqq 1/n)}$ est une martingale de carré intégrable, donc X est une semimartingale sur $]0, \infty[$. La limite X_0 existe et vaut 0, la limite $[X, X]_\infty$ existe et vaut $\sum_k D_k^2$, les mesures dA^n sont nulles—cependant, les conditions du théorème 5 ne sont pas satisfaites ($[X, X]$ n'est pas localement intégrable), et en fait X n'est pas la restriction d'une semimartingale sur $[0, \infty[$, car le processus prévisible borné H qui vaut $(-1)^k$ sur $]1/k + 1, 1/k]$ est tel que $\int_s^1 H_s dX_s \underset{s \to 0}{\to} + \infty$.

Remarque. Cet exemple (inspiré d'un exemple analogue de Lepingle et Mémin) a une signification intéressante en temps discret. On sait qu'un processus $(Y_n)_{n \in \mathbb{N}}$ est une martingale locale par rapport à une famille $(\mathscr{F}_n)_{n \in \mathbb{N}}$ si et seulement si, pour tout n, $E[Y_{n+1} | \mathscr{F}_n]$ existe et vaut Y_n. Cette caractérisation n'est plus valable pour des processus indexés par $-\mathbb{N}$. En effet, posons ici pour $n \leq 0$

$$Y_n = \sum_{k \geqq -n} D_k, \qquad \mathscr{F}_n = \sigma(D_k, k \geqq -n)$$

Alors la différence $Y_{n+1} - Y_n$ est intégrable et orthogonale à $\mathscr{F}_n$ pour tout n, mais on ne peut trouver aucun temps d'arrêt T tel que Y^T soit une martingale uniformément intégrable.

Exemple 2. Soit (B_t) un mouvement brownien réel issu de 0, relativement à sa filtration naturelle $(\mathscr{F}_t)$, rendue continue à droite et complétée. Nous désignons par (L_t) le processus du temps local de (B_t) en 0, par $(\mathscr{H}_t)$ la filtration obtenue en adjoignant à $\mathscr{F}_0$ toutes les v.a. L_t (ainsi $\mathscr{H}_t = \bigcap_{s>t} (\mathscr{F}_s \vee \sigma(L_u, u \in \mathbb{R}_+))$. Jeulin a établi le résultat suivant: pour qu'une martingale locale M de la filtration $(\mathscr{F}_t)$ reste une semimartingale pour $(\mathscr{H}_t)$, il faut et il suffit que le processus croissant $\int_0^t (1/|B_s|)|d[M, B]_s|$ soit à valeurs finies. Il en résulte que (B_t) n'est plus une semimartingale par rapport à $(\mathscr{H}_t)$, mais que $N_t = \int_0^t B_s\,dB_s = \frac{1}{2}(B_t^2 - t)$ est une semimartingale.

Nous allons exprimer ce résultat dans notre langage, et montrer que (B_t) est une semimartingale par rapport à $(\mathscr{H}_t)$ dans l'ouvert prévisible $A = \{B \neq 0\}$, qui est *maximal*: tout ouvert aléatoire A' dans lequel B est une semimartingale est contenu dans A, à un ensemble évanescent près. Cet exemple est assez suggestif: B est continu sur tout $\mathbb{R}_+$, la mesure $d[B, B]_s = ds$ se prolonge bien à $\mathbb{R}_+$, mais la "partie à variation finie" de la mesure-semimartingale dB_s dans A est égale (d'après Jeulin) à ds/B_s, et elle diverge aux extrémités des composantes connexes de A.

Pour tout $\varepsilon > 0$, soit $]R_n^\varepsilon, T_n^\varepsilon[$ le n-ième intervalle contigu au fermé $\{B = 0\}$ dont la longueur dépasse strictement ε; on sait que T_n^ε et $S_n^\varepsilon = R_n^\varepsilon + \varepsilon$ sont des temps d'arrêt de $(\mathscr{F}_t)$—d'ailleurs prévisibles, comme tous les temps d'arrêt de cette filtration. Soit $T_{np}^\varepsilon = \inf\{t > S_n^\varepsilon, |B_t| \leqq 1/p\} < T_n^\varepsilon$; le processus $K = (1/B)I]S_n^\varepsilon, T_{np}^\varepsilon]$ est prévisible/$(\mathscr{F}_t)$, borné par p, et l'intégrale $\int_0^t K_s\,dN_s$ calculée pour $(\mathscr{F}_t)$ vaut $B_t - B_{S_n^\varepsilon}$ sur $]S_n^\varepsilon, T_{np}^\varepsilon]$. Comme (N_t) est aussi une semimartingale/$(\mathscr{H}_t)$, on a le même résultat dans cette filtration, et il en résulte que B est une semimartingale/$(\mathscr{H}_t)$ dans chaque intervalle $]S_n^\varepsilon, T_{np}^\varepsilon[$, donc dans leur réunion, qui est A.

Pour établir le caractère maximal de A, nous faisons les remarques suivantes. Dans la filtration $(\mathscr{H}_t)$, l'ensemble des extrémités gauches des composantes connexes (ensemble où L est croissant à gauche, mais non à droite) est $\mathscr{H}_0$-mesurable. Comme il est à coupes dénombrables, il peut être résolu en une réunion dénombrable de graphes de v.a. $\mathscr{H}_0$-mesurables R_n. Nous posons $T_n = \inf\{t > R_n, B_t = 0\}$, qui est un temps d'arrêt de $(\mathscr{H}_t)$. Soit A' un ouvert dans lequel B (donc aussi $|B|$) est une semimartingale/$(\mathscr{H}_t)$, et soit, pour r rationnel

$$D_r = \inf\{t > r, t \notin A'\}$$

D'après le théorème 3, il existe des temps d'arrêt U_k de la filtration $(\mathscr{H}_t)$, tels que $|B|_{t \wedge U_k} - |B|_{t \wedge r}$ soit une semimartingale/$(\mathscr{H}_t)$ et que $r \leqq U_k \uparrow D_r$. D'autre part, soit J un processus prévisible/$(\mathscr{F}_t)$, borné; on a

$$\int_r^{U_k} J_s\,d|B|_s = Y_{U_k} - Y_r$$

où l'i.s. de gauche est prise dans la filtration $(\mathscr{H}_t)$, tandis que $Y = J \cdot |B|$ dans la filtration $(\mathscr{F}_t)$. En particulier, prenant $J = I_{\{B=0\}}$, on a

$$\int_r^{U_k} I_{\{B=0\}}\,d|B|_s = L_{U_k} - L_r.$$

Mais d'autre part on a, la série étant convergente en probabilité

$$\int_r^{U_k} I_{\{B\neq 0\}}\,d|B|_s = \sum_n \int_{]r,U_k]\cap]R_n,T_n[} d|B|_s = |B|_{U_k} - |B|_r$$

car la contribution de toutes les composantes connexes contenues dans $]r, U_k]$ est nulle. Par différence, on obtient

$$\int_r^{U_k} I_{\{B \neq 0\}} d|B|_s = 0$$

et donc $L_{U_k} - L_r = 0$. Comme le support du temps local est exactement l'ensemble $\{B = 0\}$, on voit que B ne s'annule pas sur $]r, U_k[$, et non plus sur $]r, D_r[$. Finalement, B ne s'annule pas sur $A' = \bigcup_r]r, D_r[$, et on a bien $A' \subset A$. Cette démonstration est inspirée de la note [1] de Barlow.

Mesures martingales locales dans $]0, +\infty[$

Les résultats de ce paragraphe sont inspirés de résultats récents de Sharpe, que nous généralisons légèrement.

Soit μ une mesure semimartingale dans l'ouvert $A =]0, +\infty[$ (cet ouvert reste fixé dans tout le paragraphe, et nous ne le mentionnons plus explicitement). Nous dirons que μ est une *mesure-martingale locale* s'il existe une suite (X^n) de martingales locales, une suite de temps d'arrêt (S_n) tendant vers 0, telles que l'on ait $\mu = dX^n$ dans $]S_n, +\infty[$.

THÉORÈME 13. *Soit μ une mesure martingale locale. Il existe une martingale X appartenant à $\mathscr{H}^1$ et un processus prévisible H partout >0, tel que pour tout temps d'arrêt $U > 0$ le processus $H^{-1}I_{]U,\infty[}$ soit localement borné, et que l'on ait $\mu = H^{-1}\,dX$ dans $[U, \infty[$.*

Démonstration. Nous reprenons les notations précédant l'énoncé. Il existe des temps d'arrêt T_n tendant vers $+\infty$ tels que $(X^n)^{T_n}$ appartienne à $\mathscr{H}^1$. Posons alors $K^N = I_{]S_n, T_n]}$, et $X = \sum_n a_n I_{K_n \setminus K_{n-1}} \cdot X^n$, où les constantes a_n $(0 < a_n \leqq 1)$ sont choisies telles que cette série converge normalement dans $\mathscr{H}^1$. Nous prenons pour H le processus prévisible borné $\sum_n a_n I_{K_n \setminus K_{n-1}}$; il est clair que $H\mu = dX$ sur $]0, +\infty[$. D'autre part, H est borné inférieurement par $c_n = \inf_{m \leqq n} a_m$ sur $K_n =]S_n, T_n]$; si U est un temps d'arrêt strictement positif, pour n assez grand $]S_n, T_n]$ contient $[U, T_n]$, donc le processus $H^{-1}I_{[U,\infty]}$ est localement borné, et on peut alors écrire $\mu = H^{-1}\,dX$ dans $[U, \infty[$.

COROLLAIRE. *Si U est un temps d'arrêt strictement positif, le processus $Y_t = \mu(]U, t])$ est une martingale locale.*

Nous nous demandons maintenant à quelle condition il existe une martingale locale N sur $[0, +\infty[$ telle que $\mu = dN$ dans $]0, +\infty[$. Nous dirons dans ce cas que μ est *prolongeable en O*. Les résultats les plus significatifs (dus à Sharpe pour l'essentiel) concernent le cas où μ ne charge aucun graphe de temps d'arrêt—nous dirons dans ce cas que μ est *continue.*

THÉORÈME 14. *Soit μ une mesure martingale locale. Pour que μ soit prolongeable en 0, il faut et il suffit que le processus croissant*

$$Z_t = \sup_{s \leqq t} |\mu(]s, t])|$$

soit localement intégrable. Si μ est continue, il faut et il suffit qu'il existe une suite (T_n) de temps d'arret >0, tendant vers 0, tel que l'on ait

$$\sup_{n,\, s \leqq t} \mu(]T_n, s])I_{\{T_n \leqq s\}} < +\infty \quad \textit{p.s. pour tout } t.^\dagger$$

Démonstration. La première condition est évidemment nécessaire. Pour voir qu'elle est suffisante, nous choisissons des temps d'arrêt $R_p \uparrow +\infty$ tels que Z_{R_p} soit intégrable. Fixons provisoirement p et posons

$$X_t^n = \mu(]1/n, t \wedge R_p])I_{\{t \wedge R_p \geqq 1/n\}}$$

qui est une martingale dont la norme *maximale* dans $\mathscr{H}^1$ est bornée, indépendamment de n, par $E[Z_{R_p}]$; comme les martingales $X^{n+1} - X^n$ sont deux à deux orthogonales au sens du lemme 3 ci-dessous, ce lemme entraîne que la suite (X^n) converge dans $\mathscr{H}^1$ vers une martingale X, nulle en 0, telle que $dX = \mu$ sur $]0, R_p]$. Notant maintenant N_p cette martingale, on vérifie immédiatement que les N_p se recollent en une martingale locale N, nulle en 0, telle que $dN = \mu$ partout.

Supposons maintenant μ continue; alors le processus croissant (Z_t) est continu, et dire qu'il est localement intégrable revient à dire qu'il est à valeurs finies. Posons cette fois

$$X_t^n = \mu(]T_n, t])I_{\{t \geqq T_n\}}.$$

Si Y est un processus, rappelons la notation usuelle Y_t^* pour $\sup_{s \leqq t} |Y_t|$, et introduisons la notation analogue $\bar{Y}_t$ pour $\sup_{s \leqq t} Y_t$ (sans valeur absolue). Rappelons l'inégalité latérale de Burkholder (dont une démonstration assez simple figure dans Yor [15]): si Y est une martingale locale continue, nulle en 0, et si α est un exposant tel que $0 < \alpha < 1$, on a

$$E[Y_t^{*\alpha}] \leqq c_\alpha E[\bar{Y}_t^\alpha].$$

Ici, l'hypothèse de l'énoncé signifie que le processus croissant continu

$$H_t = \sup_n \bar{X}_t^n$$

† Il suffit en fait que cela ait lieu pour *un* $t > 0$.

est à valeurs finies. Posons $R_p = \inf\{t : H_t \geqq p\}$. L'inégalité précédente entraîne que $E[(X^{n*}_{R_p})^\alpha \leqq c_\alpha p^\alpha$ pour tout n, et alors, d'après le lemme de Fatou

$$\liminf_n X^{n*}_{R_p} < \infty \quad \text{p.s.}$$

Mais soient s et t tels que $0 < s < t \leqq R_p$; dès que $T_n < s$ on a $|\mu(]s, t])| \leqq |\mu(]T_n, s])| + |\mu(]T_n, t])| \leqq 2X^{n*}_{R_p}$, et on a donc

$$Z_{t \wedge R_p} \leqq 2 \liminf_n X^{n*}_{R_p}.$$

Faisant tendre p vers l'infini, on voit que le processus Z est à valeurs finies, et cela achève la démonstration.

Nous dégageons le lemme suivant, qui est presque évident, extrêmement utile, mais ne semble pas être très connu (bien qu'il soit si souvent utilisé dans le cas de $\mathscr{H}^2$).

LEMME 3. *Soit (Z^n) une suite de martingales*[†] *appartenant à $\mathscr{H}^p$ $(1 \leqq p < \infty)$ et telles que $[Z^n, Z^m] = 0$ pour $m \neq n$. Pour que la suite $X^n = Z^1 + \cdots + Z^n$ converge fortement dans $\mathscr{H}^p$, il faut et il suffit que $\|X^n\|_{\mathscr{H}^p}$ soit borné en n.*

Démonstration. Supposons que $\|X^n\|_{\mathscr{H}^p}$ reste borné. Comme $[X^n, X^n]_\infty = \sum_{k=1}^n [Z^n, Z^n]_\infty$ croît avec n, cela signifie que la v.a. $U = (\sum_k [Z^k, Z^k]_\infty)^{1/2}$ appartient a L^p. Alors on a

$$\|X^{n+k} - X^n\|_{\mathscr{H}^p} = \left\| \left(\sum_{i=n+1}^{n+k} [Z^i, Z^i]_\infty \right)^{1/2} \right\|_{L^p} \leqq \left\| \left(\sum_{i=n+1}^{\infty} [Z^i, Z^i]_\infty \right)^{1/2} \right\|_{L^p}$$

qui tend vers 0 par convergence dominée lorsque $n \to +\infty$. Les X^n forment donc une suite de Cauchy dans $\mathscr{H}^p$. L'implication inverse est évidente.

Nous considérons maintenant une *martingale locale continue* (M_t) *sur* $]0, +\infty[$, c'est à dire un processus adapté, continu, tel que la mesure dM soit une mesure-martingale locale—ou encore, que pour tout temps d'arrêt $U > 0$, le processus $(M_t - M_U)I_{\{t \geqq U\}}$ soit une martingale locale ordinaire (cf. le corollaire du théorème 13). Nous pouvons redémontrer rapidement l'un des principaux résultats de Sharpe [9], comme corollaire du théorème 14.

THÉORÈME 15. *Soit M une martingale locale continue sur* $]0, +\infty[$. *Alors*

(1) *Soit A l'ensemble des $\omega \in \Omega$ tels que la limite $M_{0+}(\omega)$ existe (dans $\mathbb{R}$). Soit B l'ensemble $\{\int_0^1 d[M, M]_s < \infty\}$. Alors A est $\mathscr{F}_0$-mesurable, on a $A = B$*

[†] L'extension aux semimartingales de la classe $\mathscr{H}^p$ est immédiate.

p.s. et le processus $I_A M$ est prolongeable en une martingale locale continue sur $[0, +\infty[$.

(2) *Pour presque tout* $\omega \in A^c$, *l'une des trois éventualités suivantes est réalisée*:

$$\lim_{t\to 0} M_t(\omega) = -\infty; \qquad \lim_{t\to 0} M_t(\omega) = +\infty;$$

$$\liminf_{t\to 0} M_t(\omega) = -\infty \quad \text{et} \quad \limsup_{t\to 0} M_t(\omega) = +\infty.$$

Démonstration. (1) Nous avons défini plus haut la mesure aléatoire $d[M, M]$ pour une semimartingale dans $]0, +\infty[$. Pour tout n, on a $B = \{\int_0^{1/n} d[B, B]_s < +\infty\}$, donc B est $\mathscr{F}_{1/n}$-mesurable, et finalement $\mathscr{F}_0$-mesurable. Pour montrer que $B \subset A$ p.s., nous pouvons supposer (en nous restreignant à B si nécessaire) que $B = \Omega$, et il faut montrer qu'alors $A = \Omega$ p.s., autrement dit que M_{0+} existe p.s.. Par arrêt, nous pouvons supposer que $\int_0^1 d[M, M]_s$ est borné. On vérifie alors aussitôt (comme dans la démonstration précédente) que la suite des martingales $M_t^n = (M_t - M_{1/n}) I_{\{t \geqq 1/n\}}$ est bornée dans $\mathscr{H}^2$, donc forme une suite de Cauchy dans $\mathscr{H}^2$, et converge vers une martingale M' telle que $M_t - M_s = M'_t - M'_s$ pour $0 < s < t$; comme M'_{0+} existe, il en est de même de M_{0+}.

En sens inverse, on peut supposer que $A = \Omega$, puis (par soustraction de M_{0+} et arrêt) que M est *bornée* en valeur absolue par une constante α. Alors les martingales locales M_t^n sont bornées par 2α et, toujours par le même raisonnement, convergent dans $\mathscr{H}^2$ vers une martingale M' sur $[0, \infty[$; comme on a $d[M, M] = d[M', M']$, on voit que $B = \Omega$ p.s., le résultat cherché.

(2) Tout revient à montrer que, pour tout nombre réel a, l'ensemble

$$C = \left\{\liminf_t M_t = -\infty, \qquad a < \limsup_t M_t < +\infty\right\}$$

a une probabilité nulle. Comme C appartient à $\mathscr{F}_0$, si $P(C) > 0$ on peut conditionner par C et supposer que $P(C) = 1$. On a donc p.s. $\sup_{s \leqq 1} M_s < +\infty$ p.s. Posons $T_n = \inf\{t \geqq 1/n : M_t = a\}$; alors (T_n) est une suite décroissante de temps d'arrêt et l'on a en désignant par μ la mesure martingale locale associée à M

$$\sup_{n,\, s \leqq 1} \mu(]T_n, s]) I_{\{T_n \leqq s\}} < +\infty \quad \text{p.s.}$$

Alors le théorème 14 entraîne

$$\sup_{s \leqq 1} |\mu(s, 1)| < +\infty \quad \text{p.s.}$$

ce qui est absurde.

EXEMPLES. Soit (B_t) un mouvement brownien plan. La martingale locale $M_t = f(B_t)$ sur $]0, \infty[$ tend vers $-\infty$ lorsque $t \to 0$ si $f(z) = \log|z|$, oscille entre $-\infty$ et $+\infty$ lorsque $f(z) = \mathscr{R}(1/z)$. Enfin, la mesure martingale locale $(B_t^1\, dB_t^2 - B_t^2\, dB_t^1)/|B_t|^2$ non seulement n'est pas prolongeable en 0, mais n'est pas de la forme dM_t pour un processus adapté M.

RÉFÉRENCES

1. M. BARLOW, On the left endpoints of brownian excursions, *Sém. Probab.*, *XIII*, Lecture Notes in Mathematics, No. 721, p. 646, Springer-Verlag, Berlin and New York, 1978.
2. C. S. CHOU, P. A. MEYER, ET C. STRICKER, Sur les intégrales stochastiques de processus prévisibles non bornés, *Sém. Probab. XIV*, Lecture Notes in Mathematics, No. 784, pp. 128–139, Springer-Verlag, Berlin and New York, 1980.
3. C. DELLACHERIE, Sur l'existence de certains ess. inf de familles de processus mesurables, *Sém. Probab.*, *XII*, Lecture Notes in Mathematics, No. 649, p. 512, Springer-Verlag, Berlin and New York, 1978.
4. C. DELLACHERIE ET P. A. MEYER, "Probabilités et Potentiel A," Hermann, Paris, 1975.
5. J. JACOD, "Calcul stochastique et problèmes de martingales," Lecture Notes in Mathematics, No. 714, Springer-Verlag, Berlin and New York, 1979.
6. T. JEULIN, Semimartingales et grossissement d'une filtration, Lecture Notes in Mathematics, No. 833, Springer-Verlag, Berlin and New York, 1980.
7. P. A. MEYER, Sur un résultat de L. Schwartz, *Sém. Probab.*, *XIV*, Lecture Notes in Mathematics, No. 784, pp. 102–103, Springer-Verlag, Berlin and New York, 1980.
8. L. SCHWARTZ, "Semimartingales à valeurs dans les variétés, martingales conformes à valeurs dans les variétés complexes," Lecture Notes in Mathematics, No. 780, Springer-Verlag, Berlin and New York, 1980.
9. M. J. SHARPE, Local time and singularities of continuous local martingales, *Sém. Probab.*, *XIV*, Lecture Notes in Mathematics, No. 784, pp. 76–101, Springer-Verlag, Berlin and New York, 1980.
10. C. STRICKER, Prolongement des semi-martingales, *Sém. Probab. XIV*, Lecture Notes in Mathematics, No. 784, pp. 104–111, Springer-Verlag, Berlin and New York, 1980.
11. C. STRICKER, Quasimartingales, martingales locales et changements de probabilités, *Z.* Warsch. Verw. Gebiete **39** (1977), 55–64.
12. J.-A. YAN, Remarques sur l'intégrale stochastique de processus prévisibles non bornés, *Sém. Probab.*, *XIV*, Lecture Notes in Mathematics, No. 784, pp. 148–151, Springer-Verlag, Berlin and New York, 1980.
13. M. YOR, Remarques sur les normes H^p de semimartingales, *C. R. Acad. Sci. Paris Sér. A* **287** (1978), 461–464.
14. M. YOR, En cherchant une définition naturelle des intégrales stochastiques optionnelles, *Sém. Probab.*, *XIII*, Lecture Notes in Mathematics, No. 721, pp. 407–426, Springer-Verlag, Berlin and New York, 1979.
15. M. YOR, Les inégalités de sousmartingales comme conséquences de la relation de domination, *Stochastics* **3** (1979), 1–17.

MATHEMATICAL ANALYSIS AND APPLICATIONS, PART B
ADVANCES IN MATHEMATICS SUPPLEMENTARY STUDIES, VOL. 7B

Multiplication of Distributions

YVES MEYER

Département de Mathématiques
Université de Paris-Sud
Orsay, France

DEDICATED TO LAURENT SCHWARTZ

INTRODUCTION

We all know that multiplication of distribution does not make sense. However, there are two cases where two distributions can be multiplied.

The first one is classical: it is the case when the two distributions are, in some sense, holomorphic ones.

The second one is new and due to Bony [1]. The usual definition of the product of two functions is slightly modified into the so-called paraproduct which extends to the distributional case.

In fact, these two points of view are intimately related, at least at the proofs level, as we are going to show.

Calderón's theorem is stated (without proof) in the first section. Two proofs are given in Sections 2 and 3 (by complex interpolation or using the Littlewood–Paley theory).

In the fourth section the paraproduct of functions is defined and studied and in the two last sections we show that this notion extends to the distributional case. The spaces of distributions for which estimates are given will be the space L^p_α $(\alpha \leq 0)$ (generalizing the Sobolev spaces of negative orders) and the generalized Hardy classes H^p $(0 < p < 1)$.

1. CALDERÓN'S THEOREM

Some notations need to be fixed for stating Calderón's theorem. The space L^p is the usual $L^p(\mathbb{R}^n; dx)$ where dx is the Lebesgue measure and $1 \leq p \leq +\infty$. For a real α, L^p_α means the Banach space of all distributions $S \in \mathscr{S}'(\mathbb{R}^n)$ for which $(I - \Delta)^{\alpha/2} S \in L^p(\mathbb{R}^n)$ equipped with the norm

$$\|(I - \Delta)^{\alpha/2} S\|_p; \Delta = \partial^2/\partial x_1^2 + \cdots + \partial^2/\partial x_n^2.$$

ISBN 012-512802-9

When $\alpha = k \in \mathbb{N}$, L_k^p is the space of all functions $f \in L^p$ all of whose derivatives of order less than or equal to k belong to L^p [7].

When $p = 2$ and $\alpha = s \in \mathbb{R}$, $L_\alpha^p = H^s$, the usual Sobolev space.

Finally, if $\alpha < 0$, L_α^p is a space of distributions. If $1 < p \leq +\infty$ and q is the dual exponent, L_α^p is the dual space of $L_{-\alpha}^q$.

The Sobolev injection theorem gives the following result. Assume that $1 \leq p < q < +\infty$ and that $\varepsilon > 0$ is defined by $p^{-1} - q^{-1} = \varepsilon/n$; then $L_\alpha^p \subset L_{\alpha-\varepsilon}^q$.

When $\alpha > 0$ and $\beta \geq 0$, Hölder's inequality generalizes into the following: If $f \in L_\alpha^p$ and $q \in L_\beta^q$, then the product $fg \in L_\gamma^r$ with $\gamma = \inf(\alpha, \beta)$ and $r^{-1} = p^{-1} + q^{-1} - (|\alpha - \beta|/n)$ (p, q, and r being assumed to belong to $]1, +\infty[$).

We are going to "multiply" distributions in such a way that the product of a distribution belonging to L_α^p, $\alpha \leq 0$, and a distribution in L_β^q, $\beta \leq 0$, is a distribution in $L_{\alpha+\beta}^r$ where $r^{-1} = p^{-1} + q^{-1}$. Obviously the usual product of functions in $\mathscr{S}(\mathbb{R}^n)$ does not extend to such an operation.

DEFINITION 1. Let $\Gamma \subset \mathbb{R}^n$ be a cone defined by $|\xi| \leq M\xi_n$ where $M > 1$ is fixed and $|\xi| = (\xi_1^2 + \cdots + \xi_n^2)^{1/2}$.

A distribution $S \in \mathscr{S}'(\mathbb{R}^n)$ is called Γ-*holomorphic* if its Fourier transform $\hat{S}$ is supported by Γ.

The following statement is essentially due to Calderón [2].

THEOREM 1. *For $1 < p < +\infty$, $1 < q < +\infty$ and $1 \leq r < +\infty$ related by $r^{-1} = p^{-1} + q^{-1}$, for $\alpha \leq 0$, $\beta \leq 0$ and Γ defined above, there exists a constant C such that the following be true*

if $S \in L_\alpha^p$ and $T \in L_\beta^q$ are two Γ-holomorphic distributions, their product ST makes sense and belongs to $L_{\alpha+\beta}^r$. (1)

To be more specific, let $L_{\alpha,\Gamma}^p$ be closed subspace of L_α^p consisting in Γ-holomorphic distributions. Then there exists a continuous bilinear mapping

$$P_\Gamma : L_{\alpha,\Gamma}^p \times L_{\beta,\Gamma}^q \mapsto L_{\alpha+\beta,\Gamma}^r \tag{2}$$

such that

$$P_\Gamma(f, g) = fg,$$

when $f \in \mathscr{S}(\mathbb{R}^n)$ and $g \in \mathscr{S}(\mathbb{R}^n)$ are Γ-holomorphic. The statement given by Calderón corresponds to the case $n = 1$, $\Gamma = [0, +\infty[$, $p = q = 2$, $r = 1$, $\alpha = -1$, $\beta = 0$. We are led to three holomorphic functions $f(z)$, $g(z)$, and $h(z)$ in $\operatorname{Im} z > 0$ related by $h'(z) = f'(z)g(z)$ and tending to 0 as $\operatorname{Im} z \to +\infty$. Then $\|h\|_1 \leq C\|f\|_2\|g\|_2$. These L^p-norms are defined as

$$\sup_{y>0} \|f(x + iy)\|_{L^p(dx)}.$$

There exists a more refined statement in which p, q, and r are allowed to be any real numbers in $]0, +\infty[$. Since L^p is no longer a space of distributions when $0 < p < 1$ a substitute is needed: the generalized Hardy classes H^p. We are going to recall classical facts about H^p-spaces.

DEFINITION 2. If $p \in]0, +\infty[$ and $\varphi \in \mathscr{S}(\mathbb{R}^n)$, has integral 1, H^p is defined as the *collection of all* $S \in \mathscr{S}'(\mathbb{R}^n)$ for which $M_\varphi(S) \in L^p(\mathbb{R}^n)$ where $M_\varphi(S)(x) = \sup_{t>0} |S * \varphi_t|$ and $\varphi_t|$ and $\varphi_t(x) = t^{-n}\varphi(x/t)$.

In fact, the definition of H^p does not depend on the choice of φ and for $p > 1$, $H^p = L^p$. For $p = 1$, $H^1 \subset L^1$ and for $0 < p < 1$, H^p is a metric space where the distance between f and 0 is $\|M_\varphi(f)\|_p^p$ [6].

Let Γ be a cone as in Definition 1. Then for each $p > 0$ there exists a constant $C = C(p, \Gamma)$ such that for each Γ-holomorphic function $f \in \mathscr{S}(\mathbb{R}^n)$, one has

$$\|f\|_p \leq \|M_\varphi(f)\|_p \leq C\|f\|_p.$$

Some classical facts about pseudo-differential operators (ψdo) should be recalled.

The class $S^0_{1,0}$ consists of functions $\sigma(x,\xi) \in C^\infty(\mathbb{R}^n \times \mathbb{R}^n)$ such that

$$|\partial_\xi^\alpha \partial_x^\beta \sigma(x, \xi)| \leq C_{\alpha,\beta}(1 + |\xi|)^{-|\alpha|} \qquad \text{for} \quad \alpha \in \mathbb{N}^n, \quad \beta \in \mathbb{N}^n$$

and the corresponding ψdo. $\sigma(x, D)$ is defined by

$$\sigma(x, D)f(x) = (2\pi)^{-n} \int_{\mathbb{R}^n} e^{ix\xi}\sigma(x, \xi)\hat{f}(\xi)\, d\xi.$$

For $p \in]0, +\infty[$ and $\sigma \in S^0_{1,0}$ there exists a C such that

$$\|\sigma(x, D)f\|_p \leq C\|M_\varphi(f)\|_p.$$

Moreover, if the symbol $\sigma(x, \xi)$ does not depend on x, the corresponding ψdo maps H^p into itself.

If $p \in]0, 1]$, $\mathscr{H}^p_\alpha$ *is the metric space consisting of all distributions* $S \in \mathscr{S}'(\mathbb{R}^n)$ *such that* $(I - \Delta)^{\alpha/2} S \in H^p$

When S is Γ-holomorphic, so is $(I - \Delta)^{\alpha/2} S$ and the H^p and the L^p norms are equivalent.

THEOREM 2. *With the notations of Theorem* 1, *let* p, q, *and* r *be three numbers in* $]0, +\infty[$ *related by* $r^{-1} = p^{-1} + q^{-1}$. *Let* S *be a* Γ-*holomorphic distribution in* $\mathscr{H}^p_\alpha$ *and* T *a* Γ-*holomorphic distribution in* $\mathscr{H}^q_\alpha$ *where* α *and* β *are negative. There the product* ST *makes sense and belongs to* $\mathscr{H}^r_{\alpha+\beta}$

2. FIRST PROOF OF CALDERÓN'S THEOREM

Let us call $\mathscr{S}_\Gamma$ the subspace of $\mathscr{S}(\mathbb{R}^n)$ consisting of all $f \in \mathscr{S}(\mathbb{R}^n)$ such that $\hat{f}$ has a compact support contained in the interior of Γ.

PROPOSITION 1. *For all* $\alpha \in \mathbb{R}$ *and* $0 < p < +\infty$, $\mathscr{S}_\Gamma$ *is dense in* $\mathscr{H}^p_\alpha$

The proof is easy but of a boresome and technical nature and it will be given at the end of this section.

We are going to prove Theorem 1 by complex interpolation and use a slight variation of this argument for Theorem 2. We put $a = -\alpha/2, b = -\beta/2$, $f = (I - \Delta)^a F$, $g = (I - \Delta)^b G$ to be reduced to usual L^p-norms.

If f and g belong to $\mathscr{S}_\Gamma$, so do F and G, and we define $H \in \mathscr{S}_\Gamma$ by

$$H = (I - \Delta)^{-a-b}[((I - \Delta)^a F)((I - \Delta)^b G)].$$

Theorem 1 can be restated as the following estimate

$$\|H\|_r \le C(a, b, p, q, \Gamma)\|F\|_p\|G\|_q. \tag{3}$$

We are now going to define H_z, an entire function of z with values in $\mathscr{S}_\Gamma$, by

$$H_z = (I - \Delta)^{-(a+b)z}[((I - \Delta)^{az} F)(I - \Delta)^{bz} G)]. \tag{4}$$

For the proof of Calderón's theorem, the two following estimates will be needed:

There exist two constants C_1 and N and for each pair, $(F, G) \in \mathscr{S}_\Gamma \times \mathscr{S}_\Gamma$ a finite number $\gamma(F, G)$ such that

$$\|H_z\|_r \le (1 + |z|)^N C_1^{\operatorname{Re} z}\gamma(F, G) \qquad \text{for} \quad \operatorname{Re} z \ge 0, \tag{5}$$

and there exists another constant C_2 such that

$$\|H_{i\tau}\|_r \le C_2(1 + |\tau|)^N\|F\|_p\|G\|_q \qquad \text{for} \quad \tau \in \mathbb{R}. \tag{6}$$

In fact C_1, C_2, and N will only depend on p, q, a, b, and Γ.

Once this is achieved, the maximum modulus principle can be applied to $(1 + z)^{-N-1}C_1^{-z}H_z$ with values in the Banach space L^r. Since this function tends to 0 as $|z| \to +\infty$, $\operatorname{Re} z \ge 0$, the maximum is attained when $\operatorname{Re} z = 0$. Henceforth, $\|H\|_r \le 2^N C_1 C_2\|F\|_p\|G\|_q$.

Only in proving (5), the hypothesis that F and G are Γ-holomorphic functions is needed. Using Fourier transforms, we are led to put

$$\varphi(\xi, \eta) = \frac{(1 + |\xi|^2)^a(1 + |\eta|^2)^b}{(1 + |\xi + \eta|^2)^{a+b}}, \qquad \xi \in \mathbb{R}^n, \quad \eta \in \mathbb{R}^n,$$

and we have

$$H(x) = \frac{1}{(4\pi^2)^n}\iint e^{ix(\xi+\eta)}\varphi(\xi, \eta)\hat{F}(\xi)\hat{G}(\eta)\, d\xi\, d\eta$$

and more generally

$$H_z(x) = \frac{1}{(4\pi^2)^n}\iint e^{ix(\xi+\eta)}[\varphi(\xi, \eta)]^z\hat{F}(\eta)\hat{G}(\eta)\, d\xi\, d\eta.$$

The fundamental fact about φ is that $0 \leq \varphi(\xi, \eta) \leq C_1$ when $\xi \in \Gamma$ and $\eta \in \Gamma$ since $|\xi| \leq M\xi_n$ and $|\eta| \leq M\eta_n$ imply $|\xi| + |\eta| \leq M(\xi_n + \eta_n) \leq M|\xi + \eta|$. This integral representation of H_z can be rewritten as

$$\hat{H}_z(\xi) = \frac{1}{(2\pi)^n} \int [\varphi(\xi - \eta, \eta)]^z \hat{F}(\xi - \eta)\hat{G}(\eta)\, d\eta.$$

Since F and G belong to $\mathscr{S}_\Gamma$, both η and $\xi - \eta$ belong to a compact subset K of the interior of Γ. Hence if $\alpha \in \mathbb{N}^n$ and $|\alpha| \leq n + 1$

$$|\partial_\xi^\alpha \hat{H}_z(\xi)| \leq (1 + |z|)^{n+1} C_1^{\operatorname{Re} z} \theta(F, G). \tag{7}$$

The compact support of $\hat{H}_z(\xi)$ is contained in $K + K$ and we have

$$\int_{\mathbb{R}^n} |\partial_\xi^\alpha \hat{H}_z(\xi)|\, d\xi \leq (1 + |z|)^{n+1} C_1^{\operatorname{Re} z} \gamma(F, G).$$

Passing to Fourier transform, we obtain a similar inequality for

$$\sup_{x \in \mathbb{R}^n} |x^\alpha H_z(x)|$$

and the estimate of the L^r-norm of H_z follows easily.

This loose argument does not give any interesting information on $\gamma(F, G)$ and certainly does not allow us to take F in L^p or G in L^q.

The proof of (6) is straightforward since when $z = i\tau$ and $\tau \in \mathbb{R}$, $(I - \Delta)^{i\tau}$ is a ψdo of order 0 and classical type. The norm of $(I - \Delta)^{i\tau}: L^p \to L^p$ has a polynomial growth when $p \in]1, +\infty[$ is fixed and when $|\tau| \to +\infty$.

Let us return now to Proposition 1.

The approximation procedure of $S \in L_\alpha^p$ will be divided into three steps:

(a) multiplication by $e^{i\varepsilon x_n}$, $\varepsilon > 0$,

(b) convolution of $\hat{S}$ with an approximation to the identity

$$\varphi_k(\xi) = k^n \varphi(k\xi), \qquad \varphi \in C_0^\infty(\mathbb{R}^n), \qquad \int_{\mathbb{R}^n} \varphi(\xi)\, d\xi = 1,$$

(c) multiplication of $\hat{S}$ with a "cut-off" function $\Phi(l^{-1}\xi)$ where $\Phi(0) = 1$, $\Phi \in C_0^\infty(\mathbb{R}^n)$.

The main geometrical observation is that $e^{i\varepsilon x_n}S$ is a Γ_ε-holomorphic distribution where $\Gamma_\varepsilon = \Gamma + (0, \ldots, 0, \varepsilon)$ is strictly contained in Γ. It follows that for a small ball B, $\Gamma_\varepsilon + B \subset \Gamma$ in such a way that (b) makes sense. Obviously (c) maps Γ-holomorphic distributions into Γ-holomorphic distributions.

Starting with $S \in L_\alpha^p$ we end with a function $S_{\varepsilon,k,l}$ in $\mathscr{S}(\mathbb{R}^n)$ which belongs to $\mathscr{S}_\Gamma$ if k is large enough relatively to ε, (and if S is Γ-holomorphic). In order to prove that S is the limit in the L_α^p-norm of $S_{\varepsilon,k,l}$ we can forget that S was Γ-holomorphic. We are going to use the easy fact that $\mathscr{S}(\mathbb{R}^n)$ is dense in L_α^p for $1 < p < +\infty$ and $\alpha \in \mathbb{R}$. When $S \in \mathscr{S}(\mathbb{R}^n)$, the required convergence is obvious and we only need to prove that our operators are uniformly bounded on L_α^p.

This follows from two simple observations. The space L_α^p is invariant under multiplication by characters $e^{i\xi x}$ with a norm not exceeding $w(|\xi|)$ where w has a polynomial growth. Moreover L_α^p is isometrically invariant under translations. Theorem 1 is now completely proved.

3. Second Proof of Theorem 1 and Proof of Theorem 2

We are going to use in a systematic way the Littlewood–Paley splitting of distributions in $\mathscr{H}_\alpha^p$ and the fact that these classes are invariant under the action of classical ψdo of order 0.

Moreover the fact that $\mathscr{S}_\Gamma$ is dense in the space $L_{\alpha,\Gamma}^p$ of Γ-holomorphic distributions $S \in L_\alpha^p$ extends obviously to $\mathscr{H}_{\alpha,\Gamma}^p$. In dealing with estimates we can restrict our attention to the case where $f \in \mathscr{S}_\Gamma$ and $g \in \mathscr{S}_\Gamma$.

First, let us observe that there is nothing to be proved when $\alpha = \beta = 0$. Hölder's inequality gives $\|fg\|_r \le \|f\|_p \|g\|_g$ and the proof is concluded by the observation that L^p and $\mathscr{H}^p$ norms are equivalent on $\mathscr{S}_r$ for $p > 0$.

Let us return to the general case and introduce the Littlewood–Paley decomposition of a function. Let $\psi \in C_0^\infty(\mathbb{R}^n)$ be a function supported by $\frac{1}{3} \le |\xi| \le 1$ and such that $\sum_0^\infty \psi(2^{-j}\xi) = 1$ for $|\xi| \ge 1$. Let us call $\varphi_0 \in C_0^\infty(\mathbb{R}^n)$ a function such that $\varphi_0(\xi) + \sum_0^\infty \psi(2^{-j}\xi) = 1$ everywhere. Then an obvious computation gives

$$\varphi_0(\xi) + \sum_0^{k-1} \psi(2^{-j}\xi) = 1 - \sum_k^\infty \psi(2^{-j}\xi) = 1 - \sum_0^\infty \psi(2^{-j}2^{-k}\xi) = \varphi_0(2^{-k}\xi).$$

For $f \in \mathscr{S}(\mathbb{R}^n)$ we define $f_j (j \in \mathbb{N})$ and $S_j(f)$ by

$$[S_j(f)]\hat{} = \varphi_0(2^{-j}\xi)\hat{f}(\xi) \qquad \text{and} \qquad \hat{f}_j(\xi) = \psi(2^{-j}\xi)\hat{f}(\xi).$$

Then we have $f = S_0(f) + f_0 + \cdots + f_j + \cdots$ and

$$S_j(f) = S_0(f) + f_0 + \cdots + f_{j-1}.$$

Proposition 2. *The three following metrics are equivalent for* $0 < p < +\infty$:

$$\|f\|_{\mathscr{H}^p}, \quad \|S_0(f)\|_{\mathscr{H}^p} + \left\| \left(\sum_0^\infty |f_j|^2 \right)^{1/2} \right\|_p, \quad \textit{and} \quad \left\| \sup_{j \ge 0} |S_j(x)| \right\|_p + \|S_0(f)\|_{\mathscr{H}^p}.$$

Let us sketch the proof. We can get rid of $S_0(f)$ and restrict our attention to those f for which $S_0(f) = 0$. Then call $\omega = (\omega_k)_{k \ge 0}$ a point in $\{-1, 1\}^{\mathbb{N}} = \Omega$ equipped with the canonical probability measure (corresponding to a fair coin tossing). Let $M_\omega(D)$ be the ψdo whose symbol is $\sum_0^\infty \omega_j \psi(2^{-j}\xi)$. If

$f \in \mathscr{H}^p$ so does $M_\omega(D)(f)$ and one has

$$\|M_\omega(D)(f)\|_p \leq C_p, \tag{8}$$

since ψdo s with symbols in $S^0_{1,0}$ map $\mathscr{H}^p$ into L^p for $0 < p < +\infty$.

Raising (8) to the power p and integrating with respect to ω we obtain

$$\left\|\left(\sum_0^\infty |f_j|^2\right)^{1/2}\right\|_p \leq C_p^1 \|f\|_{\mathscr{H}^p}. \tag{9}$$

We observe that when $\varphi_{0,t}(x) = t^{-n}\varphi_0(x/t)$, we have

$$\|f\|_{\mathscr{H}^p} = \|\sup |f * \varphi_{0,t}|\|_p,$$

which in our case is easily seen to be equivalent to the same supremum with $t = 2^{-j}$. Therefore, $\|f\|_{\mathscr{H}^p}$ and $\|\sup|S_j(f)|\,\|_p$ are equivalent metrics for $p > 0$.

Finally for any Rademacher series $\sum_0^\infty c_k\omega_k$, $c_k \in \mathbb{C}$, we have that $S^*(\omega) = \sup_{m \geq 0} |\sum_0^m c_k\omega_k|$ satisfies

$$\|S^*(\omega)\|_p \leq C_p \left(\sum_0^\infty |c_k|^2\right)^{1/2}.$$

Applying this remark to $\sum_0^\infty \omega_j f_j(x)$, we obtain for every x,

$$\left\|\sup_{j \geq 0} |S_j(\omega, f)|\right\|_{L^p(\Omega)} \leq C_p \left(\sum_0^\infty |f_j(x)|^2\right)^{1/2}. \tag{10}$$

We raise (10) to the power p and then integrate in x. For each fixed ω,

$$\left\|\sum_0^\infty \omega_j f_j(x)\right\|_{\mathscr{H}^p} \leq C \left\|\sup_{j \geq 0} |S_j(\omega, f)|\right\|_p.$$

Finally classical ψdo being bounded on $\mathscr{H}^p$, we have

$$\|f\|_{\mathscr{H}^p} \leq C_p \left\|\sup_{j \geq 0} |S_j(\omega, f)|\right\|_p.$$

In order to conclude the proof of Proposition 2, it is enough to remark that the reverse inequality of (10) is also true.

PROPOSITION 3. *With the notations of Proposition 2, for $\alpha \leq 0$ the three following metrics are equivalent for $0 < p < +\infty$:*

$$\|f\|_{\mathscr{H}^p_\alpha}, \quad \|S_0(f)\|_{\mathscr{H}^p} + \left\|\left(\sum 4^{\alpha j}|f_j|^2\right)^{1/2}\right\|_p$$

and

$$\left\|\sup_{j \geq 0} 2^{\alpha j}|S_j(f)|\right\|_p + \|S_0(f)\|_{\mathscr{H}^p}.$$

This is a simple consequence of Proposition 2 and of the fact that if $m(\xi) \in S^0_{1,0}$, then $m(D)$ maps $\mathscr{H}^p$ into itself [6].

Let us return to the proof of Theorem 2. We write (as in Proposition 2) $f = S_0(f) + \sum_{j\geq 0} f_j$ and $g = S_0(g) + \sum g_k$.

The term $S_0(f)S_0(g)$ is trivial: it belongs to $\mathscr{S}_\Gamma$ and its Fourier transform is carried by $|\xi| \leq 2$. Therefore, the $L^r_{\alpha+\beta}$- and L^r-norms are equivalent ones and by Hölder

$$\|S_0(f)S_0(g)\|_r \leq \|S_0(f)\|_p \|S_0(g)\|_q \leq C_{\alpha,\beta}\|S_0(f)\|_{L^q_\beta}\|S_0(g)\|_{L^q_\beta}.$$

Then we develop the product fg into a double series. We have $fg = S_0(f)S_0(g) + S_1 + S_2 + S_3$, where

$$S_1 = \sum_{k>0} S_0(f)g_k + \sum\sum_{j\leq k-2} f_j g_k = \sum_{k\geq 0} S_{k-2}(f)g_k,$$

$$S_2 = \sum\sum_{|j-k|\leq 1} f_j g_k,$$

and

$$S_3 = \sum_{j\geq 0} S_0(g)f_j + \sum\sum_{j\geq k+2} f_j g_k = \sum_{j\geq 0} S_{j-2}(g)f_j$$

(when $j = 0$ or 1, $j-2$ should be replaced by 0).

The key observation is the fact that the spectrum of $S_{k-2}(f)$ lies inside the ball $|\xi| \leq (1/4)2^k$ while the spectrum of g_k is contained in the annulus $(1/3)2^k \leq |\xi| \leq 2^k$. It follows that the spectrum of the product $S_{k-2}(f)g_k$ is contained inside the "enlarged annulus" $(1/12)2^k \leq |\xi| \leq (5/4)2^k$ that we shall call Δ_k.

These Δ_k are not pairwise disjoint but if we select them five by five, we obtain the disjoint annuli Δ_{5k+r}, $0 \leq r \leq 4$.

Therefore we can apply to $\sum_0^\infty S_{5k+r-2}(f)g_{5k+r}$ the Littlewood–Paley estimates given by Proposition 3. The five estimates imply

$$\left\|\sum_{k\geq 0} S_{k-2}(f)g_k\right\|_{\mathscr{H}^r_{\alpha+\beta}} \leq C\left\|\left(\sum_0^\infty 4^{k(\alpha+\beta)}|S_{k-2}(f)|^2|g_k|^2\right)^{1/2}\right\|_p.$$

But for each $x \in \mathbb{R}^n$, we have the obvious inequality

$$\left(\sum_0^\infty 4^{k(\alpha+\beta)}|S_{k-2}(f)|^2|g_k|^2\right)^{1/2} \leq \sup|2^{k\alpha}S_k(f)|\left(\sum_0^\infty 4^{k\beta}|g_k|^2\right).$$

Now that we have reached this point, Proposition 3 is once more applied and yields

$$\|S_1\|_{\mathscr{H}^r_{\alpha+\beta}} \leq C\|f\|_{\mathscr{H}^p_\alpha}\|g\|_{\mathscr{H}^q_\beta}.$$

The third term S_3 is estimated the same way. The sum S_2 can also be split into three parts. We shall only examine $\sum_0^\infty f_k g_k$ since the two other sums are similar. The spectrum of each product $f_k g_k$ belongs to

$$\Gamma_k = \{\xi + \eta; (1/3)2^k \leq |\xi| \leq 2^k, (1/3)2^k \leq |\eta| \leq 2^k, \xi \in \Gamma, \eta \in \Gamma\}.$$

Since Γ is a cone, Γ_k is contained in a dyadic annulus $\{2^{k+1}/3M \leq |\xi| \leq 2^{k+1}\}$. Then the method used for S_1 applies to S_2.

4. The Paraproduct of Functions

Let $\varphi \in C^\infty(\mathbb{R}^{2n}\backslash\{0\})$ be homogeneous of degree 0 with the three following properties

$$\varphi(\eta, \xi) \in [0, 1] \quad \text{for} \quad (\eta, \xi) \neq (0, 0), \quad \eta \in \mathbb{R}^n, \quad \xi \in \mathbb{R}^n; \tag{11}$$

$$\varphi(\eta, \xi) = 0 \quad \text{whenever} \quad |\eta| \geq \tfrac{1}{10}|\xi|; \tag{12}$$

$$\varphi(\eta, \xi) = 1 \quad \text{whenever} \quad |\eta| \leq \tfrac{1}{20}|\xi|. \tag{13}$$

A bilinear operator $P_\varphi: \mathcal{S}(\mathbb{R}^n) \times \mathcal{S}(\mathbb{R}^n) \to C^\infty(\mathbb{R}^n)$ is then defined by

$$P_\varphi(f, g)(x) = \frac{1}{(4\pi^2)^n} \iint e^{i(\xi+\eta)x} \varphi(\eta, \xi) \hat{f}(\eta) \hat{g}(\xi)\, d\eta\, d\xi. \tag{14}$$

Theorem 3. *Assume that p, q and r belong to $]1, +\infty[$ and are related by $r^{-1} = p^{-1} + q^{-1}$. Then a constant $C = C(p, q, n, \varphi)$ exists such that for f and g in $\mathcal{S}(\mathbb{R}^n)$*

$$\|P_\varphi(f, g)\|_r \leq C \|f\|_p \|g\|_q. \tag{15}$$

For $a \in \mathbb{R}^n$, the translation operator $R_a: \mathcal{S}(\mathbb{R}^n) \to \mathcal{S}(\mathbb{R}^n)$ is defined by $R_a f(x) = f(x - a)$ and for $t > 0$, the dilation D_t is defined by $(D_t f)(x) = f(x/t)$. Then P_φ commutes with R_a and D_t in the following sense

$$P_\varphi(f, g) = R_a^{-1} P_\varphi(R_a f, R_a g),$$
$$P_\varphi(f, g) = D_t^{-1} P_\varphi(D_t f, D_t g).$$

Let us call $\mathcal{S}_0 \subset \mathcal{S}(\mathbb{R}^n)$ the subspace consisting of all f such that $\hat{f} = 0$ in a neighborhood of 0. Then $\mathcal{S}_0$ is dense in L^p when $1 < p < +\infty$ and we can restrict our attention to f and g in $\mathcal{S}_0$. By dilation, $\hat{f}$ and $\hat{g}$ can be assumed to vanish on $|\xi| \leq 1$.

Then the Littlewood–Paley decomposition gives $f = \sum_0^\infty f_j$ and $g = \sum_0^\infty g_k$. Then $P_\varphi(f_j g_k) = f_j g_k$ if $j \leq k - 6$ and $P_\varphi(f_j g_k)$ if $j \geq k - 1$. We have $P_\varphi(f, g) = S_2 + S_2$ where

$$S_1 = \sum_{j \leq k-6}\sum f_j g_k = \sum_6^\infty S_{k-6}(f) g_k.$$

The Littlewood–Paley theory still applies and yields

$$\|S_1\|_r \le C_r \|\sup|S_k(f)|\,\|_p \left\|\left(\sum_0^\infty |g_k|^2\right)^{1/2}\right\|_q \le C_{p,q}\|f\|_p\|g\|_q.$$

The sum S_2 equals $\sum\sum_{k-6\le j\le k-1} P_\varphi(f_j, g_k)$ and can be split into six partial sums $\sum_{j=k+m}$ where $-1 \le m \le -6$. Then the frequencies of $P_\varphi(f_{k+m}, g_k)$ belong to the enlarged dyadic annulus $(\frac{1}{3} - \frac{1}{10})2^k \le |\xi| \le (11/10)2^k$. The Littlewood–Paley theory applies and gives

$$\|S_2\| \le C \sum_{m=-6}^{-1} \left\|\left(\sum_0^\infty |P_\varphi(f_{k+m}, g_k)|^2\right)^{1/2}\right\|_r.$$

It remains to estimate $\gamma(x) = (\sum_0^\infty |P_\varphi(f_{k+m}, g_k)|^2)$. Two observations are useful.

Lemma 1. *For each $p \in]1, +\infty[$ and each bounded subset $\mathscr{B} \subset \mathscr{S}(\mathbb{R}^n)$ there exists a constant C such that for each $\varphi \in \mathscr{B}\|M_\varphi(f)\|_p \le C\|f\|_p$.*

Lemma 2. *For each bounded subset $\mathscr{B} \subset \mathscr{S}(\mathbb{R}^n)$ and for each $p \in]1, +\infty[$ there exists a C such that for all $\psi \in \mathscr{B}$ such that $\int_{\mathbb{R}^n} \psi(x)\,dx = 0$, we have*

$$\left\|\left(\sum_0^\infty |f_j(x)|^2\right)^{1/2}\right\|_p \le C_p\|f\|_p.$$

where f_j ($j \in \mathbb{N}$) is defined by $\hat{f}_j(\xi) = \psi(2^{-j}\xi)\hat{f}(\xi)$.

Let us return to estimating $\gamma(x)$. For this purpose, we remark that there exist two bounded sequences φ_ν and ψ_ν, $\nu \in \mathbb{N}$ in $\mathscr{S}(\mathbb{R}^n)$ and a rapidly decreasing sequence ω_ν of complex numbers such that

$$\varphi(\eta, \xi) = \sum_0^\infty \omega_\nu \varphi_\nu(\eta)\psi_\nu(\xi) \qquad \text{for} \quad \tfrac{1}{10} \le |\xi| \le 10 \tag{16}$$

each ψ_ν being supported by $\frac{1}{20} \le |\xi| \le 20$ and each φ_ν by $|\xi| \le 10$. Using the homogeneity of $\varphi(\eta, \xi)$, we have

$$\varphi(2^{-k}\eta, 2^{-k}\xi) = \sum_0^\infty \omega_\nu \varphi_\nu(2^{-k}\eta)\psi_\nu(2^{-k}\xi).$$

Two families $S_{\nu,k}: \mathscr{S}(\mathbb{R}^n) \to \mathscr{S}(\mathbb{R}^n)$ and $\Delta_{\nu,k}: \mathscr{S}(\mathbb{R}^n) \to \mathscr{S}(\mathbb{R}^n)$ of operators are defined by the corresponding multipliers $\varphi_\nu(2^{-k}\xi)$ and $\psi_\nu(2^{-k}\xi)$. Then

$$P_\varphi(f_{k+m}, g_k) = \sum_0^\infty \omega_\nu S_{\nu,k}(f)\Delta_{\nu,k}(g)$$

and

$$\left(\sum_0^\infty |P_\varphi(f_{k+m}, g_k)|^2\right)^{1/2} \le \sum_0^\infty |\omega_\nu| \left(\sum_{k=0}^\infty |\Delta_{\nu,k}(g)|^2\right)^{1/2} \sup_{k\ge 0} |S_{\nu,k}(f)|$$
$$= \sum_0^\infty |\omega_\nu| S_\nu(g) M_\nu(f).$$

By Lemma 1, there exists a constant C_p such that for all $\nu \in \mathbb{N} \|M_\nu(f)\|_p \le C_p\|f\|_p$. In the same way $\|S_\nu(g)\|_q \le C_q\|f\|_q$. Therefore $\|\sum_0^\infty P_\varphi(f_{k+m}, g_k)\|_r \le C_{p,q}\|f\|_p\|g\|_q$. The preceding method gives $\|P_\varphi(f,g)\|_q \le C_q\|f\|_\infty\|g\|_q$ when $1 < q < +\infty$. But the proof of the following estimate

$$\|P_\varphi(f,g)\|_p \le C_p\|f\|_p\|g\|_{\mathrm{BMO}}, \qquad 1 < p < +\infty.$$

requires a different argument. Details can be found in [5, p. 144]. Finally, if f is fixed in $L^\infty(\mathbb{R}^n)$, we define

$$K(x, y) = \int_{\mathbb{R}^n} \frac{\Omega(s, x-y)}{(|s|^2 + |x-y|^2)^n} f(x-s)\, ds, \tag{17}$$

where $\Omega(s, x)/(|s|^2 + |x|^2)^n$ is defined by its Fourier transform $\varphi(\eta, \xi)$. Therefore $\Omega \in C^\infty(\mathbb{R}^{2n}\backslash\{0\})$ is homogeneous of degree 0 and its integral on the unit sphere equals 0. It follows

$$P_\varphi(f,g) = \text{p.v.} \int_{\mathbb{R}^n} K(x, y) g(y)\, dy = \lim_{\varepsilon \downarrow 0} \int_{|x-y|\ge\varepsilon} K(x, y) g(y)\, dy.$$

The following estimates on the kernel $K(x, y)$ are easily deduced from (17).

$$|K(x, y)| \le \frac{C_n\|f\|_\infty}{|x-y|^n}, \qquad \left|\frac{\partial}{\partial x_j} K(x, y)\right| \le \frac{C_n\|f\|_\infty}{|x-y|^{n+1}}, \quad 1 \le j \le n,$$

and

$$\left|\frac{\partial}{\partial y_j} K(x, y)\right| \le \frac{C_n\|f\|_\infty}{|x-y|^{n+1}}.$$

We also have $\|P_\varphi(f,g)\|_2 \le C\|f\|_\infty\|g\|_2$.

Then the mapping $g \mapsto P_\varphi(f,g)$ is automatically bounded on L^p ($1 < p < +\infty$), maps L^1 into weak L^1 and L^∞ into BMO This follows from the general theory of [5, Chapter IV].

Moreover $P_\varphi(f,g) = 0$ whenever the function g is a constant. Then the estimate

$$\|P_\varphi(f,g)\|_{\mathrm{BMO}} \le C\|f\|_\infty\|g\|_{\mathrm{BMO}} \tag{18}$$

is obtained by the real variable methods of Calderón and A. Zygmund.

5. Paraproduct of Distributions (H^p-Classes)

The following theorem describes the case where $0 < r \leq 1$ in Theorem 3. Then L^r is replaced by H^r (the Hardy class). We can also investigate the situation where $0 < p \leq 1$ and $0 < q \leq 1$. Then we have

THEOREM 4. *Let p, q, and r be three real numbers belonging to $]0, +\infty[$ and related by $r^{-1} = p^{-1} + q^{-1}$. Then there exists a constant $C = C(p, q, \varphi)$ such that for $f \in H^p$, $g \in H^p$ we have $P_\varphi(f, g) \in H^r$ with*

$$\|P_\varphi(f,g)\|_{H^r} \leq \|f\|_{H^p}\|g\|_{H^q}.$$

The proof is identical to that of Theorem 3 and will not be given in detail.

6. Paraproduct of Distributions (L^p_α-classes, $\alpha \leq 0$)

Using the notations of Theorem 1, we have

THEOREM 5. *Assume that $\alpha \leq 0$, $\beta \leq 0$, p, q, and r belong to $]1, +\infty[$ and that $r^{-1} = p^{-1} + q^{-1}$. Then a constant C exists such that*

$$\|P_\varphi(f,g)\|_{L^r_{\alpha+\beta}} \leq C\|f\|_{L^p_\alpha}\|g\|_{L^q_\beta} \qquad \text{for all} \quad f \in L^p_\alpha \quad \text{and} \quad g \in L^q_\beta. \tag{19}$$

The proof is similar to that of Calderón's theorem. For $z \in \mathbb{C}$, we put

$$\psi_z(\eta, \xi) = (1 + |\eta|^2)^{\alpha z/2}(1 + |\xi|^2)^{\beta z/2}(1 + |\xi + \eta|^2)^{-(\alpha+\beta)z/2}\varphi(\eta, \xi).$$

Using ψ_z, the following bilinear operator is defined

$$P_z(f,g)(x) = \frac{1}{(4\pi^2)^N}\iint e^{i(\xi+\eta)x}\psi_z(\eta, \xi)\hat{f}(\eta)\hat{g}(\xi)\, d\eta\, d\xi.$$

Our goal is to prove the inequality $\|P_1(f,g)\|_r \leq C\|f\|_p\|g\|_q$ and the complex interpolation method used in Theorem 1 applies here.

7. Paraproduct of Distributions ($\mathscr{H}^p_\alpha$-classes)

When $0 < p \leq 1$, $0 < g \leq 1$, or $0 < r \leq 1$, we replace L^p_α by the already defined $\mathscr{H}^p_\alpha$-class. Then we have

THEOREM 6. *With the notations of Theorem 5, we have if p, q, and r belong to $]0, +\infty[$ and if $r^{-1} = p^{-1} + q^{-1}$*

$$\|P_\varphi(f,g)\|_{\mathscr{H}^r_{\alpha+\beta}} \leq C\|f\|_{\mathscr{H}^p_\alpha}\|g\|_{\mathscr{H}^q_\beta}. \tag{20}$$

The proof is identical to that of Theorem 2.

The "square" of the Dirac mass. We are going to compute $P_\varphi(\delta_0, \delta_0)$ where δ_0 is the Dirac mass at the origin. This measure δ_0 does not belong to H^p classes for $p < 1$ since the cancelation condition is not fulfilled. However, δ_0 belongs to L^p_α when $\alpha/n < p^{-1} - 1$. The paraproduct of δ_0 and δ_0 is then a special case of the paraproduct between two distributions in L^p_α.

For the sake of simplicity, we are going to assume that $\varphi(\eta, \xi) = \varphi_0(|\eta|/|\xi|)$ where $\varphi_0 \in C_0^\infty(\mathbb{R})$ is even, is 1 on $[-\frac{1}{20}, \frac{1}{20}]$ and is supported by $[-\frac{1}{10}, \frac{1}{10}]$.

Let S and T be two compactly supported distributions. Then $U = P_\varphi(S, T) \in \mathscr{S}'(\mathbb{R}^n)$ is defined by its Fourier transform

$$\hat{U}(\xi) = \frac{1}{(2\pi)^n} \int_{\mathbb{R}^n} \hat{S}(\eta)\hat{T}(\xi - \eta)\varphi_0\left(\frac{|\eta|}{|\xi - \eta|}\right) d\eta.$$

It should be observed that $\eta \to \varphi_0(|\eta|/|\xi - \eta|)$ is supported by a ball with volume $C_n|\xi|^n$.

Finally, if $S = T = \delta_0$, then $\hat{U}(\xi) = \gamma_n|\xi|^n(2\pi)^{-n}$, and therefore $P_\varphi(\delta_0, \delta_0) = \gamma_n(-\Delta)^{n/2}\delta_0$.

References

1. J. M. Bony, Calcul symbolique et propagation des singularités pour les equations aux dérivées partielles non linéaires, Prepublications, Dept. Math., Université Paris-Sud, Orsay, France.
2. A. P. Calderón, Commutators of singular integral operators, *Proc. Nat. Acad. Sci. U.S.A.* **53** (1965), 1092–1099
3. A. P. Calderón, Cauchy integrals on Lipschitz curves and related operators, *Proc. Nat. Acad. Sci. U.S.A.* **74** (1977), 1324–1327.
4. R. R. Coifman and Y. Meyer, Commutators of singular integrals, *Colloq. Math. Soc. Janos Bolyai, Fourier Anal., Budapest, 1976.*
5. R. R. Coifman and Y. Meyer, Audelà des opérateurs pseudo différentiels, Astérisque 57, Soc. Math. France, Paris. 1978.
6. Fefferman and E. M. Stein, H^p spaces of several variables, *Acta Math.* **129** (1972), 137–193.
7. E. M. Stein, "Singular Integrals and Differentiability Properties of Functions," Princeton Univ. Press, Princeton, New Jersey, 1970.

MATHEMATICAL ANALYSIS AND APPLICATIONS, PART B
ADVANCES IN MATHEMATICS SUPPLEMENTARY STUDIES, VOL. 7B

On the Cauchy–Kowalewski Theorem

SIGERU MIZOHATA

Department of Mathematics
Kyoto University
Kitashirakawa, Kyoto
Japan

DEDICATED TO LAURENT SCHWARTZ

1. INTRODUCTION

We are concerned here with the Cauchy problem for the linear partial differential equations with holomorphic coefficients in a neighborhood of the origin:

$$\partial_t^m u(t,x) - \sum_{j=1}^{m} a_j(t,x;\partial_x)\partial_t^{m-j}u(t,x) = f(t,x), \tag{1.1}$$

where $a_j(t,x;\zeta)$ is a polynomial in ζ, and $x = (x_1, x_2, \ldots, x_l)$. Since the coefficients are holomorphic we can define their orders in a neighborhood of the origin with respect to ζ without ambiguity. The Cauchy–Kowalewski theorem asserts that, if

$$\text{order } a_j(t,x;\zeta) \leq j, \tag{K}$$

then for any holomorphic data

$$\{\partial_t^i u(0,x)\}_{0 \leq i \leq m-1}$$

defined in a neighborhood of the origin and holomorphic $f(t,x)$, there exists a unique holomorphic solution $u(t,x)$ in a neighborhood of the origin.

Hereafter we are concerned with Eq. (1.1) where $f = 0$, namely,

$$\partial_t^m u = \sum_{j=1}^{m} a_j(t,x;\partial_x)\partial_t^{m-j}u. \tag{1.2}$$

We denote this equation by

$$L(u) = 0. \tag{1.3}$$

The purpose of this chapter is to show that the condition (K) is really necessary in order for the Cauchy–Kowalewski theorem to hold. To make

ISBN 0-12-512802-9

our problem clear, let us observe the following fact. We are always concerned with holomorphic solutions of (1.2) in a neighborhood of the origin. So, given the holomorphic data $u_i(x)$ $(0 \leq i \leq m-1)$, we can form the unique formal solution

$$u(t, x) \sim \sum_{j=0}^{\infty} u_j(x) t^j / j!,$$

where $u_j(x)$ $(j \geq m)$ are calculated from (1.2) in terms of $u_j(x)$ $(0 \leq j \leq m-1)$. The Cauchy–Kowalewski theorem claims that if (K) is assumed, then the formal solution is a convergent series. So our aim is to show that if (K) is violated, then there exists at least one holomorphic Cauchy data such that the corresponding formal solution is a divergent series.

In the original paper of Kowalewski [2], she claims that the condition (K) is necessary without giving the proof. The necessity is merely explained by taking a heat operator as an example. This author [4] proposed the problem from the general point of view. Miyake [3] showed that in the case when $m = 1$, (K) is necessary. Next, inspired by that article, this author [5] proved the following: Let

$$a_j(t, x; \zeta) = \sum_k a_{jk}(t, x; \zeta),$$

where a_{jk} is the homogeneous part of order k in ζ. Let $q(j, k)$ be the vanishing order of the polynomial $a_{jk}(t, x; \zeta)$ at $t = 0$, namely,

$$a_{jk}(t, x; \xi) = t^{q(j,k)} \mathring{a}_{jk}(t, x; \zeta), \qquad \mathring{a}_{jk}(0, x; \zeta) \not\equiv 0.$$

Then

$$\max_{j,k} \frac{k}{q(j, k) + j} \leq 1$$

is necessary for the Cauchy–Kowalewski theorem to hold. In 1976, Kitagawa and Sadamatsu [1] obtained the following, very sharp, necessary condition:

$$\max_{k > j} \frac{k}{q(j, k) + j} < \max_{k \leq j} \frac{k}{q(j, k) + j}. \tag{K–S}$$

Hereafter we assume this condition. Recently, this author [6] showed that the condition (K) is necessary in the case when the coefficients depend only on t. The argument was carried out in the following way. Let us assume (K) is violated. If we put the solution in the form $u_\zeta(t, x) = e^{\zeta x} v_\zeta(t)$, then $v_\zeta(t)$ is the solution of

$$v^{(m)}(t) = \sum a_j(t; \zeta) v^{(m-j)}(t).$$

Then we can define a family of solutions $v_\zeta(t)$ such that their initial data remain bounded, whereas there exists a number $p > 1$ such that for some positive t,

$$|v_\zeta(t)| \geq \exp(\delta|\zeta|^p t),$$

if ζ ($|\zeta| \to \infty$) is chosen appropriately. This is not compatible with the continuity property (see Proposition 2.2). This chapter extends the above method by using the microlocalization of the operator.

Finally we state our theorem.

THEOREM. *For the Cauchy–Kowalewski theorem to hold at the origin, the condition* (K) *is necessary. In other words, whenever* (K) *is violated, the Cauchy–Kowalewski theorem fails to hold.*

The content of this article was already published in [7], without giving detailed proofs.

2. PRELIMINARIES

Hereafter we assume always that the Cauchy–Kowalewski theorem holds, and that (K) is violated. Under these two assumptions, we show several fundamental facts. First, the following facts are fundamental to our argument whose proofs are given in [4,5,8].

PROPOSITION 2.1. *Let $\mathcal{O}_x$ be a connected open neighborhood of the origin in $\mathbb{C}^l$. Suppose for any initial holomorphic data $\Psi(x) = (u_0(x), \ldots, u_{m-1}(x)) \in H(\mathcal{O}_x)^m$, there is a solution $u(t,x)$ of* (1.2) *with $\Psi(x)$ at $t = 0$ such that $u(t,x) \in H(V_\Psi)$, where V_Ψ is a neighborhood (complex) of the origin, which may depend on Ψ. Then there exists a polydisk D with center at the origin such that for any $\Psi \in H(\mathcal{O}_x)^m$ there exists a unique solution u in $H(D)$.*

PROPOSITION 2.2. *The mapping*

$$\Psi(x) \in H(\mathcal{O}_x)^m \to u(t,x) \in H(D)$$

is continuous.

Let us make this assertion more concrete. Let

$$\mathcal{O}_x = \{x \in \mathbb{C}^l;\ |x_i| < r_0\ (1 \leq i \leq l)\},$$
$$D = \{(t,x) \in \mathbb{C}^{l+1};\ |t|,\ |x_i| < r_0'\ (1 \leq i \leq l)\}.$$

Take the seminorm $p_K(f) = \max_{(t,x)\in K} |f(t,x)|$, where K is a compact set in D. Now let K be the closed polydisk

$$\{(t,x) \in \mathbb{C}^{l+1};\ |t| \leq r',\ |x_i| \leq r'\},$$

where r' is an (arbitrary) positive number such that $r' < r'_0$. Proposition 2.2 implies that there exist a positive r ($<r_0$) and a constant M such that

$$\max_{|t|,\,|x_i|\le r'} |u(t,x)| \le M\left(\sum_{i=0}^{m-1} \max_{|x_i|\le r} |\partial_t^i u(0,x)|\right),$$

where $\partial_t^i u(0,x)$ $(0 \le i \le m-1)$ are holomorphic in $\mathcal{O}_x$, and all t and x_i are complex. Without loss of generality, we can assume hereafter $r, r' \le \frac{1}{2}$.

Letting

$$u(t,x) = \sum_{j\ge 0} u_j(x)t^j, \tag{2.1}$$

the above a priori estimate gives, by using the Cauchy formula and denoting r'^{-1} by K_0,

$$\max_{|x_i|\le r'} |u(t,x) \le MK_0^j \left(\sum_{i=0}^{m-1} \max_{|x_i|\le r} |\partial_t^i u(0,x)|\right), \qquad j \ge 0. \tag{2.2}$$

Now we apply this inequality to the actual situation. We assume condition (K–S), more precisely, only the condition

$$k \le q(j,k) + j.$$

Let the initial data be given by the power series around a point x_0 ($|x_0| \le r'$):

$$\partial_t^i u(0,x) = \sum_{\alpha\ge 0} a_\alpha^i (x - x_0)^\alpha, \qquad 0 \le i \le m-1.$$

Then

$$u_j(x_0) = \sum_{i=0}^{m-1} \sum_{\alpha} c_{j\alpha}^i a_\alpha^i. \tag{2.3}$$

Note that $c_{j\alpha}^i$ depends on x_0.

LEMMA 2.1. *In the above summation with respect to* α, *only the terms with* $|\alpha| \le j$ *appear. In other words,* $c_{j\alpha}^i = 0$ *when* $|\alpha| > j$. *Moreover*

$$|c_{j\alpha}^i| \le MK_0{}^j, \qquad \textit{for all} \quad i, j, \textit{and } \alpha. \tag{2.4}$$

Proof. We start from (1.2). We want to show that, if we put

$$\partial_t^{m+h} u = \sum_{j=1}^{m} a_j^{(h)}(t,x;\partial_x)\partial_t^{m-j} u,$$

then

$$\text{order}\, a_j^{(h)}(0,x;\partial_x) \le j + h.$$

We prove this in the following form: Let $a_{jk}^{(h)}(t,x;\zeta)$ be the homogeneous part of order k (in ζ) of $a_j^{(h)}$, and let $q_h(j,k)$ be its vanishing order with respect to t, then

$$k \le q_h(j,k) + j + h. \tag{2.5}$$

We show this by induction on h. This is true for $h = 0$ by assumption. By differentiating the above expression of $\partial_t^{m+h}u$ with respect to t, we have

$$\partial_t^{m+h+1}u = \sum_{j=1}^{m} a_j^{(h)}\partial_t^{m-j+1}u + \sum_{j=1}^{m} \frac{\partial}{\partial t} a_j^{(h)}\partial_t^{m-j}u.$$

Now the term $a_1^{(h)}\partial_t^m u$ should be replaced by

$$\sum_{j=1}^{m} a_1^{(h)}a_j\partial_t^{m-j}u.$$

Here

$$a_1^{(h)}a_j = \left(\sum_k a_{1k}^{(h)}\right)\left(\sum_{k'} a_{jk'}\right).$$

Take the product $a_{1k}^{(h)}(t,x;\partial_x)a_{jk'}(t,x;\partial_x)$. Its vanishing order is $q_h(1,k) + q(j,k')$. The order of the differential operator $\leqq k + k'$. By assumption,

$$k \le q_h(1,k) + 1 + h, \qquad k' \le q(j,k') + j.$$

So,

$$k + k' \le q_h(1,k) + q(j,k') + j + h + 1.$$

This shows that the inequality for $h + 1$ is satisfied. For $j \ge 2$, $a_j^{(h)}\partial_t^{m-j+1}$ satisfies the desired property. Finally,

$$\frac{\partial}{\partial t} a_j^{(h)}\partial_t^{m-j}u = \sum_k \frac{\partial}{\partial t} a_{jk}^{(h)}\partial_t^{m-j}u.$$

If $q_h(j,k) = 0$, then by assumption $k \le j + h$. So, in this case there is no problem. If $q_h(j,k) > 0$, then $(\partial/\partial t)a_{jk}^{(h)}$ has vanishing order $q_h(j,k) - 1$. So the desired property is also fulfilled.

Finally in the expression (2.3) of $u_j(x_0)$, let $\partial_t^i u(0,x) = (x - x_0)^\alpha$ and 0 for $\partial_t^j u(0,x)$ with $j \ne i$. Then $u_j(x_0) = c_{j\alpha}^i$, and (2.2) shows that the inequality (2.4) holds, since

$$\max_{|x_i| \le r} \left|(x - x_0)^\alpha\right| \le 1,$$

noting that $r < \frac{1}{2}$, $\left|x_0\right| \le r'$ $(<\frac{1}{2})$. ■

We consider a family of initial data, depending on a real large parameter n, of the form:

$$\partial_t^i u_n(0,x) = e^{n\xi_0 x}\phi_{n,i}(x), \qquad 0 \le i \le m - 1, \tag{2.6}$$

where $\xi_0 \in \mathbb{R}^l$, and $\phi_{n,i}(x)$ are entire functions satisfying

$$|\partial^\alpha \phi_{n,i}(x)| \leq c_0(nA)^{|\alpha|} \qquad \text{for all} \quad \alpha \text{ and } x \in \mathbb{R}^l.$$

Note that ξ_0 will be defined in Section 3, and $\phi_{n,i}$ in Section 6. Then, by assumption, there exists a holomorphic solution $u_n(t, x)$ of (1.2) for $|x_i|, |t| < r'_0$. Letting

$$u_n(t, x) = e^{n\xi_0 x} v_n(t, x) \tag{2.7}$$

we have

PROPOSITION 2.3 (A priori estimates). *There exist positive constants C, K, t_0 (not dependent on c_0 and A) such that*

$$|v_n(t, x)| \leq Cc_0 \exp\{n(|\xi_0| + A)K|t|\}, \tag{2.8}$$

for x real, $|x_i| \leq r'$, and $|t| \leq t_0$. This is also true for $\partial_t^i v_n(t, x)$ $(0 \leq i \leq m-1)$, if we replace C with a fixed polynomial in $n(|\xi_0| + A)K$.

Proof. Taking into account the linearity, it suffices to prove (2.8) when all $\phi_{n,i}$ except one, say, $\phi_{n,0}$, are zero. First we consider this at the point $x = 0$. In this case, we note that $u_n(t, 0) = v_n(t, 0)$. First

$$\partial_x^\alpha (e^{n\xi_0 x}\phi_{n,0})|_{x=0} = \sum_{\alpha'} C_{\alpha'}^\alpha (n\xi_0)^{\alpha'} \partial_x^{\alpha''} \phi_{n,0}(x)\Big|_{x=0}.$$

Thus, the right-hand side is estimated in absolute value by

$$c_0 \sum_k C_k^{|\alpha|} n^k |\xi_0|^k (nA)^{|\alpha| - k} \leq c_0 n^{|\alpha|}(|\xi_0| + A)^{|\alpha|}.$$

Thus, taking into account (2.3),

$$|u_j| \leq \sum_{|\alpha| \leq j} |c_{j\alpha}^0 a_\alpha^0| \leq MK_0^j \sum_{|\alpha| \leq j} |a_\alpha^0|$$

$$\leq MK_0^j c_0 \sum_{|\alpha| \leq j} \frac{n^{|\alpha|}}{\alpha!} (|\xi_0| + A)^{|\alpha|} = c_0 MK_0^j \sum_{k=0}^{j} \frac{1}{k!} (nl(|\xi_0| + A))^k.$$

Put $l(|\xi_0| + A) = A'$, then $|u_j| \leq c_0 MK_0^j \sum_{k=0}^j (nA')^k/k!$.

Define the integer j_0 by $j_0 \leq nA' < j_0 + 1$. We consider two cases of j as follows.

(1) $j \leq j_0$.

$$|u_j| \leq c_0 MK_0^j (j+1)(nA')^j/j!.$$

Since $(j+1)K_0^j \leq (2K_0)^j$,

$$|u_j| \leq c_0 M(2K_0)^j (nA')^j/j!.$$

Thus,

$$\sum_{j=0}^{j_0} |u_j t^j| \le c_0 M \sum_{j=0}^{j_0} \frac{(2K_0 nA')^j}{j!} |t|^j \le c_0 M \exp(2K_0 nA'|t|).$$

(2) $j > j_0$. In the expression u_j, the greatest term is $c_0 M K_0^j (nA')^{j_0}/j_0!$. Thus,

$$|u_j| \le c_0 M K_0^j (j+1)(nA')^{j_0}/j_0! \le c_0 M (2K_0)^j (nA')^{j_0}/j_0!.$$

$$\sum_{j > j_0} |u_j t^j| \le c_0 M \frac{(nA')^{j_0}}{j_0!} \sum_{j > j_0} (2K_0|t|)^j.$$

Thus, if we denote $t_0 = 1/4K_0$, and if $|t| \le t_0$, we have

$$\sum_{j \ge j_0} |u_j t^j| \le c_0 M \frac{(nA')^{j_0}}{j_0!} (2K_0|t|)^{j_0} < c_0 M \exp(2nA'K_0|t|).$$

Thus we have proved

$$\sum_{j=0}^{\infty} |u_j t^j| \le 2c_0 M \exp(2K_0 nA'|t|), \qquad \text{when} \quad |t| \le t_0 = 1/4K_0.$$

It is the same with the derivatives with respect to t. Consider $\sum_{j \ge 0} |j u_j t^{j-1}|$. First, when $j \le j_0$, $j|u_j| \le c_0 M K_0^j j(j+1)(nA')^j/j!$. We define a constant $C_1 = \max_j j(j+1)/2^j$. Then, $j|u_j| \le c_0 C_1 M (2K_0)^j (nA')^j/j!$. Thus,

$$\begin{aligned}\sum_{1 \le j \le j_0} |j u_j t^{j-1}| &\le c_0 C_1 M \sum_{1 \le j \le j_0} (2K_0)^j \frac{(nA')^j}{j!} |t|^{j-1} \\ &\le c_0 C_1 M (2K_0 nA') \sum_j \frac{(2K_0 nA'|t|)^{j-1}}{(j-1)!} \\ &\le c_0 C_1 M (2K_0 nA') \exp(2K_0 nA'|t|).\end{aligned}$$

In the case $j > j_0$, $j|u_j| \le c_0 C_1 M (2K_0)^j (nA')^{j_0}/j_0!$. Thus,

$$\begin{aligned}\sum_{j > j_0} |j u_j t^{j-1}| &\le c_0 C_1 M \frac{(nA')^{j_0}}{j_0!} |2K_0 t|^{j_0 - 1} 2K_0 \\ &\le c_0 C_1 M (2K_0 nA') \exp(2K_0 nA'|t|).\end{aligned}$$

Let us move on to the case when x_0 is arbitrary. Since $u_n(t, x) = e^{n\xi_0 x_0} e^{n\xi_0(x - x_0)} v_n(t, x)$, we consider $e^{n\xi_0(x - x_0)} v_n(t, x)$ around $x = x_0$, replacing c_0 by $c_0 e^{n\xi_0 x_0}$ in the above argument. The above argument will give

$$|u_n(t, x_0)| \le e^{n\xi_0 x_0} C c_0 \exp\{n(|\xi_0| + A)K|t|\}.$$

Thus, for $v_n(t, x_0) = e^{-n\xi_0 x_0} u_n(t, x_0)$, we have the desired estimate.

3. MICROLOCALIZATION

Instead of u, we consider

$$v = e^{-n\xi_0 x}u.$$

We omit the subscript n of v_n for the moment. If u is a solution of (1.3), then v satisfies

$$\tilde{L}(v) = \partial_t^m v - \sum_j a_j(t, x; n\xi_0 + \partial_x)\partial_t^{m-j} v = 0. \tag{3.1}$$

We let

$$a_j(t, x; \zeta) = \sum_{k \geq 0} a_{jk}(t, x; \zeta), \tag{3.2}$$

where a_{jk} is *homogeneous of order* k in ζ. Thus, if we let

$$a_j(t, x; n\xi_0 + i\eta) = n^j h_j(t, x; \xi_0 + i\eta'), \qquad \eta' = \eta/n, \tag{3.3}$$

then letting, for $k > j$,

$$a_{jk} = t^{q(j,k)}\mathring{a}_{jk}, \tag{3.4}$$

we have

$$h_j(t, x; \xi_0 + i\eta') = \sum_{k \leq j} a_{jk}(t, x; \xi_0 + i\eta')/n^{j-k}$$
$$+ \sum_{k > j} n^{k-j} t^{q(j,k)} \mathring{a}_{jk}(t, x; \xi_0 + i\eta'). \tag{3.5}$$

Taking into account condition (K–S) (see Section 1), for $k > j$,

$$\frac{k}{q(j, k) + j} < 1.$$

This implies, for $k > j$,

$$\max_{k > j} \frac{k - j}{q(j, k)} = \sigma_1 < 1. \tag{3.6}$$

To better understand (3.6), we draw a Newton diagram (see Fig. 1). Take the term a_{jk} where j signifies that it appears in the coefficient of ∂_t^{m-j} and k signifies its order in ∂_x. To a_{jk} we associate the point $(1 + q(j, k)/j, k/j)$. Thus the non-Kowalewskian terms are represented in the region $k/j > 1$. Then condition (K–S) says in particular that the segments joining the origin to the points in the non-Kowalewskian region ($k/j > 1$) have their slopes less than 1.

Starting from the point $P_0(1, 1)$, we draw a Newton polygon, namely, the smallest convex polygon containing all the points in the region $k/j > 1$. As we shall show, the terms which correspond to the first angular point P

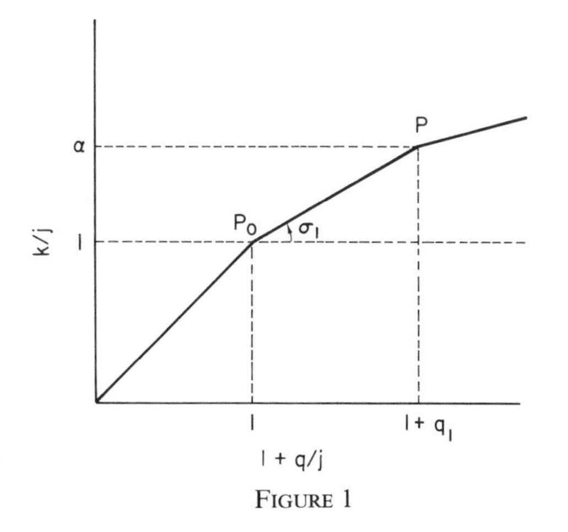

FIGURE 1

play a dominant role in our argument. We denote the coordinates of P by $(1 + q_1, \alpha)$.

Now putting

$$k - j = \sigma_1 q(j,k) - \delta_{jk}, \qquad (\delta_{jk} \geq 0), \qquad \text{for} \quad k > j,$$

the second sum on the right-hand side in (3.5) can be written

$$\sum_{k>j} (n^{\sigma_1}t)^{q(j,k)} \mathring{a}_{jk}(t, x; \xi_0 + i\eta')n^{-\delta_{jk}}.$$

The principal part will be

$$\mathring{h}_j(t, x; \xi_0 + i\eta') = a_{jj}(t, x; \xi_0 + i\eta') + \sum_{k>j}{}' (n^{\sigma_1}t)^{q(j,k)} \mathring{a}_{jk}(t, x; \xi_0 + i\eta'), \quad (3.7)$$

where $\sum'$ means the sum over the terms satisfying $\delta_{jk} = 0$, namely the terms corresponding to the points laying on the side P_0P in the Newton diagram. Now let us observe the behavior of the characteristic roots when $n^{\sigma_1}t$ becomes large. Let

$$s = n^{\sigma_1}t. \quad (3.8)$$

Let us remark that when t runs through $[0, t_0]$, s runs through $[0, n^{\sigma_1}t_0]$. In the expression of $\mathring{h}_j$, $q(j,k) \leq jq_1$, and the equal sign occurs when and only when the term is $\mathring{a}_{j,\alpha j}$ (see Fig. 1). Moreover there exists at least one j, say, j', such that $\mathring{a}_{j',\alpha j'}(0, x; \zeta) \not\equiv 0$. So we consider the *characteristic equation*:

$$\mu^m - \sum_j{}' \mathring{a}_{j,\alpha j}(0, x; \zeta)\mu^{m-j} = 0. \quad \text{(C)}$$

If eventually all $\mathring{a}_{j,\alpha j}(0, 0; \zeta)$ appearing in the coefficients of this equation are identically zero (as the polynomial in ζ), then by shifting the origin slightly in the x direction, we can assume at least one of them is not identically zero. Then, we choose $\zeta_0 = \xi_0 + i\eta_0$ in such a way that at least one of the roots, say, μ_1, when $x = 0$, satisfies

$$\operatorname{Re}\mu_1 > 0.$$

We can assume also $\eta_0 \neq 0$; $|\zeta_0| = 1$. Thus ζ_0 is fixed, and we microlocalize the differential operators $h_j(t, x; \xi_0 + \partial_x/n)$ in a Gevrey class. This is done in the following way.

Let $V(\eta_0)$ be the neighborhood of $\eta_0 : |\eta - \eta_0| < \varepsilon'$ (small). Let $\alpha(\eta)$ be a function of Gevrey class $\gamma^{(1+\varepsilon)}$, where ε is a fixed number satisfying

$$0 < \varepsilon < \sigma_1/2. \tag{3.9}$$

We take $\alpha(\eta)$ imposing the conditions:

(i) $\operatorname{supp}[\alpha(\eta)] \subset V(\eta_0)$; $\alpha(\eta) = 1$, for $|\eta - \eta_0| \leq \varepsilon'/2$
(ii) $\alpha(\eta) \in \gamma^{(1+\varepsilon)}$,
(iii) $0 \leq \alpha(\eta) \leq 1$.

In the same way, we take $\beta(x)$ so that

(i) $\operatorname{supp}[\beta(x)] \subset \{x; |x| \leq \varepsilon'\}$; $\beta(x) = 1$ for $|x| \leq \varepsilon'/2$.
(ii) $\beta(x) \in \gamma^{(1+\varepsilon)}$,
(iii) $0 \leq \beta(x) \leq 1$.

We denote $\alpha_n(\eta) = \alpha(\eta/n)$ $[= \alpha(\eta')]$, and the corresponding operator by $\alpha_n(D)$ [namely, the Fourier image of $\alpha_n(D)u$ is $\alpha_n(\eta)\hat{u}(\eta)$]. Now the localization of $\tilde{L}$ in x around $x = 0$ means the following. Let $a(x, \eta') \equiv a(x, \eta/n)$, where $a(x, \eta')$ is polynomial in η' with coefficients holomorphic in a neighborhood of the origin. We take a function $\phi(r) \in \gamma^{(1+\varepsilon)}$ for $0 \leq \phi(r) \leq 1$; $\phi(r) = 1$ for $0 \leq r \leq \varepsilon'$; $\phi(r) = 0$ for $r \geq 2\varepsilon'$. Then define

$$\tilde{a}(x, \eta') = \phi(|x|)a(x, \eta') + (1 - \phi(|x|))a(0, \eta').$$

$\tilde{a}$ is again a polynomial in η', with coefficients in $\gamma^{(1+\varepsilon)}$, defined in $\mathbb{R}^l$, which is equal to $a(x, \eta')$ for $|x| \leq \varepsilon'$. Moreover

$$\tilde{a}(x, \eta') - a(0, \eta') = \phi(|x|)(a(x, \eta') - a(0, \eta'))$$

shows that $\tilde{a}(x, \eta')$ is equal to $a(0, \eta')$ for $|x| \geq 2\varepsilon'$, so that

$$\sup_{x \in \mathbb{R}^l} |\tilde{a}(x, \eta') - a(0, \eta')| \leq \sup_{|x| \leq 2\varepsilon'} |a(x, \eta') - a(0, \eta')|.$$

This quantity becomes as small as we wish, if ε' is taken small and η' is bounded. Now if we apply $\beta(x)$ from the left to (3.1), we have the equation of the following form

$$\tilde{L}(\beta v_n) = g_n(x).$$

Here we can modify the coefficients of $\tilde{L}$ by the above method, because $\beta(x)$ has its support only in $|x| \leq \varepsilon'$. g_n has the following expression:

$$g_n = n \sum_{|p| \geq 1} \frac{(-1)^{|p|}}{p!} \sum_j h_j^{(p)}(t, x; \xi_0 + iD')n^{j-1}\partial_t^{m-j}(\beta_{(p)}v_n).$$

Here we can assume all the coefficients of $h_j^{(p)}$ are extended in the way shown above. After doing that, we apply $\alpha_n(D)$ from the left to the above equation. This time we can modify the symbol of $\tilde{L}$ in the following way. Let $a(x, \eta')$ be one of the symbols, then the localized symbol is

$$\tilde{a}(x, \eta') = a(x, \eta')\phi(|\eta' - \eta_0|) + a(x, \eta_0)(1 - \phi(|\eta' - \eta_0|)).$$

Then

$$\tilde{a}(x, \eta') - a(x, \eta_0) = (a(x, \eta') - a(x, \eta_0))\phi(|\eta' - \eta_0|).$$

Let us observe that this function (or rather this symbol) has its support in $|\eta' - \eta_0| \leq 2\varepsilon'$ and evidently $\tilde{a}(x, \eta') = a(x, \eta')$ for $|\eta' - \eta_0| \leq \varepsilon'$ [which contains the support of $\alpha_n(\eta)$]. We note

$$\sup_{x \in \mathbb{R}^l,\, \eta' \in \mathbb{R}^l} |\tilde{a}(x, \eta') - a(x, \eta_0)| \leq \sup_{x \in \mathbb{R}^l,\, |\eta' - \eta_0| \leq 2\varepsilon'} |a(x, \eta') - a(x, \eta_0)|.$$

Thus we arrive at the localized operator:

$$\tilde{L}_{\text{loc}}(\alpha_n \beta v_n) = f_n, \tag{3.10}$$

where $\tilde{L}_{\text{loc}} = \partial_t^m - n \sum_j \tilde{h}_j n^{j-1} \partial_t^{m-j}$ and $\sigma(\tilde{h}_j) = \tilde{h}_j(t, x; \xi_0 + i\eta')$ denote the microlocalized pseudo-differential operator (p.d.o.) in (x, η). f_n has the following expression:

$$\begin{aligned} \frac{1}{n} f_n = &\sum_{|p| \geq 1} \frac{(-1)^{|p|}}{p!} \sum_j \tilde{h}_j^{(p)} n^{j-1} \partial_t^{m-j}(\alpha_n \beta_{(p)} v_n) \\ &+ \sum_{p \geq 0} \frac{(-1)^{|p|}}{p!} \sum_j [\alpha_n, \tilde{h}_j^{(p)}] n^{j-1} \partial_t^{m-j}(\beta_{(p)} v_n). \end{aligned}$$

Finally, we remark that we used the usual notation: For $a(x, \eta)$,

$$a_{(\beta)}^{(\alpha)} = \partial_\eta^\alpha (i^{-1}\partial_x)^\beta a(x, \eta).$$

4. Reduction to Systems

We look at the microlocalized *homogeneous* equation (3.10) ignoring the specific form $\alpha_n \beta v_n$:

$$\tilde{L}_{\text{loc}}(w(t, x)) = 0.$$

Letting $w_{m-j} = n^{j-1}\partial_t^{m-j} w$ $(1 \le j \le m)$, and denoting

$$\mathscr{W} = {}^t(w_0, w_1, \ldots, w_{m-1}),$$

we obtain the following equivalent system

$$\frac{\partial}{\partial t}\mathscr{W} = nA\mathscr{W},$$

where

$$A = \begin{pmatrix} & 1 & & & \\ & & 1 & 0 & \\ & 0 & & \ddots & \\ & & & & 1 \\ \tilde{h}_m & \tilde{h}_{m-1} & & \cdots & \tilde{h}_1 \end{pmatrix}$$

Now we take $s\ (= n^{\sigma_1}t)$ as the new independent variable instead of t, then

$$\frac{\partial}{\partial s}\mathscr{W} = \nu A_\nu \mathscr{W}, \tag{4.1}$$

where

$$\nu = n^{1-\sigma_1}. \tag{4.2}$$

Let us note that $0 < 1 - \sigma_1 < 1$, so that ν tends to ∞ with n. Here we write A_ν instead of A to call attention to its dependency on n thus on ν.

To make clear the situation when s (hence n) is large, we make the second reduction. We put

$$\tilde{h}_j = s^{jq_1}\tilde{h}'_j, \qquad 1 \le j \le m.$$

More precisely let

$$h'_j = s^{-jq_1} \sum_{k \le j} a_{jk}(t, x; \xi_0 + i\eta') \Big/ n^{j-k} + \sum_{k > j} s^{q(j,k) - jq_1} \mathring{a}_{jk}(t, x; \xi_0 + i\eta') n^{-\delta_{jk}}.$$

Then $\tilde{h}'_j$ is a microlocalized operator in the above manner. When s is large, the first (Kowalewskian) terms become small because of the factor s^{-jq_1}. Next we divide the second (non-Kowalewskian) terms into three parts:

$$\mathring{a}_{j,\alpha j} + \sum' s^{q(j,k) - jq_1}\mathring{a}_{jk} + \sum'' s^{q(j,k) - jq_1}\mathring{a}_{jk} n^{-\delta_{jk}},$$

where the last sum is taken over the terms such that $\delta_{jk} > 0$. First, in the second summation, all terms satisfy $q(j,k) - jq_1 < 0$, so that these terms become small when s is taken large. Next, in the last summation we may have $q(j,k) - jq_1 > 0$. However, note that s takes on values up to $n^{\sigma_1}t_0$. Thus, to remove the effect of $s^{q(j,k) - jq_1}$, we restrict s up to $s_n = n^\theta$, where θ satisfies

$0 < \theta < \sigma_1$ and

$$\max_{j,k} \theta(q(j,k) - jq_1) < \delta_{jk}.$$

Thus if s_0 is taken large and we consider s only in $[s_0, s_n]$, we can regard the principal part of $\tilde{h}'_j$ as only the term $\mathring{a}_{j,\alpha j}(t, x; \xi_0 + i\eta')$. Moreover, since $\tilde{h}'_j$ is microlocalized, its symbol is close to $\mathring{a}_{j,\alpha j}(0, 0; \zeta_0)$.

Let

$$w'_{m-j} = s^{(j-1)q_1} n^{j-1} \partial_t^{m-j} w, 1 \le j \le m, \qquad \text{and} \qquad \mathscr{W}' = {}^t(w'_0, w'_1, \ldots, w'_{m-1}).$$

We obtain the following system:

$$\frac{\partial}{\partial s} \mathscr{W}' = \nu s^{q_1} A'_\nu \mathscr{W}' \qquad s \in [s_0, s_n], \tag{4.3}$$

where

$$A'_\nu = \begin{pmatrix} 1 & & & & \\ & 1 & & 0 & \\ & & 1 & & \\ & 0 & & \ddots & \\ & & & & 1 \\ \tilde{h}'_m & \tilde{h}'_{m-1} & & \cdots & \tilde{h}'_1 \end{pmatrix} + \frac{1}{\nu s^{q_1+1}} \begin{pmatrix} c_1 & & & \\ & c_2 & & 0 \\ & & \ddots & \\ & 0 & & \\ & & & c_m \end{pmatrix}$$

$$= \mathring{A}'_\nu + \frac{1}{\nu s^{q_1+1}} C$$

$[c_i = (m - 1 - i)q_1]$. Observe that A'_ν is a uniformly bounded operator acting on the L^2-space, when ν(hence n) becomes large, and $s \in [s_0, s_n]$,

Return now to (3.10): $\tilde{L}_{\text{loc}}(\alpha_n \beta v_n) = f_n$. This is expressed in the following form.

Let

$$\begin{aligned} v_{n,m-j} &= n^{j-1} \partial_t^{m-j} v_n, \qquad 1 \le j \le m, \\ V_n &= {}^t(v_{n,0}, v_{n,1}, \ldots, v_{n,m-1}); \\ v'_{n,m-j} &= n^{j-1} s^{(j-1)q_1} \partial_t^{m-j} v_n, \qquad 1 \le j \le m, \\ V'_n &= {}^t(v'_{n,0}, v'_{n,1}, \ldots, v'_{n,m-1}). \end{aligned}$$

Then (3.10) is equivalent to

$$\frac{\partial}{\partial s} (\alpha_n \beta V_n) = \nu A_\nu (\alpha_n \beta V_n) + n^{-\sigma_1} F_n, \tag{4.4}$$

$$\frac{\partial}{\partial s} (\alpha_n \beta V'_n) = \nu s^{q_1} A'_\nu (\alpha_n \beta V'_n) + n^{-\sigma_1} F_n, \tag{4.5}$$

where $F_n = {}^t(0, 0, \ldots, f_n)$.

5. Energy Inequality

We look at Eq. (4.3):

$$\frac{\partial}{\partial s}\mathscr{W}' = \nu s^{q_1} A'_\nu(s, x; D')\mathscr{W}'. \tag{5.1}$$

Let us recall that, for s and ν large, say, for $s \geq s_0$, $\nu \geq \nu_0$, the eigenvalues of $A'_\nu(s, x; \eta')$ become very close to those of $\mathring{A}'_\nu(0, 0; \eta_0)$. Now, by hypothesis, the eigenvalues $\mu_1, \mu_2, \ldots, \mu_m$ of the latter satisfy

$$\operatorname{Re}\mu_1, \ldots, \operatorname{Re}\mu_k > 0; \qquad \operatorname{Re}\mu_{k+1}, \ldots, \operatorname{Re}\mu_m \leq 0, \qquad \text{with} \quad k \geq 1.$$

Let us denote

$$\min_{1 \leq j \leq k} \operatorname{Re}\mu_i = 2\delta.$$

First we define a constant regular matrix N_0: Let

$$\mathscr{H} = \begin{pmatrix} & 1 & & & \\ & & 1 & & 0 \\ & 0 & & \ddots & \\ & & & & 1 \\ \gamma_m & \gamma_{m-1} & & \cdots & \gamma_1 \end{pmatrix},$$

where $\gamma_j = \mathring{a}_{j,\alpha j}(0, 0; \zeta_0)$. Then

$$N_0 \mathscr{H} N_0^{-1} = \begin{pmatrix} \mu_1 & & & & \\ & \mu_2 & & & 0 \\ & & \mu_3 & & \\ & b_{ij} & & \ddots & \\ & & & & \mu_m \end{pmatrix} = \mathscr{H}',$$

where we can assume $|b_{ij}| \leq \delta/8m$.

Let $A'_\nu(s, x; \eta') - \mathscr{H} = \tilde{A}_\nu(s, x; \eta')$. This is small; more precisely denoting

$$N_0 \tilde{A}_\nu(s, x; \eta') N_0^{-1} = (d_{ij}(s, x; \eta'))_{1 \leq i,\, j \leq m}$$

we have

$$\sup_{x, \eta'} |d_{ij}(s, x; \eta')| < \frac{\delta}{8m} \qquad \text{if} \quad \nu \geq \nu_0, \quad s \geq s_0.$$

Of course, for this we assume first of all the sharpness of the microlocalization, and this is always fulfilled if we take ε' small (See Section 3). Note that for the t direction, we restrict $s \in [s_0, s_n]$. Since the corresponding t_n $(= n^{-\sigma_1} s_n = n^{\theta - \sigma_1})$ tends to 0 when n tends to infinity, the sharpness in t is automatically satisfied.

Transforming (5.1) by N_0, we get

$$\frac{\partial}{\partial s} N_0 \mathscr{W}' = v s^{q_1} (\mathscr{H}' + (d_{ij})) N_0 \mathscr{W}'.$$

Now by letting

$$N_0 \mathscr{W}' = {}^t(u_1, u_2, \ldots, u_m), \tag{5.2}$$

we consider the following energy form:

$$E_n(s; \mathscr{W}') = \exp\left(-\frac{2v\delta s^{q_1+1}}{q_1+1}\right)\left(\sum_{i=1}^{k} \|u_i\|^2 - \sum_{i=k+1}^{m} \|u_i\|^2\right). \tag{5.3}$$

where $\|\cdot\|$ denotes the L^2-norm. Take the derivative of E_n, and denote $E_n(s; \mathscr{W}')$ simply by $E_n(s)$, then

$$\begin{aligned}
\exp\left(\frac{2v\delta s^{q_1+1}}{q_1+1}\right) E_n'(s) &= \sum_{i=1}^{k} 2\operatorname{Re}(u_i', u_i) - \sum_{i=k+1}^{m} 2\operatorname{Re}(u_i', u_i) \\
&\quad - 2v\delta s^{q_1}\left(\sum_{i=1}^{k} \|u_i\|^2 - \sum_{i=k+1}^{m} \|u_i\|^2\right) \\
&= \sum_{i=1}^{k} 2(\operatorname{Re}\mu_i - \delta) v s^{q_1} \|u_i\|^2 \\
&\quad - \sum_{i=k+1}^{m} 2(\operatorname{Re}\mu_i - \delta) v s^{q_1} \|u_i\|^2 \\
&\quad + \sum_{i=1}^{k} \sum_{j=1}^{m} 2 v s^{q_1} \operatorname{Re}((b_{ij} + d_{ij}) u_j, u_i) \\
&\quad - \sum_{i=k+1}^{m} \sum_{j=1}^{m} 2 v s^{q_1} \operatorname{Re}((b_{ij} + d_{ij}) u_j, u_i).
\end{aligned}$$

First, since $\sup|(b_{ij} + d_{ij})| \leq \delta/4m$, by the sharp form of Gårding's inequality, if n (hence v) is large, we get $\|b_{ij} + d_{ij}\| \leq \delta/2m$, so that

$$2\operatorname{Re}((b_{ij} + d_{ij}) u_j, u_i) \geq -(\delta/m)\|u_i\|\,\|u_j\|.$$

The right-hand side is thus estimated from below by

$$2 v s^{q_1} \delta \left(\sum_{i=1}^{m} \|u_i\|^2\right) - v s^{q_1} \frac{\delta}{m} \sum_{i=1}^{m} \sum_{j=1}^{m} \|u_i\|\,\|u_j\| \geq v s^{q_1} \delta \left(\sum_{i=1}^{m} \|u_i\|^2\right),$$

a fortiori, $\geq \nu s^{q_1}\delta(\sum_{i=1}^{k}\|u_i\|^2 - \sum_{i=k+1}^{m}\|u_i\|^2)$. Thus, $E_n'(s) \geq \nu s^{q_1}\delta E_n(s)$. Supposing $E_n(s_0) > 0$, the integration gives

$$E_n(s; \mathscr{W}') \geq \exp\{\nu(s^{q_1+1} - s_0^{q_1+1})\delta/(q_1+1)\}E_n(s_0; \mathscr{W}'). \tag{5.4}$$

Let us remark that s can be chosen arbitrarily large, more precisely $s \in [s_0, s_n]$ where $s_n = n^\theta$ (see Section 4), and δ is a positive constant, and $q_1 > 0$.

6. Choice of Initial Data

In the above inequality, we choose the initial data $\mathscr{W}'(s_0)$ or rather $\mathscr{W}_n'(s_0)$ in such a way that $E_n(s_0; \mathscr{W}_n') > 0$. For this we consider the following function $\psi(\eta')$:

$$\psi(\eta') \in \gamma^{(1+\varepsilon)}, \qquad \psi(\eta') \geq 0, \qquad \operatorname{supp}[\psi] \subset \{\eta'; |\eta' - \eta_0| < \varepsilon'/2\},$$

and

$$\int \psi(\eta')\, d\eta' = 1.$$

Let us note that $\alpha(\eta') = 1$ *on the support of* ψ. Finally denote by $\tilde{\psi}_n(x)$ the inverse Fourier image of $\psi(\eta')$:

$$\tilde{\psi}_n(x) = (2\pi)^{-l}\int e^{ix\eta}\psi(\eta')\, d\eta, \qquad \eta' = \eta/n.$$

Then

$$\mathscr{W}_n'(s_0) = N_0^{-1}\begin{pmatrix}\tilde{\psi}_n(x)\\ 0\\ 0\\ \vdots\\ 0\end{pmatrix}, \tag{6.1}$$

so that its ith component has the form

$$w_{n,i}'(s_0) = m_i\tilde{\psi}_n(x), \qquad 0 \leq i \leq m-1.$$

Let us recall the relation between $\mathscr{W}_n(s)$ and $\mathscr{W}_n'(s)$. Then

$$w_{n,i}(s_0) = s_0^{-(m-1-i)q_1}w_{n,i}'(s_0) = c_i\tilde{\psi}_n(x), \qquad 0 \leq i \leq m-1. \tag{6.2}$$

Now we are going to define the initial data at $s = 0$. The basic idea is to solve Eq. (4.1) backward starting from the data in (6.2). However we need some detailed properties of the data obtained at $s = 0$. More precisely we should know clearly at least approximately the structure of the Riemann matrix for Eq. (4.1). For this purpose we introduce the following notion and space of operators.

DEFINITION 6.1. A function $f_\nu \in L^2(\mathbb{R}^l)$ or an operator a_ν acting on $L^2(\mathbb{R}^l)$, depending on a real large parameter ν, or even a function $\phi(\nu)$ itself is called *negligible*, if for any large K, $\|f_\nu\|$, $\|a_\nu\|$, or $|\phi(\nu)|$ is estimated by $\exp(-\nu K)$ when ν is large.

DEFINITION 6.2. $\mathscr{L}^{(1+\varepsilon)}$. An operator acting on $L^2(\mathbb{R}^l)$ belongs to $\mathscr{L}^{(1+\varepsilon)}$ if and only we can decompose it in such a way that

$$a_\nu = \mathring{a}_\nu + a_{\nu,R},$$

where $a_{\nu,R}$ is negligible, and $\mathring{a}_\nu$ is a pseudo-differential operator, depending on the parameter ν, where $\mathring{a}_\nu(x,\eta')$ $(\eta' = \eta/n)$ has the following properties:

(i) There exist positive constants K and c_1 such that

$$|\partial_\eta^\alpha \partial_x^\beta \mathring{a}_\nu(x,\eta')| \le \exp(c_1\nu)\alpha!^{1+\varepsilon}\beta!^{1+\varepsilon}K^{|\alpha+\beta|}/n^{|\alpha|},$$

c_1 may depend on each a_ν.

(ii) $\mathring{a}_\nu(x,\eta') - \mathring{a}_\nu(x,\eta_0)$ has its support (as a function of η') in the set $|\eta' - \eta_0| \le 2\varepsilon'$.

Note that n and ν are connected by $\nu = n^{1-\sigma_1}$ and $0 < 1 - \sigma_1 < 1$, and ε is restricted by $0 < \varepsilon < \sigma_1/2$. We say that p.d.o. $\mathring{a}_\nu(x, D')$ is an *essential part* of a_ν. Let us note that the above decomposition is not unique.

The following fact is fundamental.

LEMMA 6.1. *Let a_ν be p.d.o. satisfying* (i), (ii) *of Definition* 6.2, *more explicitly,*

$$|a_\nu(x,\eta')^{(\alpha)}_{(\beta)}| \le f(\nu)(|\alpha+\beta|)!^{1+\varepsilon}K^{|\alpha+\beta|}/n^{|\alpha|},$$

and let

$$b_\nu = b_0 + b_1 + \cdots + b_j + \cdots + b_{j_0}, \qquad j_0 = [n^{1-2\varepsilon}],$$

where

$$|b_j(x,\eta')^{(\alpha)}_{(\beta)}| \le g(\nu)l^j \frac{(2j+|\alpha+\beta|)!^{1+\varepsilon}}{j!} \frac{K^{2j+|\alpha+\beta|}}{n^{j+|\alpha|}},$$

with $b_j(x,\eta')$ $(j \ge 1)$ having its support in $|\eta' - \eta_0| \le 2\varepsilon'$ as a function of η' and a_ν and b_0 having the property (ii).

Then denoting by $a \circ b$ the operator product of a, b, we have

$$a_\nu \circ b_\nu = c_0 + c_1 + \cdots + c_{j_0} + c_{\nu,R},$$

where c_j has the same property as b_j; more precisely,

$$|c_j(x,\eta')^{(\alpha)}_{(\beta)}| \leq c(\varepsilon)^3 f(v)g(v)l^j \frac{(2j+|\alpha+\beta|)!^{1+\varepsilon}}{j!} \times \frac{K^{2j+|\alpha+\beta|}}{n^{j+|\alpha|}},$$

and $c_{v,R}$ is negligible. More precisely, $\|c_{v,R}\| \leq \exp(-n^{1-2\varepsilon})$, *and* $f(v)$, $g(v) \leq \exp(cv)$. *Finally*

$$c(\varepsilon) = \sup_j \sum_{k=0}^{j} \frac{1}{(C^j_k)^\varepsilon} < +\infty.$$

Proof. Define

$$c_j = \sum_{|p|+k=j} \frac{1}{p!} a^{(p)}_v b_{k(p)}.^\dagger$$

Then

$$(a^{(p)}_v b_{k(p)})^{(\alpha)}_{(\beta)} = \sum_{\alpha',\beta'} C^\alpha_{\alpha'} C^\beta_{\beta'} a^{(p+\alpha')}_{v(\beta')} b^{(\alpha'')}_{k(p+\beta'')}.$$

This is estimated in absolute value by

$$f(v)g(v)\frac{K^{2j+|\alpha+\beta|}}{n^{j+|\alpha|}} l^k \sum_{\alpha',\beta'} C^\alpha_{\alpha'} C^\beta_{\beta'} |p+\alpha'+\beta'|!^{1+\varepsilon} \frac{(2k+|p+\alpha''+\beta''|)!^{1+\varepsilon}}{k!}.$$

Since $\sum_{|\alpha'|=s} C^\alpha_{\alpha'} \leq C^{|\alpha|}_s$, $\sum_{|\beta'|=t} C^\beta_{\beta'} \leq C^{|\beta|}_t$, and

$$|p+\alpha'+\beta'|!^{1+\varepsilon}(2k+|p+\alpha''+\beta''|)!^{1+\varepsilon} = (2j+|\alpha+\beta|)!^{1+\varepsilon}/(C^{2j+|\alpha+\beta|}_{|p+\alpha'+\beta'|})^{1+\varepsilon},$$

we have

$$\sum_{\alpha',\beta'} \cdots \leq \frac{(2j+|\alpha+\beta|)!^{1+\varepsilon}}{k!} \sum_{0\leq s\leq|\alpha|,\ 0\leq t\leq|\beta|} C^{|\alpha|}_s C^{|\beta|}_t/(C^{2j+|\alpha+\beta|}_{|p+\alpha'+\beta'|})^{1+\varepsilon}$$

Now

$$C^{2j+|\alpha+\beta|}_{|p|+s+t} \geq C^{2j}_{|p|} C^{|\alpha+\beta|}_{s+t} \geq C^{2j}_{|p|} C^{|\alpha|}_s C^{|\beta|}_t,$$

thus

$$\sum \cdots \leq \frac{1}{(C^{2j}_{|p|})^{1+\varepsilon}} \sum_{0\leq s\leq|\alpha|} \frac{1}{(C^{|\alpha|}_s)^\varepsilon} \sum_{0\leq t\leq|\beta|} \frac{1}{(C^{|\beta|}_t)^\varepsilon} \leq \frac{c(\varepsilon)^2}{(C^{2j}_{|p|})^{1+\varepsilon}}.$$

†In the calculus of pseudo-differential operators, especially in that of asymptotic expansions, the reader should note the difference between the two kinds of products: the operator product and the symbol product. Here the right-hand side indicates a symbol product. In the proof of Proposition 6.1, we shall use the notation of an operator product in the cases where there is the possibility of confusion and to make the notations coherent. After that, we shall not use it systematically.

So, we are led to consider

$$\sum_{|p|+k=j} \frac{l^k}{p!k!} \frac{1}{(C_{|p|}^{2j})^{1+\varepsilon}},$$

and since

$$\sum_{|\alpha|=s} \frac{1}{\alpha!} = l^s \frac{1}{s!},$$

this is equal to

$$\sum_{|p|} \frac{l^{k+|p|}}{|p|!k!} \frac{1}{(C_{|p|}^{2j})^{1+\varepsilon}} = \frac{l^j}{j!} \sum_{|p|} \frac{C_{|p|}^j}{(C_{|p|}^{2j})^{1+\varepsilon}} \leq \frac{l^j}{j!} c(\varepsilon).$$

Thus

$$|c_{j(\beta)}^{(\alpha)}| \leq c(\varepsilon)^3 f(v) g(v) \frac{(2j + |\alpha + \beta|)!^{1+\varepsilon}}{j!} \frac{K^{2j+|\alpha+\beta|} l^j}{n^{j+|\alpha|}},$$

which proves the first assertion.

Now, if we put $K_1 = 3^{2(1+\varepsilon)} K^2 l$, $K_2 = 3^{1+\varepsilon} K$, then

$$|c_{j(\beta)}^{(\alpha)}| \leq c(\varepsilon)^3 f(v) g(v) \frac{j!^{1+2\varepsilon}}{n^j} K_1^j \frac{|\alpha + \beta|!^{1+\varepsilon}}{n^{|\alpha|}} K_2^{|\alpha+\beta|}.$$

So if we let $j = j_0 = [n^{1-2\varepsilon}]$, then

$$\frac{j!^{1+2\varepsilon}}{n^j} K_1^j < \left(\frac{j^{1+2\varepsilon}}{n} K_1 \right)^j \ll \exp(-n^{1-2\varepsilon}).$$

Since $v \ll n^{1-2\varepsilon}$, and $f(v)g(v) \leq \exp(2cv)$, we have proved the desired property of $c_{v,R}$. ■

Now we return to Eq. (4.1):

$$\frac{\partial}{\partial s} \mathscr{W} = v A_v \mathscr{W}.$$

Let $\Phi_v(s, s')$ be the fundamental matrix or the Riemann matrix of Eq. (4.1) starting from s':

$$\Phi_v(s', s') = I, \qquad \frac{\partial}{\partial s} \Phi_v(s, s') = v A_v \Phi_v(s, s'), \qquad s \in [0, s_0], \quad s' \in [0, s_0].$$

The following fact plays a decisive role.

PROPOSITION 6.1 *The fundamental matrix $\Phi_v(s, s_0)$ belongs to $\mathscr{L}^{(1+\varepsilon)}$.*

Proof. Since the principle of the proof is the same, for brevity, we prove the proposition for $\Phi_v(s, 0)$. We consider only the solution of (4.1) with initial data of the form: ${}^t(\phi, 0, 0, \ldots, 0)$. We use the method of successive approximations. Namely, let

$$U_0 = {}^t(I, 0, \ldots, 0), \qquad U_1 = \nu \int_0^s A_\nu(\sigma) \circ U_0(\sigma)\, d\sigma.$$

In general

$$U_k(s) = \nu \int_0^s A_\nu(\sigma) \circ U_{k-1}(\sigma)\, d\sigma, \qquad k \geq 1.$$

Let $A = (a_{ij})$, $U_k = {}^t(u_k^{(1)}, u_k^{(2)}, \ldots, u_k^{(m)})$. First

$$u_1^{(i)} = \nu \int_0^s a_{i1}(\sigma)\, d\sigma.$$

Since

$$|a_{ij}(s, x, \eta')_{(\beta)}^{(\alpha)}| \leq c_0 |\alpha + \beta|!^{1+\varepsilon} K^{|\alpha+\beta|}/n^{|\alpha|}, \qquad s \in [0, s_0],$$

we have

$$|u_1^{(i)}(s, x, \eta')_{(\beta)}^{(\alpha)}| \leq (\nu s) c_0 |\alpha + \beta|!^{1+\varepsilon} K^{|\alpha+\beta|}/n^{|\alpha|}.$$

Next

$$u_2^{(i)}(s) = \nu \sum_{k=1}^{m} \int_0^s a_{ik}(\sigma) \circ u_1^{(k)}(\sigma)\, d\sigma, \qquad 1 \leq i \leq m.$$

By Lemma 6.1, we know that $u_2^{(i)} \in \mathscr{L}^{(1+\varepsilon)}$. More precisely, let

$$u_2^{(i)} \sim u_{2,0}^{(i)} + u_{2,1}^{(i)} + \cdots + u_{2,j}^{(i)} + \cdots.$$

Then

$$|u_{2,j}^{(i)}(s, x, \eta')_{(\beta)}^{(\alpha)}| \leq \frac{(c_1 \nu s)^2}{2!} \frac{(2j + |\alpha + \beta|)!^{1+\varepsilon}}{j!} \frac{l^j K^{2j+|\alpha+\beta|}}{n^{j+|\alpha|}},$$

where $c_1 = c(\varepsilon)^3 m c_0$. So, if we define $\mathring{u}_2^{(i)} = \sum_{j=0}^{j_0} u_{2,j}^{(i)}$, where $j_0 = [n^{1-2\varepsilon}]$, by the above lemma and if we put

$$\frac{\partial}{\partial s} \mathring{U}_2 = \nu A_\nu \circ \mathring{U}_1 + F_{2,\nu},$$

we have

$$\|F_{2,\nu}\| \leq (c_1 \nu s) \exp(-n^{1-2\varepsilon}).$$

In general, supposing $\mathring{U}_0, \ldots, \mathring{U}_{k-1}, \mathring{U}_k$ defined in such a way that

$$\mathring{U}_k = U_{k,0} + U_{k,1} + \cdots + U_{k,j_0},$$

where

(i) $$|U_{k,j}(s, x, \eta')_{(\beta)}^{(\alpha)}| \leq \frac{(c_1 \nu s)^k}{k!} \frac{(2j + |\alpha + \beta|)!^{1+\varepsilon}}{j!} \frac{l^j K^{2j+|\alpha+\beta|}}{n^{j+|\alpha|}},$$

(ii) $U_{k,0}(s, x, \eta') - U_{k,0}(s, x, \eta_0)$ has its support in $|\eta' - \eta_0| \le 2\varepsilon'$ and $U_{k,j}(s, x, \eta')$ has its support in $|\eta' - \eta_0| \le 2\varepsilon'$ for $j \ge 1$,

(iii) $(\partial/\partial s)\mathring{U}_k = \nu A_\nu \circ \mathring{U}_{k-1} + F_{k,\nu}$, where

$$\|F_{k,\nu}\| \le \frac{(c_1 \nu s)^{k-1}}{(k-1)!} \exp(-n^{1-2\varepsilon}),$$

then we can define $\mathring{U}_{k+1}$ satisfying the same properties as above. In fact, consider

$$\nu \int_0^s A_\nu(\sigma) \circ \mathring{U}_k(\sigma)\, d\sigma.$$

Since $A_\nu(\sigma) \circ \mathring{U}_k(\sigma) = A_\nu(\sigma) \circ (U_{k,0} + U_{k,1} + \cdots + U_{k,j_0})$, in view of Lemma 6.1, let

$$U_{k+1,\, j}(s) = \nu \int_0^s \sum_{i+|p|=j} \frac{1}{p!} A_\nu^{(p)}(\sigma) U_{k,i(p)}(\sigma)\, d\sigma,$$

we have

$$|U_{k+1,\, j}(s, x, \eta')_{(\beta)}^{(\alpha)}| \le \frac{(c_1 \nu s)^{k+1}}{(k+1)!} \frac{(2j + |\alpha+\beta|)!^{1+\varepsilon}}{j!} \frac{l^j K^{2j+|\alpha+\beta|}}{n^{j+|\alpha|}}.$$

Moreover letting $\mathring{U}_{k+1} = U_{k+1,\,0} + U_{k+1,\,1} + \cdots + U_{k+1,\, j_0}$ we see that

$$\frac{\partial}{\partial s} \mathring{U}_{k+1}(s) - \nu A_\nu(s) \circ \mathring{U}_k(s) = F_{k+1,\,\nu}$$

satisfies

$$\|F_{k+1,\,\nu}\| \le \frac{(c_1 \nu s)^k}{k!} \exp(-n^{1-2\varepsilon}).$$

Now we define

$$\mathring{U}_\nu = \mathring{U}_0 + \mathring{U}_1 + \cdots + \mathring{U}_{j_0}.$$

Then

$$\frac{\partial}{\partial s} \mathring{U}_\nu = \nu A_\nu \mathring{U}_\nu + \left(\sum_{k=1}^{j_0} F_{k,\nu} - \nu A_\nu \mathring{U}_{j_0} \right).$$

We show that $\nu A_\nu \mathring{U}_{j_0}$ is negligible. It suffices to show that $\mathring{U}_{j_0}$ is negligible. First, as we see in the proof of Lemma 6.1, in view of the above inequality,

$$\begin{aligned} |U_{j_0,j}(s, x, \eta')_{(\beta)}^{(\alpha)}| &\le \frac{(c_1 \nu s)^{j_0}}{j_0!} \frac{j!^{1+2\varepsilon}}{n^j} K_1^j \frac{|\alpha+\beta|!^{1+\varepsilon}}{n^{|\alpha|}} K_2^{|\alpha+\beta|} \\ &\le \left(\frac{e c_1 \nu s_0}{j_0} \right)^{j_0} \left(\frac{j^{1+2\varepsilon}}{n} K_1 \right)^j \frac{|\alpha+\beta|!^{1+\varepsilon}}{n^{|\alpha|}} K_2^{|\alpha+\beta|}. \end{aligned}$$

Now, since $j_0 = [n^{1-2\varepsilon}]$ and $\nu = n^{1-\sigma_1}$ and $2\varepsilon < \sigma_1$, we have $\nu/j_0 \ll 1$. So that the first factor $\le \exp(-n^{1-2\varepsilon})$. Moreover $\sum_{j=0}^{j_0} (j^{1+2\varepsilon} K_1/n)^j$ remains bounded (with respect to n), for $j^{1+2\varepsilon} K_1/n \ll \frac{1}{2}$ $(j \le j_0)$ if n is large. This shows that $\mathring{U}_{j_0}$ is negligible, and more precisely, $\|\nu A_\nu \mathring{U}_{j_0}\| \le \exp(-n^{1-2\varepsilon})$. Finally $\|\sum_{k=1}^{j_0} F_{k,\nu}\|$ is negligible, for $\sum_k \|F_{k,\nu}\| \le \exp(c_1 \nu s_0) \exp(-n^{1-2\varepsilon})$.

Let

$$\frac{\partial}{\partial s} \mathring{U}_\nu = \nu A_\nu \mathring{U}_\nu + F_\nu.$$

We have shown that F_ν is uniformly negligible for $s \in [0, s_0]$. Moreover $U_\nu(0) = \mathring{U}_\nu(0)$. Now

$$\frac{\partial}{\partial s} (U_\nu - \mathring{U}_\nu) = \nu A_\nu (U_\nu - \mathring{U}_\nu) - F_\nu.$$

In view of the fact that $U_\nu - \mathring{U}_\nu$ has the initial data 0, and F_ν is uniformly negligible, we see that $U_\nu - \mathring{U}_\nu$ is also uniformly negligible.

Finally,

$$\mathring{U}_\nu(s, x, \eta') = \sum_{k=0}^{j_0} \sum_{j=0}^{j_0} U_{k,j}(s, x, \eta').$$

Since

$$|U^{(\alpha)}_{k,j(\beta)}| \le \frac{(c_1 \nu s)^k}{k!} j!^{1+2\varepsilon} \frac{K_1^j |\alpha + \beta|!^{1+\varepsilon}}{n^j} \frac{K_2^{|\alpha+\beta|}}{n^{|\alpha|}}$$

then

$$|\mathring{U}_\nu(s, x, \eta')^{(\alpha)}_{(\beta)}| \le \exp(c_1 \nu s) \left(\sum_{j=0}^{j_0} \frac{j!^{1+2\varepsilon} K_1^j}{n^j} \right) \frac{|\alpha + \beta|!^{1+\varepsilon} K_2^{|\alpha+\beta|}}{n^{|\alpha|}}.$$

$\mathring{U}_\nu(s, x, D')$ has thus the desired property because the quantity in the parentheses is bounded (with respect to n). ■

We are in a position to define the initial data of $u_n(t, x)$ [hence that of $v_n(t, x)$] at $t = 0$. First by Proposition 6.1,

$$\Phi_\nu(0, s_0) = \mathring{\Phi}_\nu(0, s_0) + \Phi_{\nu,R},$$

where $\Phi_{\nu,R}$ is negligible. Let

$$W_n(s) = \mathring{\Phi}_\nu(s, s_0) \mathscr{W}_n(s_0) \qquad \text{for} \quad s \in [0, s_0]. \tag{6.3}$$

Observe that $W_n(s)$ satisfies

$$\frac{\partial}{\partial s} W_n(s) = \nu A_\nu W_n(s) + F_\nu(s), \tag{6.4}$$

where $F_\nu(s)$ is uniformly negligible.

Now take a function $\gamma(\eta')$ satisfying the following conditions: $\gamma(\eta') \in \gamma^{(1+\varepsilon)}$, $0 \le \gamma(\eta') \le 1$, $\mathrm{supp}[\gamma] \subset V(\eta_0)$, $\gamma(\eta') = 1$, *on the support of* $\psi(\eta')$. Then denoting the convolution operator by $\gamma_n(D)$ $[\gamma_n(\eta) = \gamma(\eta/n) = \gamma(\eta')]$, *we define the initial data by*

$$V_n(0) = \gamma_n(D) W_n(0) = \gamma_n(D) \mathring{\Phi}_v(0, s_0) \mathscr{W}_n(s_0). \tag{6.5}$$

More explicitly, denoting $\mathring{\Phi}_v(0, s_0) = (t_{ij,v}(x, D'))_{0 \le i,\, j \le m-1}$ and in view of (6.2) and (6.3), the ith component of $V_n(0)$ is expressed by

$$\gamma_n(D) \sum_{j=0}^{m-1} t_{ij,v}(x, D') c_j \tilde{\psi}_n(x).$$

Since the ith component of $V_n(0)$ is by definition $n^{m-1-i} \partial_t^i v_n(0, x)$, denoting

$$t_{i,v} = \sum_{j=0}^{m-1} c_j t_{ij,v}(x, D'), \tag{6.6}$$

the initial data $\phi_{n,i}(x)$ $(0 \le i \le m-1)$ in (2.6) has the form:

$$\phi_{n,i}(x) = n^{-(m-1-i)} \gamma_n(D) t_{i,v}(x, D') \tilde{\psi}_n(x). \tag{6.7}$$

Note that $\phi_{n,i}(x)$ are entire functions.

Let us see what the successive derivatives of $\phi_{n,i}(x)$ are. First we note that, in view of the above proposition, there exists some constant C (independent of v) such that

$$\|t_{i,v}(x, D')\| \le C \exp(c_1 v s_0).$$

Next put $\varphi_{n,i}(x) = t_{i,v}(x, D') \tilde{\psi}_n(x)$. Then

$$\phi_{n,i}(x) = (2\pi)^{-l} \int e^{ix\eta} \gamma(\eta') \hat{\varphi}_{n,i}(\eta)\, d\eta\, n^{-(m-1-i)}$$

so that

$$|\partial^\alpha \phi_{n,i}(x)| \le (2\pi)^{-l} \int |\eta|^{|\alpha|} \gamma(\eta') |\hat{\varphi}_{n,i}(\eta)|\, d\eta.$$

Since $\gamma(\eta')$, as a function of η, has its support in $nV(\eta_0)$, denoting by A the maximum distance from the origin to $V(\eta_0)$, we have

$$|\partial^\alpha \phi_{n,i}(x)| \le (nA)^{|\alpha|} \int \gamma(\eta') |\hat{\varphi}_{n,i}(\eta)|\, d\eta.$$

On the other hand the integral is estimated by

$$\int_{\mathrm{supp}[\gamma(\eta')]} |\hat{\varphi}_{n,i}(\eta)|\, d\eta.$$

By Schwarz and Plancherel this is estimated by the constant $n^{l/2} \|\varphi_{n,i}(x)\|$. Now $\|\varphi_{n,i}(x)\| \le \|t_{i,v}\| \, \|\tilde{\psi}_n(x)\|$. Thus we have

$$|\partial^\alpha \phi_{n,i}(x)| \le \exp(cv)(nA)^{|\alpha|}, \tag{6.8}$$

where c is a constant greater than $c_1 s_0$ (for instance, $2c_1 s_0$).

We apply Proposition 2.3 to the actual initial data (6.8). The constant c_0 there should replaced by $\exp(cv)$. Noting that $nt = vs$, we have

PROPOSITION 6.2 (A priori estimates). *$V_n(s)$ has the following estimate*

$$\sup_{|x| \leq r'} |V_n(s, x)| \leq C \exp(cv) \exp(c'vs),$$

where C is the same constant as in Proposition 2.3, c is a constant depending only on the equation and s_0, and $c' = (|\xi_0| + A)K$.

7. RELATIONS BETWEEN $V_n(s)$ AND $W_n(s)$

We compare (4.1) with (4.4). Let us recall

$$n^{-1} f_n = \sum_j \sum_{|p| \geq 1} \frac{(-1)^{|p|}}{p!} h_j^{(p)} n^{j-1} \partial_t^{m-j} (\alpha_n \beta_{(p)} v_n)$$
$$+ \sum_j \sum_{p \geq 0} \frac{(-1)^{|p|}}{p!} [\alpha_n, h_j^{(p)}] n^{j-1} \partial_t^{m-j} (\beta_{(p)} v_n).$$

Here we note that $h_j^{(p)}$ are localized in x.

The first aim is to show that $f_n(s)$ is negligible, more precisely, uniformly negligible on every compact set in $s \in [0, \infty)$, if n becomes large. This fact relies heavily on the specific form of the initial data defined at the end of the previous section.

Hereafter we are mainly concerned with functions $\alpha_n, \beta, \gamma_n$, and $t_{i,v}$. They belong to $\gamma^{(1+\varepsilon)}$ [class $(1 + \varepsilon)$ of Gevrey]. So, taking K properly, we can assume

$$|\alpha_n^{(p)}|, |\gamma_n^{(p)}| \leq p!^{1+\varepsilon} K^{|p|}/n^{|p|}; \qquad |\beta_{(p)}| \leq p!^{1+\varepsilon} K^{|p|};$$
$$|t_{i,v(q)}^{(p)}| \leq \exp(cv) p!^{1+\varepsilon} q!^{1+\varepsilon} K^{|p+q|}/n^{|p|}.$$

First we note that

$$\frac{1}{p!} [\alpha_n, h_j^{(p)}] \sim \frac{1}{p!} \sum_q \frac{1}{q!} h_{j(q)}^{(p)} \alpha_n^{(q)}$$

and that when $h_{j(q)}^{(p)}$ acts on the functions of the form $\alpha_n^{(q)} f$, its norm is estimated by

$$\left\| \frac{1}{p!q!} h_{j(q)}^{(p)} \right\| \leq C \frac{p!^{\varepsilon} q!^{\varepsilon}}{n^{|p|}} K^{|p+q|}$$
$$= C \frac{p!^{\varepsilon}}{n^{\varepsilon|p|}} \frac{q!^{\varepsilon}}{n^{\varepsilon|q|}} K^{|p+q|} n^{\varepsilon|p+q| - |p|}$$

and that $p!^{\varepsilon}n^{-\varepsilon|p|}$ is small when $|p| \le j_0 = [n^{1-2\varepsilon}]$. For later argument, it is convenient to introduce

$$d(k) = \sup_n \sum_{j=0}^{j_0} j^k(j!^{\varepsilon}K_1^j n^{-\varepsilon j}),$$

where $K_1 = 2^{1+\varepsilon}K(>K)$. We are led to consider

$$I_n(s) = \sum_{(p,q)} n^{\varepsilon|p+q|-|p|}\left(\sum_j n^{j-1}\|\alpha_n^{(q)}\beta_{(p)}\partial_t^{m-j}v_n\|\right), \tag{7.1}$$

where (p,q) is taken over $1 \le |p+q| \le j_0$. Let us remark that the term corresponding to $(p,q) = (0,0)$ does not appear in the above expression. As we shall see below $f_n(s)$ is estimated by $I_n(s)$ modulo negligible quantity when n is large. First we prove

LEMMA 7.1. *$I_n(0)$ is negligible.*

Proof. Let us recall that

$$\partial_t^i v_n(0,x) = \gamma_n(D)t_{i,v}(x,D')\tilde{\psi}_n(x)n^{-(m-1-i)}.$$

Thus we consider all the terms of the form:

$$n^{\varepsilon|p+q|-|p|}\|\alpha_n^{(q)}\beta_{(p)}\gamma_n(D)t_{i,v}\tilde{\psi}_n\|, \qquad 1 \le |p+q| \le [n^{1-2\varepsilon}].$$

First we prove that $\gamma_n(D)t_{i,v}\tilde{\psi}_n - t_{i,v}\tilde{\psi}_n$ is negligible [step (A)]. Next observe that $\alpha_n^{(q)}\tilde{\psi}_n(x) = 0$ if $|q| \ge 1$. Thus we are led to estimate the commutators $[\alpha_n^{(q)}, \beta_{(p)}]$, $[\alpha_n^{(q)}, t_{i,v}]$ [step (B)]. Finally the terms of the form $\alpha_n\beta_{(p)}t_{i,v}\tilde{\psi}_n$ are estimated [step (C)]. Hereafter we consider i as one, and we denote $t_{i,v}$ simply by t_v.

(A) By definition, $\gamma_n^{(p)}\tilde{\psi}_n = 0$ if $|p| \ge 1$, because on the support of $\psi(\eta')$, $\gamma_n^{(p)}(\eta) = 0$. Now

$$(\gamma_n t_v - t_v\gamma_n)\tilde{\psi}_n = \sum_{|p|\le j_0}\frac{1}{p!}t_{v(p)}\gamma_n^{(p)}\tilde{\psi}_n + r_n\tilde{\psi}_n.$$

The first terms are all null. To estimate r_n, let us observe

$$\|t_{v(p)}\gamma_n^{(p)}\| \le C\sum_{|\alpha+\beta|\le 2l+4}\sup|(t_{v(p)}\gamma_n^{(p)})_{(\beta)}^{(\alpha)}|.$$

Taking into account the properties of t_v and γ_n, this implies there exists a fixed monomial P depending only on l and K such that

$$\|t_{v(p)}\gamma_n^{(p)}\| \le P(|p|)\exp(cv)p!^{2+2\varepsilon}K^{2|p|}/n^{|p|}.$$

Thus

$$\|r_n\| \le Q(j_0)\exp(cv)j_0!^{1+2\varepsilon}K^{2j_0}/n^{j_0},$$

where Q is also a fixed monomial. Now the right-hand side is evidently negligible, since $(j_0^{1+2\varepsilon}K^2/n)^{j_0} \ll \exp(-j_0)$.

Now

$$\sum_{p,q} n^{\varepsilon|p+q|-|p|}\|\alpha_n^{(q)}\beta_{(p)}\| \le \sum_{p,q} n^{\varepsilon|p+q|-|p|}p!^{1+\varepsilon}q!^{1+\varepsilon}K^{|p+q|}/n^{|q|}$$
$$\le \sum_{p,q} p!^{1+\varepsilon}q!^{1+\varepsilon}K^{|p+q|}n^{-(1-\varepsilon)|p+q|},$$

where the sum is taken over $1 \le |p+q| \le j_0$. This is estimated again by

$$\left(\sum_{0\le|p|\le j_0} p!^{1+\varepsilon}K^{|p|}n^{-(1-\varepsilon)|p|}\right)^2.$$

By using Stirling's formula, it is easy to see that this remains bounded when n becomes large.

(B) We consider

$$n^{\varepsilon|p+q|-|p|}\alpha_n^{(q)}\beta_{(p)}t_v\tilde{\psi}_n(x), \qquad 1 \le |p+q| \le j_0.$$

Suppose $|q| \ge 1$. Now

$$\alpha_n^{(q)}\beta_{(p)} = \sum_r \frac{1}{r!}\beta_{(p+r)}\alpha_n^{(q+r)} + r_{p,q}.$$

In this asymptotic expansion r is taken up to $|r| = j_0 - |p+q|$. To estimate $\|r_{p,q}\|$, we estimate the general term after multiplying by $n^{\varepsilon|p+q|-|p|}$:

$$n^{\varepsilon|p+q|-|p|}r!^{-1}(p+r)!^{1+\varepsilon}(q+r)!^{1+\varepsilon}K^{|p+q+2r|}n^{-|q+r|}$$
$$\le n^{(\varepsilon-1)|p+q+r|}(r!^{\varepsilon}K_1^{|r|}n^{-\varepsilon|r|})|p+q+r|!^{1+\varepsilon}K_1^{|p+q+r|}.$$

Then letting in particular $|p+q+r| = j_0$, we see that this is estimated by

$$(r!^{\varepsilon}K_1^{|r|}n^{-\varepsilon|r|})(j_0^{1+\varepsilon}K_1/n^{1-\varepsilon})^{j_0}.$$

Thus

$$\sum_{p,q} n^{\varepsilon|p+q|-|p|}\|r_{p,q}(t_v\tilde{\psi})\| \le Q(j_0)d(l)(j_0^{1+\varepsilon}K_1/n^{1-\varepsilon})^{j_0}\exp(cv)\sum_{p,q} 1 \ll \exp(-j_0).$$

Now we estimate $n^{\varepsilon|p+q|-|p|}(1/r!)\beta_{(p+r)}\alpha_n^{(q+r)}t_v\tilde{\psi}_n$. First

$$\alpha_n^{(q+r)}t_v\tilde{\psi}_n = \sum_s \frac{1}{s!}t_{v(s)}\alpha_n^{(q+r+s)}\tilde{\psi}_n + r_{p,q,r}\tilde{\psi}_n = r_{p,q,r}\tilde{\psi}_n.$$

We take $|s| = j_0 - |p+q+r|$. The symbol of the general term is estimated by

$$\exp(cv)s!^{\varepsilon}(q+r+s)!^{1+\varepsilon}K^{|s|+|q+r+s|}n^{-|q+r+s|} \; (= A).$$

Thus $n^{\varepsilon|p+q|-|p|}\|(1/r!)\beta_{(p+r)}r_{p,q,r}\|$ is estimated by

$$n^{\varepsilon|p+q|-|p|}Q(j_0)\frac{1}{r!}|p+r|!^{1+\varepsilon}K^{|p+r|}A$$
$$\leq Q(j_0)(|r|!^{\varepsilon}K_1^{|r|}n^{-\varepsilon|r|})(|s|!^{\varepsilon}K_1^{|s|}n^{-\varepsilon|s|})\exp(cv)(j_0!^{1+\varepsilon}K_1^{j_0}n^{-(1-\varepsilon)j_0}).$$

Thus, since the number of (p, q) satisfying $|p+q| \leq j_0$ is estimated by j_0^{2l},

$$\sum_{p,q,r} n^{\varepsilon|p+q|-|p|}\left\|\frac{1}{r!}\beta_{(p+r)}r_{p,q,r}\tilde{\psi}_n\right\|$$

is estimated by

$$Q(j_0)d(l)^2 j_0^{2l}\exp(cv)(j_0^{1+\varepsilon}K/n^{1-\varepsilon})^{j_0},$$

which is evidently negligible.

(C) The case when $|p| \geq 1$, $q = 0$. In view of (A), we consider $\alpha_n\beta_{(p)}t_v\tilde{\psi}_n$.

$$\alpha_n\beta_{(p)} \sim \beta_{(p)}\alpha_n + \sum_{|r|\geq 1}\frac{1}{r!}\beta_{(p+r)}\alpha_n^{(r)}.$$

Concerning the second term, we have already treated in (B). Namely

$$\sum_{p,r} n^{(\varepsilon-1)|p|}\left\|\frac{1}{r!}\beta_{(p+r)}\alpha_n^{(r)}t_v\tilde{\psi}_n\right\|$$

is negligible. So it is enough to consider the first term:

$$\beta_{(p)}\alpha_n t_v\tilde{\psi}_n = \beta_{(p)}t_v\tilde{\psi}_n + \beta_{(p)}\left(\sum_r \frac{1}{r!}t_{v(r)}\alpha_n^{(r)}\tilde{\psi}_n + r_p\tilde{\psi}_n\right).$$

Concerning the second term of the right-hand side, we have already treated in (B). Therefore we consider only the first term:

$$\beta_{(p)}t_v\tilde{\psi}_n.$$

Now this is regarded as $(\beta_{(p)}/|x|^{2k})|x|^{2k}(t_v\tilde{\psi}_n)$. Since $\beta_{(p)}(x) = 0$ for $|x| \leq \varepsilon'/2$, we have

$$\left|\frac{\beta_{(p)}}{|x|^{2k}}\right| \leq p!^{1+\varepsilon}K^{|p|}(2\varepsilon'^{-1})^{2k}.$$

Next,

$$|x|^{2k}(t_v\tilde{\psi}_n) = (2\pi)^{-l}\int e^{ix\eta}|x|^{2k}t_v(x,\eta')\psi(\eta')\,d\eta$$
$$= (-1)^k(2\pi)^{-l}\int e^{ix\eta}\Delta_\eta^k(t_v(x,\eta')\psi(\eta'))\,d\eta$$
$$= (-1)^k(2\pi)^{-l}\sum_\alpha c_{k,\alpha}\int e^{ix\eta}\partial_\eta^{2\alpha}(t_v(x,\eta')\psi(\eta'))\,d\eta,$$

where $\sum_{|\alpha|=k} c_{k,\alpha} = l^k$. For the moment, we take k under $2k \le j_0$. Now

$$\partial_\eta^{2\alpha}(t_v\psi(\eta')) = \sum_{\beta+\beta'=2\alpha} C_\beta^{2\alpha} t_v^{(\beta)} \psi(\eta')^{(\beta')}.$$

First, the operator norm of the p.d.o. $t_v^{(\beta)}$ is estimated by

$$Q(j_0)\exp(cv)\beta!^{1+\varepsilon}K^{|\beta|}/n^{|\beta|}.$$

Next, since $\psi(\eta')^{(\beta')} = \psi^{(\beta')}(\eta')/n^{|\beta'|}$ and since, in general, $|\psi^{(\alpha)}(\eta')| \le c_0\alpha!^{1+\varepsilon}K^{|\alpha|}$, we have

$$\|\psi(\eta')^{(\beta')}\| \le \max|\psi^{(\beta')}(\eta')|n^{-|\beta'|+l/2} \le c_0 n^{l/2}\beta'!^{1+\varepsilon}K^{|\beta'|}n^{-|\beta'|}.$$

Thus denoting the inverse Fourier image of $\psi(\eta')^{(\beta')}$ by $\tilde{\psi}_{n,\beta'}(x)$, and denoting $c_0 Q(j_0)n^{l/2}$ by $C(n)$, we get

$$\sum_\beta C_\beta^{2\alpha}\|t_v^{(\beta)}\tilde{\psi}_{n,\beta'}\| \le C(n)\exp(cv)K^{2k}n^{-2k}\sum_\beta C_\beta^{2\alpha}\beta!^{1+\varepsilon}\beta'!^{1+\varepsilon}.$$

Since the summation is estimated by $(2\alpha)!^{1+\varepsilon}l^{2k}$, it follows that

$$\| |x|^{2k}(t_v\tilde{\psi}_n)\| \le C(n)\exp(cv)l^{3k}(2k)!^{1+\varepsilon}K^{2k}n^{-2k}.$$

Then

$$\begin{aligned}\|\beta_{(p)}t_v\tilde{\psi}_n\| &\le C(n)\exp(cv)p!^{1+\varepsilon}K^{|p|}(2\varepsilon'^{-1})^{2k}l^{3k}(2k)!^{1+\varepsilon}K^{2k}n^{-2k}\\ &\le C(n)\exp(cv)(|p|+2k)!^{1+\varepsilon}K'^{|p|+2k}n^{-2k},\end{aligned}$$

where $K' = 2\varepsilon'^{-1}Kl^{3/2}$ $(>K)$. Now we take k in such a way that $j_0 \le |p| + 2k < j_0 + 2$. Then

$$n^{(\varepsilon-1)|p|}\|\beta_{(p)}t_v\tilde{\psi}_n\| \le C(n)\exp(cv)(j_0^{1+\varepsilon}K'/n^{1-\varepsilon})^{j_0}.$$

Finally

$$\sum_{1\le|p|\le j_0} n^{(\varepsilon-1)|p|}\|\beta_{(p)}t_v\tilde{\psi}_n\| \le \exp(-j_0). \quad \blacksquare$$

Now we return to the beginning of this section. Let

$$S_{q',p'} = n^{\varepsilon|q'+p'|-|p'|}\sum_{j=1}^m n^{j-1}\|\alpha_n^{(q')}\beta_{(p')}\partial_t^{m-j}v_n\|. \tag{7.2}$$

In (4.4) if we replace α_n, β by $\alpha_n^{(q')}$, $\beta_{(p')}$, respectively, we have

$$\frac{\partial}{\partial s}(\alpha_n^{(q')}\beta_{(p')}V_n) = vA_v(\alpha_n^{(q')}\beta_{(p')}V_n) + n^{-\sigma_1}F_{n,q',p'}, \tag{7.3}$$

where $F_{n,q',p'} = {}^t(0,0,\cdots,f_{n,q',p'})$, and

$$\begin{aligned} n^{-1}f_{n,q',p'} &= \sum_j\sum_{|p|\ge1}\frac{(-1)^{|p|}}{p!}h_j^{(p)}n^{j-1}\partial_t^{m-j}(\alpha_n^{(q')}\beta_{(p'+p)}v_n)\\ &\quad+\sum_j\sum_{p\ge0}\frac{(-1)^{|p|}}{p!}[\alpha_n^{(q')},h_j^{(p)}]n^{j-1}\partial_t^{m-j}(\beta_{(p'+p)}v_n). \end{aligned} \tag{7.4}$$

Let us remark that

$$\|\alpha_n^{(q')}\beta_{(p')}V_n\| = \sum_{j=1}^{m} n^{j-1}\|\alpha_n^{(q')}\beta_{(p')}\partial_t^{m-j}v_n\|,$$

and that from (7.3) we have

$$\frac{d}{ds}\|\alpha_n^{(q')}\beta_{(p')}V_n\| \le c_0\nu\|\alpha_n^{(q')}\beta_{(p')}V_n\| + n^{-\sigma_1}\|F_{n,q',p'}\|. \tag{7.5}$$

Now look at (7.4).

$$[\alpha_n^{(q')}, h_j^{(p)}] \sim \sum_q \frac{1}{q!} h_{j(q)}^{(p)}\alpha_n^{(q'+q)}. \tag{7.6}$$

Note that

$$\left\|\frac{1}{p!q!} h_{j(q)}^{(p)}\right\| \le Cp!^{\varepsilon}q!^{\varepsilon}K^{|p+q|}/n^{|p|}. \tag{7.7}$$

The right-hand side is written again

$$C(p!^{\varepsilon}K^{|p|}/n^{\varepsilon|p|})(q!^{\varepsilon}K^{|q|}/n^{\varepsilon|q|})n^{\varepsilon|p+q|-|p|},$$

so that, if in (7.6) we take the expansion up to $|q| = j_0 - |p + p' + q'|$,

$$\begin{aligned}\left\|\frac{1}{n} f_{n,q',p'}\right\| \le C\sum_{p,q} \frac{p!^{\varepsilon}K^{|p|}}{n^{\varepsilon|p|}}\frac{q!^{\varepsilon}K^{|q|}}{n^{\varepsilon|q|}} \\ \times n^{\varepsilon|p+q|-|p|}\|\alpha_n^{(q'+q)}\beta_{(p'+p)}V_n\| + r_{n,q',p'},\end{aligned} \tag{7.8}$$

where (p, q) runs through $1 \le |p + q| \le j_0 - |p' + q'|$, and $r_{n,q',p'}$ corresponds to the remainder term, and C is a constant independent of (n, p', q').

Let us observe that from (7.5), (7.8) that we have

$$\begin{aligned}\frac{d}{ds} S_{q',p'} \le c_0\nu S_{q',p'} + C\nu \sum_{p,q} (p!^{\varepsilon}K^{|p|}n^{-\varepsilon|p|})(q!^{\varepsilon}K^{|q|}n^{-\varepsilon|q|})S_{q'+q,\, p'+p} \\ + n^{\varepsilon|p'+q'|-|p'|}r_{n,q',p'}.\end{aligned}$$

To estimate the remainder term, we consider the estimate of the general term, which is given by

$$n^{\varepsilon|p'+p+q'+q|-|p'+p|}\|\alpha_n^{(q'+q)}\beta_{(p'+p)}V_n\|,$$

where $|p' + p + q' + q| = j_0 + 1$. Since $j_0(= [n^{1-2\varepsilon}])$ is large, it is the same as $j_0 + 1$. In this case, we estimate by

$$Q(j_0)(q' + q)!^{1+\varepsilon}(p' + p)!^{1+\varepsilon}K^{j_0}n^{(\varepsilon-1)j_0}|V_n|_0 \ll \exp(-j_0)|V_n|_0,$$

where $|V_n|_0$ is the maximum norm of $V_n(s, x)$ on the support of β.

Summing up:

LEMMA 7.2

$$\frac{d}{ds} S_{q',p'} \leq c_0 \nu S_{q',p'} + C\nu \sum_{p,q} (p!^{\varepsilon} K^{|p|} n^{-\varepsilon|p|})(q!^{\varepsilon} K^{|q|} n^{-\varepsilon|q|}) S_{q'+q,\, p'+p}$$
$$+ \exp(-j_0)|V_n(s)|_0.$$

Since $I_n = \sum S_{q',p'}$, we sum up the above inequality over (q', p') such that $1 \leq |q' + p'| \leq j_0$, and we have the following.

LEMMA 7.3

$$\frac{d}{ds} I_n(s) \leq c'' \nu I_n(s) + \phi_n(s),$$

where $\phi_n(s)$ is uniformly negligible for $s \in [0, s_0]$, and c'' is a constant independent of ν.

Proof. In the inequality in Lemma 7.2, the coefficient of S_{q_0,p_0} is

$$((q_0 - q)!^{\varepsilon} K^{|q_0-q|}/n^{\varepsilon|q_0-q|})((p_0 - p)!^{\varepsilon} K^{|p_0-p|}/n^{\varepsilon|p_0-p|}),$$

where $q_0 - q = q'$ (≥ 0), $p_0 - p = p'$ (≥ 0). Then, in the sum of the above inequalities, the coefficient of S_{q_0,p_0} is less than

$$\sum_{q \leq q_0} \left(\frac{q!^{\varepsilon} K^{|q|}}{n^{\varepsilon|q|}}\right) \sum_{p \leq p_0} \left(\frac{p!^{\varepsilon} K^{|p|}}{n^{\varepsilon|p|}}\right).$$

Since $|q_0|, |p_0| \leq j_0$, this quantity $\leq d(l)^2$, and hence bounded when n becomes large. Finally, the number of (q', p') satisfying $|q' + p'| \leq j_0$ is of order j_0^{2l}. Thus the sum of remainder terms is estimated by $j_0^{2l} \exp(-j_0)|V_n(s)|_0$. Proposition 6.2 shows that

$$|V_n(s)|_0 \leq C \exp(c\nu) \exp(c'\nu s).$$

Now $\nu \ll j_0$ $(= [n^{1-2\varepsilon}])$, because $\sigma_1 > 2\varepsilon$. This shows that $\phi_n(s)$ is uniformly negligible. ■

Up to now, we have investigated the quantity $I_n(s)$ from (4.4). If we investigate the quantity

$$\begin{aligned} \tilde{I}_n(s) &= \sum_{1 \leq |p+q| \leq j_0} n^{\varepsilon|p+q| - |p|} \left(\sum_j n^{j-1} s^{(j-1)q_1} \|\alpha_n^{(q)} \beta_{(p)} \partial_t^{m-j} v_n\| \right) \\ &= \sum_{1 \leq |p+q| \leq j_0} n^{\varepsilon|p+q| - |p|} \|\alpha_n^{(q)} \beta_{(p)} V'_n\|, \end{aligned} \tag{7.9}$$

starting from (4.5), we have the following lemma

LEMMA 7.4

$$\frac{d}{ds}\tilde{I}_n(s) \leq c'''\nu s^{q_1}\tilde{I}_n(s) + \tilde{\phi}_n(s),$$

where $s \in [s_0, s_n]$, $\tilde{\phi}_n(s)$ *being uniformly negligible on every compact set in* $s \in [s_0, \infty)$.

Proof. Since the proof is essentially the same as that of Lemma 7.3, we restrict ourselves to showing how to modify it. First

$$n^{-\sigma_1}f_{n,q',p'} = \nu s^{q_1}\sum_p \frac{(-1)^{|p|}}{p!}\sum_j h_j'^{(p)}n^{j-1}s^{(j-1)q_1}\partial_t^{m-j}(\alpha_n^{(q')}\beta_{(p'+p)}v_n)$$
$$+ \nu s^{q_1}\sum_p \frac{(-1)^{|p|}}{p!}\sum_j [\alpha_n^{(q')}, h_j'^{(p)}]n^{j-1}s^{(j-1)q_1}\partial_t^{m-j}(\beta_{(p'+p)}v_n). \quad (7.4)'$$

Recalling the definition of $\|\alpha_n^{(q)}\beta_{(p)}V_n'\|$, we have the same type of inequality as (7.8). Namely,

$$\|n^{-\sigma_1}f_{n,q',p'}\| \leq C\nu s^{q_1}\sum_{p,q}\frac{p!^\varepsilon K^{|p|}}{n^{\varepsilon|p|}}\frac{q!^\varepsilon K^{|q|}}{n^{\varepsilon|q|}}n^{\varepsilon|p+q|-|p|}\|\alpha_n^{(q'+q)}\beta_{(p'+p)}V_n'\| + \tilde{r}_{n,q',p'}, \quad (7.8)'$$

where

$$n^{\varepsilon|q'+p'|-|p'|}\tilde{r}_{n,q',p'} \leq \exp(-j_0)s^{mq_1}|V_n(s)|_0 \qquad \text{for} \quad s \in [s_0, s_n].$$

Observe that the right-hand side is uniformly negligible on every compact set in $s \in [s_0, \infty)$, since $|V_n(s)|_0 \leq C\exp(c\nu)\exp(c'\nu s)$. More precisely there exists a sequence $s_n' = n^{\theta'}$ $(0 < \theta' < \theta)$ such that the sequence $\exp(-j_0)s^{mq_1}|V_n(s)|_0$ is uniformly negligible on $[s_0, s_n']$. ■

Now from Lemmas 7.3 and 7.4, we have the following:

PROPOSITION 7.1. *In* (4.4) *and* (4.5), $f_n(s)$ [*hence* $F_n(s)$] *is uniformly negligible on every compact set in* $s \in [0, \infty)$.

Proof. At first $I_n(s)$ is uniformly negligible for $s \in [0, s_0]$. In fact, by Lemma 7.3,

$$I_n(s) \leq I_n(0)\exp(c''\nu s) + \int_0^s \exp\{c''\nu(s-\sigma)\}\phi_n(\sigma)\,d\sigma.$$

Now Lemma 7.1 claims that $I_n(0)$ is negligible, and Lemma 7.3 claims that $\phi_n(s)$ is uniformly negligible. It follows that $I_n(s_0)$ is negligible. Hence $\tilde{I}_n(s_0)$ is also negligible, since $\tilde{I}_n(s_0) \leq s_0^{(m-1)q_1}I_n(s_0)$. Next, Lemma 7.4 shows that $\tilde{I}_n(s)$ is uniformly negligible on every compact set in $s \in [s_0, \infty)$.

Now, as we have already explained at the beginning of this section, the estimate (7.8) when $p' = q' = 0$ shows that $n^{-1}f_n$ is estimated by some constant times $I_n(s)$ except for a negligible quantity. It is the same with $\tilde{I}_n(s)$ when $s \geq s_0$. Thus $f_n(s)$ is negligible, more precisely uniformly negligible, on every compact set in $[0, \infty)$. ■

Now let us recall that

$$W_n(s) = \mathring{\Phi}_\nu(s, s_0)\mathscr{W}_n(s_0) \tag{6.3}$$

satisfies

$$\frac{\partial}{\partial s} W_n(s) = \nu A_\nu W_n(s) + F_\nu(s), \tag{6.4}$$

where $F_\nu(s)$ is uniformly negligible in $s \in [0, s_0]$. We want to show that $\alpha_n \beta W_n(s)$ satisfies the same kind of equation as (4.4). The verification can be carried out in a simplified version of the above. To see this, first apply $\beta(x)$ to (6.4) from the left:

$$\frac{\partial}{\partial s}(\beta W_n) = \nu A_\nu(\beta W_n) + \nu[\beta, A_\nu]W_n + \beta F_\nu. \tag{7.10}$$

Now

$$[\beta, A_\nu] \sim \sum_{|p| \geq 1} \frac{(-1)^{|p|}}{p!} A_\nu^{(p)} \circ \beta_{(p)},$$

$$\frac{1}{p!}\|A_\nu^{(p)}\| \leq Cp!^\varepsilon \frac{K^{|p|}}{n^{|p|}} = C(p!^\varepsilon K^{|p|} n^{-\varepsilon|p|}) n^{-(1-\varepsilon)|p|}.$$

Put

$$J_n(s) = \sum_{1 \leq |p| \leq j_0} n^{-(1-\varepsilon)|p|} \|\beta_{(p)} W_n\|.$$

We showed that $J_n(0)$ is negligible [see part (C) of the proof of Lemma 7.1]. Next, from (6.3) we know that

$$\|W_n(s)\| \leq \exp(C\nu) \qquad \text{for} \quad s \in [0, s_0].$$

In the same way as Lemma 7.3, we can show that $J_n(s)$ satisfies

$$\frac{d}{ds} J_n(s) \leq C_2 \nu J_n(s) + \gamma_n(s),$$

where $\gamma_n(s)$ is uniformly negligible. Since $J_n(0)$ is negligible, $J_n(s)$ is uniformly negligible. This implies

LEMMA 7.5. *In* (7.10), $[\beta, A_\nu]W_n(s)$ *is uniformly negligible in* $s \in [0, s_0]$.

Now we apply α_n to (7.10) from the left, then we have

LEMMA 7.6. *$W_n(s)$ satisfies*

$$\frac{\partial}{\partial s}(\alpha_n \beta W_n(s)) = v A_v(\alpha_n \beta W_n(s)) + G_v(s), \tag{7.11}$$

where $G_v(s)$ is uniformly negligible.

Proof. We start from

$$\frac{\partial}{\partial s}(\beta W_n) = v A_v(\beta W_n) + \tilde{F}_v(s),$$

where $\tilde{F}_v(s)$ is uniformly negligible in $s \in [0, s_0]$. Apply α_n from the left:

$$\frac{\partial}{\partial s}(\alpha_n \beta W_n) = v A_v(\alpha_n \beta W_n) + G_v(s),$$

where $G_v(s) = v[\alpha_n, A_v](\beta W_n) + \alpha_n \tilde{F}_v(s)$. Since

$$[\alpha_n, A_v] \sim \sum \frac{1}{p!} A_{v(p)} \alpha_n^{(p)},$$

and

$$\left\| \frac{1}{p!} A_{v(p)} \right\| \leq p!^{\varepsilon} K^{|p|} \leq (p!^{\varepsilon} K^{|p|} n^{-\varepsilon|p|}) n^{\varepsilon|p|},$$

it suffices to show that

$$\tilde{J}_n(s) = \sum_{1 \leq |p| \leq j_0} n^{\varepsilon|p|} \|\alpha_n^{(p)} \beta W_n\|$$

is uniformly negligible. First it is easy to see that $\tilde{J}_n(0)$ is negligible. In fact, denoting $\beta t_v = \tilde{t}_v$,

$$\alpha_n^{(p)} \tilde{t}_v \tilde{\psi}_n = \sum_r \frac{1}{r!} \tilde{t}_{v(r)} \alpha_n^{(p+r)} \tilde{\psi}_n + r_p \tilde{\psi}_n,$$

where $|r| \leq j_0 - |p|$, we have

$$\|r_p\| \leq Q(j_0) r!^{\varepsilon} K^{|r|} (p+r)!^{1+\varepsilon} K^{|p+r|} \exp(cv) n^{-|p+r|}.$$

Thus,

$$n^{\varepsilon|p|} \|\alpha_n^{(p)} \beta t_v\| \leq Q(j_0)(|r|!^{\varepsilon} K^{|r|} n^{-\varepsilon|r|}) j_0!^{1+\varepsilon} K^{j_0} n^{-(1-\varepsilon)j_0} \exp(cv).$$

This is obviously negligible (we put $|r| = j_0 - |p|$).

Next, considering the equation ($1 \leq |p| \leq j_0$):

$$\frac{\partial}{\partial s}(\alpha_n^{(p)}\beta W_n) = vA_v(\alpha_n^{(p)}\beta W_n) - v[A_v, \alpha_n^{(p)}](\beta W_n) + \alpha_n^{(p)}\tilde{F}_v(s),$$

we arrive at the conclusion. ■

Now we compare (4.4) with (7.11). First let us recall that $\alpha_n\beta V_n - \alpha_n\beta W_n$ is negligible at $s = 0$. [see part (A) of the proof of Lemma 7.1]. From (4.4) and (7.11), we have

$$\frac{\partial}{\partial s}(\alpha_n\beta(V_n - W_n)) = vA_v(\alpha_n\beta(V_n - W_n)) + (n^{-\sigma_1}F_n - G_v).$$

Then, since the second term of the right-hand side is uniformly negligible, we arrive at the desired conclusion:

PROPOSITION 7.2. *$\alpha_n\beta V_n(s_0) - \alpha_n\beta W_n(s_0)$ is negligible.*

8. PROOF OF THEOREM

We are now in a position to prove the Theorem in the Introduction by contradiction. Before proving this we should make some comments on the previous fairly lengthy arguments. Roughly speaking, Propositions 7.1 and 7.2 show that we can know the behavior of $\alpha_n\beta V_n(s)$ when n and s are large via that of $\mathscr{W}_n(s)$ or $\mathscr{W}'_n(s)$. Recall that $V_n(s)$, or equivalently $v_n(t, x)$, is the solution of the original equation which we supposed to exist; however, we cannot know directly its behavior when s is not large except by its a priori estimates. This is the reason we introduced $W_n(s)$.

First we observe that $W_n(s_0) = \mathscr{W}_n(s_0)$ [see (6.3)]. Thus if we take into account the relations between $\mathscr{W}_n(s_0)$ and $\mathscr{W}'_n(s_0)$ and $V_n(s_0)$ and $V'_n(s_0)$, Proposition 7.2 claims that $\alpha_n\beta\mathscr{W}'_n(s_0) - \alpha_n\beta V'_n(s_0)$ is also negligible. In view of the definition of $\mathscr{W}'_n(s_0)$ [see (6.1)], we have

$$N_0\alpha_n\beta V'_n(s_0) = {}^t(\alpha_n\beta\tilde{\psi}_n, 0, 0, \ldots, 0) + (\text{negligible}). \tag{8.1}$$

Therefore, following the definition (5.3), we have

$$E_n(s_0; \alpha_n\beta V'_n) \geq \exp\{-2v\delta s_0^{q_1+1}/(q_1+1)\}\|\alpha_n\beta\tilde{\psi}_n\|^2 - (\text{negligible}).$$

Concerning $\|\alpha_n\beta\tilde{\psi}_n\|$, we have

LEMMA 8.1. *When n is large,*

$$\|\alpha_n\beta\tilde{\psi}_n\|^2 \geq \delta_0 n^l,$$

where δ_0 is a positive constant.

Proof

$$\alpha_n\beta\tilde{\psi}_n = \beta\alpha_n\tilde{\psi}_n + [\alpha_n, \beta]\tilde{\psi}_n = \beta\tilde{\psi}_n + [\alpha_n, \beta]\tilde{\psi}_n.$$

First it is easy to see that $[\alpha_n, \beta]\tilde{\psi}_n$ is negligible. Now

$$\tilde{\psi}_n(0) = (2\pi)^{-l} \int \psi(\eta') \, d\eta = (2\pi)^{-l} n^l \int \psi(\eta') \, d\eta' = (2\pi)^{-l} n^l.$$

On the other hand $\partial_{x_i}\tilde{\psi}_n(x) = (2\pi)^{-l} \int e^{ix\eta}(i\eta_i)\psi(\eta') \, d\eta$. Thus

$$|\partial_{x_i}\tilde{\psi}_n(x)| \leq cn^{l+1}.$$

This implies that if we restrict x to $|x| \leq \delta' n^{-1}$ (δ' is a small positive constant), we have $|\tilde{\psi}_n(x)| \geq \frac{1}{2}(2\pi)^{-l}n^l$. Since $\beta(x) = 1$ on this set (if n is large), we have finally

$$\|\beta(x)\tilde{\psi}_n(x)\|^2 \geq \|\tilde{\psi}_n(x)\|^2_{(|x| \leq \delta' n^{-1})} \geq \delta_0 n^l,$$

where δ_0 is a positive constant. ∎

From this lemma, we have a fortiori

$$E_n(s_0; \alpha_n\beta V'_n) \geq \exp\{-2\nu\delta s_0^{q_1+1}/(q_1 + 1)\}, \tag{8.2}$$

for n large.

Now we apply the energy inequality to (4.5). This time, because of the presence of $n^{-\sigma_1}F_n$, (5.4) should be modified as follows:

$$E_n(s; \alpha_n\beta V'_n) \geq \exp\{\nu\delta(s^{q_1+1} - s_0^{q_1+1})/(q_1 + 1)\}E_n(s_0; \alpha_n\beta V'_n) - C\int_{s_0}^{s} \exp\left\{\nu\delta\frac{(s^{q_1+1} - \sigma^{q_1+1})}{q_1 + 1}\right\} n^{-\sigma_1}\|F_n(\sigma)\|^2 \, d\sigma, \tag{8.3}$$

for n large. Here let us recall that, by virtue of Proposition 7.1, *every time s is fixed, the second integral is negligible.*

Let us recall the definition of $E_n(s; \alpha_n\beta V'_n)$. Then Proposition 6.2 implies that there exists a constant C' such that

$$E_n(s; \alpha_n\beta V'_n) \leq C's^{2(m-1)q_1} \exp(2c\nu) \exp(2c'\nu s) \exp\{-2\nu\delta s^{q_1+1}/(q_1 + 1)\}. \tag{8.4}$$

Now we want to show that (8.3) and (8.4) are not compatible. First, observe that, if s is large, the right-hand side of (8.4) is less than $C's^{2(m-1)q_1}$. In fact, if s is large, we have

$$2\delta s^{q_1+1}/(q_1 + 1) \geq 2c + 2c's,$$

since $q_1 > 0$. Thus

$$E_n(s; \alpha_n\beta V'_n) \leq C's^{2(m-1)q_1}. \tag{8.5}$$

Next we choose s in such a way that $\frac{1}{2}s^{q_1+1} > 3s_0^{q_1+1}$. Then (8.2) and (8.3) imply

$$E_n(s; \alpha_n \beta V'_n) \geq \exp\{\tfrac{1}{2}v\delta s^{q_1+1}\} - (\text{negligible}). \tag{8.6}$$

(8.5) and (8.6) are not compatible if v is large, which proves the Theorem.

References

1. K. Kitagawa and T. Sadamatsu, A necessary condition of Cauchy–Kowalewski's theorem, *Publ. Res. Inst. Math. Sci.* **11** (1976), 523–534.
2. S. Kowalewski, Zur Theorie der partiellen Differentialgleichungen, *J. Reine Angew.* Math. **80** (1875), 1–32.
3. M. Miyake, A remark on Cauchy–Kowalewski's theorem, *Pub. Res. Inst. Math. Sci.* **10** (1974), 245–255.
4. S. Mizohata, On kowalewskian systems, *Usp. Mat. Nauk* **29** (1974), 213–227 [in Russian].
5. S. Mizohata, On Cauchy–Kowalewski's theorem; a necessary condition, *Publ. Res. Inst. Math. Sci.* **10** (1975), 509–519.
6. S. Mizohata, Une remarque sur le théorème de Cauchy–Kowalewski, *Ann. Scuola Norm. Sup. Pisa Sci. Fis. Mat.* **5** (1978), 559–566.
7. S. Mizohata, Sur le théorème de Cauchy–Kowalewski, *Sém. Goulaouic–Schwartz*, No. 19 (1979), 1–12.
8. O. A. Oleinik and E. V. Radkevich, On the analyticity of systems of linear partial differential equations, *Mat. Sb.* **90** (1973), 592–607.

MATHEMATICAL ANALYSIS AND APPLICATIONS, PART B
ADVANCES IN MATHEMATICS SUPPLEMENTARY STUDIES, VOL. 7B

Factoring the Natural Injection $\iota^{(n)}: L_n^\infty \to L_n^1$ through Finite Dimensional Banach Spaces and Geometry of Finite Dimensional Unitary Ideals[†]

A. PELCZYNSKI

Institute of Mathematics
Polish Academy of Sciences
Warsaw, Poland

AND

C. SCHÜTT

Johannes Kepler Universität Linz
Mathematisch Institut
Linz/Donau, Austria

DEDICATED TO LAURENT SCHWARTZ

For a Banach space X with $\dim X = s$ we put $\text{ké}(X) = \inf\{\|u\|\,\|v\| : u: L_s^\infty \to X, v: X \to L_s^1, vu = \iota^{(s)}\}$ where $\iota^{(s)}: L_s^\infty \to L_s^1$ is the natural injection, L_s^p denotes the space of scalar sequences $x = (x(j))_{1 \le j \le s}$ with the norm $\|x\|_p = (s^{-1}\sum_{j=1}^{q}|x(j)|^p)^{1/p}$ for $1 \le p < \infty$, $\|x\|_\infty = \max_{1 \le j \le s}|x(j)|$.

$\text{ké}(X)$—called the ké constant of X—is an invariant of local theory of Banach spaces closely related to $\text{ubc}(X)$, $\text{lust}(X)$ and the Gordon–Lewis constant $\text{gl}_H(X)$. For every X, $\text{ké}(X) \le \sqrt{\dim X}$. There is a sequence (X_s) such that $\dim X_s = s$, $\text{ké}(X_s) \ge C\sqrt{\dim X_s}$ where C is an absolute positive constant. ké constants of large subspaces of L_s and of injective and projective tensor products of L_s^p and L_s^q are studied. A characterization of Sidon sets in terms of ké constants of finite dimensional spaces spanned by characters (equipped with the sup norm) is given.

It is shown that for a unitary ideal, say α, of $n \times m$ matrices all the quantities $\text{ké}(\alpha)$, $\text{gl}_H(\alpha)$, $\text{ubc}(\alpha)$, $\text{lust}(\alpha)$ are of the same order as the Banach–Mazur distance $d(\alpha, l_{nm}^2)$. (This is a slight improvement of a result of Lewis [17] and Schütt [24].) It is also shown that for every α the Banach–Mazur distance $d(\alpha, l_{nm}^\infty)$ and the projection constant $\gamma_\infty(\alpha)$ are of the order of $\sqrt{\dim \alpha} = \sqrt{nm}$.

INTRODUCTION

In the present paper we introduce and study an invariant of local theory of Banach spaces; the invariant is called the ké constant. Roughly speaking ké(X), the ké constant of a finite dimensional Banach space X, say $\dim X = n$

[†] The research of this paper has been partially sponsored by the Sonderforschungsbereich 72 during the authors' visit at the Institut für Angewandte Mathematik der Universität Bonn.

ISBN 0-12-512802-9

is the norm of the best factorization of the natural injection $\iota^{(n)}: L_n \to L_n^1$ through X. It is shown that ké(X) is closely related to the ubc constant, lust constant, and the Gordon–Lewis constant of the same space (see definitions in Section 0) which have been the subject of intensive study [4, 6, 9, 17, 21, 24]. John [15] gives the result ké$(X) \leq \sqrt{\dim X}$. There are examples such as: for lust(X), of "random" subspaces of l_n^∞ for which ké(X) is of the worst possible order, precisely ké$(X) \geq c\sqrt{\dim X}$ where c is an absolute constant. Next we compute the ké constants of injective and projective tensor products of the spaces l_m^p and give a characterization of Sidon sets among Λ_2 sets in terms of ké constants. Section 1 ends with examples of sequences of finite dimensional spaces (even spaces with enough symmetries in the sense of Garling and Gordon [7]) whose ké constants are uniformly bounded while their Gordon–Lewis constants and therefore the lust constants tend to infinity.

The situation is different for finite dimensional unitary ideals treated in Section 2. A unitary ideal is the space of $n \times m$ matrices with a norm such that multiplications by unitary matrices from the left and from the right are isometrics of the matrix space in this norm. We show that for a unitary ideal all constants ké, lust, ubc, and the Gordon–Lewis constant are of the same order as the Banach–Mazur distance of this unitary ideal from the Hilbert space l_{nm}^2. Our technique allows also to show that the Banach–Mazur distance of every unitary ideal of $n \times m$ matrices from the space l_{nm}^∞ is of order of $\sqrt{mn}$.

0. Preliminaries

Denote by $\mathfrak{K}$ a nonspecified field of scalars: either $\mathfrak{C}$ the complex plane or $\mathfrak{R}$ the real line. Let $1 \leq p < \infty$ and let n be a fixed positive integer. Given $x = (x(j))_{1 \leq j \leq n} \in \mathfrak{K}^n$ we put $\|x\|_p = n^{-1/p}\|x\|_{l_n^p} = (n^{-1}\sum_{j=1}^n |x(j)|^p)^{1/p}$ for $1 \leq p < \infty$ and $\|x\|_\infty = \|x\|_{l_n^\infty} = \max_{1 \leq j \leq n}|x(j)|$. By L_n^p (resp., l_n^p) we denote the vector space $\mathfrak{K}^n$ with the norm $\|\cdot\|_p$ (resp., $\|\cdot\|_{l_n^p}$). For p and q with $1 \leq p$, $q \leq \infty$ the *natural injection* $\iota_{p,q}^{(n)}: L_n^p \to L_n^q$ is defined as the identity map on $\mathfrak{K}^n$ regarded as the operator from L_n^p into L_n^q. We shall often write $\iota^{(n)}$ instead of $\iota_{\infty,1}^{(n)}$.

By $\mathfrak{K}^n \otimes \mathfrak{K}^m$ we denote the algebraic tensor product of the vector spaces $\mathfrak{K}^n$ and $\mathfrak{K}^m$, i.e., the space of all $m \times n$ matrices $a = (a(j,k))_{1 \leq j \leq m,\, 1 \leq k \leq n}$. By $\mathscr{U}(n)$ [resp., $\mathscr{O}(n)$] we denote the group of all $n \times n$ unitary (resp., orthogonal) matrices. A *norm* α on $\mathfrak{C}^n \otimes \mathfrak{C}^m$ is called *unitary* provided $\alpha(a) = \alpha(v \cdot a \cdot u)$ for every $a \in \mathfrak{C}^n \otimes \mathfrak{C}^m$, $v \in \mathscr{U}(m)$, $u \in \mathscr{U}(n)$. The Banach space $(\mathfrak{C}^n \otimes \mathfrak{C}^m, \alpha)$ is called a *unitary ideal* provided α is a unitary norm. We shall often denote the unitary ideal $(\mathfrak{C}^n \otimes \mathfrak{C}^m, \alpha)$ also by α. Recall that the Hilbert–Schmidt norm of a matrix $a \in \mathfrak{C}^n \otimes \mathfrak{C}^m$ is defined by hs$(a) = (\sum_{j=1}^m \sum_{k=1}^n |a(j,k)|^2)^{1/2}$. The unitary ideal hs $= (\mathfrak{C}^n \otimes \mathfrak{C}^m, \text{hs})$ is obviously a Hilbert space with the inner product $\langle a,b\rangle = \mathrm{tr}(b^*a) = \sum_{j=1}^{m}\sum_{k=1}^{n} \overline{b(k,j)}a(j,k)$. Here for $b \in \mathfrak{C}^n \otimes \mathfrak{C}^m$ the *transposed* $b^* \in \mathfrak{C}^m \otimes \mathfrak{C}^n$ is defined by $b^*(k,j) = \overline{b(j,k)}$ $(j = 1,2,\ldots,m;\ k = 1,2,\ldots,n)$ and for an $n \times n$ matrix, say c, the *trace* $\mathrm{tr}(c)$ is defined by $\mathrm{tr}(c) = \sum_{k=1}^{n} c(k,k)$. If $(\mathfrak{C}^n \otimes \mathfrak{C}^m, \alpha)$ and $(\mathfrak{C}^n \otimes \mathfrak{C}^m, \beta)$ are unitary ideals then the natural injection $I_{\alpha,\beta}:\alpha \to \beta$ is defined to be the identity map on $\mathfrak{C}^n \otimes \mathfrak{C}^m$ regarded as the operator from α into β. Given a unitary norm α on $\mathfrak{C}^n \otimes \mathfrak{C}^m$ we define the unitary norm α^* on $\mathfrak{C}^m \otimes \mathfrak{C}^n$ by $\alpha^*(b) = \sup\{|\mathrm{tr}(ba)| : a \in \mathfrak{C}^n \otimes \mathfrak{C}^m,\ \alpha(a) \le 1\}$. The unitary ideal $(\mathfrak{C}^m \otimes \mathfrak{C}^n, \alpha^*)$ can be identified with the dual space of $(\mathfrak{C}^n \otimes \mathfrak{C}^m, \alpha)$; the duality is given by the trace.

The uniform norm $|||a|||$ of an $a \in \mathfrak{C}^n \otimes \mathfrak{C}^m$ is defined by

$$|||a||| = \sup\left\{\left|\sum_{j=1}^{m}\sum_{k=1}^{n} a(j,k)x(k)y(j)\right| : x \in \mathfrak{C}^n,\ y \in \mathfrak{C}^m \right.$$
$$\left.\text{with } \sum_{k=1}^{n} |x(k)|^2 \sum_{j=1}^{m} |y(j)|^2 \le 1\right\};$$

The sequence $(s_\nu(a))_{1\le\nu\le\infty}$ of the *s numbers* of a is defined by

$$s_\nu(a) = \inf\{|||a - b||| : b \in \mathfrak{C}^n \otimes \mathfrak{C}^m \text{ rank } b \le \nu - 1\}.$$

Clearly $s_1(a) = |||a||| \ge s_2(a) \ge \cdots$ and $s_\nu(a) = 0$ for $\nu > \min(n,m)$. Recall that

0.I. *Given $a \in \mathfrak{C}^n \otimes \mathfrak{C}^m$, there are $v \in \mathscr{U}(m)$ and $u \in \mathscr{U}(n)$ such that $a = v\Lambda(a)u$ where $\Lambda(a) \in \mathfrak{C}^n \otimes \mathfrak{C}^m$ is the diagonal operator defined by $\Lambda(a)(j,j) = s_j(a)$ for $j \le \min(n,m)$, $\Lambda(a)(j,k) = 0$ for $j \ne k$ and for $j = k > \min(n,m)$.*

0.II. *Let n, m be positive integers and let $d = \min(n,m)$. Then α is a unitary norm on $\mathfrak{C}^n \otimes \mathfrak{C}^m$ iff there is a norm p on $\mathfrak{C}^d$ which is symmetric (i.e. $p((t(j))) = p((\varepsilon_j t(\pi(j))))$ for every $(t(j)) \in \mathfrak{C}^d$, $|\varepsilon_j| = 1$ for $j = 1,2,\ldots,d$ and for every permutation $\pi:\{1,2,\ldots,d\} \to \{1,2,\ldots,d\}$) such that $\alpha(a) = p((s_\nu(a))_{1\le\nu\le d})$ for $a \in \mathfrak{C}^n \otimes \mathfrak{C}^m$.*

The proof of 0.II for $n = m$ can be found, e.g., in [8, Section 6]. The general case of arbitrary n and m reduces to the previous one via 0.I.

If X is a Banach space, X^* denotes the dual of X and id_X the identity operator on X. A *basis* for a finite dimensional Banach space X, say $\dim X = n$ is a biorthogonal sequence $(x_j, x_j^*)_{1\le j\le n}$, i.e., $x_j \in X$, $x_j^* \in X^x$, $x_k^*(x_j) = \delta_j^k$ for $j, k = 1, 2, \ldots, n$. We put

$$\mathrm{ubc}(x_j, x_j^*) = \sup\left\{\sum_{j=1}^{n} |x^*(x_j)x_j^*(x)| : x \in X,\ x^* \in X^* \text{ with } \|x\|\,\|x^*\| \le 1\right\},$$
$$\mathrm{ubc}(X) = \inf \mathrm{ubc}(x_j, x_j^*),$$

where the infimum extends on all bases for X.

Next recall that

$$\operatorname{lust}(X) = \inf \|u\| \, \|v\|$$

where the infimum extends on all finite dimensional Banach spaces Y with $\operatorname{ubc}(Y) = 1$ and all pairs of operators $u: X \to Y$, $v: Y \to X$ with $vu = \mathrm{id}_X$.

Let $u: X \to Y$ be a (bounded linear) operator acting between Banach spaces X and Y. Then $u^*: Y^* \to X^*$ denotes the adjoint of u, $u|_{X_0}$ denotes the restriction of u to a subspace X_0 of X, $\|u\|$ stands for *operator norm of* u, $\pi_p(u)$ for *p-absolutely summing norm of* u ($1 \leq p < \infty$), $\mathbf{n}(u)$ for the *nuclear norm* of u, and $\gamma_p(u)$ for the *norm* of *factorization* of u *through* L_p*-spaces* (see, e.g. [20] for the definitions).

By $d(X, Y)$ we denote the *Banach–Mazur distance* of Banach spaces X and Y with $\dim X = \dim Y < \infty$. Recall that

$$d(X, Y) = \inf\{\|u\| \, \|u^{-1}\| : u: X \to Y \text{ isomorphism}\}.$$

Finally recall that the *Gordon–Lewis constant* of a finite dimensional Banach space X in symbols $\mathrm{gl}_H(X)$ is defined by

$$\mathrm{gl}_H(X) = \inf\{\gamma_1(T): T: X \to \mathrm{l}_m^2, \pi_1(T) \leq 1, m = 1, 2, \ldots\}.$$

1. The ké Constants

DEFINITION. Let n be a positive integer and let X be a Banach space with $n \leq \dim X = s < \infty$. Let us put

$$\mathrm{k\acute{e}}_n(X) = \inf\{\|u\| \, \|v\| : u: L_n^\infty \to X, v: X \to L_n^1, vu = \imath^{(n)}\},$$
$$\mathrm{k\acute{e}}(X) = \mathrm{k\acute{e}}_{\dim X}(X).$$

The numbers $\mathrm{k\acute{e}}_n(X)$, called sometimes the ké *constants*, are invariants of the local theory of Banach spaces, precisely we have

PROPOSITION 1.1. *If* $u: X \to Y$, $v: Y \to X$ *are operators with* $vu = \mathrm{id}_X$ *and* $\|u\| \, \|v\| \leq d$ *then* $\mathrm{k\acute{e}}_n(Y) \leq d \, \mathrm{k\acute{e}}_n(X)$; *in particular if* X *and* Y *are isomorphic then* $\mathrm{k\acute{e}}_n(X) \leq d(X, Y) \, \mathrm{k\acute{e}}_n(Y)$.

The proof of Proposition 1.1 is trivial. Less obvious are the following estimates

PROPOSITION 1.2. *Let* $s = \dim X \geq n \geq 1$. *Then*

(a) $\mathrm{k\acute{e}}_n(X) \leq \operatorname{ubc}(X)$,
(b) $\mathrm{k\acute{e}}_n(X) \leq \sqrt{n}$, $n = 1, 2, \ldots, \dim X$.

Proof. (a) By Proposition 1.1 it is enough to show that if X has a basis, say (x_j, x_j^*), with $\mathrm{ubc}(x_j, x_j^*) = 1$ then $\mathrm{k\acute{e}}_n(X) = 1$. Let

$$X_n = \mathrm{span}\{x_j : j = 1, 2, \ldots, n\}$$

and let $P: X \to X_n$ be the projection with $\ker P = \mathrm{span}\{x_j : j > n\}$. By a result of Lozanovskiĭ [18], and Jamison and Ruckle [14] there are $z \in X_n$ and $z^* \in X_n^*$ such that $\|z\| = \|z^*\| = 1$ and $z^*(x_j)x_j^*(z) = 1/n$ for $j = 1, 2, \ldots, n$. Define $u: L_n^\infty \to X$ and $v: X \to L_n^1$ by

$$u(f) = \sum_{j=1}^{n} f(j)x_j^*(z)x_j \qquad \text{for} \quad f \in L_n^\infty,$$

$$v(x) = (nx_j^*(Px)z^*(x_j))_{1 \le j \le n} \qquad \text{for} \quad x \in X.$$

Remembering that $\|z\| = \|z^*\| = \|P\| = \mathrm{ubc}(x_j, x_j^*) = 1$ we easily get

$$\begin{aligned} \|u\| &= \sup\left\{\left\|\sum_{j=1}^{n} f(j)x_j^*(z)x_j\right\| : \|f\|_\infty \le 1\right\} \\ &\le \sup\left\{\sum_{j=1}^{n} |x_j^*(z)|\,|x^*(x_j)| : \|x^*\| \le 1\right\} \\ &\le \mathrm{ubc}(x_j, x_j^*) = 1. \\ \|v\| &= \sup\left\{n^{-1}\sum_{j=1}^{n} |nx_j^*(Px)|\,|z^*(x_j)| : \|x\| \le 1\right\} \\ &= \sup\left\{\sum_{j=1}^{n} |x_j^*(Px)|\,|z^*(x_j)| : \|x\| \le 1\right\} \\ &\le \mathrm{ubc}(x_j, x_j^*)\|P\|\,\|z^*\| \le 1. \end{aligned}$$

Finally a direct checking shows that $vu = \imath^{(n)}$.

(b) Let Y be any n-dimensional subspace of X. By a result of John [15], there is a factorization of the identity $\mathrm{id}_{L_n^2}: L_n^2 \xrightarrow{u} Y \xrightarrow{v} L_n^2$ such that $\|u\| = 1$ and $\pi_2(v) \le \sqrt{n}$ (see, e.g. [1, 2]). Thus there is an extension $\tilde{v}: X \to L_n^2$ of v with $\pi_2(\tilde{v}) = \pi_2(v)$ (see, e.g. [20]). Thus if $\imath_{\infty,2}^{(n)}. L_n^\infty \to L_n^2$ and $\imath_{2,1}^{(n)}: L_n^2 \to L_n^1$ are natural injections then $\imath^{(n)} = \imath_{2,1}^{(n)}\,\tilde{v}u\,\imath_{\infty,2}^{(n)}$ and $\|\imath_{2,1}^{(n)}\tilde{v}\|\,\|u\imath_{\infty,2}^{(n)}\| \le \pi_2(\tilde{v}) \le \sqrt{n}$.

Our next aim is to give some estimates for $\mathrm{k\acute{e}}_n(X)$ from below. They are closely related to estimates of $\mathrm{gl}_H(X)$. They are based on the classical result due to Horn (see [8, Section 4]), a simple consequence of which is the following fact

(#) *Let $u: H \to H$ be a bounded operator acting in a Hilbert space H. Assume that there is another Hilbert space H_1 and nuclear operators $A: H \to H_1$*

and $B: H_1 \to H$ such that $u = BA$. Then the s numbers of u satisfy the inequality

$$\left(\sum_j [s_j(u)]^{1/2}\right)^2 \leq \mathbf{n}(A)\mathbf{n}(B).$$

In particular, if u is an orthogonal projection on a k-dimensional subspace or more specific the identity on L_k^2 then

$$k^2 \leq \mathbf{n}(A)\mathbf{n}(B).$$

Now we are ready to state the main result of this section.

PROPOSITION 1.3. *Let n and s be integers with $0 < n \leq s$ and let X be an s-dimensional Banach space. Then*

(i) $\text{ké}_n(X) \geq \sup\left\{\dfrac{n}{\pi_1(u)\pi_1((u^{-1})^*)} : u: X \to l_s^2,\ u \text{ invertible}\right\}$

(ii) $\text{ké}_n(X) \geq \sup\left\{\dfrac{a^2}{2}\dfrac{\sqrt{n}}{\pi_1(u)} \inf\|u|_Y\|\right\}$

where a with $0 < a < \frac{1}{2}$ is an absolute constant independent of X, n, s, and the supremum is taken over all invertible operators $u: X \to l_s^2$ and the infimum over all linear subspaces $Y \subset X$ with dim $Y \geq an$.

Proof of (i). Consider the commutative diagram

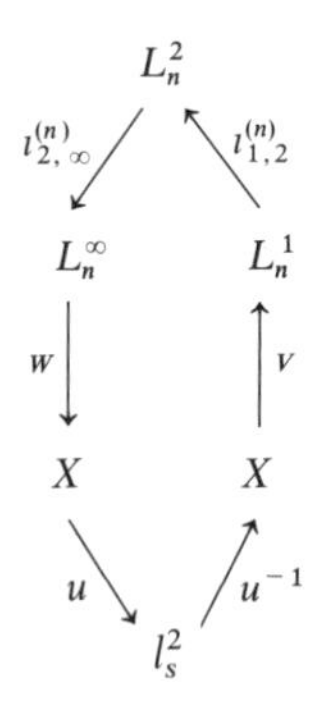

where $vw = \iota^{(n)}$, $\iota_{2,\infty}^{(n)}: L_n^2 \to L_n^\infty$, and $\iota_{1,2}^{(n)}: L_n^1 \to L_n^2$ are natural injections $u: X \to l_s^2$ is an arbitrary invertible operator. Clearly

$$\text{id}_{L_n^2} = (\iota_{1,2}^{(n)} v u^{-1})(u w \iota_{2,\infty}^{(n)}).$$

Thus, by (#),

$$n^2 \leq \mathbf{n}(\iota_{1,2}^{(n)} v u^{-1})\mathbf{n}(u w \iota_{2,\infty}^{(n)}).$$

Clearly $\|\iota_{1,2}^{(n)}\| = \|\iota_{2,\infty}^{(n)}\| = \sqrt{n}$. Using the fact that for every operator from L_n^∞ its nuclear norm equals its 1-absolutely summing norm we get

$$\mathbf{n}(uw\iota_{2,\infty}^{(n)}) \leq \|\iota_{2,\infty}^{(n)}\|\mathbf{n}(uw) = \|\iota_{2,\infty}^{(n)}\|\boldsymbol{\pi}_1(uw) \leq \|\iota_{2,\infty}^{(n)}\| \|w\|\boldsymbol{\pi}_1(u) = \sqrt{n}\|w\|\boldsymbol{\pi}_1(u).$$

Similarly

$$\begin{aligned}\mathbf{n}(\iota_{1,2}^{(n)}vu^{-1}) &= \mathbf{n}((\iota_{1,2}^{(n)}vu^{-1})^*) \leq \|(\iota_{1,2}^{(n)})^*\| \|v^*\|\boldsymbol{\pi}_1((u^{-1})^*) \\ &= \|\iota_{1,2}^{(n)}\| \|v\|\boldsymbol{\pi}_1((u^{-1})^*) = \sqrt{n}\|v\|\boldsymbol{\pi}_1((u^{-1})^*).\end{aligned}$$

Thus

$$n^2 \leq n\|w\| \|v\|\boldsymbol{\pi}_1(u)\boldsymbol{\pi}_1((u^{-1})^*).$$

Therefore for arbitrary factorization (w, v) of $\iota^{(n)}$ through X one has $\|w\| \|v\| \geq n/\boldsymbol{\pi}_1(u)\pi_1((u^{-1})^*)$ which yields (i).

(ii) We begin with a lemma due to Figiel and Johnson [4].

LEMMA 1.1. *Let $v: X \to L_n^1$ and $u: X \to l_s^2$ be linear operators. Assume that u is invertible and there is a subspace X_0 of X with* $\dim X_0 = n$ *such that $v|_{X_0}$ is invertible. Then there is a subspace Y of X_0 such that* $\dim Y \geq an$ *and $\boldsymbol{\pi}_2(vu^{-1}P) \leq (2\|v\|/\|u|_Y\|)\sqrt{n}$ where $P: l_s^2 \to l_s^2$ is the orthogonal projection onto $u^{-1}(Y)$ and a, with $0 < a < \frac{1}{2}$, is an absolute constant independent of s, n, X, and u.*

Proof. The cotype 2 constants of L_n^1 are uniformly bounded for $n = 1, 2, \ldots$ (in fact by $\sqrt{2}$). Thus a result of Figiel *et al.* [5] yields the existence of a subspace G of $v(X_0)$, with $\dim G \geq 2a \dim v(X_0) = 2an$, which admits an isomorphism $\varphi: G \to l_{\dim G}^2$ with $\|\varphi^{-1}\| \|\varphi\| \leq 2$ (a is an absolute constant). Let $X_1 = v^{-1}(G)$. Consider on X_1 two inner products induced by operators $u|_{X_1}$ and $\varphi v|_{X_1}$. Then there is a subspace, say Y, of X_1 with $\dim Y \geq \frac{1}{2}\dim X_1 \geq an$ such that the inner products restricted to this subspace are proportional. (This is a restatement of a well-known geometric fact that every n-dimensional ellipsoid has an $(n+1)/2$-dimensional spherical section). Thus there is a $c > 0$ such that $\|\varphi v(y)\| = c\|u(y)\|$ for $y \in Y$. (c is positive because φ, $v|_Y$, and u are invertible). Pick $y_0 \in Y$ so that $\|y_0\| = 1$ and $\|u|_Y\| = \|u(y_0)\|$. Then $c\|u|_Y\| = \|\varphi v(y_0)\| \leq \|\varphi\| \|v\|$. Hence $c \leq \|v\| \|\varphi\|/\|u|_Y\|$. Now let $P: l_s^2 \to l_s^2$ be the orthogonal projection onto $u^{-1}(Y)$. Consider the operator $\varphi vu^{-1}P: l_s^2 \to l_{\dim Y}^2$. Clearly $(\varphi vu^{-1}P)(\xi) = 0$ for $\xi \in [u^{-1}(Y)]^\perp = \ker P$, while for $\xi \in u^{-1}(Y)$ one has

$$\|(\varphi vu^{-1}P)(\xi)\| = \|\varphi vu^{-1}(\xi)\| = c\|u(u^{-1}(\xi))\| = c\|\xi\|.$$

Thus we can estimate the Hilbert–Schmidt norm of $\varphi v u^{-1} P$; we have

$$\mathrm{hs}(\varphi v u^{-1} P) = c\sqrt{\dim Y} \le c\sqrt{\dim X_0} = c\sqrt{n} \le \frac{\|\varphi\|\,\|v\|}{\|u|_Y\|}\sqrt{n}.$$

Finally using the fact that for operators between Hilbert spaces the Hilbert–Schmidt norm equals the 2-absolutely summing norm we get

$$\begin{aligned}\pi_2(vu^{-1}P) &= \pi_2(\varphi^{-1}\varphi v u^{-1}P) \le \|\varphi^{-1}\|\pi_2(\varphi v u^{-1} P)\\ &\le \|\varphi^{-1}\|\,\|\varphi\|\frac{\|v\|}{\|u|_Y\|}\sqrt{n} \le \frac{2\|v\|}{\|u|_Y\|}\sqrt{n}.\end{aligned}$$

Proof of (ii). Let $w: L_n^\infty \to X$ and $v: X \to L_n^1$ be operators with $vw = \iota^{(n)}$. Put $X_0 = w(L_n^\infty)$. Clearly $\dim X_0 = n$ and $v|_{X_0}$ is invertible. Let $u: X \to l_s^2$ be any invertible operator. Then, by Lemma 1.1, there is a subspace $Y \subset X_0$ with $\dim Y \ge an$ and $\pi_2(vu^{-1}P) \le 2\|v\|/\|u|_Y\|\sqrt{n}$ where $P: l_s^2 \to l_s^2$ is the orthogonal projection onto $u^{-1}(Y)$. Consider the operator

$$Q: L_n^2 \xrightarrow{\iota_{2,\infty}^{(n)}} L_n^\infty \xrightarrow{w} X \xrightarrow{u} l_s^2 \xrightarrow{P} l_s^2 \xrightarrow{u^{-1}} X \xrightarrow{v} L_n^1 \xrightarrow{\iota_{1,2}^{(n)}} L_n^2.$$

A moment of reflection shows that Q is the orthogonal projection with $\dim Q(L_n^2) = \dim Y \ge an$. Thus, by (#), $a^2n^2 \le (\dim Q(L_n^2))^2 \le \mathbf{n}(A)\mathbf{n}(B)$ where $A = uw\iota_{2,\infty}^{(n)}$ and $B = \iota_{1,2}^{(n)} v u^{-1} P$. Clearly

$$\mathbf{n}(A) \le \|\iota_{2,\infty}^{(n)}\|\mathbf{n}(uw) = \|\iota_{2,\infty}^{(n)}\|\pi_1(uw) = \|\iota_{2,\infty}^{(n)}\|\,\|w\|\pi_1(u) = \sqrt{n}\|w\|\pi_1(u).$$

Next we have $\mathbf{n}(B) \le \pi_2(\iota_{1,2}^{(n)})\pi_2(vu^{-1}P)$ (see, e.g. [20, 24.6.5]). It is well known and easy to evaluate that $\pi_2(\iota_{1,2}^{(n)}) = \sqrt{n}$ (see, e.g. [20, 22.4.8]). Finally, by Lemma 1.1, $\pi_2(vu^{-1}\varphi) \le 2(\|v\|/\|u|_Y\|)\sqrt{n}$. Hence $\mathbf{n}(B) \le (2\|v\|/\|u|_Y\|)n$. Thus $a^2n^2 \le n^{3/2}(2\|v\|\,\|w\|/\|u|_Y\|)\pi_1(u)$. Hence $\|w\|\,\|v\| \ge (a^2/2)(\|u|_Y\|\sqrt{n}/\pi_1(u))$ which yields (ii).

COROLLARY 1.1. *In the notation of Proposition* 1.3 *for every s-dimensional Banach space X*

$$\text{ké}(X) \ge \sup\left\{\frac{s}{\pi_1(u)\pi_1((u^{-1})^*)} : u: X \to l_s^2,\ u\text{-invertible}\right\}; \tag{j}$$

$$\text{ké}(X) \ge \sup\left\{\frac{a^2\sqrt{s}}{2\pi_1(u)}\inf\|u|_Y\|\right\}, \tag{jj}$$

here the infimum is taken over all subspaces Y of X with $\dim Y \ge as$.

It is interesting to compare Corollary 1.1 with similar estimates for $\mathrm{gl}_H(X)$. Namely one has

PROPOSITION 1.4. *For every s-dimensional Banach space X*

$$\mathrm{gl}_H(X) = \sup\left\{\frac{\mathrm{tr}(uv)}{\pi_1(u)\pi_1(v^*)} : u: X \to l^2,\, v: X \to l^2,\, u \neq 0,\, v \neq 0\right\}, \quad \text{(GL)}$$

$$\mathrm{gl}_H(X) \geq \sup\left\{\frac{a}{\sqrt{2\pi}}\frac{\sqrt{s}}{\pi_1(u)}\{\inf\|u|_Y\| : Y \subset X \dim Y \geq as\}\right\}, \quad \text{(FJ)}$$

where the absolute constant a with $0 < a < \frac{1}{2}$ and the sets on which the "sup" and the "inf" are extended are the same as in Proposition 1.3.

Proof. (GL) is due to Gordon and Lewis [9]. The equality in (GL) follows from the duality between the ideal of 1-absolutely summing operators and the ideal of L_1-factorable operators.

(FJ) is essentially due to Figiel and Johnson [4]. It can be deduced from Lemma 1.1 as follows:

Let an invertible $u: X \to l_s^2$ be given. By definition of $\mathrm{gl}_H(X)$ there are $w: L^1 \to l_s^2$ and $v: X \to L^1$ with $wv = u$ and $\|v\|\,\|w\| \leq \mathrm{gl}_H(X)\pi_1(u)$. Since u is invertible, so is v. Thus, by Lemma 1.1, there is a subspace $Y \subset X$ with $\dim Y \geq as$ such that $\pi_2(vu^{-1}P) \leq (2\|v\|/\|u|_Y\|)\sqrt{s}$ where $P: l_s^2 \to l_s^2$ is the orthogonal projection onto $u^{-1}(Y)$. Thus

$$as \leq \dim u^{-1}(Y) = \mathbf{n}(wvu^{-1}P) \leq \pi_2(w)\pi_2(vu^{-1}P).$$

Since every operator from an L_1-space into a Hilbert space is 2-absolutely summing, $\pi_2(w) \leq K\|w\|$ where K is an absolute constant. (In fact for the best possible $K: K = \sqrt{\pi/2}$ for real spaces and $K = 2/\sqrt{\pi} \leq \sqrt{\pi/2}$ for complex spaces; see Grothendieck [10, pp. 50–52].) Thus

$$as \leq 2\sqrt{\frac{\pi}{2}}\|v\|\,\|w\|\sqrt{s}\|u|_Y\|^{-1} \leq \frac{\mathrm{gl}_H(X)\pi_1(u)}{\|u|_Y\|}\sqrt{s}\sqrt{2\pi},$$

which yields the desired inequality.

Let us observe that the right-hand side of (GL) is greater than or equal to the right-hand side of Corollary 1.1(j) [because if $u: X \to l_s^2$ is invertible and $v = u^{-1}$ then $\mathrm{tr}(uv) = \mathrm{tr}(uu^{-1}) = s$], and the right-hand side of (FJ) is greater than or equal to the right-hand side of Corollary 1.1(jj) (because $0 < a < \frac{1}{2}$). This observation and Proposition 1.2a justify the following:

Conjecture. $\mathrm{k\acute{e}}(X) \leq C\,\mathrm{gl}_H(X)$ for every finite dimensional Banach space X; C is an absolute constant independent of X.

Spaces with large ké constants. Our estimations of ké constants from below base on estimating from below the right-hand sides of formulas (i) and (ii) of Proposition 1.3. Hence we got similar effects as for gl_H constants (cf. [4,6,9]).

We begin with "large" subspaces of L_N^∞ spaces.

Let E be a subspace of $\mathfrak{C}^N$ (resp., $\mathfrak{R}^N$), $N < \infty$. Put $E_p = (E, \|\cdot\|_p)$ for $1 \le p \le \infty$, i.e., E_p is E regarded as a subspace of L_N^p. Let $\imath_{p,q}^E : E_p \to E_q$ denote the natural injection $\imath_{p,q}^{(n)}$ restricted to E_p.

Under the above notation we have

PROPOSITION 1.5. *Let n, s, N be positive integers with $n \le s \le N$. Let E be an s-dimensional subspace of $\mathfrak{C}^N$ (resp. $\mathfrak{R}^N$). Then*

$$\text{ké}_n(E_\infty) \ge \frac{a^{5/2}}{2\|\imath_{1,2}^E\|} \frac{n}{\sqrt{N}},$$

where a is the absolute constant appearing in Proposition 1.3.

Proof. We apply formula (ii) of Proposition 1.3 for $X = E_\infty$. Put $u = \imath_{\infty,2}^E : E_\infty \to E_2$. (Clearly E_2 being an s-dimensional subspace of L_N^2 can be identified with l_s^2). Consider a subspace F of E with $\dim F \ge an$ and put $Y = F_\infty$. Next we need

LEMMA 1.2. $\|\imath_{\infty,2}^F\| \ge \sqrt{\dim F/N}$.

For the proof of Lemma 1.2 see Figiel and Johnson [4, Corollary 1.2] Applying Lemma 1.2 we get $\|\imath_{\infty,2}^E|_Y\| = \|\imath_{\infty,2}^F\| \ge \sqrt{an/N}$. On the other hand $\imath_{\infty,2}^E = \imath_{\infty,1}^E \imath_{1,2}^E$ and obviously $\pi_1(\imath_{\infty,1}^E) \le \pi_1(\imath_{\infty,1}^{(N)}) = 1$. Thus $\pi_1(\imath_{\infty,2}^E) \le \|\imath_{1,2}^E\|$. Therefore, by Proposition 1.3(ii)

$$\begin{aligned}\text{ké}_n(E_\infty) &\ge \sup\left\{\frac{a^2}{2}\frac{\sqrt{n}}{\pi_1(u)}\inf\|u|_Y\|\right\} \ge \frac{a^2\sqrt{n}}{2\pi_1(\imath_{\infty,2}^E)}\inf\{\|\imath_{\infty,2}^F\| : F \subset E, \dim F \ge an\} \\ &\ge \frac{a^2\sqrt{a}}{2\|\imath_{2,1}^E\|}\frac{n}{\sqrt{N}}.\end{aligned}$$

COROLLARY 1.2. *There exists a sequence (X_s) of Banach spaces such that $\dim X_s = s$ and $\text{ké}_n(X_s) \ge C(n/\sqrt{s})$ for $1 \le n \le s$, $s = 1, 2, \ldots$, where C is an absolute constant; in particular $\text{ké}(X_s) \ge C\sqrt{s}$ is of the worst possible order (because by Proposition 1.2, $\text{ké}(X_s) \le \sqrt{s}$).*

Proof. Fix $b > 1$. By a result of Kashin [16] (cf. [26]) there is an absolute constant $d(b)$ such that there is an s-dimensional subspace of $\mathfrak{C}^{[bs]}$ (resp., $\mathfrak{R}^{[bs]}$), say, $E^{(s)}$, such that $\|\imath_{1,2}^{E^{(s)}}\| = \sup\{(\|x\|_2/\|x\|_1) : 0 \ne x \in E^{(s)}\} \le d(b)$. Put

$X_s = E^{(s)}_\infty$. Then, by Proposition 1.5,

$$\text{ké}_n(X_s) \geq \frac{a^{5/2}}{2d(b)} \frac{n}{\sqrt{[sb]}} \geq \frac{a^{5/2}}{2d(b)\sqrt{b}} \frac{n}{\sqrt{s}}.$$

We put $C = a^{5/2}(2d(b)b)^{-1}$.

Remarks. 1°. We do not know whether $n/\sqrt{s}$ is the worst possible order for factorization of $\iota^{(n)}$ through an s-dimensional space, precisely:

Problem 1. Does there exist a constant $C > 0$ such that for every s-dimensional space X, $\text{ké}_n(X) \geq Cn/\sqrt{s}$?

In fact we do not know even whether there is a function $n \to s(n)$ and a constant C_1 such that for every $s(n)$-dimensional space X, $\text{ké}_n(X) \leq C_1$. It follows from recent results due to Benyamini and Gordon [1, Section 5] that under additional assumptions on X (like K-convexity) the desired uniformly bounded factorizations exist.

2°. If (Y_s) is a sequence of subspace of the sequence (X_s), i.e., $Y_s \subset X_s$, for $s = 1, 2, \ldots$, and if (n_s) is a sequence of integers with $1 \leq n_s \leq \dim Y_s$, for $s = 1, 2, \ldots$, then

$$\text{ké}_{n_s}(Y_s) \geq Cn_s/\sqrt{s}$$

where C is the same constant as in Corollary 1.2. (The proof is the same as that of Corollary 1.2. On the other hand, it follows from recent results due to Figiel and Johnson [4, Section 6] that there is an absolute constant C such that X_s contains a subspace, say Z_s, with $d(Z_s, l^\infty_{\dim Z_s}) \leq C$ and $\dim Z_s = [\sqrt{s}]$. Hence $\text{ké}_{[\sqrt{s}]}(X_s) \leq C$. This fact gives evidences for a positive solution of Problem 1.

3°. Proposition 1.5 allows to estimate from below by the fourth root of dimension for the $\text{ké}_{n/2}$ constant of special $n/2 = 2^{k-1}$-dimensional subspaces of L^∞_n:

(a) the space spanned by the fixed points of the involution of $\mathfrak{C}^n$ (resp., $\mathfrak{R}^n$) which in the unit vector basis is represented by a $2^k \times 2^k$ Walsh matrix (= the Sobczyk spaces),

(b) the spaces spanned by some Walsh functions,

(c) the spaces spanned by these exponents e^{imt} that the coefficients of the nth Rudin–Shapiro polynomials at these exponents equal $+1$.

For all these spaces, say E_n, we have $\|\iota^{E_n}_{1,2}\| \leq c\sqrt[4]{n}$ (see [19, Lecture 10] for details).

Proposition 1.3(i) yields a convenient criterion for estimating ké constants from below which we shall formulate next.

Let X be an s-dimensional space; let $(e_j, e_j^*)_{1 \leq j \leq s}$ be a basis for X. Clearly $(e_j^*, e_j)_{1 \leq j \leq s}$ is the basis for X^*. Denote by $\mathfrak{T}^s$ the s-dimensional torus. For each $\omega = (\omega(j))_{1 \le j \le s} \in \mathfrak{T}^s$ define a linear operator $T_\omega : X \to X$ by

$$T_\omega(x) = \sum_{j=1}^{s} \omega(j) e_j^*(x) e_j \qquad \text{for} \quad x \in X.$$

Clearly for the adjoint operator T^* we have

$$T_\omega^*(x^*) = \sum_{i=1}^{s} \omega(j) x^*(e_j) e_j^* \qquad \text{for} \quad x^* \in X^*.$$

Let us put

$$\rho = \rho((e_j)_{1 \le j \le s}) = \inf\left\{\left\| \sum_{j=1}^{s} \omega(j) e_j \right\| : \omega \in \mathfrak{T}^s\right\},$$

$$\rho^* = \rho((e_j^*)_{1 \le j \le s}) = \inf\left\{\left\| \sum_{j=1}^{s} \omega(j) e_j^* \right\| : \omega \in \mathfrak{T}^s\right\}.$$

Now we are ready to formulate our criterion

PROPOSITION 1.6. *Let X, $(e_j, e_j^*)_{1 \le j \le s}$ be as above. Assume that there is a subset $G \subset \mathfrak{T}^s$ and a probability measure m on Borel subsets of G such that for some constants c_G and M_G with $0 < c_G < \infty, 0 < M_G \le 1$ one has*

$$\|T_\sigma\| \le M_G \qquad \textit{for} \quad \sigma \in G, \tag{1}$$

$$\int_G \left| \sum_{j=1}^{s} \sigma(j) c_j \right| m(d\sigma) \ge c_G \left(\sum_{j=1}^{s} |c_j|^2 \right)^{1/2} \tag{2}$$

for arbitrary scalars $c_1, c_2, \ldots, c_s$. Then, for $n = 1, 2, \ldots, s$,

$$\text{ké}_n(X) \ge \frac{n}{\rho \cdot \rho^*} (c_G/M_G)^2.$$

Proof. By Proposition 1.3(i) it is enough to show that

$$\pi_1(u) \le \frac{M_G}{c_G} \rho^*; \qquad \pi_1(u^{-1})^* \le \frac{M_G}{c_G} \rho \tag{3}$$

where $u : X \to l_s^2$ is defined by $u(x) = (e_j^*(x))_{1 \le j \le s}$. We shall check the first inequality in (3). The proof of the second is the same; one has to observe that $(u^{-1})^*(x^*) = (x^*(e_j))_{1 \le j \le s}$ and replace in the argument below the $e_j - s$ by the $e_j^* - s$ and vice versa.

Fix $\omega \in \mathfrak{T}^s$ and pick $x_1, x_2, \ldots, x_N \in X$ arbitrarily. Then, for every $\sigma = (\sigma(j)) \in G$,

$$\sup_{\|x^*\| \leq 1} \sum_{\nu=1}^{N} |x^*(x_\nu)| \geq \left\| \sum_{j=1}^{s} \omega(j) e_j^* \right\|^{-1} M_G^{-1} \sum_{\nu=1}^{N} \left| \left(\sum_{j=1}^{s} \omega(j) e_j^* \right) (T_\sigma(x_\nu)) \right|$$

$$= \left\| \sum_{j=1}^{s} \omega(j) e_j^* \right\|^{-1} M_G^{-1} \sum_{\nu=1}^{N} \left| \sum_{j=1}^{s} \omega(j) \sigma(j) e_j^*(x_\nu) \right|.$$

Integrating against $m(d\sigma)$ and using (2) we get

$$\left\| \sum_{j=1}^{s} \omega(j) e_j^* \right\| M_G \sup_{\|x^*\| \leq 1} \sum_{\nu=1}^{N} |x^*(x_\nu)| \geq \int_G \sum_{\nu=1}^{N} \left| \sum_{j=1}^{s} \sigma(j) \omega(j) e_j^*(x_\nu) \right| m(d\sigma)$$

$$= \sum_{\nu=1}^{N} \int_G \left| \sum_{j=1}^{s} \sigma(j) \omega(j) e_j^*(x_\nu) \right| m(d\sigma)$$

$$\geq c_G \sum_{\nu=1}^{N} \left(\sum_{j=1}^{s} |\omega(j) e_j^*(x_\nu)|^2 \right)^{1/2}$$

$$= c_G \sum_{\nu=1}^{N} \left(\sum_{j=1}^{s} |e_j^*(x_\nu)|^2 \right)^{1/2}$$

$$= c_G \sum_{\nu=1}^{N} \|u(x_\nu)\|.$$

Since the above estimate holds for arbitrary sequence $x_1, x_2, \ldots, x_N$ in X and for $N = 1, 2, \ldots$, it follows that $\pi_1(u) \leq (M_G/c_G) \|\sum_{j=1}^{s} \omega(j) e_j^*\|$. Since the last inequality holds for all $\omega \in \mathfrak{T}^s$ we get the first inequality of (3).

Next we give two applications of Proposition 1.6.

As usual $p^* = 1/(p-1)$ for $1 < p < \infty$, $p^* = \infty$ for $p = 1$, $p^* = 1$ for $p = \infty$. The symbols $X \hat{\otimes} Y$ and $X \hat{\hat{\otimes}} Y$ stand for projective and injective tensor products, respectively.

COROLLARY 1.3. *There is an absolute constant $C > 0$ such that for $m = 1, 2, \ldots$ and $n = 1, 2, \ldots, m^2$ one has*

$$\text{ké}_n(l_m^{p^*} \hat{\otimes} l_m^{q^*}) = \text{ké}_n(l_m^p \hat{\hat{\otimes}} l_m^q) \geq C(n/m^{a(p,q)})$$

where

$$a(p,q) = \begin{cases} \frac{3}{2} & \textit{for} \quad 1 \leq p, q \leq 2, \\ 2 - 1/\max(p,q) & \textit{for} \quad \max(p,q) \geq 2. \end{cases}$$

Proof. Since $\text{ké}_n(X) = \text{ké}_n(X^*)$ for every finite dimensional Banach space and since $(l_m^p \hat{\hat{\otimes}} l_m^q)^* = (l_m^{p^*} \hat{\otimes} l_m^{q^*})$, the equality between ké constants in the assertion of the Corollary is obvious. To estimate from below the constant

$\text{ké}_n(l_m^p \hat{\hat{\otimes}} l_m^q)$, we consider the basis in $l_m^p \hat{\hat{\otimes}} l_m^q$ consisting of the tensors $(\delta_k \otimes \eta_l)_{1 \le k, l \le m}$ where $(\delta_k)_{1 \le k \le m}$ and $(\eta_l)_{1 \le l \le m}$ are the unit vector basis in l_m^p and l_m^q, respectively. Represent the torus $\mathfrak{T}^{m^2}$ as the tensor product $\mathfrak{T}^m \otimes \mathfrak{T}^m$. Let G be the set of all $\omega = (\omega(k,l)) \in \mathfrak{T}^m \otimes \mathfrak{T}^m$ such that $\omega(k,l) = \omega_1(k)\omega_2(l)$ for some $\omega_1, \omega_2 \in \mathfrak{T}^m$. Let $\lambda = \lambda_1 \times \lambda_2$ be the product measure of the Haar measures for $\mathfrak{T}^m$. (Clearly there is a natural identification of G with $\mathfrak{T}^m \times \mathfrak{T}^m$). Then $M_G = 1$, and applying twice the Khinchine inequality for the Steinhaus variables we infer that $c_G \le (\sqrt{2})^2 = 2$. To estimate the quantities ρ and ρ^* for the basis $(\delta_k \otimes \eta_l)_{1 \le k, l \le m}$ we consider the elements

$$x = \sum_{k=1}^{m} \sum_{l=1}^{m} e^{2\pi i k l/m} \delta_k \otimes \eta_l \qquad \text{and} \qquad y = \sum_{k=1}^{m} \sum_{l=1}^{m} \delta_k \otimes \eta_l = \sum_{k=1}^{m} \delta_k \otimes \sum_{l=1}^{m} \eta_l.$$

Clearly $\|y\|_{l_m^{p^*} \hat{\otimes} l_m^{q^*}} = m^{2-(1/p)-(1/q)}$. Furthermore it is well known and easy to establish that there are absolute positive constants (independent of m, p, q), say c_1 and c_2, such that $c_1 m^{b(p,q)} \le \|x\|_{l_m^p \hat{\hat{\otimes}} l_m^q} \le c_2 m^{b(p,q)}$, where

$$b(p,q) = \begin{cases} 1/p + 1/q - \frac{1}{2} & \text{for} \quad 1 \le p, q \le 2, \\ 1/\min(p,q) & \text{for} \quad \max(p,q) \ge 2. \end{cases}$$

Thus $\rho\rho^* \le c_2 m^{a(p,q)}$. The desired conclusion follows now immediately from Proposition 1.6.

Remark. For $n = m^2$ the estimate in Corollary 1.3 is of exact order, i.e., there exists an absolute constant C_0 such that $\text{ké}(l_m^p \hat{\hat{\otimes}} l_m^q) \le C_0 m^{2-a(p,q)}$. By Proposition 1.2 to obtain the last inequality it is enough to show that $\text{ubc}(l_m^p \hat{\hat{\otimes}} l_m^q) \le C_0 m^{2-a(p,q)}$. Let $p = \max(p,q)$. Observe that $\text{ubc}(l_m^\infty \hat{\hat{\otimes}} l_m^q) = 1$ because the space $l_m^\infty \hat{\hat{\otimes}} l_m^q$ is isometrically isomorphic with the space $(l_m^q \times l_m^q \times \cdots \times l_m^q)_{l_m^\infty}$. Next, by a result due to Gurarii *et al.* [11], $d(l_m^\infty, l_m^p) \le C_0 m^{\min(1/p, 1/2)}$. Thus

$$d(l_m^\infty \hat{\hat{\otimes}} l_m^q, l_m^p \hat{\hat{\otimes}} l_m^q) \le C_0 m^{\min(1/p,1/2)} = C_0 m^{2-a(p,q)}.$$

Hence $\text{ubc}(l_m^p \hat{\hat{\otimes}} l_m^q) \le C_0 m^{2-a(p,q)}$.

Our last application is a characterization of Sidon sets in terms of ké constants. We recall standard notation. G stands for a compact abelian group, Γ for its dual, and m is the normalized Haar measure on G. For $f \in L^1(G,m) = L^1(G)$ and for $\gamma \in \Gamma$, we put $\hat{f}(\gamma) = \int_G f(g)\bar{\gamma}(g)\, dg$. If $S \subset \Gamma$, S nonempty, then $\mathcal{T}_S$ denotes the set of all trigonometric polynomials generated by S, i.e., the set of all finite linear combinations of elements of S. For $f \in \mathcal{T} = \mathcal{T}_\Gamma$ we put

$$\|f\|_\infty = \sup\{|f(g)| : g \in G\},$$

$$\|f\|_p = \left(\int_G |f(g)|^p m(dg) \right)^{1/p} \qquad \text{for} \quad 1 \le p < \infty.$$

For $F \subset S$, F, finite nonempty, we put $C_F = (\mathscr{T}_F, \|\cdot\|_\infty)$, $L_F^p = (\mathscr{T}_F, \|\cdot\|_p)$. Under the above notation we have

PROPOSITION 1.7. *For a nonempty subset $S \subset \Gamma$ the following conditions are equivalent*:

(+) *S is a Sidon set, i.e. there is a $c_1 > 0$ such that*

$$\sum_{\gamma \in S} |\hat{f}(\gamma)| \le c_1 \|f\|_\infty \qquad \text{for} \quad f \in \mathscr{T}_S$$

(++) *S has the following properties*:

(a) *S is a Λ_2-set, i.e., there is a $c_2 > 0$ such that $\|f\|_2 \le c_2 \|f\|_1$ for $f \in \mathscr{T}_S$,*

(b) *There is a $c_3 > 0$ such that for arbitrary nonempty finite subset B of S,* $\text{ké}(C_B) \le c_3$.

Proof. (+)$\Rightarrow$(++). The fact that (+)$\Rightarrow$(a) is classic (see, e.g. [13, Section 37]). It follows from (+) that $d(C_B, l^1_{|B|}) \le c_1$. Clearly $\text{ké}(l^1_{|B|}) = 1$. Hence $\text{ké}(C_B) \le c_1$.

(++)$\Rightarrow$(+). Fix a nonempty $B \subset S$ and consider the basis $(\gamma)_{\gamma \in B}$ in C_B. The biorthogonal functionals $(\gamma^*)_{\gamma \in B}$ are given by $\gamma^*(f) = \hat{f}(\gamma)$ for $f \in C_B$. Now consider the map $u: C_{|B|} \to l^2_{|B|}$ defined by $u(f) = (\hat{f}(\gamma))_{\gamma \in B}$. By (a) we have $\|u(f)\|_{l^2_{|B|}} = \|f\|_2 \le c_2 \int_G |f(g)| m(dg)$ for $f \in \mathscr{T}_B$. Thus $\pi_1(u) \le c_2$. Next for $g \in G$ define $\omega_g \in \mathfrak{T}^{|B|}$ by $\omega_g(\gamma) = \gamma(g)$ for $\gamma \in B$. In that way we identify G with a subset of $\mathfrak{T}^{|B|}$, and we consider this subset with the Haar measure m on G. We are now in the situation of Proposition 1.6. First observe that $M_G = 1$. Indeed for $f \in \mathscr{T}_B$ and $\sigma \in G$ we have

$$(T_\sigma f)(g) = \sum_{\gamma \in B} \gamma(\sigma)\hat{f}(\gamma)\gamma(g) = \sum_{\gamma \in B} \hat{f}(\gamma)\gamma(g + \sigma) = f(g + \sigma).$$

Thus $\|T_\sigma\| = 1$. Next we may put $c_G = c_2^{-1}$. Indeed, by (a),

$$\int_G \left| \sum_{\gamma \in B} \gamma(g) t_\gamma \right| m(dg) \ge \frac{1}{c_2} \sqrt{\int_G \left| \sum_{\gamma \in B} \gamma(g) t_\gamma \right|^2 m(dg)} = \frac{1}{c_2} \sqrt{\sum_{\gamma \in B} |t_\gamma|^2}$$

for an arbitrary sequence of scalars $(t_\gamma)_{\gamma \in B}$. Hence it follows from the proof of Proposition 1.6 that $\pi_1(u^{-1})^* \le (M_G/c_G)\rho = c_2 \rho$, where

$$\rho = \rho((\gamma)_{\gamma \in B}) = \inf\left\{ \left\| \sum_{\gamma \in B} \omega(\gamma)\gamma \right\|_\infty : \omega = (\omega(\gamma)) \in \mathfrak{T}^B \right\}.$$

Hence combining (b) with Proposition 1.3(i) we get

$$c_3 \ge \text{ké}(C_B) \ge |B| / \pi_1(u)\pi_1(u^{-1})^* \ge |B|/c_2^2.$$

Thus we have shown that for every nonempty finite set $B \subset S$, for every sequence $(\omega(\gamma))_{\gamma \in B}$ with $|\omega(\gamma)| = 1$ for $\gamma \in B$, one has $\|\sum_B \omega(\gamma)\gamma\|_\infty \ge (1/c_2^2 c_3)|B|$. Hence, by a complex result due to Pisier [22], S is a Sidon set.

We end Section 1 by two examples of sequences of finite dimensional spaces whose ké constants are uniformly bounded while their Gordon–Lewis constants tend to infinity.

EXAMPLE 1. Given $s = N^3$ there is a Banach space X_N with $\dim X = s$, $\mathrm{ke}(X_N) \le 2$, $\mathrm{gl}_H(X_N) \ge Cs^{1/3}$ where C is an absolute constant independent of s.

Proof. Let Y_N by any N^2-dimensional space with $\mathrm{gl}_H(Y_N) \ge CN$; C an absolute constant (for the existence of such a Y_N, cf. [4,6]). Put $X_N = (Y_N \times L^2_{N^3-N^2})_2$. Clearly $\dim X_N = N^3$ and $\mathrm{gl}_H(X_N) \ge \mathrm{gl}_H(Y_N) \ge CN$. By the result of John [15] there is an isomorphism $u: X_N \to L^2_{N^3}$ with $\|u\| = \|u^{-1}\| \le \sqrt{\dim Y_N} = N$ which takes Y_N onto the span of the first N^2 unit vectors of $L^2_{N^3}$, say E_2, and u restricted to $L^2_{N^3-N^2}$ is an isometry onto the orthogonal complement of the span. Let E_∞ denote E_2 equipped with the norm $\|\cdot\|_\infty$ and let $\imath^s_{\infty,2}: L^\infty_s \to L^2_s$ and $\imath^{(s)}_{2,1}: L^2_s \to L^1_s$ be natural injections. Observe that

$$\|\imath^{(s)}_{\infty,2}|_{E_\infty}\| = \sqrt{N^2/N^3} = N^{-1/2} = \|\imath^{(s)}_{2,1}|_{E_2}\|. \tag{*}$$

Put $v = u^{-1}\imath^s_{\infty,2}$ and $w = \imath^s_{1,2}u$. Then $\imath^{(s)} = wv$, and by the construction of u, $\|v\| \le \sqrt{2}$. (Indeed given $f \in L^\infty_s$ with $\|f\|_\infty = 1$ let f_1 and f_2 be orthogonal projections of f onto E_2 and its orthogonal complement, respectively. We have

$$\begin{aligned}\|u^{-1}\imath^{(s)}_{\infty,2}f\|_{X_N} &= \|u^{-1}(f_1 + f_2)\|_{X_N} = (\|u^{-1}f_1\|^2_{X_N} + \|u^{-1}f_2\|^2_{L^2_{N^3-N^2}})^{1/2}\\ &\le (\|u^{-1}\|^2\|f_1\|^2_2 + \|f_2\|^2)^{1/2} \le \sqrt{2}\end{aligned}$$

because (*) yields that $\|f_1\|_2 \le N^{-1/2}$). Similarly $\|w\| \le \sqrt{2}$. Thus $\|v\|\,\|w\| \le 2$.

The sequence (X_N) constructed in Example 1 consists of spaces each of which is a product of two factors of different nature. The next example is free of this defect. Recall that a Banach space $(X, \|\cdot\|)$ has enough symmetries [7] if the group G of isometries of X has the property: The only operators $T: X \to X$ which commute with all g in G (i.e., $Tg = gT$ for $g \in G$) are the scalar multiples of identity (i.e., $T = \lambda\,\mathrm{id}_X$ for some scalar λ).

EXAMPLE 2. There is a sequence $(F_n)_{n=1}^\infty$ of spaces with enough symmetries such that

$$\dim F_n = 2^{2^{n+1}}, \qquad \mathrm{gl}_H(F_n) \ge C_1 n, \qquad \mathrm{k\acute{e}}(F_n) \le C_2,$$

(Here and in the proof below $C_1, C_2, \ldots$ denote absolute constants independent of n.)

Proof. Let E_n denote the $N = 2^{2^n}$-dimensional symmetric space with the norm defined by

$$\|x\| = \sup\left\{\sum_{j\in A} |x(j)| : A \subset \{1, 2, \ldots, 2^{2^n}\}, |A| = n\right\}$$

($|A|$ denotes the cardinality of a set A). Let $F_n = E_n \hat{\hat{\otimes}} E_n$ the injective tensor product of E_n. Clearly $\dim F_n = (\dim E_n)^2 = 2^{2^{n+1}}$. Since E_n contains a norm-one complemented subspace isometrically isomorphic to l_n^1, F_n contains a norm-one complemented subspace isomorphic to $l_n^1 \hat{\hat{\otimes}} l_n^1$. Hence [9, Theorem 3.5] $\mathrm{gl}_H(F_n) \geq \mathrm{gl}_H(l_n^1 \hat{\hat{\otimes}} l_n^1) \geq C_1\sqrt{n}$. It is well known and easy to check that if E is any symmetric space then the tensor product $E \hat{\hat{\otimes}} E$ has enough symmetries. Hence each of the spaces F_n has enough symmetries.

Next we shall prove that $\text{ké}(F_n) \leq C$ for all n. We show that if $u: L_{N^2}^\infty \to F_n$ and $v: F_n \to L_{N^2}^1$ are natural injections then $\|u\|\,\|v\| \leq C$ where C is chosen so that

$$C > 2 \qquad \text{and} \qquad \frac{(C-2)N^2}{4n} > 2\frac{N^2}{2^n} + N2^{n^2}e^{2^n n^2 N^{-1}} \qquad \text{for all} \quad n; \quad (\diamondsuit)$$

obviously $vu = \iota^{(N^2)}$. Recall that u can be defined as follows. Represent $L_{N^2}^\infty$ as the tensor product $L_N^\infty \hat{\hat{\otimes}} L_N^\infty$ and put $u = I_{\infty,E_n} \otimes I_{\infty,E_n}$, the tensor product of the natural injections from L_N^∞ into E_n. It follows immediately from the definition of the norm in E_n that $\|I_{\infty,E_n}\| \leq n$. Thus $\|u\| \leq \|I_{\infty,E_n}\|^2 \leq n^2$. It remains to show that $\|v\| \leq Cn^{-2}$. Assume to the contrary that $\|v\| > Cn^{-2}$. Then for some $x \in E_n \hat{\hat{\otimes}} E_n$ with $\|x\|_{E_n \hat{\hat{\otimes}} E_n} = 1$ one would have

$$N^2\|v(x)\|_{L_{N^2}^1} = \sum_{j,k=1}^{N} |x(j,k)| > C\frac{N^2}{n^2},$$

Let $A = \{(j,k), 1 \leq j, k \leq N: |x(j,k)| > 2/n^2$ and let as usual $|A|$ denote the cardinality of A. Clearly the condition $\|x\|_{E_n \hat{\hat{\otimes}} E_n} = 1$ implies $|x(j,k)| \leq 1$ for $j, k = 1, 2, \ldots, N$. Thus

$$C\left(\frac{N}{n}\right)^2 < \sum_{j,k=1}^{N} |x(j,k)| \leq |A| + (N^2 - |A|)\frac{2}{n^2} < |A| + \frac{N^2}{n^2}2.$$

Thus $|A| \geq (N/n)^2(C-2)$. Replacing, if necessary x by either $-x$ or $\pm ix$, one may assume without loss of generality that $|A_+| \geq (N/2n)^2(C-2)$ where

$$A_+ = \{(j,k), 1 \leq j, k \leq N: x(j,k) > 1/n^2\}.$$

Now applying Lemma 1.3 stated below we infer that there exist sets $\Delta_\nu \subset \{1, 2, \ldots, N\}$, $\nu = 1, 2$ such that

$$|\Delta_1| = |\Delta_2| = n; \qquad \Delta_1 \times \Delta_2 \subset A_+. \qquad (+)$$

Define for $\nu = 1, 2$, the projections $P_\nu: E_n \to E_n$ by $(P_\nu\xi)(j) = \xi(j)$ for $j \in \Delta_\nu$, $(P_\nu\xi)(j) = 0$ for $j \notin \Delta_\nu$. Clearly $\|P_\nu\| = 1$ and $P_\nu(E_n)$ is naturally isometrically isomorphic to l_n^1. Hence $\|P_1 \otimes P_2\| = 1$ and $P_1(E_n) \hat{\hat{\otimes}} P_2(E_n)$ is naturally isometrically isomorphic to $l_n^1 \hat{\hat{\otimes}} l_n^1$. Observe that for a $z \in l_n^1 \hat{\hat{\otimes}} l_n^1$ one has

$$\|z\|_{l_n^1 \hat{\hat{\otimes}} l_n^1} = \sup\left\{ \sum_{1 \leq s,t \leq n} z(s,t)\varepsilon_s\eta_t : |\varepsilon_s| = |\eta_t| = 1\right\} \geq \left|\sum_{1 \leq s,t \leq n} z(s,t)\right|.$$

Thus remembering that for $(j,k) \in \Delta_1 \times \Delta_2 \subset A_+$, $\operatorname{Re} x(j,k) > n^{-2}$ we get

$$\|x\|_{E_n \hat{\hat{\otimes}} E_n} \geq \|(P_1 \otimes P_2)(x)\|_{E_n \hat{\hat{\otimes}} E_n} \geq \left| \sum_{(j,k) \in \Delta_1 \times \Delta_2} x(j,k) \right| > n^2 \frac{1}{n^2} = 1,$$

which contradicts the assumption $\|x\|_{E_n \hat{\hat{\otimes}} E_n} = 1$.

To complete the proof of Example 2 we need to prove Lemma 1.3 stated below which is a corollary to the following

COMBINATORIAL LEMMA. *Let $1 < n < N$ be positive integers and let $M = (M(j,k))$ be an $N \times N$ matrix with the entries equal either 0 or 1. Assume that for some integer p, with $p > n - 1$, $\sum_{1 \leq j,\, k \leq N} M(j,k) > a(N,n,p)$ where*

$$a(N,n,p) = \frac{n-1}{\binom{p}{n-1}} \binom{N}{n} + \frac{(p+1)(n-1)}{n} N.$$

Then there are sets $\Delta_\nu \subset \{1,2,\ldots,N\}$ with $|\Delta_\nu| = n$ for $\nu = 1, 2$, such that $M(j,k) = 1$ whenever $(j,k) \in \Delta_1 \times \Delta_2$.

For the proof see Roman [23, Theorem 1].

LEMMA 1.3. *Let $n > 1$ and let $N = 2^{2^n}$ let A_+ be a subset of $\{1,2,\ldots,N\} \times \{1,2,\ldots,N\}$ with $|A_+| > \frac{1}{4}(C-2)(N/n)^2$, where C satisfies ($\diamond$). Then there are sets $\Delta_\nu \subset \{1,2,\ldots,N\}$ for $\nu = 1, 2$ which satisfy (+).*

Proof. Define the $N \times N$ matrix M by $M(j,k) = 1$ for $(j,k) \in A_+$ and $M(j,k) = 0$ for $(j,k) \notin A_+$. Pick $p = N/2^n = 2^{2^{n-1}}$. Then

$$a(N,n,p) < |A_+| = \sum_{1 \leq j,\, k \leq N} M(j,k).$$

Indeed for $N = 2^{2^n}$ and $p = N/2^n$ one has $(p+1)(n-1)/n \leq 2p$ and

$$\begin{aligned}
\frac{n-1}{\binom{p}{n-1}} \binom{N}{n} &= \frac{(n-1)N!(p-n+1)!(n-1)!}{p!n!(N-n)!} \\
&= \frac{n-1}{n} N \frac{(N-1)(N-2)\cdots(N-n+1)}{p(p-1)\cdots(p-n+2)} \\
&= \frac{n-1}{n}(N-n+1)\frac{(N-1)\cdots(N-n+2)}{(N-2^n)\cdots(N-(n-2)2^n)} 2^{n(n-1)} \\
&\leq N 2^{n^2} \prod_{k=1}^{n-2} \frac{N-k}{N-k2^n} \leq N 2^{n^2} \frac{1}{(1-n2^n/N)^{n-2}} \\
&\leq N 2^{n^2} e^{2^n n^2/N}
\end{aligned}$$

Thus

$$a(N,n,p) \leq N2^{n^2}e^{2^n n^2/N} + 2Np$$
$$= 2\,\frac{N^2}{2^n} + N2^{n^2}e^{2^n n^2/N} < \frac{C-2}{4}\,\frac{N^2}{n^2}.$$

2. Finite Dimensional Unitary Ideals

The starting point for our investigation is the following fact which is essentially due to Lewis [17].

Proposition 2.1. *For every unitary ideal $\alpha = (\mathfrak{C}^n \otimes \mathfrak{C}^m, \alpha)$ one has*

$$C\pi_1(I_{\alpha,\mathrm{hs}}) \leq \frac{\sqrt{nm}}{\|I_{\mathrm{hs},\alpha}\|} = \pi_2(I_{\alpha,\mathrm{hs}}) \tag{1}$$

where C is an absolute constant independent of n, m and α.

Proposition 2.1 is an easy consequence of the following noncommutative analog of the Khinchine inequality which is a nonessential modification of a result of Helgason [12] and Figa-Talamanca and Rider [3] concerning the integral $\int_{\mathcal{U}(n)} |\mathrm{tr}(au)|\,du$.

Lemma 2.1. *For arbitrary $a \in \mathfrak{C}^n \otimes \mathfrak{C}^m$ and $b \in \mathfrak{C}^m \otimes \mathfrak{C}^n$ one has*

$$\left(\int_{\mathcal{U}(n)}\int_{\mathcal{U}(m)} |\mathrm{tr}(aubv)|^2\,dv\,du\right)^{1/2} = \frac{\mathrm{hs}(a)\mathrm{hs}(b)}{\sqrt{nm}}, \tag{2}$$

$$\int_{\mathcal{U}(n)}\int_{\mathcal{U}(m)} |\mathrm{tr}(aubv)|\,dv\,du \geq C\,\frac{\mathrm{hs}(a)\mathrm{hs}(b)}{\sqrt{nm}}, \tag{3}$$

where $C > 0$ is an absolute constant independent of a, b, n and m; the integration is taken with respect to the normalized Haar measures of $\mathcal{U}(n)$ and $\mathcal{U}(m)$ respectively.

First assuming the lemma we derive Proposition 2.1

Proof of Propositions 2.1. Fix $a_1, a_2, \ldots, a_k$ in $\mathfrak{C}^n \otimes \mathfrak{C}^m$ and put

$$M = \sup\left\{\sum_{j=1}^{k} |\mathrm{tr}(a_j b')| : b' \in \mathfrak{C}^m \otimes \mathfrak{C}^n,\ \alpha^*(b') \leq 1\right\}.$$

By (3), for $b \in \mathfrak{C}^m \otimes \mathfrak{C}^n$ with $\alpha^*(b) = 1$ we have

$$\sum_{j=1}^{k} \mathrm{hs}(a_j) \leq \frac{\sqrt{nm}}{C\,\mathrm{hs}(b)} \int_{\mathcal{U}(n)}\int_{\mathcal{U}(m)} \sum_{j=1}^{k} |\mathrm{tr}(a_j ubv)|\,dv\,du \leq \frac{\sqrt{nm}}{C\,\mathrm{hs}(b)} M.$$

Hence

$$\sum_{j=1}^{k} \mathrm{hs}(a_j) \leq \inf\left\{\frac{\sqrt{nm}}{C\,\mathrm{hs}(b)} M : b \in \mathfrak{C}^m \otimes \mathfrak{C}^n,\ \alpha^*(b) = 1\right\}$$

$$= \frac{\sqrt{nm}}{C\|I_{\alpha^*,\mathrm{hs}}\|} M = \frac{\sqrt{nm}}{C\|I_{\mathrm{hs},\alpha}\|} M,$$

because $I_{\alpha^*,\mathrm{hs}} = (I_{\mathrm{hs},\alpha})^*$. Thus

$$\pi_1(I_{\alpha,\mathrm{hs}}) \leq \frac{nm}{C\|I_{\mathrm{hs},\alpha}\|}.$$

A similar argument which uses (2) instead of (3) yields $\pi_2(I_{\alpha,\mathrm{hs}}) \leq nm/\|I_{\mathrm{hs},\alpha}\|$. Finally fix $a \in \mathfrak{C}^n \otimes \mathfrak{C}^m$ with $\mathrm{hs}(a) = 1$. Taking into account that the integral in (2) can be regarded as a limit of convex combinations of numbers $|\mathrm{tr}(aubv)|^2 = |\mathrm{tr}(vaub)|^2$ for $u \in \mathscr{U}(n)$ and $v \in \mathscr{U}(m)$ we get

$$1 = \mathrm{hs}(a)^2 = \int_{\mathscr{U}(n)} \int_{\mathscr{U}(m)} |\mathrm{hs}(vau)|^2\, dv\, du$$

$$\leq \pi_2^2(I_{\alpha,\mathrm{hs}}) \sup\left\{\int_{\mathscr{U}(n)} \int_{\mathscr{U}(m)} |\mathrm{tr}(vaub)|^2\, dv\, du : b \in \mathfrak{C}^m \otimes \mathfrak{C}^n,\ \alpha^*(b) = 1\right\}$$

$$= \pi_2^2(I_{\alpha,\mathrm{hs}}) \sup\left\{\frac{\mathrm{hs}(b)^2}{nm} : b \in \mathfrak{C}^m \otimes \mathfrak{C}^n,\ \alpha^*(b) = 1\right\}$$

$$= \frac{1}{nm} \pi_2^2(I_{\alpha,\mathrm{hs}})\|I_{\alpha^*,\mathrm{hs}}\|^2 = \frac{1}{nm} \pi_2^2(I_{\alpha,\mathrm{hs}})\|I_{\mathrm{hs},\alpha}\|^2.$$

Thus $nm/\|I_{\mathrm{hs},\alpha}\| \leq \pi_2(I_{\alpha,\mathrm{hs}})$. This completes the proof of the equality in (1).

Proof of Lemma 2.1. Let us put

$$A = \int_{\mathscr{U}(n)} \int_{\mathscr{U}(m)} |\mathrm{tr}(aubv)|^2\, dv\, du = \int_{\mathscr{U}(n)} \int_{\mathscr{U}(m)} \mathrm{tr}(aubv)\,\overline{\mathrm{tr}(aubv)}\, dv\, du.$$

Writing the trace as the sum of matrix elements, performing multiplication and interchanging the signs $\sum$ and $\int$ we get

$$A = \sum a(j,k) b(l,r) \overline{a(j',k')} \overline{b(l',r')} \int_{\mathscr{U}(n)} u(k,l)\overline{u(k',l')}\, du \int_{\mathscr{U}(m)} v(r,j) v(r',j')\, dv,$$

where the sum is extended over all sequences of indices (j,k,l,r,j',k',l',r') with $1 \leq j, j', r, r' \leq m$ and $1 \leq k, k', l, l' \leq n$. By the orthogonality relations [13, Section 27]

$$\int_{\mathscr{U}(d)} w(\nu,\mu)\overline{w(\nu',\mu')}\, dw = \begin{cases} 1/d & \text{for} \quad \nu = \nu' \text{ and } \mu = \mu', \\ 0 & \text{otherwise.} \end{cases}$$

Thus

$$A = \sum_{\substack{1 \le j, r \le m \\ 1 \le k, l \le n}} \overline{a(j,k)} a(j,k) b(l,r) \overline{b(l,r)} \frac{1}{n} \cdot \frac{1}{m}$$
$$= \sum_{\substack{1 \le j, r \le m \\ 1 \le k, l \le n}} \frac{1}{nm} |a(j,k)|^2 |b(l,r)|^2 = \left(\frac{\mathrm{hs}(a)\mathrm{hs}(b)}{\sqrt{nm}} \right)^2.$$

This proves the formula (2).

The proof of the inequality (3) requires some preparation. Recall that if $w \to (w(j,k))$ is the natural representation of $\mathcal{U}(d)$ then

$$\int_{\mathcal{U}(d)} |w(j,k)|\, dw = \int_{S^{d-1}} |\langle x, e_k \rangle|\, dx = c_d, \qquad j, k = 1, 2, \ldots, d. \tag{4}$$

Here S^{d-1} denotes the unit sphere of the d-dimensional complex Hilbert space l_d^2, e_k is the kth unit vector in l_d^2 $(k = 1, 2, \ldots, d)$, and $\int_{S^{d-1}} \cdots dx$ is the integral with respect to the normalized Lebesgue measure of S^{d-1}.

Indeed S^{d-1} can be identified via the map $x \to [x] = \{w \in \mathcal{U}(d): we_j = x\}$ (here j is fixed) with the homogeneous space $\mathcal{U}(d)/\mathcal{U}(d-1)$ of all cosets of $\mathcal{U}(d)$ with respect to the stationary subgroup of e_j which is identified with $\mathcal{U}(d-1)$. Hence

$$\int_{\mathcal{U}(d)} |w(j,k)|\, dw = \int_{\mathcal{U}(d)} |\langle we_j, e_k \rangle|\, dw$$
$$= \int_{\mathcal{U}(d-1)} \int_{\mathcal{U}(d)/\mathcal{U}(d-1)} |\langle [x] ve_j, e_k \rangle| d[x]\, dv$$
$$= \int_{\mathcal{U}(d-1)} \int_{S^{d-1}} |\langle x, e_k \rangle|\, dx\, dv$$
$$= \int_{S^{d-1}} |\langle x, e_k \rangle|\, dx.$$

It is well known that $c_1 = 1$, $c_d = (2 \cdot 4 \cdots 2d-2)/(3 \cdot 5 \cdots 2d-1)$ for $d \ge 2$. Hence

$$1 = \sqrt{1} c_1 \ge \sqrt{2} c_2 \ge \sqrt{3} c_3 \ge \cdots \ge \lim_d c_d \sqrt{d} = \frac{\sqrt{\pi}}{2}.$$

We shall identify the tours group $\mathfrak{T}^d$ with the subgroup of $\mathcal{U}(d)$ consisting of $d \times d$ diagonal matrices $\varepsilon = (\varepsilon_j) = (\varepsilon_j \delta_j^k)$ with $|\varepsilon_j| = 1$ $(j = 1, 2, \ldots, d)$. By $E_{\mathfrak{T}^d} \cdots d\varepsilon$ we denote the integral with respect to the normalized Haar measure of $\mathfrak{T}^d$. Recall the Orlicz inequality for Steinhaus variables (see, e.g. [25]); if (Ω, Σ, P) is a probability space then for every $f_1, f_2, \ldots, f_n \in L^1(\Omega, \Sigma, P)$ $(n = 1, 2, \ldots)$ one has for $d \ge n$

$$E_{\mathfrak{T}^d} \int_\Omega \left| \sum_{j=1}^d \varepsilon_j f_j(\omega) \right| P(d\omega)\, d\varepsilon \ge \frac{1}{\sqrt{2}} \left(\sum_{j=1}^d \left[\int_\Omega |f_j(\omega)| P(d\omega) \right]^2 \right)^{1/2} \tag{5}$$

The constant $1/\sqrt{2}$ is not the best possible; the best constant is related to the constant in the Khinchine inequality for Steinhaus variables which is unknown.

Now we are ready to prove (3). Given $a \in \mathfrak{C}^n \otimes \mathfrak{C}^m$ and $b \in \mathfrak{C}^m \otimes \mathfrak{C}^n$ there are $u_a, u_b \in \mathcal{U}(n)$ and $v_a, v_b \in \mathcal{U}(m)$ such that $a = v_a \Lambda(a) u_a$, $b = u_b \Lambda(b) v_b$, where $\Lambda(a) = (s_j(a)\delta_j^k) \in \mathfrak{C}^n \otimes \mathfrak{C}^m$, $\Lambda(b) = (s_l(b)\delta_l^r) \in \mathfrak{C}^m \otimes \mathfrak{C}^n$. $(s_j(a))$ and $(s_l(b))$ are the sequences of s-numbers of a and b, respectively.

Let us put

$$B = \int_{\mathcal{U}(n)} \int_{\mathcal{U}(m)} |\mathrm{tr}(aubv)| \, dv \, du .$$

Using the cyclic property of the trace and the translation invariance of the Haar measures of $\mathcal{U}(n)$ and $\mathcal{U}(m)$, for arbitrary $\varepsilon \in \mathfrak{T}^m$ and $\eta \in \mathfrak{T}^n$ we obtain

$$\begin{aligned} B &= \int_{\mathcal{U}(n)} \int_{\mathcal{U}(m)} |\mathrm{tr}(v_a \varepsilon^{-1} \varepsilon \Lambda(a) u_a u u_b \eta^{-1} \eta \Lambda(b) v_b v)| \, dv \, du \\ &= \int_{\mathcal{U}(n)} \int_{\mathcal{U}(m)} |\mathrm{tr}(\varepsilon \Lambda(a) u_a u u_b \eta^{-1} \eta \Lambda(b) v_b v v_a \varepsilon^{-1})| \, dv \, du \\ &= \int_{\mathcal{U}(n)} \int_{\mathcal{U}(m)} |\mathrm{tr}(\varepsilon \Lambda(a) u \eta \Lambda(b) v)| \, dv \, du. \end{aligned}$$

Averaging over $\mathfrak{T}^n$ and $\mathfrak{T}^m$, writing the trace as the sum of matrix elements, and using twice the Orlicz inequality we obtain

$$\begin{aligned} B &= EE \int_{\mathfrak{T}^m \mathfrak{T}^n \mathcal{U}(n)} \int_{\mathcal{U}(m)} \left| \sum_{\substack{1 \le j,\, r \le m \\ 1 \le k,\, l \le n}} \varepsilon_j s_j(a) \delta_j^k u(j,k) \eta_l s_l(b) \delta_l^r v(r,j) \right| dv \, du \, d\eta \, d\varepsilon \\ &= EE \int_{\mathfrak{T}^m \mathfrak{T}^n \mathcal{U}(n)} \int_{\mathcal{U}(m)} \left| \sum_{1 \le j,\, l \le \min(n,m)} \varepsilon_j s_j(a) u(j,l) \eta_l s_l(b) v(l,j) \right| dv \, du \, d\eta \, d\varepsilon \\ &\ge \frac{1}{\sqrt{2}} \left[\sum_{1 \le j \le \min(n,m)} \left(E \int_{\mathfrak{T}^m \mathcal{U}(n)} \int_{\mathcal{U}(m)} \left| \sum_{1 \le l \le \min(n,m)} s_j(a) u(j,l) \right. \right. \right. \\ &\qquad \left. \left. \left. \times \, \eta_l s_l(b) v(l,j) \right| dv \, du \, d\eta \right)^2 \right]^{1/2} \\ &\ge \frac{1}{2} \left\{ \sum_{1 \le j,\, l \le \min(n,m)} [s_j(a)]^2 [s_l(b)]^2 \left[\int_{\mathcal{U}(n)} |u(j,l)| \, du \int_{\mathcal{U}(m)} |v(l,j)| \, dv \right]^2 \right\}^{1/2} \\ &= \frac{1}{2} \frac{\mathrm{hs}(a)\mathrm{hs}(b)}{\sqrt{nm}} \sqrt{n c_n} \sqrt{m c_m} \ge \frac{\pi}{8} \frac{\mathrm{hs}(a)\mathrm{hs}(b)}{\sqrt{nm}} \end{aligned}$$

{because of (4), (5), and the identities $[\mathrm{hs}(a)]^2 = \sum_{1 \le j \le \min(n,m)} s_j(a)^2$, $[\mathrm{hs}(b)]^2 = \sum_{1 \le l \le \min(n,m)} s_l(b)^2$}. This completes the proof of (3).

Remarks. 1° Using an argument due to Helgason [12] (cf. [13, p. 146]) one can show that

$$\int_{\mathcal{U}(n)}\int_{\mathcal{U}(m)}|\mathrm{tr}(aubv)|^4\,dv\,du \le 4\left(\int_{\mathcal{U}(n)}\int_{\mathcal{U}(m)}|\mathrm{tr}(aubv)|^2\,dv\,du\right)^2.$$

This yields via a standard convexity argument that

$$\int_{\mathcal{U}(n)}\int_{\mathcal{U}(m)}|\mathrm{tr}(aubv)|\,dv\,du \ge 2^{-1}\left(\int_{\mathcal{U}(n)}\int_{\mathcal{U}(m)}|\mathrm{tr}(aubv)|^2\,dv\,du\right)^{1/2}.$$

Thus in that way we get an alternative proof of (3) with the constant 2^{-1} which is better than $\pi/8 \approx (2,55)^{-1}$.

2° The identity (2) and the inequality (3) (with a different constant) are valid if one replaces the unitary groups $\mathcal{U}(n)$ and $\mathcal{U}(m)$ by the orthogonal groups $\mathcal{O}(n)$ and $\mathcal{O}(m)$. The proof is essentially the same.

Next we state some corollaries to Proposition 2.1. The first is essentially known (cf. [17, 24]).

COROLLARY 2.1. *For arbitrary unitary ideal* $\alpha = (\mathfrak{C}^n \otimes \mathfrak{C}^m, \alpha)$ *one has*

$$\mathrm{gl}_H(\alpha) \overset{①}{\le} \mathrm{lust}(\alpha) \overset{②}{\le} \mathrm{ubc}(\alpha) \overset{③}{\le} d(\alpha, \mathrm{hs}) \overset{④}{\le} \frac{1}{C^2}\,\frac{nm}{\pi_1(I_{\alpha,\mathrm{hs}})\pi_1(I_{\alpha^*,\mathrm{hs}})} \overset{⑤}{\le} \frac{1}{C^2}\,\mathrm{gl}_H(\alpha)$$

where C *is the absolute constant appearing in the inequality* (1) *of Proposition* 2.1.

Proof. The inequalities ①, ② are standard (cf. [9, 19]); ③ is an immediate consequence of the facts that $\mathrm{ubc}(\mathrm{hs}) = 1$ and $\mathrm{ubc}(\alpha) \le d(\alpha, \mathrm{hs})\mathrm{ubc}(\mathrm{hs})$. For ④ observe that, by (1) applied twice to α and $\alpha^* = (\mathfrak{C}^m \otimes \mathfrak{C}^n, \alpha^*)$,

$$\|I_{\mathrm{hs},\alpha}\|\,\|I_{\mathrm{hs},\alpha^*}\| \le \frac{1}{C^2}\,\frac{nm}{\pi_1(I_{\mathrm{hs},\alpha})\pi_1(I_{\alpha^*,\mathrm{hs}})}.$$

Thus

$$\begin{aligned} d(\alpha, \mathrm{hs}) &\le \|I_{\mathrm{hs},\alpha}\|\,\|I_{\alpha,\mathrm{hs}}\| = \|I_{\mathrm{hs},\alpha}\|\,\|(I_{\alpha,\mathrm{hs}})^*\| \\ &= \|I_{\mathrm{hs},\alpha}\|\,\|I_{\mathrm{hs},\alpha^*}\| \le C^{-2}nm/\pi_1(I_{\alpha,\mathrm{hs}})\pi_1(I_{\alpha^*,\mathrm{hs}}). \end{aligned}$$

Finally for ⑤ observe that, by the trace duality between the ideals of L_1-factorable operators and adjoints to one-absolutely summing operators, one has (cf. [19, 20])

$$\mathrm{gl}_H(\alpha) = \sup\left\{\frac{\mathrm{tr}(v^*u)}{\pi_1(u)\pi_1(v)} : u : \alpha \to l^2,\ v : \alpha^* \to l^2\right\}.$$

Specifying $u = I_{\alpha,\mathrm{hs}}$, $v = I_{\alpha^*,\mathrm{hs}}$ one gets $v^*u = I_{\mathrm{hs},\alpha}I_{\alpha,\mathrm{hs}} = \mathrm{id}_{\mathrm{hs}}$, thus $\mathrm{tr}(v^*u) = nm$. Hence

$$\mathrm{gl}_H(\alpha) \geq \frac{nm}{\pi_1(I_{\alpha,\mathrm{hs}})\pi_1(I_{\alpha^*,\mathrm{hs}})}.$$

COROLLARY 2.2. *For arbitrary unitary ideal* $\alpha = (\mathfrak{C}^n, \otimes \mathfrak{C}^m, \alpha)$ *one has*

$$\text{ké}_{nm}(\alpha) \leq d(\alpha, \mathrm{hs}) \leq \frac{1}{C^2}\,\text{ké}_{nm}(\alpha),$$

where C is the absolute constant appearing in (1).

Proof. Clearly $\text{ké}_{nm}(\mathrm{hs}) = 1$ and $\text{ké}(\alpha) \leq \text{ké}(\mathrm{hs})d(\alpha, \mathrm{hs})$. Thus the left-hand side inequality is trivial. For the right-hand side one observes that, by Corollary 2.1,

$$d(\alpha, \mathrm{hs}) \leq \frac{1}{C^2}\frac{nm}{\pi_1(I_{\alpha,\mathrm{hs}})\pi_1(I_{\alpha^*,\mathrm{hs}})}.$$

Next, by Corollary 1.1,

$$\text{ké}_{nm}(\alpha) \geq \sup\left\{\frac{\mathrm{tr}(vu)}{\pi_1(u)\pi_1(v^*)} : u : \alpha \to l^2_{nm}, v = u^{-1}\right\}.$$

Thus specifying $u = I_{\alpha,\mathrm{hs}}$ one gets $v = I_{\mathrm{hs},\alpha}$, $v^* = I_{\alpha^*,\mathrm{hs}}$, hence

$$\frac{\mathrm{tr}(vu)}{\pi_1(u)\pi_1(v^*)} = \frac{nm}{\pi_1(I_{\mathrm{hs},\alpha})\pi_1(I_{\alpha^*,\mathrm{hs}})}.$$

Therefore

$$\text{ké}_{nm}(\alpha) \geq \frac{nm}{\pi_1(I_{\mathrm{hs},\alpha})\pi_1(I_{\alpha^*,\mathrm{hs}})} \geq C^2 d(\alpha, \mathrm{hs}).$$

Our next result provides rather satisfactory information on the projection constant $\gamma_\alpha(\mathrm{id}_\alpha)$ of an arbitrary unitary ideal $\boldsymbol{\alpha}$.

COROLLARY 2.3. *For arbitrary unitary ideal* $\alpha = (\mathfrak{C}^n \otimes \mathfrak{C}^m, \alpha)$ *one has*

$$\sqrt{nm} \leq \pi_1(\mathrm{id}_\alpha) \leq \frac{1}{C}\sqrt{nm}, \qquad C\sqrt{nm} \leq \gamma_\infty(\mathrm{id}_\alpha) \leq \sqrt{nm}.$$

Proof. By, Proposition 2.1

$$\pi_1(\mathrm{id}_\alpha) \leq \pi_1(I_{\alpha,\mathrm{hs}})\|I_{\mathrm{hs},\alpha}\| \leq \frac{\sqrt{nm}}{C}.$$

On the other hand, by a result of Garling and Gordon [7], $\sqrt{nm} = \pi_2(\mathrm{id}_\alpha) \leq \pi_1(\mathrm{id}_{\boldsymbol{\alpha}})$.

The second chain of inequalities is an immediate consequence of the first one and the Garling–Gordon [7] formula for spaces X with enough symmetries, $\gamma_\infty(\mathrm{id}_X)\pi_1(\mathrm{id}_X) = \dim X$.

Clearly for every Banach space X, $\gamma_\infty(\mathrm{id}_X) \leq d(X, l^\infty_{\dim X})$. Our next result improves the second chain of inequalities in Corollary 2.3.

THEOREM 2.1. *For every unitary ideal* $\alpha = (\mathfrak{C}^n \otimes \mathfrak{C}^m, \alpha)$ *one has*

$$C^{-1}\sqrt{nm} \leq d(\alpha, l^\infty_{nm}) \leq \sqrt{2}\sqrt{nm} \tag{6}$$

where C *is an absolute constant appearing in* (1).

Proof. The left-hand side of inequality (6) is an immediate consequence of Corollary 2.3. To prove the right-hand side of inequality (6) we need the following

PROPOSITION 2.2. *Let* n, m *be positive integers and let* $\mu = \min(\eta, m)$. *Given a sequence* $\lambda_1 \geq \lambda_2 \cdots \geq \lambda_\mu \geq 0$ *with* $\sum_{j=1} \lambda_j^2 = 1$, *there exists an* $a \in \mathfrak{C}^n \otimes \mathfrak{C}^m$ *such that* $s_j(a) \leq \sqrt{2}\lambda_j$ $(j = 1, 2, \ldots)$, $\sum_{j=1} s_j(a)^2 = 1, |a(s,t)| = 1/\sqrt{nm}$ *for* $1 \leq s \leq m$, $1 \leq t \leq n$.

Proof. Put $k_j^* = [2n\lambda_j^2]$ $(j = 1, 2, \ldots, \mu)$. Clearly

$$k_1^* \geq k_2^* \geq \ldots \geq k_\mu^* \geq 0 \qquad \text{and} \qquad \sum_{j=1}^{\mu} k_j^* \geq \sum_{j=1}^{\mu} (2n\lambda_j^2 - 1) = 2n - \mu \geq n.$$

If $\sum_{j=1}^{\mu} k_j^* > n$ we replace k_μ^* by the integers $s = k_\mu^* - 1, k_\mu^* - 2, \ldots, 0$ until either we reach zero or the sum $\sum_{j=1}^{\mu-1} k_j^* + s = n$; if the first case occurs then $\sum_{j=1}^{\mu-1} k_j^* > n$ and we continue replacing $k_{\mu-1}^*$ by smaller integers $s = k_{\mu-1}^* - 1, k_{\mu-1}^* - 2, \ldots, 0$ until either we reach zero or the sum $\sum_{j=1}^{\mu-2} k_j^* + s = n$; if the first case occurs we continue. Finally we find the index ν with $1 \leq \nu < \mu$ and an integer s with $k_{\nu+1}^* \geq s \geq 0$ such that $\sum_{j=1}^{\nu} k_j^* + s = n$. We put $k_j = k_j^*$ for $1 \leq j \leq \nu$, $k_{\nu+1} = s$ and $k_j = 0$ for $\mu \geq j > \nu$. Then clearly $k_1 \geq k_2 \geq \cdots \geq k_\mu \geq 0$ $\sum_{j=1}^{\mu} k_j = n$; $k_j/n \leq 2\lambda_j^2$ $(j = 1, 2, \ldots, \mu)$. For conveniency we put $k_0 = 0$, $k_j = 0$ for $j > \mu$. Let $j_0 = \max\{j : k_j \neq 0\}$. Next we define the orthonormal sequence of "normalized characteristic functions" $[(1/\sqrt{k_j})\chi_j]_{1 \leq j \leq j_0}$, where

$$\chi_j(t) = \begin{cases} 1 & \text{for } \sum_{l=0}^{j-1} k_l < t \leq \sum_{l=0}^{j} k_l \\ 0 & \text{otherwise} \end{cases}$$

Finally we define the desired matrix $a \in \mathfrak{C}^n \otimes \mathfrak{C}^m$ by

$$a(s,t) = \sum_{j=1}^{j_0} \frac{1}{\sqrt{nm}} e^{2\pi i s j m^{-1}} \chi_j(t), \qquad 1 \leq s \leq m, \quad 1 \leq t \leq n.$$

Since $\sum_{j=1}^{j_0} k_j = n$, for every t with $1 \le t \le n$ there is exactly one $j(t)$ with $1 \le j(t) \le j_0$ such that $\chi_j(t) = 0$ for $j \ne j(t)$ and $\chi_{j(t)}(t) = 1$. Thus $a(s,t) = (1/\sqrt{nm})e^{2\pi i s j(t) m^{-1}}$ and consequently $|a(s,t)| = (nm)^{-1/2}$ for $1 \le s \le m$. Next observe that $a = v\Lambda u$ where $v \in \mathscr{U}(m)$ is defined by $v(s,j) = m^{-1/2}e^{2\pi i s j m^{-1}}$, $u \in \mathscr{U}(n)$ is any $n \times n$ unitary matrix with $u(l,t) = (1/\sqrt{k_l})\chi_l(t)$ for $\sum_{j=0}^{l-1} k_j < t \le \sum_{j=0}^{l} k_j$ and for $1 \le l \le j_0$, and $\Lambda \in \mathfrak{C}^n \otimes \mathfrak{C}^m$ is the "diagonal" matrix $\Lambda(j,l) = \sqrt{k_j/n}\,\delta_j^l\,(1 \le j \le m;\ 1 \le l \le n)$. Since $k_1 \ge k_2 \ge \cdots k_m \ge 0$, we have $s_j(\Lambda) = \sqrt{k_j/n}$ for $j = 1, 2, \ldots$. Hence $s_j(a) = s_j(v\Lambda u) = s_j(\Lambda) = \sqrt{k_j/n} \le 2\lambda_j$ $(j = 1, 2, \ldots)$. This completes the proof of the Proposition 2.2.

Remark. The infinite version of Proposition 2.2 is even more elegant.

For every nonnegative nonincreasing sequence $(\lambda_j)_{1 \le j < \infty}$ *with* $\sum \lambda_j^2 = 1$ *there exists a Hilbert–Schmidt operator* $\tilde{a}: L^2[0,1] \to L^2[0,1]$ *defined by* $\tilde{a}(f) = \int_0^1 f(s)a(s,t)\,ds$ *for* $f \in L^2[0,1]$ *where* $a:[0,1] \times [0,1] \to \mathfrak{R}$ *is a measurable function with* $|a(s,t)| = 1$ *for* $(s,t) \in [0,1] \times [0,1]$, *and the sequence of s-numbers of* $\tilde{a}$ *satisfies* $s_j(a) = \lambda_j$ $(j = 1, 2, \ldots)$.

Proof. Let $j_0 = \sup\{j : \lambda_{j-1} > 0\}$. Clearly j_0 may be finite or infinite. Let $(A_j)_{1 \le j < j_0}$ be a sequence of mutually disjoint subsets of $[0,1]$ whose measures satisfy $|A_j| = \lambda_j^2$ for $j < j_0$. The desired function $a:[0,1] \times [0,1] \to \mathfrak{R}$ is defined by

$$a(s,t) = \sum_{j<j_0} w_j(s)\chi_{A_j}(t) = \sum_{j<j_0} \lambda_j w_j(s)\frac{\chi_{A_j}}{\|\chi_{A_j}\|_2},$$

where $(w_j)_{1 \le j < \infty}$ denotes the Walsh orthonormal system and χ_A denotes the characteristic function of a subset A of $[0,1]$.

Now we are ready to complete the proof of Theorem 2.1. To this end it is clearly enough to prove the "dual" inequality $d(((\mathfrak{C}^n \otimes \mathfrak{C}^m), \alpha), l_{nm}^1) \le \sqrt{2nm}$ (changing the role of the indices n and m is clearly inessential).

Fix an $a_0 \in \mathfrak{C}^n \otimes \mathfrak{C}^m$ with $\mathrm{hs}(a_0) = \sum_{j=1}^{\mu}(s_j(a_0))^2 = 1$ and construct for $s_j(a_0) = \lambda_j$ $(j = 1, 2, \ldots)$ an $a \in \mathfrak{C}^n \otimes \mathfrak{C}^m$ satisfying the assertion of Proposition 2.2. By 0.II, $\alpha(a) \le 2\alpha(a_0)$. It is convenient for us to identify the space l_{nm}^2 with hs $= (\mathfrak{C}^n \otimes \mathfrak{C}^m, \mathrm{hs})$ via the map which takes the unit vectors of l_{nm}^2 onto the unit matrices $e_{p,q}$ where $e_{p,q}(s,t) = \delta_p^s\delta_q^t$ $(1 \le p_1\xi \le m;\ 1 \le q, t \le n)$. Let $A:\mathrm{hs} \to \mathrm{hs}$ be the unitary operator which takes the unit matrices $e_{p,q}$ onto the matrices $a_{p,q}$ defined by

$$a_{p,q}(s,t) = e^{2\pi i p s m^{-1}}a(s,t)e^{2\pi i q t n^{-1}}, \qquad 1 \le s \le m, \quad 1 \le t \le n,$$
$$1 \le p \le m, \quad 1 \le q \le n.$$

Recall that $\langle a, b\rangle = \mathrm{tr}(b^*a)$ is the inner product of elements a and b in hs. Then remembering that $|a(s,t)| = 1/nm$ for $1 \le s \le m$, $1 \le t \le n$ we have

$$\langle a_{p,q}, a_{p',q'} \rangle = \operatorname{tr}(a^*_{p',q'} a_{p,q})$$
$$= \sum_{1 \le s \le m} \sum_{1 \le t \le n} e^{2\pi i (p-p') s m^{-1}} |a(s,t)|^2 e^{2\pi i (q-q') t n^{-1}} = \delta_p^{p'} \delta_q^{q'}.$$

This proves that the operator $A: \mathrm{hs} \to \mathrm{hs}$ is unitary. Now we define $Q: l^1_{nm} \to (\mathfrak{C}^n \otimes \mathfrak{C}^m, \alpha)$ as the unique operator which takes the unit vectors of l^1_{nm} onto the matrices $a_{p,q}$ ($1 \le p \le m$, $1 \le q \le n$). Clearly $Q^{-1} = J_{2,1} A^{-1} I_{\alpha,\mathrm{hs}}$ where $J_{2,1}: \mathrm{hs} = l^2_{nm} \to l^1_{nm}$ is the natural injection. Observe that $a_{p,q} = d_p^{(m)} a d_q^{(n)}$ where $d_p^{(m)}$ (resp., $d_q^{(n)}$) are $m \times m$ (resp., $n \times n$) unitary diagonal matrices, $d_p^{(m)}(x, y) = e^{2\pi i p x m^{-1}} \delta_x^y$ ($1 \le x, y \le m$). Hence $\alpha(a_{p,q}) = \alpha(a)$ for $1 \le p \le m$, $1 \le q \le n$. Thus $\|Q\| = \max_{1 \le p \le m,\, 1 \le q \le n} \alpha(a_{p,q}) = \alpha(a)$. Since $\|J_{2,1}\| = \sqrt{nm}$ and $\|A^{-1}\| = 1$, we have

$$\|Q^{-1}\| \le \sqrt{nm} \|I_{\alpha,\mathrm{hs}}\|.$$

Thus $\|Q\| \, \|Q^{-1}\| \le \sqrt{nm} \|I_{\alpha,\mathrm{hs}}\| \alpha(a) \le \sqrt{2nm} \|I_{\alpha,\mathrm{hs}}\| \alpha(a_0)$. Now we specify the matrix $a_0 \in \mathfrak{C}^n \otimes \mathfrak{C}^m$ so that

$$\alpha(a_0) = \inf\{\alpha(b): b \in \mathfrak{C}^n \otimes \mathfrak{C}^m, \mathrm{hs}(b) = 1\} = \|I_{\alpha,\mathrm{hs}}\|^{-1}.$$

Then $\|Q\| \, \|Q^{-1}\| \le \sqrt{2nm}$. This completes the proof of the theorem.

Remark. The authors are indebted to Michal Misiurewicz and Nicole Tomczak-Jaegermann for the proofs of Proposition 2.2 and Theorem 2.1, respectively presented here. The original argument of the authors was more complicated.

Our next corollary is also a consequence of Proposition 2.2. It completes Corollary 2.1. Roughly speaking, it says that the unconditionality of the natural basis of a unitary ideal $\alpha = (\mathfrak{C}^n \otimes \mathfrak{C}^m, \alpha)$ and therefore the distance $d(\alpha, \mathrm{hs})$ can be measured looking only at elements whose coefficients with respect to the natural basis (= the entries) have constant absolute value.

Let us set

$$A(n,m) = \{a \in \mathfrak{C}^n \otimes \mathfrak{C}^m : |a(s,t)| = 1 \ (1 \le s \le m, 1 \le t \le n)\},$$
$$M_\alpha = \sup\{\alpha(a): a \in A(n,m)\}, \qquad m_\alpha = \inf\{\alpha(a): a \in A(n,m)\}.$$

COROLLARY 2.4 *For arbitrary unitary ideal $\alpha = (\mathfrak{C}^n \otimes \mathfrak{C}^m, \alpha)$ one has*

$$\frac{1}{\sqrt{2}} \frac{M_\alpha}{m_\alpha} \le d(\alpha, \mathrm{hs}) \le \sqrt{2} \frac{M_\alpha}{m_\alpha},$$

$$\frac{M_\alpha M_{\alpha^*}}{nm} \le d(\alpha, \mathrm{hs}) \le 2 \frac{M_\alpha M_{\alpha^*}}{nm},$$

$$\frac{nm}{2 m_\alpha m_{\alpha^*}} \le d(\alpha, \mathrm{hs}) \le \frac{nm}{m_\alpha m_{\alpha^*}},$$

Proof. It is well known that $d(\alpha, \mathrm{hs}) = \|I_{\alpha,\mathrm{hs}}\| \|I_{\mathrm{hs},\alpha}\|$ (cf. [1,24]). Observe that

$$\|I_{\mathrm{hs},\alpha}\| = \frac{\sup\{\alpha(a): a \in \mathfrak{C}^n \otimes \mathfrak{C}^m, \mathrm{hs}(a) = \sqrt{nm}\}}{\sqrt{nm}},$$

$$\|I_{\alpha,\mathrm{hs}}\| = \frac{\sqrt{nm}}{\inf\{\alpha(a): a \in \mathfrak{C}^n \otimes \mathfrak{C}^m, \mathrm{hs}(a) = \sqrt{nm}\}}.$$

Clearly if $a \in A(n,m)$ then $\mathrm{hs}(a) = \sqrt{nm}$. Thus, by Proposition 2.2, we get

$$\frac{M_\alpha}{\sqrt{nm}} \le \|I_{\mathrm{hs},\alpha}\| \le \frac{\sqrt{2}M_\alpha}{\sqrt{nm}}, \qquad \frac{\sqrt{nm}}{\sqrt{2}m_\alpha} \le \|I_{\alpha,\mathrm{hs}}\| \le \frac{\sqrt{nm}}{m_\alpha}. \tag{7}$$

Multiplying these inequalities we get the first desired inequality. To obtain the remaining two inequalities it is enough to use (7) with α replaced by α^* and to note that $\|I_{\mathrm{hs},\alpha}\| = \|I_{\alpha^*,\mathrm{hs}}\|$, $\|I_{\alpha,\mathrm{hs}}\| = \|I_{\mathrm{hs},\alpha^*}\|$.

The rest of this section is devoted to investigation of subspaces of the unitary ideal $(\mathfrak{C}^n \otimes \mathfrak{C}^m, |||\cdot|||) = l_n^2 \hat{\hat{\otimes}} l_m^2$.

The crucial role is played by the following analog of Lemma 1.2. Here $I_{\alpha,\beta}^E = I_{\alpha,\beta}|E$.

LEMMA 2.2. *If E is an s-dimensional subspace of $\mathfrak{C}^n \otimes \mathfrak{C}^m$ then*

$$\|I^E_{|||\cdot|||,\mathrm{hs}}\| \ge \sqrt{\frac{s}{n+m}}.$$

Lemma 2.2 is an immediate consequence of the next one.

LEMMA 2.3. *If E is a subspace of $\mathfrak{C}^n \otimes \mathfrak{C}^m$ with $\dim E = s$, then there exists an a in E such that $s_j(a) = 1$ for $j = 1, 2, \ldots, [s/(n+m)] + 1$.*

Proof. Let r be the largest positive integer such that there exists an $a \in E$ with $s_j(a) = 1$ for $j = 1, 2, \ldots, r$. Clearly $r \le \min(n,m)$, because $s_j(b) = 0$ for $j > \min(n,m)$ for every $b \in \mathfrak{C}^n \otimes \mathfrak{C}^m$. Thus there are $u \in \mathscr{U}(n)$, $v \in \mathscr{U}(m)$ and diagonal operator $\Lambda(a) = (s_j(a)\delta_j^k)_{1 \le j \le m,\, 1 \le k \le n}$ such that $a = v\Lambda u$. Put

$$B = \{b \in \mathfrak{C}^n \otimes \mathfrak{C}^m : b(j,k) = 0 \text{ for } \min(j,k) \le r\}.$$

Now assume to the contrary that $r \le [s/(n+m)]$. Let $E_0 = E \cap vBu$. Then $\dim E_0 \ge \dim E + \dim B - nm = s + (n-r)(m-r) - nm = s - (n+m)r + r^2 > 0$. Thus there is a $b_0 \in B$ with $|||b_0||| = 1$ such that $vb_0u \in E$. Define the diagonal operator $\Delta \in B$ by $\Delta(j,k) = 0$ for $\min(j,k) \le r$ and $\Delta(j,k) = \Lambda(a)(j,k)$ for $\min(j,k) > r$. Observe that $|||\Delta||| = s_{r+1}(a) < 1$ because of definition of r. Consider the continuous function $p(t) = |||\Delta + tb_0|||$ for $t \ge 0$. Then $p(0) < 1$ and $p(1 + |||\Delta|||) \ge |||(1 + |||\Delta|||)b_0||| - |||\Delta||| = 1$. Hence there is a $t_0 \in [0, 1 + |||\Delta|||]$ such that $|||\Delta + t_0 b_0||| = 1$. Now consider the element $a_0 = a + t_0 v b_0 u = v(\Lambda(a) + t_0 b_0)u$. Clearly $s_j(a_0) = s_j(\Lambda(a) + t_0 b_0)$ for all j.

A moment of reflection shows that $s_j(\Lambda(a) + t_0 b_0) = s_j(\Lambda(a)) = 1$ for $j = 1, 2, \ldots, r$ and $s_{r+1}(\Lambda(a) + t_0 b_0) = s_1(\Delta + t_0 b_0) = |||\Delta + t_0 b_0||| = 1$. Thus the first $(r + 1)$ s-numbers of $a_0 \in E$ equal one. This contradicts the definition of r. Thus the assumption $r \leq [s/(n + m)]$ leads to a contradiction. Hence $r > s/(n + m)$.

Our next result is analogous to similar results for random subspaces of l_n^∞ (see Figiel and Johnson [4], and Proposition 1.5 in the present paper).

PROPOSITION 2.3. *Let E be an s-dimensional subspace of* $\mathfrak{C}^n \otimes \mathfrak{C}^m$. *Then*

$$\mathrm{gl}_H((E, |||\cdot|||)) \geq C_1 \frac{s}{\sqrt{(n + m)nm}}, \qquad \mathrm{k\acute{e}}_t((E, |||\cdot|||)) \geq C_2 \sqrt{\frac{ts}{(n + m)nm}},$$

in particular,

$$\mathrm{k\acute{e}}((E, |||\cdot|||)) \geq C_2 \frac{s}{\sqrt{(n + m)nm}}.$$

Proof. By Proposition 2.1,

$$\pi_1(I^E_{|||\cdot|||,\mathrm{hs}}) \leq \pi_1(I_{|||\cdot|||,\mathrm{hs}}) \leq \frac{1}{C} \frac{\sqrt{nm}}{\|I_{\mathrm{hs},|||\cdot|||}\|} = \frac{1}{C} \sqrt{nm},$$

because for every $a \in \mathfrak{C}^n \otimes \mathfrak{C}^m$, $\mathrm{hs}(a) \geq |||a|||$. Next, by Lemma 2.2, for every subspace $Y \subset E$ with $\dim Y \geq as$ [where a is the absolute constant appearing in Proposition 1.3(ii)] we have

$$\|I^E_{|||\cdot|||,\mathrm{hs}}|_Y\| = \|I^Y_{|||\cdot|||,\mathrm{hs}}\| \geq \sqrt{\frac{s}{n + m}}.$$

The desired conclusion follows now by direct application either of Proposition 1.4 (to obtain the first inequality) or of Proposition 1.3(ii) (to obtain the second and third inequalities).

For $n = m$ we obtain

COROLLARY 2.5. *If E is an s-dimensional subspace of* $\mathfrak{C}^n \otimes \mathfrak{C}^n$, *then* $\mathrm{gl}_H(E, |||\cdot|||) \geq (C_1/\sqrt{2})(s/n^{3/2})$; $\mathrm{k\acute{e}}_t(E) \geq (C_2/\sqrt{2})(\sqrt{st}/n^{3/2})$ *for* $1 \leq t \leq s$.

The first part of this corollary was obtained by Figiel and Johnson [4]. Their argument is different.

Remark. Nonessential changes are required to extend principal results of Section 2 on real matrix spaces. An *orthogonal ideal* is defined to be a real Banach space $(\mathfrak{R}^n \otimes \mathfrak{R}^m, \alpha)$ provided the norm α satisfies the condition $\alpha(vau) = \alpha(a)$ for every $a \in \mathfrak{R}^n \otimes \mathfrak{R}^m$ and every $v \in \mathcal{O}(m)$, $u \in \mathcal{O}(n)$. We describe briefly some changes of the argument.

The proofs of Proposition 2.1 and Corollaries 2.1, 2.2, 2.3 do not change except replacing Lemma 2.1 by its real analog for the orthogonal groups (cf. Remark 2° after the proof of Lemma 2.1). The real analog of Proposition 2.2 holds for n, m diadic, i.e., $n = 2^h$, $m = 2^g$ $(h, g = 0, 1, \ldots)$; in the proof we replace the matrices of nth and mth roots of unity by corresponding Walsh matrices. This allows us to prove Theorem 2.1 for diadic n and m [with the same constant $\sqrt{2}$ in the right-hand side of (6)]; again in the proof we use Walsh matrices instead of matrices of roots of unity. The general case of Theorem 2.1 is completed as follows. Fix arbitrary a and m and write the diadic expansions $n = \sum_{\kappa=1}^{K} d_\kappa$, $m = \sum_{\eta=1}^{N} D_\eta$; $d_1 < d_2 < \cdots < d_k$, $D_1 < D_2 < \cdots < D_N$ are diadic numbers. Next we divide the pairs of indices of $n \times m$ matrix onto $K \cdot N$ diadic rectangles, say $\Delta_{\kappa,\eta}$ with the lengths of sides d_κ and D_η, respectively. Let $I_{\kappa,\eta}: \mathfrak{R}^{d_\kappa} \otimes \mathfrak{R}^{D_\eta} \to \mathfrak{R}^n \otimes \mathfrak{R}^m$ be the natural embedding which assigns to each $d_\kappa \times D_\eta$ matrix the $n \times m$ matrix whose entries are zero outside the rectangle $\Delta_{\kappa,\eta}$ and inside $\Delta_{\kappa,\eta}$ are equal to the corresponding entries of the original $d_\kappa \times D_\eta$ matrix; let $P_{\kappa,\eta}: \mathfrak{R}^n \otimes \mathfrak{R}^m \to \mathfrak{R}^{d_\kappa} \otimes \mathfrak{R}^{D_\eta}$ be the natural projection defined by $P_{\kappa,\eta} I_{\kappa,\eta} = \mathrm{id}_{(\mathfrak{R}^{d_\kappa} \otimes \mathfrak{R}^{D_\eta})}$; let $\alpha_{\kappa,\eta} = \alpha P_{\kappa,\eta}$. Since the real analog of Theorem 2.1 has been already established for diadic numbers, there are isomorphisms

$$U_{\kappa,\eta}: (\mathfrak{R}^{d_\kappa} \otimes \mathfrak{R}^{D_\eta}, \alpha_{\kappa,\eta}) \to l^\infty_{d_\kappa D_\eta} \quad \text{with} \quad \|U_{\kappa,\eta}\| = \|U_{\kappa,\eta}^{-1}\| \leqslant C_0 \sqrt[4]{d_\kappa D_\eta}.$$

Define $U: (\mathfrak{R}^n \otimes \mathfrak{R}^m, \alpha) \to l^\infty_{nm}$ by

$$U = \sum_{\kappa=1}^{K} \sum_{\eta=1}^{N} I_{\kappa,\eta} U_{\kappa,\eta} p_{\kappa,\eta}.$$

Then

$$\begin{aligned} \|U\|\,\|U^{-1}\| &\leq \sum_{\kappa=1}^{K} \sum_{\kappa'=1}^{K} \sum_{\eta=1}^{N} \sum_{\eta'=1}^{N} \|U_{\kappa,\eta}\|\,\|U_{\kappa',\eta'}^{-1}\| C_0^2 \\ &\leq 2 \sum_{p=0}^{[\log_2 n]} \sum_{p'=0}^{[\log_2 n]} \sum_{q=0}^{[\log_2 m]} \sum_{q'=0}^{[\log_2 m]} 2^{(p+p'+q+q')/4} C_0^2 \\ &\leq C_0^2 C_1 \sqrt{nm} \end{aligned}$$

where $C_1 \leq \sqrt{22(2^{1/4} - 1)^{-4}}$, $C_0^2 = \sqrt{2C}$, C is the constant in real analogue of Lemma 2.1

Proposition 2.3 and Corollary 2.5 extend to orthogonal ideals automatically.

References

1. Y. Benyamini and Y. Gordon, Random factorizations of operators between Banach spaces, *J. Analyse Math.* (1981), to appear.
2. W. J. Davis, V. Milman, and N. Tomczak-Jaegermann, The diameter of the space of n-dimensional spaces, *Israel J. Math.*, to appear.

3. A. FIGA-TALAMANCA AND D. RIDER, A theorem of Littlewood and lacunary series for compact groups, *Pacific J. Math.* **16** (1966), 505–514.
4. T. FIGIEL AND W. B. JOHNSON, Large subspaces of l_∞^n and estimates of the Gordon–Lewis constant, Israel *Math. J.* **37** (1980), 92–112.
5. T. FIGIEL, J. LINDENSTRAUSS, AND V. MILMAN, The dimension of almost spherical sections of convex bodies, *Acta Math.* **139** (1977), 53–94.
6. T. FIGIEL, S. KWAPIEŃ, AND A. PEŁCZYŃSKI, Sharp estimates for the constant of local unconditional structure of Minkowski spaces, *Bull. Acad. Polon. Sci. Sér. Math. Astronom. Phys.* **25** (1977), 1221–1226.
7. D. J. H. GARLING AND Y. GORDON, Relations between some constants associated with finite dimensional Banach spaces, *Israel J. Math.* **9** (1971), 346–361.
8. I. C. GOHBERG AND M. G. KREN, "Introduction to the Theory of Linear Nonselfadjoint Operators in Hilbert Space," Moscow, 1965 [in Russian]; Providence, 1969; Paris, 1971 [in French].
9. Y. GORDON AND D. R. LEWIS, Absolutely summing operators and local unconditional structures, *Acta Math.* **133** (1974), 24–48.
10. A. GROTHENDIECK, Résumé de la théorie métrique des produits tensoriels topologiques, *Bol. Soc. Mat. Sao Paulo* **8** (1956), 1–79.
11. V. I. GURARII, M. I. KADEC, AND V. I. MACAEV, Distances between finite dimensional analogs of the L_p-spaces, *Mat. Sb.* **70**, No. 112 (1966), 24–29 [in Russian].
12. S. HELGASON, Topologies of group algebras and a theorem of Littlewood, *Trans. Amer. Math. Soc.* **86** (1957), 269–283.
13. E. HEWITT AND K. A. ROSS, "Abstract Harmonic Analysis, II," Springer-Verlag, Berlin and New York, 1970.
14. R. I. JAMISON AND W. H. RUCKLE, Factoring absolutely converging series, *Math. Ann.* **224** (1976), 143–148.
15. F. JOHN, "Extremum Problems with Inequalities as Subsidiary Conditions," R. Courant Aniversary Volume, pp. 187–204, Wiley (Interscience), New York, 1948.
16. B. S. KASHIN, Diameters of some finite dimensional sets of some classes of smooth functions, *Lzv. Akad. Navk SSSR, Ser. Mat.* **41** (1977), 334–351 [in Russian].
17. D. R. LEWIS, An isomorphic characterization of the Schmidt class, *Compositio Math.* **30** (1975), 293–297.
18. G. J. LOZANOVSKIĬ, On some Banach lattices I; III, *Sibirsk. Mat. Ž.* **10**, No. 3 (1969), 584–599; **13**, No. 6 (1972), 1304–1313 [in Russian].
19. A. PEŁCZYŃSKI, Geometry of finite dimensional Banach spaces and operator ideals, *in* "Notes in Banach Spaces" (H. E. Lacey, ed.), pp. 81–181, Univ. of Texas Press, Austin and London 1980.
20. A. PIETSCH, "Operator Ideals," VEB, Berlin, 1978.
21. G. PISIER, Some results, on Banach spaces without unconditional structures, *Compositio Math.* **37** (1978), 3–19.
22. G. PISIER, De nouvelles caractérisations des ensembles de Sidon, this volume, Chapter 29.
23. S. ROMAN, A problem of Zarankiewicz, *J. Combin. Theory Ser.* **18** (1975), 207–212.
24. O. SCHÜTT, Unconditionality in tensor products, *Israel J. Math.* **31** (1978), 209–216.
25. S. J. SZAREK, On the best constants in the Khinchine inequality, *Studia Math.* **58** (1976), 197–208.
26. S. J. SZAREK, On Kashin's almost euclidean orthogonal decomposition of l_n^1, *Bull. Acad. Polon. Sci. Sér. Math. Astronom. Phys.* **26** (1978), 691–694.

MATHEMATICAL ANALYSIS AND APPLICATIONS, PART B
ADVANCES IN MATHEMATICS SUPPLEMENTARY STUDIES, VOL. 7B

De nouvelles caractérisations des ensembles de Sidon

GILLES PISIER[†]

Centre de Mathématiques
École Polytechnique
Palaiseau, France

DÉDIÉ À LAURENT SCHWARTZ

Let G be a compact abelian group with dual group Γ. This paper shows that Sidon sets may be characterized by their behavior in certain Orlicz spaces, namely, the Orlicz spaces $L^{\psi_q}(G)$ associated with the functions $\psi_q(x) = \exp|x|^q - 1$ $(2 \leq q < \infty)$. Fix p, q with $1 < p < 2 < q < \infty$ and $p^{-1} + q^{-1} = 1$; it is proved that $\Lambda \subset \Gamma$ is a Sidon set if and only if the Fourier transform (resp., the inverse Fourier transform) is bounded from $l_p(\Lambda)$ into $L_\Lambda^{\psi_q}$ [resp., from $L_\Lambda^{\psi_q}$ into $l_{p,\infty}(\Lambda)$]. Moreover, it is shown that, in the inequalities which are known to characterize Sidon sets, it is enough to consider functions which have Fourier coefficients equal to ± 1 or 0. These results yield an arithmetic characterization of Sidon sets in the particular case when $G = \prod_{j=1}^{\infty} \mathbb{Z}(p_j)$ and (p_j) is a *bounded* sequence of integers. Moreover, we prove an analog for Sidon sets of the quantitative version of Dvoretzky's theorem (cf. [31]). The paper also includes a theorem of interpolation (by the Lions–Peetre method) between the spaces of bounded multipliers from $L^2(G)$ into $L^{\psi_q}(G)$ $(2 \leq q \leq \infty)$. Finally, the nonabelian and the noncompact case are considered. When the group G is not assumed abelian, we prove that the "Sidonicity" of Λ is characterized by the behavior of the *central* functions in $L_\Lambda^p(G)$ when $p \to \infty$. (This is rather surprising in view of known results.)

Contents

[†] Laboratoire de Recherche Associé au C.N.R.S. No. 169.

ISBN 0-12-512802-9

0. Notations générales

Tout au long de cet article nous conservons les notations suivantes.

G est un groupe compact de mesure de Haar normalisée m. Si G est abélien, on note Γ le groupe dual. On notera $\mathscr{P}(G)$ l'espace des polynômes trigonométriques sur G (i.e., l'espace des fonctions dont la transformée de Fourier est à support fini). On pose

$$\psi_q(x) = \exp|x|^q - 1 \qquad \text{et} \qquad \varphi_q(x) = |x|(1 + \mathrm{Log}(1 + |x|))^{1/q}.$$

Pour toute fonction $\Phi: \mathbb{R}_+ \to \mathbb{R}_+$ croissante, convexe et nulle à l'origine, l'espace $L^\Phi(G)$ est l'espace des fonctions $f \in L^0(G)$ telles que $\exists c > 0$ avec

$$\int \Phi\left(\left|\frac{f}{c}\right|\right) dm < \infty;$$

on peut le munir de la norme

$$\|f\|_\Phi = \inf\left\{c > 0 \int \Phi\left(\left|\frac{f}{c}\right|\right) dm \le \Phi(1)\right\}.$$

Les espaces $L^{\varphi_q}(G)$ et $L^{\psi_q}(G)$ sont en dualité (cf. [13]); on vérifie en effet classiquement que $L^{\psi_q}(G)$ s' identifie à $L^{\varphi_q}(G)'$ par la dualité usuelle. Précisément, si l'on pose

$$\|f\|_{\varphi_q^*} = \sup\left\{\int fg\, dm \,\middle|\, \|g\|_{\varphi_q} \le 1\right\},$$

il existe une constante $\Delta_q > 0$ telle que:

$$\forall f \in L^{\psi_q}(G) \qquad (1/\Delta_q)\|f\|_{\psi_q} \le \|f\|_{\varphi_q^*} \le \Delta_q\|f\|_{\psi_q}. \tag{0.1}$$

Pour alléger, on notera le plus souvent L^Φ au lieu de $L^\Phi(G)$.

Soit Λ une partie de Γ et soit $\mathscr{F}(G)$ un certain ensemble de fonctions sur G [par exemple, $C(G)$, $L^2(G)$, $L^{\psi_q}(G)$, $\mathscr{P}(G)$]. On notera $\mathscr{F}_\Lambda(G)$ ou pour alléger simplement $\mathscr{F}_\Lambda$ le sous-ensemble de $\mathscr{F}(G)$ formé des fonctions à spectre dans Λ: $\mathscr{F}_\Lambda = \{f \in \mathscr{F} \mid \hat{f}(\gamma) = 0\ \forall \gamma \notin \Lambda\}$. Pour tout ensemble fini A, on note $|A|$ le cardinal de A.

Dans tout cet article, on notera $(\varepsilon_\gamma)_{\gamma \in \Gamma}$ une suite de variables aléatoires de Bernoulli indexées par Γ; c'est-a-dire, une suite de variables indépendantes et prenant les valeurs $+1$ et -1 avec probabilité $\frac{1}{2}$. On note $(\Omega, \mathscr{B}, \mathbb{P})$ l'espace de probabilité sur lequel sont définies les variables $(\varepsilon_\gamma)_{\gamma \in \Gamma}$.

Rappelons une inégalité classique: il existe une constante—notée B_2—telle que $\forall(\alpha_\gamma) \in l^2(\Gamma)$, on a

$$\left\|\sum \alpha_\gamma \varepsilon_\gamma\right\|_{L^{\psi_2(d\mathbb{P})}} \le B_2\left(\sum |\alpha_\gamma|^2\right)^{1/2}. \tag{0.2}$$

Enfin rappelons la

DÉFINITION 0.1. Un ensemble $\Lambda \subset \Gamma$ est un ensemble de Sidon s'il existe une constante C telle que

$$\forall f \in C_\Lambda(G) \qquad \sum_{\gamma \in \Lambda} |\hat{f}(\gamma)| \leq C \|f\|_{C(G)}. \tag{0.3}$$

On note $S(\Lambda)$ la plus petite constante C vérifiant (0.3).

1. INTRODUCTION

Dans [22], les ensembles de Sidon ont été caractérisés de la manière suivante: une partie $\Lambda \subset \Gamma$ est un ensemble de Sidon si (et seulement si) toute fonction $f \in L^2(G)$ à spectre dans Λ vérifie nécessairement

$$\int \exp|f|^2 \, dm < \infty.$$

En d'autres termes, Λ est un ensemble de Sidon si et seulement si $L^{\psi_2}_\Lambda(G)$ est isomorphe à $l^2(\Lambda)$. D'autre part, par définition, Λ est de Sidon si et seulement si (ssi) la transformation de Fourier définit un isomorphisme de $C_\Lambda(G)$ sur $l^1(\Lambda)$. Il est donc naturel de chercher si des résultats analogues subsistent "entre" $C(G)$ et $L^{\psi_2}(G)$. En particulier, on peut se demander s'il existe des conditions sur $L^{\psi_q}_\Lambda(G)$ pour $2 < q < \infty$ caractérisant les ensembles de Sidon. La section 2 ci-dessous répond à cette question: tout d'abord, on peut montrer (par des méthodes classiques) que, si Λ est un ensemble de Sidon, alors la transformation de Fourier établit un isomorphisme entre $L^{\psi_q}_\Lambda$ et $l^{p,\infty}(\Lambda)$ où

$$1 < p < 2 < q < \infty \qquad \text{et} \qquad \frac{1}{p} + \frac{1}{q} = 1. \tag{1.1}$$

Dans le théorème 2.3 ci-dessous, nous démontrons que pour chaque choix de p, q vérifiant (1.1), les ensembles de Sidon sont les seuls ensembles Λ possédant la propriété précédente. Plus généralement, si la transformation de Fourier (resp., son inverse) est bornée de $l^p(\Lambda)$ dans $L^{\psi_q}_\Lambda$ [resp., de $L^{\psi_q}_\Lambda$ dans $l^{p,\infty}(\Lambda)$], alors Λ est nécessairement un ensemble de Sidon.

La démonstration de ce résultat nous a conduit à d'autres conditions équivalentes montrant que les ensembles de Sidon sont caractérisés par le comportement des polynômes trigonométriques dont les coefficients non nuls sont égaux à ± 1. En particulier, Λ est de Sidon ssi

$$\sup \left\{ |A|^{-1/2} \left\| \sum_{\gamma \in A} \gamma \right\|_{\psi_2} \Big| A \subset \Lambda, |A| < \infty \right\} < \infty.$$

D'autre part, pour que Λ soit un ensemble de Sidon il suffit que (0.3) soit vérifié par tous les polynômes trigonométriques $f \in \mathscr{P}_\Lambda$ tels que

$$\hat{f}(\gamma) \in \{-1, 0, 1\}.$$

Enfin, un ensemble Λ est de Sidon si toute partie finie A de Λ contient elle-même un sous-ensemble B tel que $|B| \geq \delta |A|$ et $S(B) \leq C$, où $\delta > 0$ et C sont des constantes indépendantes de A.

Dans le cas où $\Gamma = \mathbb{Z}$, l'un des principaux problèmes ouverts est de caractériser les ensembles de Sidon par des propriétés arithmétiques. La dernière propriété que nous avons donnée permet d'espérer quelques progrès sur cette question. En effet, dans la section 3, nous donnons (a l'aide de cette propriété) une caractérisation arithmétique des ensembles de Sidon dans le cas où $G = \prod_{j=1}^{\infty} \mathbb{Z}(p_j)$ où $\{p_j | j = 1, \ldots\}$ est une suite *bornée* d'entiers positifs. En particulier, dans le cas $p_j = p$ avec p premier, nous retrouvons un résultat de Malliavin-Brameret et Malliavin [17] (sans utiliser le lemme de Rado–Horn).

Dans la section 4 (pratiquement indépendante des précédentes) nous démontrons un théorème d'interpolation concernant l'espace—note M_{2,ψ_q}—des multiplicateurs de $L^2(G)$ dans $L^{\psi_q}(G)$. Précisément nous montrons que si $2 \leq q_0 \leq q_1 \leq \infty$, $0 < \theta < 1$ et $1/q_\theta = (1-\theta)/q_0 + \theta/q_1$, alors on a

$$[M_{2,\psi_{q_0}}, M_{2,\psi_{q_1}}]_{\theta,\infty} = M_{2,\psi_{q_\theta}}$$

avec équivalence des normes correspondantes. (Cela devient faux pour $q_0 < 2$.) Dans les sections 5 et 6, on considère brièvement des généralisations au cas non abélien et au cas non compact. Enfin, dans la Section 7, nous démontrons un analogue pour les ensembles de Sidon d'un théorème de [31]; soit A une partie finie de Γ, posons $M(A) = \mathbb{E}\|\sum_{\gamma \in A} \varepsilon_\gamma \gamma\|_{C(G)}$, alors, il existe $B \subset A$ tel que $|B| \geq \alpha M(A)^2 |A|^{-1}$ et $S(B) \leq C$ où $\alpha > 0$ et C sont des constantes numériques.

2. Caractérisations des ensembles de Sidon

Soit Λ un ensemble dénombrable.

Soit $(\alpha_\gamma)_{\gamma \in \Lambda}$ une suite de scalaires. On note (α_n^*) la suite obtenue en réarrangeant en ordre décroissant la suite $(|\alpha_\gamma|)_{\gamma \in \Lambda}$. On dit que la suite $(\alpha_\gamma)_{\gamma \in \Lambda}$ est dans $l^{p,\infty}(\Lambda)$ si $\sup_{n \geq 1} n^{1/p} \alpha_n^* < \infty$ et l'on pose

$$\|\{\alpha_\gamma\}_{\gamma \in \Lambda}\|_{p,\infty} = \sup_{n \geq 1} n^{1/p} \alpha_n^*.$$

Il est bien connu que $\| \; \|_{p,\infty}$ est équivalent à une norme sur $l^{p,\infty}(\Lambda)$ pour tout $p > 1$. On notera $l_0^{p,\infty}(\Lambda)$ la fermeture dans $l^{p,\infty}(\Lambda)$ de l'espace des suites à support fini.

Soit $(r_n)_{n \geq 1}$ la suite des fonctions de Rademacher sur $[0,1]$ muni de la mesure de Lebesgue. Il est bien connu que, pour tout $p < \infty$, cette suite engendre dans $L^p([0,1])$ un espace isomorphe à l^2; plus précisément, pour tout $q \leq 2$, on sait que la suite (r_n) engendre dans $L^{\psi_q}([0,1])$ un sous-espace isomorphe à l^2. Il est aussi bien connu que ce dernier résultat n'est plus vrai pour $q > 2$; mais, curieusement, les auteurs classiques ne semblent pas s'être attachés à savoir quel était le sous-espace engendré par (r_n) dans $L^{\psi_q}([0,1])$ pour $q > 2$. Pourtant, on peut démontrer, par des arguments entièrement classiques, que cet espace n'est autre que $l_0^{p,\infty}(\mathbb{N})$ où p est l'exposant conjugué de $q(1/p + 1/q = 1)$.

Il est étonnant que ce fait ne soit pas mieux connu. Je suis reconnaissant à J. Lindenstrauss de m'avoir signalé l'article [27] de Rodin–Semyonov, où ce résultat (que j'avais démontré indépendamment, mais après eux) est mentionné. V. Milman m'a aussi informé oralement que le résultat était sans doute connu de Mačaev. Donnons-en l'énoncé précis:

PROPOSITION 2.1. *Soit $(\alpha_n)_{n \geq 1}$ une suite de carré sommable, de sorte que $S = \sum_{n=1}^{\infty} \alpha_n r_n$ converge presque surement (p.s.) Soit $1 < p < 2 < q < \infty$ avec $p^{-1} + q^{-1} = 1$. Alors, S appartient à $L^{\psi_q}([0,1])$—c'est-à-dire qu'il existe $\delta > 0$ tel que $\int \exp \delta |S|^q \, dt < \infty$—si et seulement si $(\alpha_n)_{n \geq 1} \in l^{p,\infty}$. De plus, il existe des constantes A_p* et *B_p telles que*

$$A_p^{-1} \|(a_n)\|_{p,\infty} \leq \|S\|_{\psi_q} \leq B_p \|(\alpha_n)\|_{p,\infty}. \tag{2.1}$$

Démonstration. Il suffit clairement de démontrer (2.1) en supposant que $(|\alpha_n|)_n$ est une suite décroissante. Montrons d'abord le côté droit de (2.1).

Supposons que $\sup_n |\alpha_n| n^{-1/p} \leq 1$ et posons $S_n = \sum_{k=1}^{n} \alpha_k r_k$. On note μ le mesure de Lebesgue sur $[0,1]$. On a:

$$\forall c > 0 \qquad \mu(\{|S| > 2c\}) \leq \mu(\{|S_n| > c\}) + \mu(\{|S - S_n| > c\}).$$

Choisissons $c = qn^{1/q} \geq \sum_{k=1}^{n} k^{-1/p}$ de sorte que

$$\mu(\{|S_n| > c\}) = 0;$$

d'autre part, par une inégalité classique [11, p.43] on a

$$\mu(\{|\sum \alpha_m r_m| > c\}) \leq 2 \exp - \left\{ \frac{c^2}{2} (\sum \alpha_m^2)^{-1} \right\},$$

d'où

$$\mu(\{|S - S_n| > c\}) \leq 2 \exp - \left(c^2 \Big/ 2 \sum_{k > n} k^{-2/p} \right)$$

soit

$$\mu(\{|S| > 2qn^{1/q}\}) \leq 2 \exp - \{\tfrac{1}{2} q(q-2)\, n\}$$

pour tout entier n. On en déduit immédiatement

$$\|S\|_{\psi_q} \leq B_p,$$

où B_p est une constante ne dépendant que de p (ou de q). Par homogénéité, cela démontre le coté droit de (2.1).

Pour démontrer l'inégalité inverse, on note que

$$\left\|\sum_{k=1}^{n} r_k\right\|_{\psi_q} \geq \left(\frac{2}{A_p}\right) n^{1/p} \tag{2.2}$$

où A_p est une constante indépendante de n.

En effet, on vérifie sans peine que si $\pi_n = \prod_{k=1}^{n} (1 + r_k)$, alors $\|\pi_n\|_{\varphi_q} \leq A'_q n^{1/q}$, où A'_q est une constante. On a donc, d'après (0.1):

$$n = \int \left(\sum_1^n r_k\right) \pi_n \, d\mu \leq \Delta_q \left\|\sum_1^n r_k\right\|_{\psi_q} \|\pi_n\|_{\varphi_q}$$

d'où

$$\left\|\sum_{k=1}^{n} r_k\right\|_{\psi_q} \geq (\Delta_q A'_q)^{-1} n^{1/p},$$

ce qui établit (2.2) avec $A_p = 2^{-1} A'_q \Delta_q$.

Démontrons maintenant le côté gauche de (2.1): de nouveau, on peut (sans restreindre la généralité) supposer que la suite $(|\alpha_n|)_{n \geq 1}$ est décroissante. On écrit alors

$$\left\|\sum_{k=1}^{\infty} \alpha_k r_k\right\|_{\psi_q} \geq \left\|\sum_{k=1}^{n} \alpha_k r_k\right\|_{\psi_q} \geq \frac{1}{2} \left\|\sum_{k=1}^{n} r_k\right\|_{\psi_q} \inf_{k \leq n} |\alpha_k|$$
$$\geq A_p^{-1} n^{1/p} a_n^*,$$

d'où le côté gauche de (2.1).

Remarque: Le côté droit de (2.1) résulte aussi d'un théorème d'interpolation de Bennett [2, Corollaire B] qui assure que l'on a $[L^{\psi_2}, L^{\infty}]_{\theta,\infty} = L^{\psi_q}$, au sens de l'interpolation de Lions–Peetre, avec $q^{-1} = (1 - \theta)/2$.

Par une technique bien connue (introduite dans [28]), on déduit de la proposition 2.1 un résultat analogue pour n'importe quel ensemble de Sidon:

PROPOSITION 2.2. *Tout ensemble de Sidon $\Lambda = \{\gamma_n | n \geq 1\} \subset \Gamma$ vérifie nécessairement, pour tout $q > 2$, la propriété suivante: pour toute suite (α_n) de nombres complexes dans l^2, on a*

$$(2S(\Lambda)A_p)^{-1} \|(\alpha_n)\|_{p,\infty} \leq \left\|\sum_{n=1}^{\infty} \alpha_n \gamma_n\right\|_{\psi_q} \leq 2S(\Lambda) B_p \|(\alpha_n)\|_{p,\infty} \tag{2.3}$$

(*avec* $1/p + 1/q = 1$).

Démonstration. L'argument est standard, rappelons-le brièvement: on déduit facilement de la définition que: $\forall t \in [0,1]$ $\exists \mu_t \in M(G)$ tel que

$$\hat{\mu}_t(\gamma_n) = r_n(t)$$

et

$$\|\mu_t\|_{M(G)} \leq S(\Lambda). \tag{2.4}$$

Posons $f = \sum_{n=1}^{\infty} \alpha_n \gamma_n$. On a [d'après (2.4)]

$$\|\mu_t * f\|_{\psi_q} \leq S(\Lambda)\|f\|_{\psi_q} \tag{2.5}$$

et

$$\|f\|_{\psi_q} = \|\mu_t * \mu_t * f\|_{\psi_q} \leq S(\Lambda)\|\mu_t * f\|_{\psi_q}. \tag{2.6}$$

Posons $F(t,x) = \sum_{n=1}^{\infty} r_n(t)\, \alpha_n \gamma_n(x) = \mu_t * f(x)$. On voit aisément que, quelle que soit la fonction $F(t,x)$ (et quel que soit $q \geq 1$), on a

$$\begin{aligned} &\tfrac{1}{2}\max\{\|F(t,x)\|_{L^1_{dt}(L^{\psi_q}_{dm})}, \|F(t,x)\|_{L^1_{dm}(L^{\psi_q}_{dt})}\} \\ &\leq \|F(t,x)\|_{L^{\psi_q}(dt \otimes dm)} \\ &\leq \inf\{\|F(t,x)\|_{L^{\infty}_{dt}(L^{\psi_q}_{dm})}, \|F(t,x)\|_{L^{\infty}_{dm}(L^{\psi_q}_{dt})}\}. \end{aligned} \tag{2.7}$$

[Le côté gauche de (2.7) résulte du fait que $x^{-1}\psi_q(x)$ est croissante pour $x \geq 0$.] On déduit donc de (2.5), (2.7), et (2.1)

$$(2A_p)^{-1}\|(\alpha_n)\|_{p,\infty} \leq S(\Lambda)\|f\|_{\psi_q}.$$

De même, on obtient à partir de (2.6), (2.7), et (2.1):

$$\|f\|_{\psi_q} \leq 2S(\Lambda)B_p\|(\alpha_n)\|_{p,\infty}. \quad \blacksquare$$

Les résultats principaux du présent article sont en fait des réciproques à la proposition précédente. Nous pouvons les réunir de la manière suivante:

THÉORÈME 2.3. Soient G, Γ, m *comme ci-dessus. Soit* $\Lambda = \{\gamma_n | n \in \mathbb{N}\}$ *une partie de* Γ.
Les propriétés suivantes de Λ *sont équivalentes:*

(i) Λ *est un ensemble de Sidon.*
(ii) *il existe une constante C telle que, pour toute partie finie A de* Λ *on a:*

$$\left\|\sum_{\gamma \in A} \gamma\right\|_{\psi_2} \leq C|A|^{1/2}.$$

(iii) *Il existe p avec* $1 < p < 2$, *et une constante C tels que pour toute partie finie A de* Λ *on a:*

$$\left\|\sum_{\gamma \in A} \gamma\right\|_{\psi_q} \leq C|A|^{1/p},$$

où $q^{-1} + p^{-1} = 1$.

(iv) *Il existe $\delta > 0$ et C tels que pour toute partie finie A de Λ, on puisse trouver une sous-partie B de A avec $|B| \geq \delta|A|$ et $S(B) \leq C$.* [*$S(B)$ est la constante de "sidonicité" de B, cf. Définition* 0.1].

(v) *Il existe $\delta > 0$ et p avec $1 < p < 2$ tels que, pour toute partie finie A de Λ, on a*

$$\mathbb{E}\left\|\sum_{\gamma \in A} \varepsilon_\gamma \gamma\right\|_{\psi_q} \geq \delta|A|^{1/p}.$$

(vi) *Il existe $\delta > 0$ tel que pour toute partie finie A de Λ, on a*

$$\mathbb{E}\left\|\sum_{\gamma \in A} \varepsilon_\gamma \gamma\right\|_{C(G)} \geq \delta|A|.$$

Remarques: Il est démontré dans [22] que Λ est un ensemble de Sidon ssi il existe une constante C telle que: $\forall(\alpha_\gamma)_{\gamma \in \Lambda} \in l^2(\Lambda)$, on a

$$\left\|\sum_{\gamma \in \Lambda} \alpha_\gamma \gamma\right\|_{\psi_2} \leq C\left(\sum |\alpha_\gamma|^2\right)^{1/2}.$$

L'équivalence (i)$\Leftrightarrow$(ii) du théorème améliore ce résultat.

D'autre part, d'aprés [26], Λ est de Sidon ssi il existe $\delta > 0$ tel que, pour toute suite $(\alpha_\gamma)_{\gamma \in \Gamma}$ de scalaires dont un nombre fini seulement sont non nuls, on a

$$\mathbb{E}\left\|\sum \varepsilon_\gamma \alpha_\gamma \gamma\right\|_{C(G)} \geq \delta \sum |\alpha_\gamma|.$$

L'équivalence (i)$\Leftrightarrow$(vi) du théorème 2.3 améliore ce résultat.

Enfin, on peut remarquer que les équivalences (i)$\Leftrightarrow$(iii)$\Leftrightarrow$(v) constituent une réciproque à la proposition 2.2. En effet, il en résulte que tout ensemble vérifiant des inégalités analogues à (2.3) est nécessairement un ensemble de Sidon; il suffit d'ailleurs qu'il vérifie l'une ou l'autre des inégalités (2.3) pour qu'il soit un ensemble de Sidon.

Nous reviendrons sur les équivalences (ii)$\Leftrightarrow$(iv)$\Leftrightarrow$(vi) à la fin du paragraphe 4.

Notations Supplémentaires

Soit $(\alpha_\gamma)_{\gamma \in \Gamma}$ une famille de scalaires telle que $\alpha_\gamma \to 0$ à l'infini. On note encore $(\alpha_n^*)_n$ la suite obtenue en réarrangeant en ordre décroissant la famille $(|\alpha_\gamma|)_{\gamma \in \Gamma}$. L'espace de Lorentz $l^{p,1}(\Gamma)$ est défini habituellement comme l'espace des suites $(\alpha_\gamma)_{\gamma \in \Gamma}$ telles que

$$\sum_{n=1}^{\infty} \alpha_n^* n^{-1/q} < \infty, \qquad \frac{1}{p} + \frac{1}{q} = 1.$$

Nous utiliserons en fait une autre définition de cet espace: on définit la norme $\| \|_{p,1}$ sur l'espace des suites à support fini, de la manière suivante:

$$\|(\alpha_\gamma)_{\gamma\in\Gamma}\|_{p,1} = \inf \sum_{i=1}^{N} |\lambda_i|,$$

où l'infimum porte sur toutes les représentations de la suite à support fini $(\alpha_\gamma)_{\gamma\in\Gamma}$ de la forme

$$\alpha_\gamma = \sum_{i=1}^{N} \frac{\xi_\gamma^i 1_{A_i}(\gamma)}{|A_i|^{1/p}} \lambda_i$$

avec $\lambda_i \in \mathbb{C}$ et où $(A_i)_{i=1}^N$ est une famille de parties finies de Γ, 1_{A_i} est la fonction indicatrice de A_i, et ξ_γ^i sont des nombres complexes de module 1. En d'autres termes l'ensemble des suites $(\alpha_\gamma)_{\gamma\in\Gamma}$ à support fini telles que $\|(\alpha_\gamma)\|_{p,1} \leq 1$ est l'enveloppe convexe de l'ensemble des suites $(\alpha_\gamma)_{\gamma\in\Gamma}$ telles que

$$|\alpha_\gamma| = |A|^{-1/p} \quad \text{si } \gamma \in A \quad \text{et} \quad |\alpha_\gamma| = 0 \quad \text{sinon,}$$

où A est le support de $(\alpha_\gamma)_{\gamma\in\Gamma}$.

On peut définir alors $l^{p,1}(\Gamma)$ comme le complété de l'espace des suites à support fini pour la norme $\| \|_{p,1}$.

Il est bien connu que cette définition équivaut à celle donnée ci-dessus (cf. [10, (1.5) et (2.5)]).

Dans [22], un théorème dû à Dudley et Fernique sur les processus gaussiens, joue un rôle crucial. Ce théorème donne une condition nécessaire et suffisante pour la continuité p.s. des trajectoires d'un processus gaussien stationnaire [7]. Cette condition porte sur l'entropie d'un espace métrique associé au processus. Cette notion joue aussi un grand rôle dans le présent article, ce qui nous oblige à introduire les notations suivantes:

Soit $f \in L^2(G)$ et soit p avec $1 < p \leq 2$. Nous noterons $d_{p,1}^f(s,t)$ la (pseudo-) métrique définie sur G par:

$$\forall s, t \in G \qquad d_{p,1}^f(s,t) = \|\{\hat{f}(\gamma)(\gamma(t) - \gamma(s))\}_{\gamma\in\Gamma}\|_{p,1}.$$

Cette métrique est bien définie si l'on suppose que $\hat{f} \in l^{p,1}(\Gamma)$. Soit d une pseudo-métrique sur G.

On notera $N_d(\varepsilon)$ le plus petit nombre de boules de rayon ε pour la métrique d qui suffisent à recouvrir G. On pose, pour simplifier, $N_{p,1}(\varepsilon, f) = N_{d_{p,1}^f}(\varepsilon)$. Enfin, on notera $J_{p,1}(f)$ l'intégrale suivante:

$$J_{p,1}(f) = \int_0^\infty (\operatorname{Log} N_{p,1}(\varepsilon, f))^{1/q}\, d\varepsilon \qquad \text{avec} \quad \frac{1}{p} + \frac{1}{q} = 1.$$

Tout à fait similairement, on pose

$$d_p^f(s,t) = (\sum |\hat{f}(\gamma)|^p |\gamma(s) - \gamma(t)|^p)^{1/p},$$

on définit $N_p(\varepsilon, f)$ comme ci-dessus:

$$N_p(\varepsilon, f) = N_{d_p^f}(\varepsilon).$$

Enfin, on pose

$$J_p(f) = \int_0^\infty (\operatorname{Log} N_p(\varepsilon, f))^{1/q}\, d\varepsilon \qquad \text{avec} \quad \frac{1}{p} + \frac{1}{q} = 1.$$

Au cours de la démonstration du principal résultat de [22], on a établi en fait le résultat suivant:

LEMME 2.4. *Soit $\Lambda \subset \Gamma$ une partie de Γ. Supposons*:

(i)′ *il existe une constante C telle que, pour tout f dans L^2_Λ on a*:

$$\sum_{\gamma \in \Lambda,\ \gamma \neq 0} |\hat{f}(\gamma)| \leq C J_2(f).$$

Alors Λ est un ensemble de Sidon. (*La réciproque est aussi vraie.*)

Pour plus de détails, cf. [19, 22, 24]. Dans le cours de la démonstration du théorème 2.3 nous verrons que les propriétés (i) à (v) sont en fait équivalentes à:

(iii)′ il existe p avec $1 < p < 2$ et une constante C tels que, pour tout $f \in \mathscr{P}_\Lambda$ on a

$$\sum_{\gamma \in \Gamma,\ \gamma \neq 0} |\hat{f}(\gamma)| \leq C J_{p,1}(f).$$

Pour utiliser (iii)′, le lemme suivant sera très utile, nous le démontrons à la fin de la section 2.

LEMME 2.5. *Supposons que $1 < r < p < 2$. Soit θ tel que $p^{-1} = (1-\theta)/r + \theta/2$. Alors il existe une constante K_θ telle que, pour tout f dans $\mathscr{P}_\Lambda$, on a*:

$$J_{p,1}(f) \leq K_\theta J_2(f)^\theta J_r(f)^{1-\theta}.$$

Nous aurons aussi besoin d'un corollaire d'une version modifiée d'un théorème dû à Dudley [6], (nous pourrions utiliser aussi les résultats de [25]; dans le cas particulier que nous considérons, cela revient au même).

LEMME 2.6 *Soit d une pseudo-métrique sur G. On suppose que*

$$\int_0^\infty (\operatorname{Log} N_d(\varepsilon))^{1/q}\, d\varepsilon < \infty$$

avec $0 < q < \infty$. Soit $f \in L^{\psi_q}(G)$. On suppose que f vérifie

$$\forall t, s \in G \qquad \|f_t - f_s\|_{\psi_q} \leq d(t, s)$$

où l'on a noté f_t la fonction $f_t(x) = f(x+t)$. Dans ces conditions f est continue et l'on a:

$$\sup_{x,\, y \in G} |f(x) - f(y)| \leq D_q \int_0^\infty (\operatorname{Log} N_d(\varepsilon))^{1/q}\, d\varepsilon$$

où D_q est une constante.

On peut démontrer ce lemme exactement comme en [22] (voir aussi [19]) où la démonstration est donnée pour $q = 2$.

Remarque. Bien entendu, on peut prendre en particulier dans le lemme 2.6 $d(s, t) = \|f_t - f_s\|_{\psi_q}$. C'est d'ailleurs le meilleur choix de d.

Nous utiliserons le lemme suivant qui est tout à fait élémentaire:

LEMME 2.7. *Soit q, q_1 tels que $0 < q_1 < q < \infty$. Posons $\theta = q_1/q$. Alors, pour toute fonction f de $L^\infty(G)$, on a:*

$$\|f\|_{\psi_q} \leq \|f\|_{\psi_{q_1}}^{\theta} \|f\|_\infty^{1-\theta}.$$

La démonstration est évidente.

Nous aurons également besoin du lemme suivant qui est essentiellement dû à Paley–Zygmund [11, pp. 43–44]. Pour une démonstration détaillée, voir [24] ou [19].

LEMME 2.8. *Si $h \in L^2(G)$ et $k \in L^{\varphi_2}(G)$ alors la série $\sum_{\gamma \in \Gamma} \varepsilon_\gamma \hat{h}(\gamma)\hat{k}(\gamma)$ est presque sûrement continue et l'on a:*

$$\mathbb{E}\left\| \sum_{\gamma \in \Gamma} \varepsilon_\gamma \hat{h}(\gamma)\hat{k}(\gamma)\gamma \right\|_{C(G)} \leq R\|h\|_2 \|k\|_{\varphi_2}$$

où R est une constante numérique.

Nous aurons enfin besoin du résultat suivant qui n'est qu'une reformulation d'un théorème dû à Marcus [18, Theorem 1].

LEMME 2.9. *Soit $f \in C(G)$ tel que $\sum_{\gamma \in \Gamma} |\hat{f}(\gamma)| < \infty$ et soit r avec $1 < r < \infty$. Alors $J_r(f) < \infty$. De plus, il existe une constante C_r telle que:*

$$J_r(f) \leq C_r \sum_{\gamma \neq 0} |\hat{f}(\gamma)|. \tag{2.8}$$

Démonstration. Soit $\{a_k \mid k \in \mathbb{N}\}$ le réarrangement décroissant de $\{|\hat{f}(\gamma)| \mid \gamma \in \Gamma\}$. On pose

$$B\{a_k\} = \{\{x_k\} \in l^r(\mathbb{N}) \mid |x_k| \leq a_k \ \forall k \in \mathbb{N}\}.$$

Soit $N(\varepsilon)$ le plus petit nombre de boules de rayon ε pour la norme de $l^r(\mathbb{N})$ suffisant à recouvrir $B\{a_k\}$. Marcus [18, Theorem 1] a montré que si $\sum a_k < \infty$

alors $\int_0^\infty (\operatorname{Log} N(\varepsilon))^{1/s}\, d\varepsilon < \infty$ $(r^{-1} + s^{-1} = 1)$. En réalité, sa démonstration établit l'existence d'une constante C'_r telle que

$$\int_0^\infty (\operatorname{Log} N(\varepsilon))^{1/s}\, d\varepsilon \le C'_r \sum a_k.$$

(L'existence de cette constante résulte aussi d'un argument a priori à partir du résultat énoncé par Marcus.)

Le lemme 2.9 en résulte immédiatement, car il est très facile de vérifier que

$$\forall \varepsilon > 0 \qquad N_r(2\varepsilon, f) \le N(\varepsilon). \tag{2.9}$$

En effet, soit $\{\gamma_k\}$ un réarrangement de l'ensemble $\{\gamma \in \Gamma \mid \hat{f}(\gamma) \neq 0\}$ de telle sorte que $|\hat{f}(\gamma_k)| = a_k$; l'application $F: G \to l^r(\mathbb{N})$ définie par $F(t) = \{\gamma_k(t)\hat{f}(\gamma_k)\} \in l^r(\mathbb{N})$ est une isométrie de G muni de la métrique d_r^f dans $l^r(\mathbb{N})$ muni de sa norme. L'inégalité (2.9) est alors évidente.

Le lecteur notera enfin que $J_r(f)$ est indépendant de $\hat{f}(0)$, ce qui permet d'écrire (2.8) en supprimant $\hat{f}(0)$ dans le second membre.

Nous pouvons maintenant conclure:

Démonstration du théorème 2.3. (i) $\Rightarrow$ (ii): C'est un affaiblissement du résultat classique de Rudin [28] qui affirme que tout ensemble de Sidon Λ vérifie $L^2_\Lambda \subset L^{\psi_2}_\Lambda$.

(ii) $\Rightarrow$ (iii): Supposons (ii), soit p quelconque tel que $1 < p < 2$, et soit q tel que $p^{-1} + q^{-1} = 1$. D'après le lemme 2.7, on a:

$$\Big\|\sum_{\gamma \in A} \gamma\Big\|_{\psi_q} \le \Big\|\sum_{\gamma \in A} \gamma\Big\|_{\psi_2}^{\theta} \Big\|\sum_{\gamma \in A} \gamma\Big\|_{\infty}^{1-\theta}$$

où $\theta = 2/q$. On a donc

$$\Big\|\sum_{\gamma \in A} \gamma\Big\|_{\psi_q} \le (C|A|^{1/2})^{\theta} |A|^{1-\theta} = C^{\theta} |A|^{1/p}.$$

Ce qui prouve que (ii) $\Rightarrow$ (iii).

(iii) $\Rightarrow$ (iii)'. Soient p et C comme en (iii). On a évidemment (inégalité triangulaire) encore:

$$\forall \xi_\gamma = \pm 1 \qquad \Big\|\sum_{\gamma \in A} \xi_\gamma \gamma\Big\|_{\psi_q} \le 2C|A|^{1/p};$$

il en résulte immédiatement, quel que soit $\xi_\gamma \in \mathbb{C}$ avec $|\xi_\gamma| = 1$:

$$\Big\|\sum_{\gamma \in A} \xi_\gamma \gamma\Big\|_{\psi_q} \le 4C|A|^{1/p},$$

par conséquent, un argument de convexité immédiat nous donne:

$$\forall f \in \mathscr{P}_\Lambda \qquad \|f\|_{\psi_q} \leq 4C\|\{\hat{f}(\gamma)\}\|_{l^{p,1}(\Gamma)}. \tag{2.10}$$

Considérons alors la fonction

$$g = \sum |\hat{f}(\gamma)|\gamma.$$

On a, d'après (2.10):

$$\forall f \in \mathscr{P}_\Lambda, \forall s, t \in G \qquad \|g_t - g_s\|_{\psi_q} \leq 4Cd^f_{p,1}(s, t).$$

D'où l'on déduit, par le lemme 2.6:

$$\sup_{x,\, y \in G} |g(x) - g(y)| \leq 4CD_q J_{p,1}(f).$$

Mais on a trivialement

$$\sum_{\gamma \neq 0} |\hat{f}(\gamma)| = \sum_{\gamma \neq 0} |\hat{g}(\gamma)| = \sup_{x \in G} \left|g(x) - \int g(y)\, dm(y)\right|$$
$$\leq \sup_{x,\, y \in G} |g(x) - g(y)|.$$

On conclut donc que

$$\sum_{\gamma \neq 0} |\hat{f}(\gamma)| \leq 4CD_q J_{p,1}(f),$$

ce qui prouve que (iii) $\Rightarrow$ (iii)′.

(iii)′ $\Rightarrow$ (i)′. Cette implication est une conséquence du lemme 2.5. En effet, supposons (iii)′. On a donc:

$$\forall f \in \mathscr{P}_\Lambda \qquad \sum_{\gamma \in \Lambda,\, \gamma \neq 0} |\hat{f}(\gamma)| \leq CJ_{p,1}(f)$$

soit d'après le lemme 2.5:

$$\sum_{\gamma \in \Lambda,\, \gamma \neq 0} |\hat{f}(\gamma)| \leq CK_\theta J_2(f)^\theta J_r(f)^{1-\theta}$$

avec $1 < r < p < 2$ et $p^{-1} = (1 - \theta)/r + \theta/2$.

Mais d'après le lemme 2.9 on en déduit

$$\sum_{\gamma \neq 0} |\hat{f}(\gamma)| \leq CK_\theta J_2(f)^\theta \left(C_r \sum_{\gamma \neq 0} |\hat{f}(\gamma)|\right)^{1-\theta}$$

soit encore (après division):

$$\forall f \in \mathscr{P}_\Lambda \qquad \sum_{\gamma \neq 0} |\hat{f}(\gamma)| \leq (CK_\theta C_r^{1-\theta})^{1/\theta} J_2(f),$$

ce qui prouve que (iii)′ $\Rightarrow$ (i)′. D'après le lemme 2.4 on a (i)′ $\Rightarrow$ (i) et on conclut donc que (i), (ii), (iii), et (iii)′ sont des propriétés équivalentes.

Passons aux implications restantes:

(i) $\Rightarrow$ (iv) est triviale.

(iv) $\Rightarrow$ (v) est facile: si (iv) est vérifiée, on a:

$$\mathbb{E}\left\|\sum_{\gamma\in A}\varepsilon_\gamma\gamma\right\|_{\psi_q}\geq\mathbb{E}\left\|\sum_{\gamma\in B}\varepsilon_\gamma\gamma\right\|_{\psi_q}$$

soit d'après la proposition 2.2:

$$\begin{aligned}&\geq(2A_pS(B))^{-1}|B|^{1/p}\\&\geq\delta^{1/p}(2A_pC)^{-1}|A|^{1/p},\end{aligned}$$

ce qui prouve que (iv) $\Rightarrow$ (v).

(v) $\Rightarrow$ (vi) est aussi facile, supposons

$$\delta|A|^{1/p}\leq\mathbb{E}\left\|\sum_{\gamma\in A}\varepsilon_\gamma\gamma\right\|_{\psi_q};$$

alors, d'après le lemme 2.7, on a:

$$\begin{aligned}\delta|A|^{1/p}&\leq\mathbb{E}\left\{\left\|\sum_{\gamma\in A}\varepsilon_\gamma\gamma\right\|_{\psi_2}^{2/q}\left\|\sum_{\gamma\in A}\varepsilon_\gamma\gamma\right\|_{C(G)}^{1-2/q}\right\}\\&\leq\left(\mathbb{E}\left\|\sum_{\gamma\in A}\varepsilon_\gamma\gamma\right\|_{\psi_2}\right)^{2/q}\left(\mathbb{E}\left\|\sum_{\gamma\in A}\varepsilon_\gamma\gamma\right\|_{C(G)}\right)^{1-2/q};\end{aligned}\tag{2.11}$$

d'autre part d'après (2.7) et (0.2)

$$\mathbb{E}\left\|\sum_{\gamma\in A}\varepsilon_\gamma\gamma\right\|_{\psi_2}\leq2\left\|\sum_{\gamma\in A}\varepsilon_\gamma\gamma\right\|_{L^{\psi_2}(d\mathbb{P}\otimes dm)}\leq2B_2|A|^{1/2}.$$

On déduit donc de (2.11) que

$$\mathbb{E}\left\|\sum_{\gamma\in A}\varepsilon_\gamma\gamma\right\|_{C(G)}\geq\delta_1|A|\qquad\text{avec}\quad\delta_1=\{\delta(2B_2)^{-2/q}\}^{q/q-2}.$$

Ce qui prouve que (v) $\Rightarrow$ (vi).

Il reste donc à démontrer que (vi) $\Rightarrow$ (ii).

Pour cela, on remarque que l'on a, d'après le lemme 2.8:

$$\forall k\in L^{\varphi_2}(G),\quad\forall A\subset\Lambda\qquad\mathbb{E}\left\|\sum_{\gamma\in A}\varepsilon_\gamma\gamma\hat{k}(\gamma)\right\|_{C(G)}\leq R|A|^{1/2}\|k\|_{\varphi_2}.$$

D'autre part, il est évident que, si l'on suppose (vi), on a:

$$\delta|A|\inf_{\gamma\in A}|\hat{k}(\gamma)|\leq2\mathbb{E}\left\|\sum_{\gamma\in A}\varepsilon_\gamma\hat{k}(\gamma)\gamma\right\|_{C(G)},$$

d'où

$$|A|^{1/2}\inf_{\gamma\in A}|\hat{k}(\gamma)|\leq2R\delta^{-1}\|k\|_{\varphi_2}.\tag{2.12}$$

Soit $\{\gamma_1, \ldots, \gamma_N\}$ une énumération de A telle que $N = |A|$ et telle que $|\hat{k}(\gamma_j)|$ est une fonction décroissante de j. Si l'on applique (2.12) en remplaçant A par $\{\gamma_1, \ldots, \gamma_j\}$, on trouve, pour chaque $j = 1, 2, \ldots, N$:

$$j^{1/2}|\hat{k}(\gamma_j)| \leq 2R\delta^{-1}\|k\|_{\varphi_2}$$

d'où

$$\sum_{j=1}^{N} |\hat{k}(\gamma_j)| \leq 2R\delta^{-1}\left(\sum_{j=1}^{N} j^{-1/2}\right)\|k\|_{\varphi_2}$$

soit

$$\left|\sum_{\gamma\in A} \hat{k}(\gamma)\right| \leq 4R\delta^{-1}\sqrt{N}\|k\|_{\varphi_2}.$$

En d'autres termes, on a:

$$\left\|\sum_{\gamma\in A} \gamma\right\|_{\varphi_2^*} \leq 4R\delta^{-1}\sqrt{N}$$

soit finalement d'après (0.1):

$$\left\|\sum_{\gamma\in A} \gamma\right\|_{\psi_2} \leq 4R\delta^{-1}\Delta_2|A|^{1/2},$$

et on conclut bien que (vi) $\Rightarrow$ (ii). Ce qui achève la démonstration du théorème 2.3.

Démonstration du lemme 2.5: Le point de départ est l'inégalité bien connue suivante: si $1 \leq r < p < 2$, $p^{-1} = (1-\theta)/r + \theta/2$, on a pour toute suite de scalaires $\{\alpha_\gamma\}_{\gamma\in\Gamma}$:

$$\|\{\alpha_\gamma\}\|_{l^{p,1}(\Gamma)} \leq K'_\theta(\sum|\alpha_\gamma|^r)^{(1-\theta)/r}(\sum|\alpha_\gamma|^2)^{\theta/2},$$

où K'_θ est une constante.
On a donc

$$\forall s, t \in G \qquad d^f_{p,1}(s,t) \leq K'_\theta d^f_r(s,t)^{1-\theta} d^f_2(s,t)^\theta.$$

Fixons f dans $\mathscr{P}(G)$ et posons pour simplifier:

$$\sigma_{p,1}(t) = d^f_{p,1}(t,0), \qquad \sigma_r(t) = d^f_r(t,0), \qquad \text{et} \qquad \sigma_2(t) = d^f_2(t,0).$$

On a

$$\sigma_{p,1}(t) \leq K'_\theta\sigma_r(t)^{1-\theta}\sigma_2(t)^\theta. \tag{2.13}$$

Pour plus de clarté, nous introduisons, pour chaque pseudo-métrique d sur G, le nombre $M_d(\varepsilon)$ comme le plus petit entier n pour lequel il existe une partition $A_1, \ldots, A_n$ de G en ensembles de d-diamètre inférieur a ε. On vérifie facilement que

$$M_d(2\varepsilon) \leq N_d(\varepsilon) \leq M_d(\varepsilon). \tag{2.14}$$

Soient d_0, d_1, d_θ trois pseudo-métriques sur G telles que

$$d_\theta(s,t) \le d_0(s,t)^{1-\theta} d_1(s,t)^\theta.$$

Posons pour simplifier $M_\theta(\varepsilon) = M_{d_\theta}(\varepsilon)$ et de même pour d_0, d_1. On a alors

$$\forall \varepsilon_0, \varepsilon_1 \in \mathbb{R}_+ \qquad M_\theta(\varepsilon_0^{1-\theta}\varepsilon_1^\theta) \le M_0(\varepsilon_0) M_1(\varepsilon_1). \tag{2.15}$$

En effet si $(A_j^0)_{j \le M_0(\varepsilon_0)}$ [resp., $(A_k^1)_{k \le M_1(\varepsilon_1)}$] est une partition de G en ensembles de d_0-diamètre (resp., d_1-diamètre) inférieur à ε_0 (resp., ε_1), alors $(A_j^0 \cap A_k^1)$ est une partition de G en ensembles de d_θ-diamètre inférieur à $\varepsilon_0^{1-\theta}\varepsilon_1^\theta$.

Supposons maintenant que d est une pseudo-distance invariante par translation sur G. On pose

$$\sigma(t) = d(t,0) \qquad \text{et} \qquad \mu_\sigma(\varepsilon) = m(\{x \in G \mid \sigma(x) < \varepsilon\}).$$

On voit facilement que

$$\forall \varepsilon > 0 \qquad \frac{1}{\mu_\sigma(\varepsilon)} \le N_d(\varepsilon) \le \frac{1}{\mu_\sigma(\varepsilon/2)}. \tag{2.16}$$

Posons pour simplifier

$$J_\alpha(\sigma) = \int_0^\infty (\operatorname{Log} N_d(\varepsilon))^{1/\alpha}\, d\varepsilon \qquad \text{et} \qquad K_\alpha(\sigma) = \int_0^\infty \left(\operatorname{Log} \frac{1}{\mu_\sigma(\varepsilon)}\right)^{1/\alpha} d\varepsilon.$$

On déduit évidemment de (2.16) que

$$K_\alpha(\sigma) \le J_\alpha(\sigma) \le 2K_\alpha(\sigma). \tag{2.17}$$

Posons $\bar\sigma(t) = \sup\{y \mid \mu_\sigma(y) < t\}$ pour $0 \le t \le 1$. On peut considérer la fonction $\bar\sigma : [0,1] \to \mathbb{R}_+$ comme la réarrangée croissante de la fonction $\sigma : G \to \mathbb{R}_+$. En effet, $\bar\sigma$ a la même distribution sur $([0,1], dt)$ que σ considérée comme fonction sur (G, m). Pour plus de détails sur ces questions, voir [19, Chapitre 2].

Posons $\sigma_0(t) = d_0(t,0)$, $\sigma_1(t) = d_1(t,0)$, $\sigma_\theta(t) = d_\theta(t,0)$. On peut alors retranscrire (2.15) dans le cas où d_0, d_1, et d_θ sont invariantes par translation, on trouve

$$\bar\sigma_\theta(ts) \le 4(\bar\sigma_0(t))^{1-\theta}(\bar\sigma_1(s))^\theta, \tag{2.18}$$

en effet, on déduit de (2.14), (2.15) et (2.16):

$$\frac{1}{\mu_{\sigma_\theta}(\varepsilon_0^{1-\theta}\varepsilon_1^\theta)} \le \frac{1}{\mu_{\sigma_0}(\varepsilon_0/4)} \frac{1}{\mu_{\sigma_1}(\varepsilon_1/4)}$$

d'où (2.18) résulte immédiatement.

Considérons alors l'intégrale

$$I_\alpha(\sigma) = \int_0^1 \frac{\bar\sigma(t)}{t(\operatorname{Log} t^{-1})^{1-1/\alpha}}\, dt.$$

Vérifions tout d'abord l'inégalité suivante

$$I_q(\sigma_{p,1}) \le K''_\theta [I_s(\sigma_r)]^{1-\theta} [I_2(\sigma_2)]^\theta \tag{2.19}$$

avec $p^{-1} + q^{-1} = 1$, $r^{-1} + s^{-1} = 1$, où K''_θ est une constante. On a en effet par un changement de variable:

$$I_q(\sigma_{p,1}) = \int_0^1 \frac{\bar{\sigma}_{p,1}(t^2)}{t^2(\operatorname{Log} t^{-2})^{1/p}}\, dt^2 = 2^{1/q} \int_0^1 \frac{\bar{\sigma}_{p,1}(t^2)}{t(\operatorname{Log} t^{-1})^{1/p}}\, dt$$

soit d'après (2.13) et (2.18):

$$I_q(\sigma_{p,1}) \le 4.2^{1/q} K'_\theta \int_0^1 \frac{\bar{\sigma}_r(t)^{1-\theta} \bar{\sigma}_2(t)^\theta}{t(\operatorname{Log} t^{-1})^{1-\theta/r+\theta/2}}\, dt$$

d'où par l'inégalité de Hölder:

$$I_q(\sigma_{p,1}) \le 4.2^{1/q} K'_\theta \left(\int_0^1 \frac{\bar{\sigma}_r(t)}{t(\operatorname{Log} t^{-1})^{1/r}}\, dt \right)^{1-\theta} \left(\int_0^1 \frac{\bar{\sigma}_2(t)}{t(\operatorname{Log} t^{-1})^{1/2}}\, dt \right)^\theta,$$

ce qui démontre (2.19) avec $K''_\theta = 4.2^{1/q} K'_\theta$.

Pour terminer la démonstration du lemme 2.5, il nous reste à vérifier l'équivalence des fonctionnelles $I_\alpha(\sigma)$ et $J_\alpha(\sigma)$. Précisément, on va montrer que

$$\alpha K_\alpha(\sigma) = I_\alpha(\sigma),$$

ce qui, compte-tenu de (2.17), achèvera la démonstration du lemme 2.5. En effet, on a, par un changement de variable et une intégration par parties

$$\begin{aligned} K_\alpha(\sigma) &= \int_0^\infty \left(\operatorname{Log} \frac{1}{\mu_\sigma(\varepsilon)} \right)^{1/\alpha} d\varepsilon = \int_0^1 \left(\operatorname{Log} \frac{1}{t} \right)^{1/\alpha} d\bar{\sigma}(t) \\ &= \left[\bar{\sigma}(t) \left(\operatorname{Log} \frac{1}{t} \right)^{1/\alpha} \right]_0^1 + \frac{1}{\alpha} \int_0^1 \frac{\bar{\sigma}(t)}{t(\operatorname{Log} t^{-1})^{1-1/\alpha}}\, dt. \end{aligned}$$

D'où $K_\alpha(\sigma) = \alpha^{-1} I_\alpha(\sigma)$ puisque le premier terme de l'intégration par parties est nul si $I_\alpha(\sigma)$ existe. ■

Remarques. Le théorème de Marcus [18] utilisé au lemme 2.9 a été généralisé récemment par Carl [4].

D'autre part, l'inégalité (2.18) est très voisine de certains résultats connus sur le comportement de l'entropie vis-à-vis de l'interpolation [29, pp. 116–117].

On peut prolonger facilement les résultats donnés au théorème 2.3 dans plusieurs directions. Par exemple:

COROLLAIRE 2.10. *Λ est un ensemble de Sidon ssi*

(vii) *il existe q, $2 \le q < \infty$, tel que la suite $\{\gamma | \gamma \in \Lambda\}$ soit une base inconditionnelle de l'espace $L_\Lambda^{\psi_q}$.*

Démonstration. Dire que $\{\gamma | \gamma \in \Lambda\}$ est une base inconditionnelle de $L_\Lambda^{\psi_q}$ signifie qu'il existe une constante C telle que: $\forall f \in \mathscr{P}_\Lambda(G)$, et $\forall \varepsilon_\gamma = \pm 1$ on a

$$\frac{1}{C} \|f\|_{\psi_q} \leq \left\| \sum_{\gamma \in \Lambda} \varepsilon_\gamma \gamma \hat{f}(\gamma) \right\|_{\psi_q} \leq C \|f\|_{\psi_q}. \tag{2.20}$$

Il est clair que tout ensemble de Sidon vérifie (2.20) (cf. proposition 2.2). Réciproquement, si (2.20) est vérifiée, on a:

$$\|f\|_{\psi_q} \leq C \int \|\textstyle\sum \varepsilon_\gamma \gamma \hat{f}(\gamma)\|_{\psi_q} d\mathbb{P}(\{\varepsilon_\gamma\})$$

soit d'après (2.7) et (2.1): $\|f\|_{\psi_q} \leq 2CB_p \|\{\hat{f}(\gamma)\}\|_{l^{p,\infty}(\Lambda)}$ si $q > 2$. L'implication (iii) $\Rightarrow$ (i) du théorème 2.3 montre donc que Λ est de Sidon. Le cas $q = 2$ est similaire.

Remarque. L'implication (vi) $\Rightarrow$ (i) du théorème 2.3 répond positivement à une question posée par Pełczyński, dans un contexte tout à fait différent, cf. [21].

3. Conditions arithmétiques pour qu'un ensemble soit de Sidon

L'implication (iv) $\Rightarrow$ (i) dans le théorème 2.3 conduit à une nouvelle condition suffisante de "sidonicité". Pour l'énoncer, nous aurons besoin d'une notion déjà considérée (avec des variantes) par divers auteurs (voir [16] et les références qui y sont citées).

Nous dirons qu'une partie finie B de Γ est "quasi-indépendante" si la seule égalité du type $\sum_{\gamma \in B} n_\gamma \gamma = 0$ avec $n_\gamma \in \{-1, 0, +1\}$ est l'égalité triviale correspondant à $n_\gamma = 0$ pour tout $\gamma \neq 0$. On pose

$$[B] = \left[\sum_{\gamma \in B} n_\gamma \gamma \,|\, n_\gamma \in \{-1, 0, 1\} \right\}.$$

Nous pouvons énoncer le critère suivant:

Théorème 3.1. *Soit Λ une partie de Γ. Supposons qu'il existe une constante K telle que:*

Pour toute partie quasi-indépendante $B \subset \Lambda$, on a:

$$|\Lambda \cap [B]| \leq K|B|. \tag{3.1}$$

Alors, Λ est un ensemble de Sidon.

Démonstration. On va montrer que (3.1) entraîne la condition (iv) du théorème 2.3.

Soit $A \subset \Lambda$ une partie finie quelconque de Λ.

Soit $B \subset A$ une partie quasi-indépendante maximale.

On a nécessairement $A \subset [B]$, car sinon $\exists \gamma \in A \dot{-} [B]$ et alors $B \cup \{\gamma\}$ est encore une partie quasi-indépendante, ce qui contredit la maximalité de B. On a donc $A \subset [B]$, d'ou, si l'on suppose (3.1):

$$|A| \leq |\Lambda \cap [B]| \leq K|B|$$

soit finalement $|B| \geq K^{-1}|A|$. Pour conclure, il suffit donc d'utiliser l'implication (iv) $\Rightarrow$ (i) du théorème 2.3 ainsi que le lemme (bien connu des spécialistes) suivant:

LEMME 3.2. *Il existe une constante numérique C telle que pour tout ensemble quasi-indépendant $B \subset \Gamma$, on a*

$$S(B) \leq C.$$

Démonstration. On peut évidemment supposer que $0 \notin B$, et que B est fini. Pour tout choix de signes $\varepsilon_\gamma = \pm 1$, on pose

$$\Phi(\{\varepsilon_\gamma\}) = \prod_{\gamma \in B} \left(1 + \varepsilon_\gamma \frac{\gamma + \bar{\gamma}}{2}\right).$$

Puisque $\Phi(\{\varepsilon_\gamma\}) \geq 0$, on a

$$\|\Phi(\{\varepsilon_\gamma\})\|_{L_1(G)} = \int \Phi(\{\varepsilon_\gamma\})\, dm = 1$$

quel que soit (ε_γ); la dernière égalité résulte de la quasi-indépendance de B.

Soit $\mathbb{P}$ la probabilité uniforme sur $\{-1, +1\}^\Gamma$. On peut écrire

$$\forall f \in \mathscr{P}_B \qquad \frac{1}{2} \sum_{\gamma \in B} |\hat{f}(\gamma)| \leq \int \left\langle \sum_{\gamma \in B} |\hat{f}(\gamma)| \varepsilon_\gamma \gamma, \Phi\{\varepsilon_\gamma\} \right\rangle d\mathbb{P}(\{\varepsilon_\gamma\})$$

$$\leq \int \left\| \sum_{\gamma \in B} |\hat{f}(\gamma)| \varepsilon_\gamma \gamma \right\|_{C(G)} d\mathbb{P}(\{\varepsilon_\gamma\})$$

$$\leq 2 \int \left\| \sum_{\gamma \in B} \varepsilon_\gamma \hat{f}(\gamma) \gamma \right\|_{C(G)} d\mathbb{P}(\{\varepsilon_\gamma\}).$$

On conclut donc que $S(B) \leq C$ (où C est une constante numérique) en appliquant un théorème dû à Rider [26]. (On pourrait aussi appliquer un argument analogue à celui donné dans [16, p. 28].)

Remarque. Dans le cas particulier du groupe $G = \mathbb{Z}(p)^{\mathbb{N}}$ avec p premier, le théorème 3.1 fournit une nouvelle démonstration d'un résultat de [17]. En effet, dans ce cas particulier, Γ peut être considéré comme un espace vectoriel sur le corps $\mathbb{Z}(p)$ et la condition:

Pour tout sous-espace vectoriel W de Γ, on a

$$|\Lambda \cap W| \leq K \dim W, \qquad \text{où } K \text{ est une constante} \tag{3.2}$$

est nécessaire et suffisante pour que Λ soit un ensemble de Sidon. Le fait que cette condition (3.2) est suffisante est démontré dans [17] à partir d'un lemme d'algèbre dû à Rado–Horn dont la démonstration est relativement délicate. Nous pouvons démontrer le même résultat sans faire appel à ce lemme. En effet, il est immédiat que (3.2) entraîne (3.1), car si W est le sous-espace engendré par B, on a $[B] \subset W$ et $\dim W \leq |B|$. L'avantage de cette nouvelle preuve est qu'elle se généralise sans difficulté au cas où p n'est pas premier, ce qui permet de répondre à une question posée dans [30] (*cf.* [14, Chapitre 6, p. 82]).

COROLLAIRE 3.3. *Considérons le cas particulier* $G = \prod_{j=1}^{\infty} \mathbb{Z}(p_j)$ *où* p_j *est une suite bornée d'entiers. Alors une partie* $\Lambda \subset \Gamma$ *est un ensemble de Sidon si et seulement si il existe une constante K telle que*

Pour tout sous-groupe fini $H \subset \Gamma$, *on a*

$$|\Lambda \cap H| \leq K \operatorname{rang}(H), \tag{3.3}$$

où le rang de H est défini comme le plus petit nombre de générateurs de H.

Démonstration. La partie "seulement si" est connue [14, Chapitre 5]. Pour démontrer la partie "si", il suffit de remarquer (comme dans la remarque précédente) que si B est quasi-indépendant alors le sous-groupe H engendré par B vérifie

$$[B] \subset H \qquad \text{et} \qquad \operatorname{rang}(H) \leq |B|.$$

Remarque. Dans le cas $G = \prod_{j=1}^{\infty} \mathbb{Z}(p_j)$ avec $\sup_j p_j < \infty$, on voit donc que la condition suffisante donnée au théorème 3.1 est aussi nécessaire. Malheureusement, ce n'est pas le cas en général (cf. Proposition 7.3).

Dans le cas général, la "condition de maille" considérée par exemple dans [12, p. 54] (cf. [8, p. 340]) est toujours une condition nécessaire; mais, a priori cette condition entraîne seulement l'existence d'une constante K telle que

$$|\Lambda \cap [B]| \leq K|B|(1 + \operatorname{Log}|B|)$$

quel que soit $B \subset \Lambda$.

Par conséquent, cette condition entraîne une propriété a priori plus faible que (iv): toute partie finie $A \subset \Lambda$ contient une partie quasi-indépendante $B \subset A$ telle que $|B| \geq |A|/(1 + \operatorname{Log}|A|)$ où $\delta > 0$ est une constante. Malheureusement, la présence de ce facteur logarithmique ne nous permet pas de conclure en dehors des cas ci-dessus.

D'autre part, dans le cas $G = \mathbb{Z}(p)^{\mathbb{N}}$ avec p premier, on sait, d'après [17], que tout ensemble de Sidon dans Γ est réunion d'un nombre fini d'ensembles indépendants. Les méthodes du présent article ne permettent apparemment pas de retrouver ce résultat.

4. SUR CERTAINS ESPACES DE MULTIPLICATEURS

Soit q tel que $2 \leq q < \infty$.

On note M_{2,ψ_q} l'espace des opérateurs bornés de $L^2(G)$ dans $L^{\psi_q}(G)$ qui commutent avec les translations de G. On munit cet espace de la norme naturelle

$$\|T\|_{M_{2,\psi_q}} = \sup\{\|Tx\|_{\psi_q} \mid \|x\|_2 \leq 1\}.$$

Il est bien connu que l'on peut construire un prédual de M_{2,ψ_q} de la manière suivante:

on note A_{2,φ_q} l'espace formé des fonctions f de $L^2(G)$ de la forme $f = \sum_{n=1}^{\infty} h_n * k_n$ avec $h_n \in L(G)$, $k_n \in L^{\varphi_q}(G)$ et $\sum \|h_n\|_2 \|k_n\|_{\varphi_q} < \infty$. On peut munir cet espace de la norme naturelle:

$$\|f\|_{A_{2,\varphi_q}} = \inf\left\{\sum_{n=1}^{\infty} \|h_n\|_2 \|k_n\|_{\varphi_q}\right\}$$

où l'infimum porte sur toutes les représentations possibles de f. On voit facilement que si $T \in M_{2,\psi_q}$ et $f \in A_{2,\varphi_q}$ alors $T(f) \in C(G)$, ce qui permet de mettre ces deux espaces en dualité en posant

$$\langle T, f \rangle = T(f)(0).$$

On démontre, par un argument classique (voir, e.g. [8, Theorem 10.2.16] dans le cas des espaces L^p) que M_{2,ψ_q} s'identifie au dual de A_{2,φ_q}, la dualité étant définie comme ci-dessus. Précisément, si l'on pose

$$\|T\|_{A'_{2,\varphi_q}} = \sup\{|\langle T, f \rangle| \mid \|f\|_{A_{2,\varphi_q}} \leq 1\},$$

alors les inégalités (0.1) entraînent

$$\forall T \in M_{2,\psi_q} \qquad \Delta_q^{-1}\|T\|_{M_{2,\psi_q}} \leq \|T\|_{A'_{2,\varphi}} \leq \Delta_q \|T\|_{M_{2,\psi_q}}. \tag{4.0}$$

L'espace A_{2,φ_2} a été étudié en détail dans [24] (cf. aussi [19]), où l'on démontre que $f \in A_{2,\varphi_2}$ si et seulement si $J_2(f) < \infty$.

Nous allons faire une étude comparable pour A_{2,φ_q} pour $q > 2$. Soit donc $f \in L^2(G)$. On pose (avec les notations de la section 2)

$$E_q(f) = \int_0^{\infty} (\operatorname{Log} N_2(\varepsilon, f))^{1/q} \, d\varepsilon.$$

On note que, si $q = 2$, $E_2(f)$ coïncide avec l'intégrale $J_2(f)$ définie à la section 2.

Rappelons brièvement la définition de l'espace d'interpolation obtenu par la méthode de Lions–Peetre (cf. [15, 29]). Soit (A_0, A_1) un couple d'interpolation. On pose, pour $0 < t < \infty$, et pour $a \in A_0 + A_1$:

$$K_t(a; A_0, A_1) = \inf\{\|a_0\|_{A_0} + t\|a_1\|_{A_1} \mid a_0 \in A_0,\, a_1 \in A_1,\, a = a_0 + a_1\}.$$

Soient θ et q tels que $0 < \theta < 1$ et $1 \leq q \leq \infty$. On définit l'espace $[A_0, A_1]_{\theta,q}$ comme l'espace formé des éléments a de $A_0 + A_1$ tels que:

$$\|a\|_{[A_0,A_1]_{\theta,q}} = \left(\int_0^\infty (t^{-\theta} K_t(a; A_0, A_1))^q \frac{dt}{t} \right)^{1/q} < \infty,$$

(avec la convention usuelle pour $q = \infty$).

Muni de cette norme, l'espace $[A_0, A_1]_{\theta,q}$ est un espace de Banach. Bennett [2, Corollaire B] a démontré que si $0 < \alpha_0 < \alpha_1$ et si $\alpha_\theta^{-1} = (1 - \theta)/\alpha_0 + \theta/\alpha_1$, alors

$$[L^{\varphi_{\alpha_0}}, L^{\varphi_{\alpha_1}}]_{\theta,1} = L^{\varphi_{\alpha_\theta}}$$

avec équivalence des normes correspondantes.

Nous pouvons maintenant énoncer le principal résultat de ce paragraphe:

THÉORÈME 4.1. *Supposons que $2 < q < \infty$; soit $\theta = 2/q$.*

(i) *L'espace A_{2,φ_q} peut être identifié (avec équivalence des normes correspondantes) avec l'espace $[L^2(G), A_{2,\varphi_2}]_{\theta,1}$.*

(ii) *De plus, $f \in A_{2,\varphi_q}$ si et seulement si il existe une suite f_n dans $L^2(G)$ telle que $f = \sum_{n=1}^\infty f_n$ et $\sum_{n=1}^\infty E_q(f_n) < \infty$.*

(iii) *Par dualité à partir de (i), on a aussi l'identification $M_{2,\psi_q} = [L^2(G), M_{2,\psi_2}]_{\theta,\infty}$ avec équivalence des normes correspondantes.*

Rappelons tout d'abord un résultat de [24] (cf. [19, Chapitre 6]).

Soit $f \in L^2(G)$, la fonction f est dans A_{2,φ_2} si et seulement si $E_2(f) < \infty$ et il existe une constante D telle que

$$D^{-1}(E_2(f) + |\hat{f}(0)|) \leq \|f\|_{A_{2,\varphi_2}} \leq D(E_2(f) + |\hat{f}(0)|). \tag{4.1}$$

Ce résultat montre en particulier que si l'on se restreint aux fonctions f telles que $\hat{f}(0) = 0$, la fonctionnelle positivement homogène $f \to E_2(f)$ est équivalente à une norme. Nous ignorons si cela se généralise à $E_q(f)$ pour $q > 2$, ce qui explique la formulation de (ii) dans le théorème 4.1. [Voir la note (2) rajoutée sur les épreuves.]

Pour la démonstration, nous aurons besoin du lemme suivant:

LEMME 4.2. *Soit $f \in L^2(G)$ tel que $E_q(f) < \infty$ $(0 < q < \infty)$, alors $f \in A_{2,\varphi_q}$ et l'on a*

$$\|f\|_{A_{2,\varphi_q}} \leq \beta_q\{|\hat{f}(0)| + E_q(f)\} \tag{4.2}$$

où β_q est une constante ne dépendant que de q.

Démonstration. La démonstration est la même que celle donnée pour $q = 2$ dans [24] (cf. [19, Chapitre 6]). Pour la commodité du lecteur, nous en donnons le principe:

On peut remarquer qu'il suffit de démontrer (4.2) pour $f \in \mathscr{P}(G)$. D'autre part, d'après (4.0), $M_{2,\psi_q} \approx A'_{2,\varphi_q}$ et il suffit de montrer:

$$\forall f \in \mathscr{P}(G), \quad \forall T \in M_{2,\psi_q} \qquad |T(f)(0)| \leq \beta_q \Delta_q^{-1}\{|\hat{f}(0)| + E_q(f)\}. \tag{4.3}$$

Or, l'inégalité (4.3) résulte du lemme 2.6, en effet si l'on pose $g = T(f)$, on a

$$\forall t, s \in G \qquad \|g_t - g_s\|_{\psi_q} \leq \|T\|_{M_{2,\psi_q}} \|f_t - f_s\|_2$$

d'où (cf. lemme 2.6):

$$\sup_{x,\, y \in G} |g(x) - g(y)| \leq D_q \|T\|_{M_{2,\psi_q}} E_q(f).$$

D'autre part, on a trivialement:

$$|g(0) - \hat{g}(0)| = \left| g(0) - \int g(y)\, dm(y) \right| \leq \sup_{x,\, y \in G} |g(x) - g(y)|$$

d'où

$$|T(f)(0)| = |g(0)| \leq D_q \|T\|_{M_{2,\psi_q}} E_q(f) + |\hat{T}(0)\hat{f}(0)|$$

et comme

$$|\hat{T}(0)| \leq \|T\|_{M_{2,\psi_q}},$$

(4.3) en résulte.

Le lemme 4.2 entraîne le résultat suivant qui est le point crucial de la démonstration:

LEMME 4.3. *Soit $2 < q < \infty$ et soit $f \in A_{2,\varphi_2}$. On a:*

$$\|f\|_{A_{2,\varphi_q}} \leq R_\theta \|f\|_2^{1-\theta} \|f\|_{A_{2,\varphi_2}}^\theta, \tag{4.4}$$

où $\theta = 2/q$ et R_θ est une constante ne dépendant que de θ.

Démonstration. Il est clair que l'on peut (sans restreindre la généralité) supposer $\hat{f}(0) = 0$. On a alors, d'après (4.1) et puisque $\operatorname{Log} N_2(\varepsilon, f) = 0$ si $\varepsilon > 2\|f\|_2$:

$$\begin{aligned}
\|f\|_{A_{2,\varphi_q}} &\leq \beta_q E_q(f) = \beta_q \int_0^{2\|f\|_2} (\operatorname{Log} N_2(\varepsilon, f))^{\theta/2}\, d\varepsilon \\
&\leq \beta_q (2\|f\|_2)^{1-\theta} \left\{ \int_0^{2\|f\|_2} (\operatorname{Log} N_2(\varepsilon, f))^{1/2}\, d\varepsilon \right\}^\theta \\
&\leq \beta_q (2\|f\|_2)^{1-\theta} (E_2(f))^\theta
\end{aligned}$$

soit, d'après (4.1)

$$\leq \beta_q (2\|f\|_2)^{1-\theta} (D\|f\|_{A_{2,\varphi_2}})^\theta,$$

ce qui établit (4.4).

L'inclusion $A_{2,\varphi_q} \subset [L^2, A_{2,\varphi_2}]_{\theta,1}$ est en fait la conséquence d'un théorème général bien connu sur l'interpolation entre espaces de multiplicateurs. Nous donnons brièvement la démonstration pour la commodité du lecteur.

PROPOSITION 4.4. *On a* $A_{2,\varphi_q} \subset [L^2, A_{2,\varphi_2}]_{\theta,1}$ *et* $[L^2, M_{2,\psi_q}]_{\theta,\infty} \subset M_{2,\psi_q}$, *avec les inégalités associées, où* $\theta = 2/q$.

Démonstration: Il suffit de montrer la première inclusion, la seconde en résulte par un théorème général de dualité [29, Section 1.11.2]. Pour démontrer que $A_{2,\varphi_q} \subset [L^2, A_{2,\varphi_2}]_{\theta,1}$, il suffit évidemment de considérer les éléments de A_{2,φ_q} de la forme $f = h * k$ avec $h \in L^2(G)$ et $k \in L^{\varphi_q}(G)$. D'après un résultat (déjà cité) de [2], on a

$$L^{\varphi_q} = [L^1, L^{\varphi_2}]_{\theta,1}$$

avec équivalence des normes correspondantes. Il existe donc une constante C'_q telle que:

$$\forall k \in L^{\varphi_q} \qquad \int_0^\infty t^{-\theta} K_t(k; L^1, L^{\varphi_2}) \frac{dt}{t} \leq C'_q \|k\|_{\varphi_q}. \tag{4.5}$$

On peut remarquer que

$$K_t(h * k; L^2, A_{2,\varphi_2}) \leq \|h\|_2 K_t(k; L^1, L^{\varphi_2}). \tag{4.6}$$

En effet, si $K_t(k; L^1, L^{\varphi_2}) < 1$, alors il existe $k_0 \in L^1$, $k_1 \in L^{\varphi_2}$ tels que $k = k_0 + k_1$ et

$$\|k_0\|_{L^1} + t\|k_1\|_{\varphi_2} < 1.$$

On a donc $h * k = h * k_0 + h * k_1$ et

$$\|h * k_0\|_2 + t\|h * k_1\|_{A_{2,\varphi_2}} < \|h\|_2$$

ce que établit (4.6) par homogénéité.

On obtient donc, à l'aide de (4.5) et (4.6);

$$\|h * k\|_{[L^2, A_{2,\varphi_2}]_{\theta,1}} \leq C'_q \|h\|_2 \|k\|_{\varphi_q},$$

d'où l'on déduit immédiatement l'inclusion annoncée $A_{2,\varphi_q} \subset [L^2, A_{2,\varphi_2}]_{\theta,1}$.

Remarque. La proposition 4.4 est un cas particulier dun phénomène très général [19, Section 1.19.5].

Remarque. Le lecteur aura sans doute noté que l'on a en fait montré ci-dessus l'inégalité

$$K_t(h * k; L^2, A_{2,\varphi_2}) \leq \|h\|_2 K_t(k; \mathscr{F} l^\infty(\Gamma), L^{\varphi_2})$$

qui est a priori meilleure que (4.6).

A posteriori, il en résulte que, si l'on pose $I_q = [\mathscr{F} c_0(\Gamma), L^{\varphi_2}]_{\theta,1}$ alors il revient au même de dire que $f \in A_{2,\varphi_q}$ ou de dire $\exists h_n \in L^2$, $k_n \in I_q$ tels que $f = \sum h_n * k_n$ et $\sum \|h_n\|_2 \|k_n\|_{I_q} < \infty$.

Par dualité, on peut identifier M_{2,ψ_q} avec l'espace des opérateurs bornés de L^2 dans $(I_q)' = [\mathscr{F} l^1(\Gamma), L^{\psi_2}]_{\theta,\infty}$ qui commutent aux translations. Ce phénomène mériterait sans doute d'être éclairci.

Démonstration du théorème 4.1: D'après un théorème bien connu [29, Section 1.10.1] l'inégalité (4.4) implique que

$$[L^2, A_{2,\varphi_2}]_{\theta,1} \subset A_{2,\varphi_q}.$$

L'inclusion inverse résulte de la proposition 4.4. On a donc démontré (i). (Il est clair que l'on a démontré en même temps l'équivalence des normes correspondantes.)

Montrons (ii): Notons provisoirement $\mathscr{A}_q$ l'espace des fonctions f de L^2 qui s'écrivent

$$f = \sum_{n=1}^{\infty} f_n \qquad \text{avec} \qquad \sum_{n=1}^{\infty} E_q(f) < \infty.$$

On peut munir cet espace de la norme

$$\mathscr{E}_q(f) = |\hat{f}(0)| + \inf\left\{\sum_1^{\infty} E_q(f_n)\right\}$$

où l'infimum porte sur toutes les représentations de f.

Il est clair que $\mathscr{A}_q$, muni de la norme $\mathscr{E}_q(\cdot)$, est un espace de Banach. D'après le lemme 4.2, on a $\mathscr{A}_q \subset A_{2,\varphi_q}$ et

$$\forall f \in \mathscr{A}_q \qquad \|f\|_{A_{2,\varphi_q}} \leq \beta'_q \mathscr{E}_q(f) \tag{4.7}$$

où β'_q est une constante.

Inversement, on a (cf. la démonstration du lemme 4.3):

$$\begin{aligned}\mathscr{E}_q(f) &\leq |\hat{f}(0)| + E_q(f) \leq |\hat{f}(0)| + (2\|f\|_2)^{1-\theta}(E_2(f))^{\theta} \\ &\leq |\hat{f}(0)| + 2^{1-\theta}D^{\theta}\|f\|_2^{1-\theta}\|f\|_{A_{2,\varphi_2}}^{\theta} \leq (1 + 2^{1-\theta}D^{\theta})\|f\|_2^{1-\theta}\|f\|_{A_{2,\varphi_2}}^{\theta}.\end{aligned}$$

D'après un résultat classique [29, Section 1.10.1] il en résulte, puisque $\mathscr{E}_q$ est une norme, que l'on a:

$$\mathscr{E}_q(f) \leq C_q'''\|f\|_{[L^2, A_{2,\varphi_2}]_{\theta,1}} \tag{4.8}$$

où C_q''' est une constante. [Voir (2) à la fin de cet article.]

Finalement, on déduit de (4.7) et (4.8) que $\mathscr{A}_q$ coïncide avec $[L^2, A_{2,\varphi_2}]_{\theta,1}$ et que les normes de ces espaces sont équivalentes.

Remarque. Il est naturel d'identifier (par convention) L^{ψ_∞} avec L^∞ et L^{φ_∞} avec L^1; avec cette convention, on peut identifier M_{2,ψ_∞} et A_{2,φ_∞} avec L^2, de sorte que les espaces M_{2,ψ_q}, $2 \leq q \leq \infty$, forment une échelle d'interpolation. On notera que, par le théorème de réitération [29, Section 1.10.2] on a: Si $2 \leq q_0 < q < q_1 \leq \infty$

$$M_{2,\psi_q} = [M_{2,\psi_{q_0}}, M_{2,\psi_{q_1}}]_{\theta,\infty}$$

avec $q^{-1} = (1-\theta)/q_0 + \theta/q_1$.

Il est facile de voir (considérer $M_{2,\psi_q\Lambda}$ où Λ est un ensemble de Sidon) que le résultat précédent n'est plus valable si $q_0 < 2$.

Le lecteur aura remarqué que le point crucial de la démonstration précédente est l'inégalité (4.4). Curieusement, nous ne connaissons pas de démonstration directe de cette inégalité. En particulier, nous ignorons si l'analogue de cette inégalité est valable pour des groupes non compacts et en particulier pour $G = \mathbb{R}$. De même, nous ignorons si l'on a encore $[L^2(G), M_{2,\psi_2}]_{\theta,\infty} = M_{2,\psi_q}$ $(\theta = 2/q < 1)$ dans le cas où $G = \mathbb{R}$.

Dans les assertions équivalentes du théoreme 2.3, il est important de considérer *toutes* les parties finies de Λ. Néanmoins, le résultat suivant subsiste si l'on fixe une partie A:

PROPOSITION 4.5. *Soit $A \subset \Gamma$ une partie finie de Γ. Soient p, q tels que* $1 < p \leq 2 \leq q < \infty$ et $p^{-1} + q^{-1} = 1$.

(a) *Supposons que*

$$\left\| \sum_{\gamma \in A} \gamma \right\|_{\psi_q} \leq C|A|^{1/p}, \tag{4.9}$$

alors, il existe $\delta = \delta(C, q) > 0$ tel que:

$$\mathbb{E}\left\| \sum_{\gamma \in A} \varepsilon_\gamma \gamma \right\|_{C(G)} \geq \delta|A|. \tag{4.10}$$

(b) *Supposons que A vérifie* (4.10) *pour un $\delta > 0$, alors il existe $B \subset A$ avec $|B| \geq \alpha\delta|A|$ et $S(B) \leq K_1$, où $K_1 = K_1(\delta)$ est une constante ne dépendant que de δ et où $\alpha > 0$ est une constante numérique.*

La démonstration utilisera le lemme élémentaire suivant:

LEMME 4.6. *Soit $q > 0$, il existe une constante γ_q telle que, si f et g sont dans L^{ψ_q} et vérifient*

$$\forall \gamma \in \Gamma \qquad |\hat{f}(\gamma)| \leq \hat{g}(\gamma),$$

alors on a

$$\|f\|_{\psi_q} \leq \gamma_q \|g\|_{\psi_q}.$$

Démonstration. Il est bien connu qu'une fonction f est dans L^{ψ_q} ssi $\|f\|_p \in O(p^{1/q})$ quand $p \to \infty$. (Pour le voir, on développe $\exp x^q$ en série de Taylor et on utilise la formule de Stirling).

Par conséquent, on voit facilement qu'il existe une constante γ'_q telle que

$$\forall f \in \mathscr{P}(G) \qquad \frac{1}{\gamma'_q}\|f\|_{\psi_q} \leq \sup_{n \in \mathbb{N}_+} n^{-1/q}\|f\|_{2n} \leq \gamma'_q\|f\|_{\psi_q}.$$

Or, il est bien connu que l'on a pour tout entier $n \geq 1$: si $|\hat{f}(\gamma)| \leq \hat{g}(\gamma)$, $\forall \gamma \in \Gamma$, alors $\|f\|_{2n} \leq \|g\|_{2n}$.

(Pour plus de détails sur cette question, voir [1].) On en déduit donc immédiatement, sous la même hypothèse:

$$\|f\|_{\psi_q} \leq (\gamma'_q)^2\|g\|_{\psi_q}. \quad \blacksquare$$

Démonstration de la proposition 4.5. Montrons d'abord (a). Si (4.9) est vérifié, alors le lemme 4.6 montre que l'on a, quels que soient les coefficients complexes (α_γ):

$$\left\|\sum_{\gamma \in A} \alpha_\gamma \gamma\right\|_{\psi_q} \leq C'|A|^{1/p} \sup_{\gamma \in A} |\alpha_\gamma| \tag{4.11}$$

avec $C' = \gamma_q C$.

D'autre part, on a trivialement:

$$\left\|\sum_{\gamma \in A} \alpha_\gamma \gamma\right\|_{\infty} \leq \sum_{\gamma \in A} |\alpha_\gamma|. \tag{4.12}$$

Par un résultat classique sur l'interpolation complexe, on a: $[L^{\psi_q}, L^\infty]_{1/2} \subset L^{\psi_{2q}}$ avec égalité des normes correspondantes [3, Section 13.6]. En interpolant entre (4.11) et (4.12), on trouve donc:

$$\left\|\sum_{\gamma \in A} \alpha_\gamma \gamma\right\|_{\psi_{2q}} \leq (C')^{1/2}|A|^{1/2p}\left(\sum_{\gamma \in A} |\alpha_\gamma|^2\right)^{1/2},$$

d'où

$$\left\|\sum_{\gamma \in A} \alpha_\gamma \gamma\right\|_{M_{2,\psi_{2q}}} \leq (C')^{1/2}|A|^{1/2p} \sup_{A} |\alpha_\gamma|.$$

Cette dernière inégalité entraîne par dualité:

$$\forall (\alpha_\gamma) \in \mathbb{C}^A \qquad \sum_{\gamma \in A} |\alpha_\gamma| \leq \Delta_{2q}(C')^{1/2}|A|^{1/2p}\left\|\sum_{\gamma \in A} \alpha_\gamma \gamma\right\|_{A_{2,\varphi_{2q}}}.$$

Posons $f = \sum_{\gamma \in A} \alpha_\gamma \gamma$; on a, d'après (4.4),

$$\sum_{\gamma \in A} |\hat{f}(\gamma)| \leq R_\theta \Delta_{2q}(C')^{1/2} |A|^{1/2p} (\|f\|_2^{1-\theta} \|f\|_{A_{2,\varphi_2}}^\theta)$$

avec $\theta = 1/q$.

On en déduit donc

$$\delta' |A|^{-q/2p} \left(\sum_{\gamma \in A} |\hat{f}(\gamma)| \right)^q \Big/ \|f\|_2^{q-1} \leq \|f\|_{A_{2,\varphi_2}} \tag{4.13}$$

avec $\delta' = \{\Delta_{2q}(C')^{1/2} R_{1/q}\}^{-q}$.

Dans le cas particulier où $\alpha_\gamma = 1$ pour $\gamma \in A$, on trouve:

$$\delta' |A| \leq \left\| \sum_{\gamma \in A} \gamma \right\|_{A_{2,\varphi_2}}. \tag{4.14}$$

D'après [19] (théorèmes 6.1. et 1.1.5), on a

$$\|f\|_{A_{2,\varphi_2}} \leq \lambda \mathbb{E} \left\| \sum_{\gamma \in A} \varepsilon_\gamma \hat{f}(\gamma) \gamma \right\|_{C(G)}, \tag{4.15}$$

où λ est une constante numérique. On déduit donc de (4.14)

$$\mathbb{E} \left\| \sum_{\gamma \in A} \varepsilon_\gamma \gamma \right\|_{C(G)} \geq \delta' \lambda^{-1} |A|,$$

ce qui montre que (4.9) entraine (4.10).

Passons à la seconde partie. Supposons maintenant que l'on a (4.10). Alors, d'après le lemme 2.8

$$\delta |A| \leq R \left\| \sum_{\gamma \in A} \gamma \right\|_{A_{2,\varphi_2}};$$

par conséquent, on a [cf. (4.0)]:

$$\delta |A| \leq R \Delta_2 \sup \left\{ \left| \sum_{\gamma \in A} \hat{T}(\gamma) \right| \|T\|_{M_{2,\psi_2}} \leq 1 \right\}.$$

Posons $\delta'' = \delta (R \Delta_2)^{-1}$. Soit T tel que $\|T\|_{M_{2,\psi_2}} \leq 1$ et

$$\delta'' |A| \leq \left| \sum_{\gamma \in A} \hat{T}(\gamma) \right|.$$

On a

$$\sup_{\gamma \in A} |\hat{T}(\gamma)| \leq \|T\|_{M_{2,\psi_2}},$$

donc si

$$B = \{\gamma \in A \mid |\hat{T}(\gamma)| \geq \delta''/2\}$$

on a

$$\delta''|A| \leq \delta''/2|A - B| + |B|$$

d'où

$$|B| \geq \tfrac{1}{2}\delta''|A|.$$

D'autre part, puisque $|\hat{T}(\gamma)| \geq \delta''/2$, $\forall \gamma \in B$, on a

$$\frac{\delta''}{2}\left\|\sum_{\gamma \in B} \gamma\right\|_{M_{2,\psi_2}} \leq \left\|\sum_{\gamma \in B} \hat{T}(\gamma)\gamma\right\|_{M_{2,\psi_2}} \leq \|T\|_{M_{2,\psi_2}} \leq 1,$$

par conséquent, $\forall f \in \mathscr{P}_B(G)$, on a

$$\|f\|_{\psi_2} \leq (2/\delta'')\|f\|_2,$$

ce qui implique, d'après le principal résultat de [22], que $S(B) \leq K_1$ où K_1 est une constante ne dépendant que de δ'' (et donc que δ). ■

5. Extension au cas non commutatif

Soit G un groupe compact non nécessairement abélien. On note encore m la mesure de Haar normalisée de G. On note Σ l'"objet dual" de G, c'est-à-dire l'ensemble des classes d'équivalence des représentations irréductibles de G; puisque G est compact, ces dernières sont de dimension finie. Pour chaque i dans Σ, on note U_i une représentation irréductible de G (sur un espace de Hilbert H_i), appartenant à la classe définie par i. On note d_i la dimension de H_i. D'après le théorème de Peter–Weyl [9, Section 27], toute fonction $f \in L^2(G)$ admet un développement (analogue aux séries de Fourier) de la forme

$$f(x) = \sum_{i \in \Sigma} d_i \operatorname{tr} U_i(x)\hat{f}(i)$$

où la série converge dans $L^2(G)$ avec

$$\|f\|_2 = \left(\sum_{i \in \Sigma} d_i \operatorname{tr}|\hat{f}(i)|^2\right)^{1/2},$$

et où le "coefficient" $\hat{f}(i)$ de f est l'opérateur linéaire sur H_i défini par

$$\hat{f}(i) = \int U_i(-x)f(x)\,dm(x).$$

On pose $\chi_i(x) = \operatorname{tr} U_i(x)$; χ_i est appelé le "caractère" de la représentation U_i. On notera $\mathscr{P}(G)$ l'espace des fonctions $f \in L^2(G)$ telles que l'ensemble $\{i \in \Sigma \mid \hat{f}(i) \neq 0\}$ est fini.

DÉFINITION 5.1. On dit qu'une partie $\Lambda \subset \Sigma$ est de Sidon s'il existe une constante C telle que

$$\forall f \in \mathscr{P}_\Lambda(G) \qquad \Sigma\, d_i |\hat{f}(i)| \leq C \|f\|_{C(G)}. \tag{5.1}$$

On note $S(\Lambda)$ la plus petite constante C possédant cette propriété.

Pour plus de détails sur toutes ces questions, voir [9].

Les techniques des articles [22–24] utilisées dans la Section 2 sont présentées en détail dans le cadre non abélien dans [19, Chapitres 5 et 6], où la théorie des séries de Fourier aléatoires "non commutatives" est développée. Dans ce cadre, les variables de Bernoulli sont définies de la manière suivante: on note $(\varepsilon_i)_{i\in\Sigma}$ une suite de variables aléatoires indépendantes, telles que pour chaque i dans Σ, ε_i prend ses valeurs dans le groupe orthogonal $O(d_i)$ et est uniformément distribuée sur ce groupe [i.e., ε_i a pour loi la mesure de Haar normalisée sur $O(d_i)$]. A priori, ε_i est une matrice aléatoire, néanmoins, si l'on suppose donnée une base orthonormale de H_i, on pourra aussi identifier ε_i avec l'opérateur aléatoire sur H_i correspondant à cette matrice.

Nous allons tout d'abord généraliser (en partie) les propositions 2.1 et 2.2. Pour cela, nous devons préciser plusieurs points: pour tout opérateur compact A sur un espace de Hilbert H, on note $\lambda_n(|A|)$ la suite des valeurs propres de $|A| = (A^*A)^{1/2}$ rangées en ordre décroissant et répétées suivant leur multiplicité. On note $C_p(H)$ l'espace des opérateurs compact $A: H \to H$ tels que $\operatorname{tr}|A|^p < \infty$. Muni de la norme $\|A\|_p = (\operatorname{tr}|A|^p)^{1/p}$, cet espace est un espace de Banach pour tout $p \geq 1$. On notera que $\|A\|_p = (\sum_n \lambda_n(|A|)^p)^{1/p}$. L'espace $C_2(H)$ n'est autre que l'espace (hibertien) des opérateurs de Hilbert–Schmidt sur H. On note $C_{p,\infty}(H)$ l'espace des opérateurs compacts A tels que $\|A\|_{p,\infty} = \sup_{n\geq 1} n^{1/p}\lambda_n(|A|)$. Il est bien connu que la quasi-norme $\|\ \|_{p,\infty}$ est équivalente à une norme sur $C_{p,\infty}(H)$ pour tout $p > 1$.

Soit f dans $L^2(G)$. On notera $\alpha_n^*(f)$ la suite obtenue en réarrangeant en ordre décroissant l'ensemble des nombres

$$\{\lambda_j(|\hat{f}(i)|) \mid 1 \leq j \leq d_i,\ i \in \Sigma\}$$

en ayant pris soin de répéter chacun des nombres $\lambda_j(|\hat{f}(i)|)\, d_i$ fois. Avec cette convention, on a

$$\forall p \geq 1 \qquad \sum_n \alpha_n^*(f)^p = \sum_{i\in\Sigma} d_i \operatorname{tr}|\hat{f}(i)|^p.$$

Soit $\mathscr{H}$ la somme directe hilbertienne suivante: $\mathscr{H} = \bigoplus_{i\in\Sigma} c_2(\mathscr{H}_i)$. Le lecteur vérifiera aisément que si l'on note T_f l'opérateur de $\mathscr{H}$ dans lui-même défini par $T_f((A_i)_{i\in\Sigma}) = (\hat{f}(i)A_i)_{i\in\Sigma}$, alors on a $\alpha_n^*(f) = \lambda_n(|T_f|)$. Pour tout f dans $\mathscr{P}(G)$, on pose

$$\|f\|_{X^p} = \left(\sum \alpha_n^*(f)^p\right)^{1/p} = \|T_f\|_p,$$

et on note X^p l'espace obtenu en complétant $\mathscr{P}(G)$ muni de cette norme. Soit $X^{p,\infty}$ l'espace formé des $f \in X^\infty$ tels que

$$\|f\|_{X^{p,\infty}} = \sup_{n \geq 1} n^{1/p} \alpha_n^*(f) = \|T_f\|_{p,\infty} < \infty.$$

On munit cet espace de la quasi norme correspondante.

Les espaces X^p et $X^{p,\infty}$ s'identifient isométriquement (par la correspondance $f \to T_f$) à des sous-espaces notés Λ_p et $\Lambda_{p,\infty}$ des espaces $C_p(\mathscr{H})$ et $C_{p,\infty}(\mathscr{H})$; on vérifie sans difficulté que la projection orthogonale de $C_2(\mathscr{H})$ sur Λ_2 est en fait bornée sur $C_p(\mathscr{H})$ [resp., $C_{p,\infty}(\mathscr{H})$] pour tout $p \geq 1$ (resp., $p > 1$).

Il en résulte, par des arguments standard (cf. [29, Sections 1.19.7 et 1.17.1]) que l'on a

$$[X^2, X^1]_{\theta,\infty} = X^{p,\infty} \tag{5.2}$$

avec $(1-\theta)/2 + (\theta/1) = p^{-1}$, et les normes correspondantes sont équivalentes. On sait d'autre part [2] que

$$[L^{\psi_2}, L^\infty]_{\theta,\infty} = L^{\psi_q}. \tag{5.3}$$

On en déduit donc immédiatement

PROPOSITION 5.2. *Soient $1 < p < 2 < q$, avec $p^{-1} + q^{-1} = 1$. Soit $\Lambda \subset \Sigma$ un ensemble de Sidon. Alors, on a $X_\Lambda^{p,\infty} \subset L^{\psi_q}(G)$; de plus, il existe une constante B'_p telle que:*

$$\forall f \in X_\Lambda^{p,\infty} \qquad \|f\|_{\psi_q} \leq B'_p S(\Lambda) \|f\|_{X^{p,\infty}}.$$

J' ignore si l'inclusion et l'inégalité inverses sont vraies.

Démonstration. D'après un résultat bien connu dû à Figa–Talamanca et Rider (voir [9, Théorème 36.2]), on a

$$X_\Lambda^2 \subset L^{\psi_2}. \tag{5.4)'$$

D'autre part, on a trivialement $X^1 \subset L^\infty$, donc:

$$X_\Lambda^1 \subset L^\infty. \tag{5.4)''$$

De plus, (5.2) implique évidemment

$$[X_\Lambda^2, X_\Lambda^1]_{\theta,\infty} = X_\Lambda^{p,\infty}. \tag{5.5}$$

Il suffit donc pour démontrer la proposition 5.2 d'interpoler entre (5.4)' et (5.4)'' en utilisant (5.3) et (5.5).

Les remarques données à la fin du paragraphe précédent ont une généralisation particulièrement intéressante dans le cas non abélien.

PROPOSITION 5.3. *Soit $A \subset \Sigma$ une partie finie de Σ.*

(a) *Supposons qu'il existe $1 < p \leq 2$ et C tels que*

$$\left\| \sum_{i \in A} d_i \chi_i \right\|_{\psi_q} \leq C \left(\sum_{i \in A} d_i^2 \right)^{1/p}, \qquad \frac{1}{p} + \frac{1}{q} = 1. \tag{5.6}$$

Alors, il existe $\delta = \delta(C, p) > 0$ tel que

$$\mathbb{E} \left\| \sum_{i \in A} d_i \operatorname{tr} U_i \varepsilon_i \right\|_{C(G)} \geq \delta \sum_{i \in A} d_i^2. \tag{5.7}$$

(b) *Supposons maintenant que A vérifie* (5.7); *alors, il existe une constante numérique $\alpha > 0$, une constante $C = C(\delta)$, et un ensemble $B \subset A$ tel que $\sum_{i \in B} d_i^2 \geq \alpha\delta \sum_{i \in A} d_i^2$* et *$S(B) \leq C$.*

Démonstration. La démonstration de (a) est tout-à-fait analogue à celle donnée à la fin de la Section 4. Signalons seulement que l'on utilise l'identité $[X^\infty, X^1]_{1/2} = X_2$ [29, Section 1.19.7] et que l'analogue non abélien du lemme 4.6 résulte du théorème 4.1 de [20].

Pour démontrer (b), si l'on procède comme dans le cas abélien, on obtient $B \subset A$ tel que $\sum_{i \in B} d_i^2 \geq \alpha\delta \sum_{i \in A} d_i^2$ et T dans l'espace M_{2,ψ_2} (défini dans [19, Chapitre 6]) tel que $\|T\|_{M_{2,\psi_2}} \leq 1$ et

$$\forall i \in B \qquad \operatorname{tr} \hat{T}(i) \geq \alpha\delta d_i$$

ou $\alpha > 0$ est une constante numérique.

Puisque la projection centrale $T \to \sum_{i \in \Sigma}(\operatorname{tr} \hat{T}(i))\chi_i$ est un opérateur de norme ≤ 1 sur M_{2,ψ_2}, il en résulte que $\alpha\delta \|\sum_{i \in B} d_i \chi_i\|_{M_{2,\psi_2}} \leq \|T\|_{M_{2,\psi_2}} \leq 1$, ce qui permet de conclure exactement comme dans le cas abélien que $S(B) \leq C(\delta)$.

Remarque. Il est bon de souligner que la proposition précédente est non triviale même si $|A| = 1$, contrairement au cas abélien. En effet, on peut prendre dans ce cas $B = A$, explicitons-le:

COROLLAIRE 5.4. *Soit $i \in \Sigma$ fixé. Si $\|\chi_i\|_{\psi_2} \leq C$ alors il existe $\delta = \delta(C) > 0$ (indépendant de la dimension d_i) tel que, quel que soit l'opérateur A sur H_i, on a*

$$\sup_{x \in G} |\operatorname{tr} U_i(x) A| \geq \delta \operatorname{tr} |A|. \tag{5.8}$$

Inversement, on sait (*cf.* [9, Théorème 36.2]) *que si U_i vérifie* (5.8), *alors on a* (5.9) *$\|\chi_i\|_{\psi_2} \leq C'$ où $C' = C'(\delta)$ est une constante indépendante de la dimension de U_i.*

Les inégalités (5.8) et (5.9) sont donc essentiellement équivalentes. Bien entendu, *ce résultat ne subsiste pas si U_i n'est pas supposée irréductible.* Il est clair que les autres résultats de la Section 4 s'étendent sans difficulté au cas non abélien. Enoncons pour finir l'analogue du théorème 2.3:

THÉORÈME 5.5. *Soit Λ une partie de Σ. Les propriétés suivantes sont équivalentes*:

(i) *Λ est un ensemble de Sidon.*

(ii) *Il existe p, q avec $1 < p \leq 2 \leq q$, $p^{-1} + q^{-1} = 1$, et une constante C tels que toute partie finie A de Λ vérifie* (5.6).

(iii) *Il existe une constante C telle que*:

$$\forall (\alpha_i)_{i \in \Lambda} \in l^2(\Lambda) \qquad \left\| \sum_{i \in \Lambda} \alpha_i \chi_i \right\|_{\psi_2} \leq C \left(\sum |\alpha_i|^2 \right)^{1/2}.$$

(iv) *Il existe $\delta > 0$ tel que toute partie finie A de Λ vérifie* (5.7).

(v) *Il existe $\delta > 0$ et C tels que toute partie finie A de Λ contienne une sous-partie B avec $\sum_{i \in B} d_i^2 \geq \delta \sum_{i \in A} d_i^2$ et $S(B) \leq C$.*

Remarque. L'équivalence (i) $\Leftrightarrow$ (iii) est assez surprenante; en effet, elle montre que les ensembles de Sidon sont caractérisées par le comportement dans $L^{\psi_2}(G)$ des fonctions *centrales* à spectre dans Λ. Par contre, on sait depuis la thèse de W. Parker (voir [16, Section 10.14] et les références qui y sont citées) qu'il existe des ensembles Λ qui sont centralement de Sidon [c'est-à-dire tels que (5.1) est vérifiée seulement par les fonctions centrales à spectre dans Λ] mais qui ne sont pas des ensembles de Sidon.

6. EXTENSION AU CAS NON COMPACT

Dans ce paragraphe G est un groupe localement compact abélien de dual Γ. On note encore m la mesure de Haar. Pour toute partie compacte K de mesure positive, on note m_K la mesure de Haar restreinte à K et normalisée. On suppose de plus que Γ est métrisable.

Nous allons énoncer les généralisations des résultats de la Section 2 au cas des ensembles de Sidon topologiques étudiés dans [5].

DÉFINITION 6.1. Soit $\Lambda = \{\gamma_n | n \in \mathbb{N}\}$ une partie dénombrable de Γ. On dit que Λ est un ensemble de Sidon topologique s'il existe un compact $K \subset G$ et une constante C tels que:

$$\forall n \in \mathbb{N} \quad \forall (\alpha_i) \in \mathbb{C}^n \qquad \sum_1^n |\alpha_i| \leq C \left\| \sum_1^n \alpha_i \gamma_i \right\|_{C(K)}. \tag{6.1}$$

On note $S(B, K)$ la plus petite constante C vérifiant (6.1). On dit que Λ est de pas strictement positif s'il existe $\alpha > 0$ et un compact K tels que

$$\forall \lambda' \neq \lambda'' \in \Lambda \qquad \|\lambda' - \lambda''\|_{C(K)} \geq \alpha.$$

Les résultats de la Section 2 se généralisent sans difficulté à ce nouveau cadre. Nous ne ferons qu'énoncer le résultat, et nous renvoyons le lecteur aux articles [23] et [19] où il trouvera les ingrédients permettant d'adapter la démonstration détaillée à la Section 2.

THÉORÈME 6.2. *Soit $\Lambda = \{\gamma_n | n \in \mathbb{N}\}$ une partie de Γ de pas strictement positif. Les propriétés suivantes sont équivalentes:*

(i) *Λ est un ensemble de Sidon.*

(ii) *Il existe p avec $1 < p \leq 2$, une constante C et un compact $K \subset G$ de mesure positive, tels que, pour toute partie finie A de Λ, on a:*

$$\left\|\sum_{\gamma \in A} \gamma\right\|_{L^{\psi_q}(dm_K)} \leq C|A|^{1/p}, \qquad \frac{1}{p} + \frac{1}{q} = 1.$$

(iii) *Il existe p avec $1 \leq p < 2$, une constante $\delta > 0$ et un compact $K \subset G$ de mesure positive tels que, pour toute partie finie A de Λ, on a*

$$\mathbb{E}\left\|\sum_{\gamma \in A} \varepsilon_\gamma \gamma\right\|_{L^{\psi_q}(dm_K)} \geq \delta|A|^{1/p}, \qquad \frac{1}{p} + \frac{1}{q} = 1.$$

[Dans le cas $p = 1$, on identifie $L^{\psi_q}(dm_K)$ avec $L^\infty(dm_K)$.]

(iv) Il existe $\delta > 0$, un compact $K \subset G$ et une constante C tels que toute partie finie A de Λ contienne un sous-partie $B \subset A$ telle que

$$|B| \geq \delta|A| \qquad \text{et} \qquad S(B, K) \leq C.$$

De plus, si l'une des propriétés équivalentes ci-dessus est vérifiée, alors quel que soit le compact $K \subset G$ de mesure positive, et quel que soient p, q avec $1 < p < 2 < q$ et $p^{-1} + q^{-1} = 1$, il existe des constantes $A_p(K)$ et $B_p(K)$ telles que:

$$\forall (\alpha_i) \in l^2 \qquad A_p(K)^{-1}\|(\alpha_i)\|_{l^{p,\infty}} \leq \left\|\sum_1^\infty \alpha_i \gamma_i\right\|_{L^{\psi_q}(m_K)} \leq B_p(K)\|(\alpha_i)\|_{l^{p,\infty}}.$$

7. SOUS-ENSEMBLES DE SIDON D'UNE PARTIE DONNÉE DE Γ

Les résultats donnés à la fin de la Section 4 permettent de démontrer un théorème général donnant une estimation du cardinal des sous-ensembles de Sidon dans une partie finie A d'un groupe *discret abélien* Γ

THÉORÈME 7.1. *Soit $A \subset \Gamma$ une partie finie de Γ. Posons:*

$$\mu = \mathbb{E}\left\|\sum_{\gamma \in A} \varepsilon_\gamma \gamma\right\|_{C(G)}$$

alors il existe $B \subset A$ tel que

$$|B| \geq \alpha\mu^2|A|^{-1}, \qquad et \qquad S(B) \leq C$$

où $\alpha > 0$ et C sont des constantes numériques.

Remarque. L'ordre de grandeur de $|B|$ dans l'énoncé précédent est en accord avec les résultats généraux de l'article [31].

Démonstration. Il suffit d'après la proposition 4.5 d'exhiber une partie B de A telle que

$$\left\|\sum_{\gamma \in B} \gamma\right\|_{\psi_2} \leq C_1\sqrt{|B|} \qquad \text{et} \qquad |B| \geq \alpha_1\mu^2|A|^{-1}$$

où $\alpha_1 > 0$ et C_1 sont des constantes numériques. On va montrer que si l'on choisit "au hasard" d'une manière convenable l'ensemble B, alors il a les propriétés voulues. Tout d'abord rappelons que l'on a, d'après le lemme 2.8:

$$\forall f \in \mathscr{P}(G) \qquad \mathbb{E}\left\|\sum_{\gamma \in \Gamma} \varepsilon_\gamma \hat{f}(\gamma)\gamma\right\|_{C(G)} \leq R\|f\|_{A_{2,\varphi_2}}$$

donc, d'apprès (4.0)

$$\leq R\Delta_2 \sup\{\textstyle\sum \hat{f}(\gamma)\hat{T}(\gamma) \mid T \in M_{2,\psi_2} \|T\|_{M_{2,\psi_2}} \leq 1\}.$$

On en déduit donc qu'il existe $T \in M_{2,\psi_2}$ tel que $\|T\|_{M_{2,\psi_2}} \leq 1$ et

$$\left|\sum_{\gamma \in A} \hat{T}(\gamma)\right| \geq (R\Delta_2)^{-1}\mu. \tag{7.1}$$

Remarquons que l'on a aussi

$$\|T\|_{M_{2,\psi_2}} = \left\|\sum_{\gamma \in A} |\hat{T}(\gamma)|\gamma\right\|_{M_{2,\psi_2}} \leq 1,$$

par conséquent:

$$\left\|\sum_{\gamma \in A} |\hat{T}(\gamma)|\gamma\right\|_{\psi_2} \leq |A|^{1/2}. \tag{7.2}$$

On pose

$$p_\gamma = |\hat{T}(\gamma)|\left(\sum_{\gamma \in A} |\hat{T}(\gamma)|\right)^{-1} \qquad \forall \gamma \in A.$$

Soit $(\Omega, \mathscr{B}, \mathbb{P})$ un espace de probabilité. Soit $\gamma_1^\omega, \ldots, \gamma_k^\omega$ une suite de k variables aléatoires indépendantes sur $(\Omega, \mathscr{B}, \mathbb{P})$, à valeurs dans A ayant la même distribution déterminée de la manière suivante:

$$\forall j = 1, 2, \ldots, k, \quad \forall \gamma \in A \qquad \mathbb{P}(\{\omega \,|\gamma_j^\omega = \gamma\}) = p_\gamma.$$

on remarque que

$$\forall j \leq k \qquad \mathbb{E}\gamma_j^\omega = \sum_{\gamma \in A} p_\gamma \gamma$$

donc

$$\|\mathbb{E}\gamma_j^\omega\|_{\psi_2} = \left\|\sum_{\gamma \in A} p_\gamma \gamma\right\|_{\psi_2}$$

soit, d'après (7.2):

$$\leq \left(\sum_{\gamma \in A} |\hat{T}(\gamma)|\right)^{-1} |A|^{1/2},$$

d'où d'après (7.1):

$$\|\sum p_\gamma \gamma\|_{\psi_2} \leq R\Delta_2 \mu^{-1} |A|^{1/2}. \tag{7.3}$$

Nous allons montrer que l'on a

$$\mathbb{E}\left\|\sum_{j=1}^k \gamma_j^\omega\right\|_{L^{\psi_2}(dm)} \leq C_2\sqrt{k} + R\Delta_2 k\mu^{-1}|A|^{1/2}, \tag{7.4}$$

où C_2 est une constante numérique.

On va considérer la somme $\sum_{j=1}^k \gamma_j^\omega - \gamma_j^{\omega'}$ considérée comme variable aléatoire sur $(\Omega, \mathscr{B}, \mathbb{P}) \times (\Omega, \mathscr{B}, \mathbb{P})$. Notons $\mathbb{E}_{\omega'}$ l'espérance relative à la variable ω'. Pour tout ω fixé, on peut écrire (puisque $\| \; \|_{\psi_2}$ est une norme):

$$\left\|\sum_{j=1}^k \gamma_j^\omega - \mathbb{E}_{\omega'}\gamma_j^{\omega'}\right\|_{\psi_2} \leq \mathbb{E}_{\omega'}\left\|\sum_{j=1}^k \gamma_j^\omega - \gamma_j^{\omega'}\right\|_{\psi_2}$$

d'où

$$\mathbb{E}_\omega\left\|\sum_{j=1}^k \gamma_j^\omega - k\sum_{\gamma \in A} p_\gamma \gamma\right\|_{\psi_2} \leq \mathbb{E}_\omega\mathbb{E}_{\omega'}\left\|\sum_{j=1}^k \gamma_j^\omega - \gamma_j^{\omega'}\right\|_{\psi_2}. \tag{7.5}$$

Soit (r_j) les fonctions de Rademacher sur $[0, 1]$. Par symétrie, on voit que,

$$\forall t \in [0, 1], \qquad \mathbb{E}_\omega\mathbb{E}_{\omega'}\left\|\sum_{j=1}^k (\gamma_j^\omega - \gamma_j^{\omega'})r_j(t)\right\|_{\psi_2} = \mathbb{E}_\omega\mathbb{E}_{\omega'}\left\|\sum_{j=1}^k \gamma_j^\omega - \gamma_j^{\omega'}\right\|_{\psi_2}.$$

D'où

$$\mathbb{E}_\omega\mathbb{E}_{\omega'}\left\|\sum_{j=1}^k \gamma_j^\omega - \gamma_j^{\omega'}\right\|_{\psi_2} \leq \sup_{\omega, \omega'} \int dt \left\|\sum_{j=1}^k r_j(t)(\gamma_j^\omega - \gamma_j^{\omega'})\right\|_{\psi_2}$$

d'où, d'après (2.7):

$$\leq 2 \sup_{\substack{\omega,\omega' \\ x \in G}} \left\| \sum_{j=1}^{k} r_j(t)(\gamma_j^{\omega}(x) - \gamma_j^{\omega'}(x)) \right\|_{L_{dt}^{\psi_2}},$$

soit, d'après (0.2):

$$\leq 4B_2 k^{1/2}.$$

On déduit donc de (7.5) que l'on a

$$\mathbb{E}_\omega \left\| \sum_{j=1}^{k} \gamma_j^{\omega} \right\|_{\psi_2} \leq k \left\| \sum_{\gamma \in A} p_\gamma \gamma \right\|_{\psi_2} + 4B_2 k^{1/2}$$

ce qui donne bien le résultat annoncé en (7.4) si l'on tient compte de (7.3).

Choisissons k de sorte que

$$k^{1/2} \mu^{-1} |A|^{1/2} \leq 1. \tag{7.6}$$

L'inégalité (7.4) montre alors qu'il existe $\gamma_1, \ldots, \gamma_k$ éléments de A tels que

$$\left\| \sum_{j=1}^{k} \gamma_j \right\|_{\psi_2} \leq C_3 k^{1/2} \tag{7.7}$$

où C_3 est une constante numérique.

C'est presque le résultat cherché, si ce n'est que les $(\gamma_j)_{j \leq k}$ ne sont pas nécessairement distincts. Soit donc B_1 l'ensemble des caractères apparaissant dans $\{\gamma_1, \ldots, \gamma_k\}$, et pour chaque γ dans B_1, notons m_γ le nombre de fois que γ apparait dans $\{\gamma_1, \ldots, \gamma_k\}$. On a

$$\sum_{\gamma \in B_1} m_\gamma = k, \qquad m_\gamma \geq 1 \quad \forall \gamma \in B_1, \qquad \text{et} \qquad \sum_{\gamma \in B_1} m_\gamma \gamma = \sum_{j=1}^{k} \gamma_j.$$

Donc

$$\sum_{\gamma \in B_1} m_\gamma^2 = \left\| \sum_{j=1}^{k} \gamma_j \right\|_2^2 \leq (C_4)^2 \left\| \sum_{j=1}^{k} \gamma_j \right\|_{\psi_2}^2 \leq (C_4 C_3)^2 k$$

où C_4 est une constante numérique. Posons $K = (C_4 C_3)^2$. D'après ce qui précède, on a:

$$\sum_{\gamma \in B_1} m_\gamma^2 \leq K \sum_{\gamma \in B_1} m_\gamma = Kk.$$

Soit alors $B = \{\gamma \in B_1 \mid m_\gamma \leq 2K\}$. On a

$$2K \sum_{\gamma \notin B} m_\gamma \leq \sum_{\gamma \notin B} m_\gamma^2 \leq Kk$$

d'où $\sum_{\gamma \notin B} m_\gamma \leq k/2$ et par conséquent: $\sum_{\gamma \in B} m_\gamma \geq k/2$. On a donc $2K|B| \geq \sum_{\gamma \in B} m_\gamma \geq k/2$, c'est-à-dire,

$$|B| \geq k(4K)^{-1}. \tag{7.8}$$

De plus, on a d'après le lemme 4.6:

$$(\gamma_2)^{-1}\left\|\sum_{\gamma\in B}\gamma\right\|_{\psi_2}\leq\left\|\sum_{j=1}^{k}\gamma_j\right\|_{\psi_2}$$

d'où d'après (7.7): $\leq C_3k^{1/2}$, soit d'après (7.8): $\leq C_3(4K)^{1/2}|B|^{1/2}$.

L'ensemble B a donc bien les propriétés annoncées au début de la démonstration, et l'on a donc établi le théorème 7.1, puisque l'on peut choisir k vérifiant (7.6) et aussi $k\geq\frac{1}{2}\mu^2|A|^{-1}$, il suffit de prendre k égal à la partie entière de $\mu^2|A|^{-1}$.

Donnons un exemple d'application du théorème 7.1 et de la proposition 4.5:

Considérons le cas où $G=\mathbb{T}^{\mathbb{N}}$. On note $(\omega_n)_{n\in\mathbb{N}}$ la suite des coordonnées sur $\mathbb{T}^{\mathbb{N}}$.

Soit $\pi_n=\prod_{k=1}^{n}(1+\omega_k)$. On note $A_n\subset\mathbb{Z}^{\mathbb{N}}$ le support de $\hat{\pi}_n$ (ou encore le spectre de π_n). Notons que l'on a $|\Lambda_n|=2^n$.

D'après un résultat classique de Littlewood et Salem (voir [11, p. 57]), on a

$$\mathbb{E}\left\|\sum_{\gamma\in A_n}\varepsilon_\gamma\gamma\right\|_{C(G)}\leq C(2^n n\operatorname{Log}(n+1))^{1/2},$$

ou C est une constante numérique.

On peut vérifier que l'inégalité inverse est aussi vraie, c'est-a-dire,

$$\mathbb{E}\left\|\sum_{\gamma\in A_n}\varepsilon_\gamma\gamma\right\|_{C(G)}\geq\delta(2^n n\operatorname{Log}(n+1))^{1/2}\tag{7.9}$$

où $\delta>0$ est une constante numérique.

Pour vérifier (7.9) il suffit de vérifier que l'on a:

$$\left\|\sum_{\gamma\in A_n}\gamma\right\|_{\psi_2}\leq\beta 2^n(n\operatorname{Log}(n+1))^{-1/2},\tag{7.10}$$

où β est une constante numérique.

En effet, (7.10) entraine (7.9) comme le montre clairement le lemme suivant:

LEMME 7.2. *Il existe une constante numérique $\delta_1>0$ telle que, pour toute partie finie A de Γ, on ait*:

$$\mathbb{E}\left\|\sum_{\gamma\in A}\varepsilon_\gamma\gamma\right\|_{C(G)}\geq\delta_1\frac{|A|^{3/2}}{\left\|\sum_{\gamma\in A}\gamma\right\|_{\psi_2}}.$$

Démonstration. Il suffit d'appliquer les inégalités (4.13) et (4.15) de la Section 4 dans le cas particulier où $p=q=2$, $C'=\gamma_2C=\gamma_2\|\sum_{\gamma\in A}\gamma\|_{\psi_2}|A|^{-1/2}$, et $f=\sum_{\gamma\in A}\gamma$.

On en déduit immédiatement (7.11).

Il reste donc à vérifier (7.10); pour cela, on rappelle (cf. la démonstration du lemme 4.6) que l'on a

$$\Big\|\sum_{\gamma\in A_n}\gamma\Big\|_{\psi_2}\le\gamma_2'\sup_{N\ge1}N^{-1/2}\Big\|\sum_{\gamma\in A_n}\gamma\Big\|_{2N}.$$

Or, on vérifie aisément que:

$$\begin{aligned}\Big\|\sum_{\gamma\in A_n}\gamma\Big\|_{2N}&=\|\pi_n\|_{2N}=\|1+\omega_1\|_{2N}^n=2^n\left(\frac{2}{\pi}\int_0^{\pi/2}|\cos t|^{2N}\,dt\right)^{n/2N}\\&=2^n\left(\prod_{k=1}^{k=n}\left(1-\frac{1}{2k}\right)\right)^{n/2N},\end{aligned}$$

et un calcul élémentaire permet alors de vérifier que

$$\sup_{N\ge1}\Big\|\sum_{\gamma\in A_n}\gamma\Big\|_{2N}N^{-1/2}\le\beta'2^n(n\operatorname{Log}(n+1))^{-1/2},$$

où β' est une constante. ■

En appliquant (7.9) et le théorème 7.1 on obtient

PROPOSITION 7.3. *Il existe $\delta>0$ et C (indépendants de n) et pour chaque n, il existe une partie $B_n\subset A_n$ telle que $|B_n|\ge\delta\, n\operatorname{Log} n$ et $S(B_n)\le C$.*

La proposition précédente montre que la condition suffisante pour qu'un ensemble soit de Sidon donnée au théorème 3.1 n'est pas nécessaire en général.

On peut vérifier que le théorème 7.1 donne le bon ordre de grandeur de $|B|$ dans tous les cas où l'on sait calculer μ, par exemple si A est une progression arithmétique, ou bien si A est le produit de deux (ou plusieurs) ensembles de Sidon.

8. REMARQUES ET PROBLÈMES

Revenons ici au cas d'un groupe G compact et abélien.

8.1 Notons $C_{\text{p.s.}}$ (resp., $L^{\psi_q}_{\text{p.s.}}$) l'espace formé des fonctions $f\in L^2(G)$ telles que la série de Fourier aléatoire

$$\sum_{\gamma\in\Gamma}\varepsilon_\gamma(\omega)\hat f(\gamma)\gamma$$

représente ω-presque sûrement une fonction de $C(G)$ [resp., de $L^{\psi_q}(G)$]. On peut munir cet espace de la norme

$$\|f\|_{C_{\text{p.s.}}}=\mathbb{E}\Big\|\sum_{\gamma\in\Gamma}\varepsilon_\gamma\hat f(\gamma)\gamma\Big\|_{C(G)}\qquad\left[\text{resp., }\|f\|_{\psi^q_{\text{p.s.}}}=\mathbb{E}\Big\|\sum_{\gamma\in\Gamma}\varepsilon_\gamma\hat f(\gamma)\gamma\Big\|_{\psi_q}\right].$$

Muni de cette norme, cet espace est un espace de Banach.

L'espace $C_{\text{p.s.}}$ est étudié en détail dans [24] et [19]; en particulier (cf. [19, 24]) on sait que $C_{\text{p.s.}}$ coïncide avec l'espace A_{2,φ_2} considéré à la Section 4.

D'autre part, on voit aisément [e.g., en utilisant les inégalités (2.7)] que $L^{\psi_2}_{\text{p.s.}}$ coincide avec L^2.

On vérifie aisément que $[L^{\psi_2}_{\text{p.s.}}, C_{\text{p.s.}}]_\theta \subset L^{\psi_q}_{\text{p.s.}}$ si $q^{-1} = (1-\theta)/2, 0 < \theta < 1$, mais on peut voir que l'inclusion inverse n'est pas vraie.

Problème 8.1. Caractériser les fonctions f appartenant à $L^{\psi_q}_{\text{p.s.}}$ quand $2 < q < \infty$. Comparer $L^{\psi_q}_{\text{p.s.}}$ avec l'espace $[L^{\psi_2}_{\text{p.s.}}, C_{\text{p.s.}}]_{\theta,\infty}$ pour $(1-\theta)/2 = q^{-1}$. On notera que si $(\hat{f}(\gamma))_{\gamma\in\Gamma} \in l^{p,\infty}(\Gamma)$ pour $p = q/(q-1) < 2$, alors $f \in L^{\psi_q}_{\text{p.s.}}$.

8.2 Notons $\mathscr{P}_0$ l'espace des polynômes trigonométriques f tels que $\hat{f}(0) = 0$. Nous savons que la fonctionnelle $f \to E_q(f)$ est équivalente à une norme sur $\mathscr{P}_0$ si $q = 2$, mais nous ignorons si cela est encore vrai pour $q > 2$. [Voir la note (2) ci-dessous.] Par contre on peut voir que cela devient faux pour $q < 2$. Il me parait plausible également que, si $1 < p < \infty$, la fonctionnelle $f \to J_p(f)$ (considérée à la Section 2) est équivalente à une norme sur $\mathscr{P}_0$, mais je ne sais pas le démontrer si $p \neq 2$.

8.3 Soit Λ une partie de Γ. Supposons qu'il existe $1 < p < \infty$ et une constante C tels que:

$$\forall f \in \mathscr{P}_\Lambda \qquad \sum_{\substack{\gamma\in\Lambda \\ \gamma\neq 0}} |\hat{f}(\gamma)| \leq C J_p(f)$$

alors Λ est un ensemble de Sidon (et inversement tout ensemble de Sidon vérifie cette propriété).

En effet, on peut remarquer que l'on a (même démonstration que pour le lemme 2.5) si $1 < p_0 < p_\theta < p_1 < \infty$ et $p_\theta^{-1} = (1-\theta)/p_0 + \theta/p_1$

$$\forall f \in \mathscr{P} \qquad J_p(f) \leq C_\theta J_{p_0}(f)^{1-\theta} J_{p_1}(f)^\theta$$

où C_θ est une constante ne dépendant que de θ, p_0 et p_1. On peut donc démontrer l'assertion ci-dessus exactement comme on a démontré l'implication (iii)$'$ $\Rightarrow$ (i)$'$ au théorème 2.3. Comme nous l'avons vu plus haut (cf. la démonstration du lemme 2.5) $J_p(f)$ est équivalente à

$$I_p(f) = \int_0^1 \frac{\bar{\sigma}_p(f,t)}{t(\operatorname{Log} t^{-1})^{1/p}}\, dt$$

où l'on a posé $\sigma_p(f,t) = d_p^f(t,0)$ avec les notations précédant le lemme 2.4.

Problème 8.3. Supposons qu'il existe une constante C telle que:

$$\forall f \in \mathscr{P}_\Lambda \qquad \sum_{\gamma\neq 0} |\hat{f}(\gamma)| \leq C I_\infty(f) = C \int_0^1 \frac{\bar{\sigma}_\infty(f,t)}{t}\, dt,$$

est-ce que Λ est un ensemble de Sidon? (On peut voir que tout ensemble de Sidon vérifie cette propriété.)

8.4 On peut montrer (cf. un article en préparation en collaboration avec M. B. Marcus) que l'on a:

$$\forall f \in \mathscr{P}(G) \qquad J_2(f) \leq K_p J_p(f) \qquad \text{si} \quad 1 < p < 2$$

où K_p est une constante ne dépendant que de p. D'où la question suivante

Problème 8.4. Soient $2 \leq p_0 < p_1 \leq \infty$. Existe-t-il une constante $K = K(p_0, p_1)$ telle que:

$$\forall f \in \mathscr{P}(G) \qquad J_{p_1}(f) \leq K J_{p_0}(f)?$$

8.5 On sait que si Λ est un ensemble de Sidon, la projection orthogonale P de L^2 sur L^2_Λ, est aussi bornée de L^{ψ_2} sur $L^{\psi_2}_\Lambda$; on sait aussi que cette projection n'est bornée ni sur L^∞, ni sur L^1. De plus, on vérifie aisément que P n'est pas bornée non plus sur L^{ψ_q} quand $q > 2$. Ces observations nous ont conduit à formuler la question suivante:

Problème 8.5. Soit q tel que $2 < q < \infty$. Soit $P: L^{\psi_q}(G) \to L^{\psi_q}(G)$ une projection bornée sur $L^{\psi_q}(G)$, et commutant avec les translations. Est-ce que P est nécessairement bornée sur $C(G)$ et $L^1(G)$? En d'autres termes, P est-elle un opérateur de convolution par une mesure bornée idempotente?

Notes ajoutées sur les épreuves: (1) Les résultats de Carl mentionnés en [4] et cités à la section 2 se trouvent dans un article à paraître de B. Carl intitulé: "Entropy numbers, s-numbers, and eigenvalue problems."

Nous aurions dû signaler que nous avions besoin de ces résultats (au lieu de ceux de Marcus [18]) pour démontrer le théorème 5.5 et la proposition 5.3.

(2) Alors que cet article était à l'imprimerie, je me suis aperçu que la fonctionnelle $f \to E_q(f)$ (considérée à la section 4 et en 8.2) est équivalente à une norme sur $\mathscr{P}_0$ pour $2 < q < \infty$. On peut donc améliorer un peu le théorème 4.1: Une fonction f est dans A_{2,φ_q} ssi $E_q(f) < \infty$ (pour $2 < q < \infty$). Les détails paraîtront ailleurs.

Remerciements

Je remercie M. Déchamps-Gondim et N. Th. Varopoulos pour des conversations stimulantes sur la Section 3.

Références

1. G. Bachelis, On the upper and lower majorant properties in $L^p(G)$, *Quart. J. Math. Oxford* **24** (1973), 119–128.
2. C. Bennett, Intermediate spaces and the class L Log$^+$ L, *Ark. Mat.* **11** (1973), 215–228.
3. A. P. Calderón, Intermediate spaces and interpolation: the complex method, *Studia Math.* **24** (1964), 113–190.
4. B. Carl, A collection of recent results concerning entropy numbers, s-numbers, eigenvalues, Notes manuscrites.
5. M. Déchamps-Gondim, Ensembles de Sidon topologiques, *Ann. Inst. Fourier* (*Grenoble*) **22**, No. 3 (1972), 51–79.

6. R. M. DUDLEY, The sizes of compact subsets of Hilbert space and continuity of Gaussian processes, *J. Funct. Anal.* **1** (1967), 290–330.
7. X. FERNIQUE, Régularité des trajectoires des processus gaussiens, Ecole d'Eté de St. Flour, 1974, Springer Lecture Notes in Mathematics, No. 480, Springer-Verlag, Berlin and New York, 1975.
8. C. GRAHAM ET O. C. MC GEHEE, "Essays in Commutative Harmonic Analysis," Grundlehren der mathematischen Wissenschaften, No. 238. Springer-Verlag, Berlin and New York, 1979.
9. E. HEWITT ET K. ROSS, "Abstract Harmonic Analysis, I and II," Springer-Verlag, Berlin and New York, 1970.
10. R. A. HUNT, On $L(p,q)$ spaces, *Enseign. Math.* **12** (1966), 249–275.
11. J. P. KAHANE, "Some Random Series of Functions," Heath, Lexington, Massachusetts, 1968.
12. J.P. KAHANE, "Séries de Fourier absolument convergentes," Ergebnisse der Mathematik, No. 50, Springer-Verlag, Berlin and New York, 1970.
13. M. KRASNOSELSKII ET Y. RUTIČKII, "Convex Functions and Orlicz Spaces," Noordhoff, Gröningen, 1961.
14. L. Å. LINDHAL ET F. POULSEN, EDS., "Thin Sets in Harmonic Analysis," Lecture Notes in Pure and Applied Mathematics, No. 2, Dekker, New York, 1971.
15. J. L. LIONS ET J. PEETRE, Sur une classe d'espaces d'interpolation, *Pub. Math. I.H.E.S.* No. 19 (1964).
16. J. M. LÓPEZ ET K. A. ROSS, "Sidon Sets," Lecture Notes in Pure and Applied Mathematics, No. 13, Dekker, New York, 1975.
17. M. P. MALLIAVIN-BRAMERET ET P. MALLIAVIN, Caractérisation arithmétique des ensembles de Helson, *C. R. Acad. Sci. Paris Sér. A* **264** (1967), 192–193.
18. M. B. MARCUS, The ε-entropy of some compact subsets of l^p, *J. Approx. Theory* **10** (1974), 304–312.
19. M. B. MARCUS ET G. PISIER, "Random Fourier Series with Applications to Harmonic Analysis," Annals of Maths. Studies, Princeton Univ. Press. A paraître.
20. D. OBERLIN, The derived space of L (G), *Symp. Math.* **22** (1977), 231–247.
21. A. Pełczyński et. C. Shütt, Factoring the natural injection $i^{(n)}$: $L_n^\infty \to L_n^1$ through finite dimensional Banach spaces and geometry of finite dimensional unitary ideals, *in* "Mathematical Analysis and Applications," (L. Nachbin ed.), Part B, Advances in Mathematics Supplementary Studies, Vol. 7B, Academic Press, New York 1981.
22. G. PISIER, Ensembles de Sidon et processus gaussiens, *C. R. Acad. Sci. Paris Sér. A* **286** (1978), 671–674.
23. G. PISIER, Lacunarité et processus gaussiens, *C. R. Acad. Sci. Paris Ser. A* **286** (1978), 1003–1006.
24. G. PISIER, Sur l'espace de Banach des séries de Fourier aléatoires presque sûrement continues, *Séminaire sur la geometrie des espaces de Banach, Ecole Polytechnique, Palaiseau*, Exposé No. 17–18, 1977–1978.
25. C. PRESTON, Banach spaces arising from some integral inequalities, *Indiana Univ. Math. J.* **20** (1971), 997–1015.
26. D. RIDER, Randomly continuous functions and Sidon sets, *Duke Math. J.* **42** (1975), 759–764.
27. V. A. RODIN ET E. M. SEMYONOV, Rademacher series in symmetric spaces, *Anal. Math.* **1** (1975), 207–222.
28. W. RUDIN, Trigonometric series with gaps, *J. Math. Math. Mech.* **9** (1960), 203–227.
29. H. TRIEBEL, "Interpolation Theory, Function Spaces, Differential Operators," North-Holland Publ. Amsterdam, 1978.
30. N. T. VAROPOULOS, Some combinatorial problems in harmonic analysis, Notes de l'Ecole d'Eté de Warwick (rédigées par D. Salinger), 1968.
31. T. FIGIEL, J. Lindenstrauss, ET V. MILMAN, The dimension of almost spherical sections of convex bodies, *Acta Math.* **139** (1977), 53–94.

MATHEMATICAL ANALYSIS AND APPLICATIONS, PART B
ADVANCES IN MATHEMATICS SUPPLEMENTARY STUDIES, VOL. 7B

Topological Algebras of Vector-Valued Continuous Functions

JOÃO B. PROLLA

Instituto de Matemática
Universidade Estadual de Campinas
Campinas, Brazil

DEDICATED TO LAURENT SCHWARTZ ON THE OCCASION OF HIS 65TH BIRTHDAY

1. INTRODUCTION

Throughout this paper X stands for a completely regular Hausdorff space, and E stands for a nonzero locally convex algebra over $\mathbb{K}$ (where $\mathbb{K} = \mathbb{R}$ or $\mathbb{K} = \mathbb{C}$); that is, E is an algebra over $\mathbb{K}$ equipped with a topology in such a way that E is a locally convex space and the multiplication

$$(u, v) \in E \times E \mapsto uv \in E$$

is *jointly continuous.*

The space $\mathscr{C}(X;E)$ of all continuous E-valued functions on X, under pointwise operations becomes an algebra over $\mathbb{K}$ too, and we are interested in subalgebras of $\mathscr{C}(X;E)$ which are locally convex algebras when equipped with suitable topologies, namely, *weighted topologies*. To describe precisely those subalgebras we shall assume, throughout this paper, that V is an upper directed set of nonnegative real-valued functions on X satisfying the following two conditions

(a) *each $v \in V$ is upper semicontinuous;*
(b) *given $x \in X$, there is a $v \in V$ such that $v(x) > 0$.*

Under these assumptions, $\mathscr{C}V_\infty(X;E)$ denotes the vector subspace of all $f \in \mathscr{C}(X;E)$ such that vf vanishes at infinity, for each $v \in V$. Hence, for every $f \in \mathscr{C}V_\infty(X;E)$ and $v \in V$, given $\varepsilon > 0$ and a continuous seminorm p on E, there is a compact set $K \subset X$ such that $v(x)p(f(x)) < \varepsilon$ for all $x \notin K$. The space $\mathscr{C}V_\infty(C; E)$ is endowed with the topology τ_V generated by the family of seminorms

$$f \mapsto \|f\|_{p,v} = \sup\{v(x)p(f(x)); x \in X\},$$

ISBN 0-12-512802-9

where v ranges over V and p over a set of continuous seminorms on E which generates the topology of E. We call the locally convex space $\mathscr{C}V_\infty(X; E)$ a *Nachbin space*. If it is a locally convex algebra, we call it a *Nachbin algebra*.

A very simple sufficient condition for $\mathscr{C}V_\infty(X; E)$ to be a Nachbin algebra is the following:

(c) *$v \in V$ implies that $\sqrt{v} \in V$.*

Indeed, let f and g be two elements of $\mathscr{C}V_\infty(X; E)$. We claim that fg belongs to $\mathscr{C}V_\infty(X; E)$ too. Let then $v \in V$, $\varepsilon > 0$ and a continuous seminorm p on E be given. Since E is a locally convex algebra, there exist $\delta > 0$ and q, a continuous seminorm on E, such that $q(u) < \delta$ and $q(t) < \delta$ implies $p(ut) < \varepsilon$. Let

$$K = \{x \in X; \sqrt{v(x)}q(f(x)) \geq \delta\} \qquad \text{and} \qquad K' = \{x \in X; \sqrt{v(x)}q(g(x)) \geq \delta\}.$$

Then $K \cup K'$ is compact, and $v(x)p(f(x)g(x)) < \varepsilon$ whenever $x \notin K \cup K'$. Hence $fg \in \mathscr{C}V_\infty(X; E)$. Moreover $\|f\|_{w,q} < \delta$ and $\|g\|_{w,q} < \delta$ imply that $\|fg\|_{v,p} < \varepsilon$, where $w = \sqrt{v}$. Hence multiplication in $\mathscr{C}V_\infty(X; E)$ is jointly continuous, i.e., $\mathscr{C}V_\infty(X; E)$ is a locally convex topological algebra.

Let us give examples of sets of weights satisfying condition (c).

EXAMPLE 1. Let V be the set of all characteristic functions of compact subsets of X. Then $\mathscr{C}V_\infty(X; E)$ is just $\mathscr{C}(X; E)$ with the *compact–open topology* κ.

EXAMPLE 2. Let V be the set of all nonnegative constant functions on X. Then $\mathscr{C}V_\infty(X; E)$ is the algebra $\mathscr{C}_\infty(X; E)$ of all continuous E-valued functions that vanish at infinity, equipped with the *uniform topology* σ.

EXAMPLE 3. Let V be the set of all nonnegative elements of $\mathscr{C}_\infty(X; \mathbb{R})$. Then $\mathscr{C}V_\infty(X; E)$ is the algebra $\mathscr{C}_b(X; E)$ of all continuous E-valued functions which are bounded on X, equipped with the *strict topology* β.

Remarks. In Examples 2 and 3 we assume that X is *locally compact*.

In all three examples the Nachbin algebra $\mathscr{C}V_\infty(X; E)$ satisfies the following important property:

(d) *given $(x, y, u, t) \in X \times X \times E \times E$, with $u = t$ if $x = y$, there exists a function $f \in \mathscr{C}V_\infty(X; E)$ such that $f(x) = u$ and $f(y) = t$.*

Clearly, (d) implies the following:

(e) *for every $x \in X$, given $u \in E$, there exists $f \in \mathscr{C}V_\infty(X; E)$ such that $f(x) = u$.*

When $\mathscr{C}V_\infty(X; E)$ satisfies property (d), we say that it is *strongly separating on* X; when $\mathscr{C}V_\infty(X; E)$ satisfies property (e) we say that it is *essential.*

A final remark: It is to be noted that *any Nachbin space* $\mathscr{C}V_\infty(X; E)$ *is a* $\mathscr{C}_b(X; \mathbb{K})$*-module*; that is, given $f \in \mathscr{C}V_\infty(X; E)$ and $\varphi \in \mathscr{C}_b(X; \mathbb{K})$ then $x \mapsto \varphi(x)f(x)$ belongs to $\mathscr{C}V_\infty(X; E)$.

2. The Spectrum of $\mathscr{C}V_\infty(X; E)$

If E is any locally convex algebra over $\mathbb{K}$, its *spectrum* is the set $\Delta(E)$ of all nonzero continuous algebra homomorphisms of E onto $\mathbb{K}$, equipped with the relative $\sigma(E', E)$ topology.

Let $\mathscr{C}V_\infty(X; E)$ be a Nachbin algebra. For each $x \in X$, the map

$$f \mapsto \delta_x(f) = f(x),$$

defined on $\mathscr{C}V_\infty(X; E)$ is an algebra homomorphism of $\mathscr{C}V_\infty(X; E)$ into E. We claim that δ_x is continuous. Indeed, let p be a continuous seminorm of E. By condition (b), Section 1, there is a $v \in V$ such that $v(x) > 0$. Therefore

$$p(\delta_x(f)) = p(f(x)) \leq [1/v(x)]\|f\|_{v,p}$$

for all $f \in \mathscr{C}V_\infty(X; E)$.

Assume now that $\mathscr{C}V_\infty(X; E)$ is *essential*, that is, condition (e), Section 1, is verified, and let $h \in \Delta(E)$ be given. Choose $u \in E$ with $h(u) = 1$, and then choose $f \in \mathscr{C}V_\infty(X; E)$ with $f(x) = u$. Therefore $(h \circ \delta_x)(f) = 1$ and $h \circ \delta_x \in \Delta(\mathscr{C}V_\infty(X; E))$. This shows that the map G defined by

$$(x, h) \mapsto h \circ \delta_x,$$

for all $(x, h) \in X \times \Delta(E)$, takes values in $\Delta(\mathscr{C}V_\infty(X; E))$, whenever $\mathscr{C}V_\infty(X; E)$ is essential.

We would like the map G to be one to one. Let then $(x, h) \neq (y, k)$ be given in $X \times \Delta(E)$. If $x = y$, then $h \neq k$. Choose $u \in E$ such that $h(u) \neq k(u)$. Since $\mathscr{C}V_\infty(X; E)$ is essential, choose now $f \in \mathscr{C}V_\infty(X; E)$ such that $f(x) = u$. It follows that

$$(h \circ \delta_x)(f) = h(u) \neq k(u) = (k \circ \delta_x)(f).$$

If $x \neq y$, choose $\varphi \in \mathscr{C}_b(X; \mathbb{K})$ such that $\varphi(x) = 0$ and $\varphi(y) = 1$. This is possible because X is a completely regular Hausdorff space. Choose now $u \in E$ with $k(u) = 1$. Let $f \in \mathscr{C}V_\infty(X; E)$ be such that $f(y) = u$. Then $g = \varphi f$ belongs to $\mathscr{C}V_\infty(X; E)$ and we have

$$(h \circ \delta_x)(g) = h(g(x)) = h(0) = 0,$$

but $(k \circ \delta_y)(g) = k(g(y)) = k(u) = 1$.

The above argument shows that *the map G is indeed one to one, whenever $\mathscr{C}V_\infty(X; E)$ is essential.*

In fact, we are interested in conditions under which the map G is a *homeomorphism* between $X \times \Delta(E)$ and $\Delta(\mathscr{C}V_\infty(X; E))$.

When X is *compact*, and E is a *Banach algebra*, Examples 1 and 2 of Section 1 coincide with the Banach algebra $\mathscr{C}(X; E)$ with the sup-norm. In this case the spectrum of E [resp., $\mathscr{C}(X; E)$] is homeomorphic with the space of all regular maximal two-sided ideals in E [resp., $\mathscr{C}(X; E)$] topologized in the Gelfand sense. With this formulation, Hausner [2, p. 248] proved that $X \times \Delta(E)$ and $\Delta(\mathscr{C}(X; E))$ are homeomorphic, in the case of a *commutative* Banach algebra E. Hausner's theorem was an improvement of an earlier result of Yood [9, Lemma 5.1], in which E has a unit, and it was assumed that every maximal ideal in $\mathscr{C}(X; E)$ was of the form $\{f \in \mathscr{C}(X; E); f(x) \in M\}$, with $x \in X$ and M a maximal ideal in E. In turn, Dietrich [1, Theorem 4] extended the Hausner–Yood theorem to the case of a completely regular *k-space*, and of a *complete* locally convex algebra E for which $\Delta(E)$ is locally equicontinuous. On the other hand, Mallios [4, Theorem 5.1] had extended the Hausner–Yood theorem to the case of a *locally compact* space X, of a *complete* locally m-convex Q-algebra E, and $\mathscr{C}_\infty(X; E)$ with the topology of uniform convergence on X. Both Dietrich and Mallios used the technique of tensor products. They had to rely on the representation of $\mathscr{C}(X; E)$ and $\mathscr{C}_\infty(X; E)$ as tensor products $\hat{\otimes}_\varepsilon$ of $\mathscr{C}(X; \mathbb{K})$ and $\mathscr{C}_\infty(X; \mathbb{K})$, respectively, with E. That is why X has to be a k-space in [1], so that $\mathscr{C}(X; \mathbb{K})$ with the compact–open topology is complete, and E has to be complete in both [1] and [4]. While working in the similar problem for normed algebras over valued fields, in non-Archimedean analysis (see Prolla [8]), we proved that the map G is a homeomorphism, without using tensor product techniques. Our proof relies on characterizing the closed regular two-sided maximal ideals of $\mathscr{C}V_\infty(X; E)$ by exploiting the fact that they are $\mathscr{C}_b(X; \mathbb{K})$-modules, and using an approximation theorem valid for $\mathscr{C}_b(X; \mathbb{K})$-submodules of $\mathscr{C}V_\infty(X; E)$.

Hence our proof shows that, for $X \times \Delta(E)$ and $\Delta(\mathscr{C}V_\infty(X; E))$ to be homeomorphic, there is no need to assume that X is a k-space or that E is complete. Concerning the hypothesis of local equicontinuity of $\Delta(E)$, it is automatically satisfied in the case of a *normed algebra* E, since in this case $\Delta(E)$ is a subset of the unit ball of E'.

We now state our main result.

THEOREM 2.1. *Let X be a completely regular Hausdorff space, let E be a locally convex algebra over $\mathbb{K}$, and let $\mathscr{C}V_\infty(X; E)$ be a Nachbin algebra. Assume that $\mathscr{C}V_\infty(X; E)$ is essential, and that $\Delta(E)$ is locally equicontinuous. Then the* *map G defined by*

$$(x, h) \mapsto h \circ \delta_x$$

is a homeomorphism between $X \times \Delta(E)$ *and* $\Delta(\mathscr{C}V_\infty(X; E))$.

We have already seen that G is one to one. We will postpone the proof that G is onto $\Delta(\mathscr{C}V_\infty(X; E))$, and that G and G^{-1} are continuous to Section 4. Let us derive some consequences of Theorem 2.1.

COROLLARY 2.2. *Let* X *be a completely regular Hausdorff space, and let* E *be a locally convex algebra over* $\mathbb{K}$ *such that* $\Delta(E)$ *is locally equicontinuous. Then the map* G, *defined by*

$$(x, h) \mapsto h \circ \delta_x$$

is a homeomorphism between $X \times \Delta(E)$ *and* $\Delta(A)$, *in the following cases*:

(a) $A = \mathscr{C}(X; E)$ *with the compact-open topology.*

(b) X *is locally compact and* $A = \mathscr{C}_\infty(X; E)$ *with the topology of uniform convergence on* X.

(c) X *is locally compact and* $A = \mathscr{C}_b(X; E)$ *with the strict topology* β.

Proof. One has only to remark that in all three cases A is an essential Nachbin algebra as remarked in Section 1.

COROLLARY 2.3. *Let* X *be a completely regular Hausdorff space, and let* $\mathscr{C}V_\infty(X; \mathbb{K})$ *be an essential Nachbin algebra. Then* $\Delta(\mathscr{C}V_\infty(X; \mathbb{K}))$ *is homeomorphic to* X *under the map* $x \in X \mapsto \delta_x$.

COROLLARY 2.4. *Assume the hypothesis of Theorem 2.1. Suppose that* $\mathscr{C}V_\infty(X; E)$ *is a commutative locally m-convex algebra. Then,* $W \subset \mathscr{C}V_\infty(X; E)$ *is a proper closed regular maximal ideal if, and only if,* W *is of form*

$$W = \{f \in \mathscr{C}V_\infty(X; E); h(f(x)) = 0\}$$

for some $x \in X$ *and* $h \in \Delta(E)$.

Proof. Clearly, since $h \circ \delta_x \in \Delta(\mathscr{C}V_\infty(X; E))$ any set of the form $\{f \in \mathscr{C}V_\infty(X; E)); h(f(x)) = 0\}$ is a proper closed regular maximal ideal.

Conversely, assume that $\mathscr{C}V_\infty(X; E)$ is a commutative locally m-convex algebra, and let W be a proper closed regular maximal ideal in $\mathscr{C}V_\infty(X; E)$. By Michael [6, Corollary 2.10], W is the kernel of some $H \in \Delta(\mathscr{C}V_\infty(X; E))$. Since the map G is onto $\Delta(\mathscr{C}V_\infty(X; E))$, there is some $(x, h) \in X \times \Delta(E)$ such that $H = h \circ \delta_x$. Hence $W = H^{-1}(0)$ is of the form $\{f \in \mathscr{C}V_\infty(X; E); h(f(x)) = 0\}$.

COROLLARY 2.5. *Assume the hypothesis of Theorem 2.1. Suppose that both* E *and* $\mathscr{C}V_\infty(X; E)$ *are commutative locally m-convex algebras. Then a subset*

$W \subset \mathscr{C}V_\infty(X;E)$ is a proper closed regular maximal ideal if, and only if, W is of the form

$$W = \{f \in \mathscr{C}V_\infty(X;E); f(x) \in M\}$$

for some $x \in X$ and $M \subset E$ a proper closed regular maximal ideal in E.

Proof. By Michael [6, Corollary 2.10], any $M \subset E$ which is a proper closed regular maximal ideal in E is the kernel of some $h \in \Delta(E)$. Hence $\{f \in \mathscr{C}V_\infty(X;E);\ f(x) \in M\} = \{f \in \mathscr{C}V_\infty(X;E);\ h(f(x)) = 0\} = (h \circ \delta_x)^{-1}(0)$, and it is a proper closed regular maximal ideal in $\mathscr{C}V_\infty(X;E)$.

Conversely, by Corollary 2.4, any proper closed regular maximal ideal W in $\mathscr{C}V_\infty(X;E)$ is of the form

$$W = \{f \in \mathscr{C}V_\infty(X;E); h(f(x)) = 0\}$$

for some $x \in X$ and $h \in \Delta(E)$. Now set $M = h^{-1}(0)$. Then M is a proper closed regular maximal ideal in E, and

$$W = \{f \in \mathscr{C}V_\infty(X;E); f(x) \in M\}.$$

Remark. Sometimes the hypothesis that E is a commutative locally m-convex algebra already implies that $\mathscr{C}V_\infty(X;E)$ is a commutative locally m-convex algebra too. This is the case of Examples 1 and 2 of Section 1. Hence we have the following.

COROLLARY 2.6. *Let X be a completely regular space and let E be a commutative locally m-convex algebra such that $\Delta(E)$ is locally equicontinuous. Let $A = \mathscr{C}(X;E)$ with the compact–open topology, or let $A = \mathscr{C}_\infty(X;E)$ with the uniform topology (assuming in this case that X is locally compact). Then a subset $W \subset A$ is a proper closed regular maximal ideal if, and only if, W is of the form stated in Corollary 2.5.*

3. IDEALS IN NACHBIN ALGEBRAS

We start with our

APPROXIMATION LEMMA 3.0. *Let $\mathscr{C}V_\infty(X; E)$ be a Nachbin space, and let $W \subset \mathscr{C}V_\infty(X;E)$ be a $\mathscr{C}_b(X;\mathbb{K})$-submodule. For every $f \in \mathscr{C}V_\infty(X;E)$, $f \in \overline{W}$ if, and only if, $f(x) \in W(x)$ for each $x \in X$.*

Proof. Since $A = \mathscr{C}_b(X;\mathbb{K})$ is a separating self-adjoint subalgebra of $\mathscr{C}(X;\mathbb{K})$, and every $\varphi \in \mathscr{C}_b(X;\mathbb{K})$ is bounded on X, our result follows from [7, Theorem 5.20]. However, the direct proof is so simple that we prefer to present it, rather than derive the approximation lemma from very general results.

Since $V > 0$, the condition is necessary. Conversely, assume that $f(x) \in \overline{W(x)}$, for every $x \in X$. Let $v \in V$, $\varepsilon > 0$ and p a continuous seminorm on E be given. For each $x \in X$, there is some $g_x \in W$ such that

$$v(x)p(f(x) - g_x(x)) < \varepsilon.$$

Let $K_x = \{t \in X; v(t)p(f(t) - g_x(t)) \geq \varepsilon\}$. Then K_x is compact and $x \notin K_x$. It follows that K_x is a *closed* subset of the Stone–Cech compactification βX of X and $\bigcap \{K_x; x \in X\} = \emptyset$. By compactness of βX, there are $x_1, x_2, \ldots, x_n$ in X such that $K_{x_1} \cap K_{x_n} = \emptyset$. Let $U_i = \beta X \backslash K_{x_i}$ for each $i = 1, \ldots, n$. Let $\varphi_1, \ldots, \varphi_n$ be a partition of unity in βX subordinated to $U_1, \ldots, U_n$; that is, each φ_i belongs to $\mathscr{C}(\beta X; \mathbb{R})$, $0 \leq \varphi_i \leq 1$, φ_i vanishes outside of U_i, and $\sum_{i=1}^n \varphi_i = 1$ on βX. Let $h_i = \varphi_i | X$; then $h_i \in \mathscr{C}_b(X; \mathbb{R})$, $0 \leq h_i \leq 1$, h_i vanishes on K_{x_i} and $\sum_{i=1}^n h_i = 1$ on X. Let $g \in W$ be the function $\sum_{i=1}^n h_i g_{x_i}$. We claim that

$$v(x)p(f(x) - g(x)) < \varepsilon$$

for all $x \in X$. Indeed,

$$v(x)p(f(x) - g(x)) \leq \sum_{i=1}^{n} v(x)h_i(x)p(f(x) - g_{x_i}(x)).$$

Let $J_x \subset \{1, \ldots, n\}$ be the set of all $1 \leq i \leq n$ such that $x \notin K_{x_i}$. Then $J_x \neq \emptyset$, and for any $1 \leq i \leq n$, with $i \notin J_x$, $h_i(x) = 0$. Hence the above sum is equal to $\sum_{i \in J_x} v(x)h_i(x)p(f(x) - g_{x_i}(x))$. Now

$$v(x)h_i(x)p(f(x) - g_{x_i}(x)) < \varepsilon h_i(x)$$

for all $i \in J_x$, and therefore

$$\sum_{i \in J_x} v(x)h_i(x)p(f(x) - g_{x_i}(x)) < \varepsilon \sum_{i \in J_x} h_i(x) = \varepsilon.$$

This shows that $f \in \overline{W}$, thus ending the proof.

LEMMA 3.1. *Let $\mathscr{C}V_\infty(X; E)$ be a Nachbin algebra. Then every regular right (resp., left) ideal in $\mathscr{C}V_\infty(X; E)$ is a $\mathscr{C}_b(X; \mathbb{K})$-module.*

Proof: Let $W \subset \mathscr{C}V_\infty(X; E)$ be a regular right ideal. Let $f \in \mathscr{C}V_\infty(X; E)$ be a left identity modulo W; that is $fg - g \in W$ for all $g \in \mathscr{C}V_\infty(X; E)$. Let $\varphi \in \mathscr{C}_b(X; \mathbb{K})$ and $h \in W$ be given. Then $\varphi f \in \mathscr{C}V_\infty(X; E)$ and $(\varphi f)h = f(\varphi h)$. Hence $f(\varphi h) \in W$. Now $f(\varphi h) - \varphi h \in W$, and therefore $\varphi h \in W$, i.e., W is a $\mathscr{C}_b(X; \mathbb{K})$-module.

The proof for a left ideal is similar.

Combining Lemmas 3.0 and 3.1 we get the following.

LEMMA 3.2. *Let $\mathscr{C}V_\infty(X; E)$ be a Nachbin algebra, let $W \subset \mathscr{C}V_\infty(X; E)$ be a regular right (resp., left) ideal, and let $f \in \mathscr{C}V_\infty(X; E)$. Then f belongs to* *the closure of W if, and only if, $f(x)$ belongs to the closure of $W(x)$ in E, for every $x \in X$.*

THEOREM 3.3. *Let $\mathscr{C}V_\infty(X; E)$ be an essential Nachbin algebra. Any closed regular right (resp., left) ideal W in $\mathscr{C}V_\infty(X; E)$ has the following form: For every $x \in X$ a closed regular right (resp., left) ideal W_x in E is given in such a way that we can find $g \in \mathscr{C}V_\infty(X; E)$ such that $g(x)$ is a left (resp., right) identity modulo W_x in E for each $x \in X$; and W consists of all those $f \in \mathscr{C}V_\infty(X; E)$ such that $f(x) \in W_x$ for each $x \in X$.*

Proof. Let us assume that, for each $x \in X$, a closed regular right (resp., left) ideal W_x in E has been given with the property stated above. Let $W = \{f \in \mathscr{C}V_\infty(X; E); f(x) \in W_x\}$. Using condition (b), Section 1, we see that W is closed. W is clearly a right (resp., left) ideal, and g is a left (resp., right) identity modulo W in $\mathscr{C}V_\infty(X; E)$, that is W is regular.

Conversely, let W be a closed regular right (resp., left) ideal in $\mathscr{C}V_\infty(X; E)$, and let $g \in \mathscr{C}V_\infty(X;E)$ be a left (resp., right) identity modulo W. Let W_x be the closure of $W(x)$ in E, where $W(x) = \{f(x); f \in W\}$. Then W_x is a closed right (resp., left) ideal in E. We claim that W_x is regular, and that $g(x)$ is a left (resp., right) identity modulo W_x. Indeed, let $u \in E$. Since $\mathscr{C}V_\infty(X; E)$ is essential, choose $f \in \mathscr{C}V_\infty(X; E)$ such that $f(x) = u$. Then $gf - f \in W$. Hence $g(x)u - u \in W(x) \subset W_x$. Let us define

$$J = \{f \in \mathscr{C}V_\infty(X; E); f(x) \in W_x \text{ for all } x \in X\}.$$

Clearly, $W \subset J$. Conversely, if $f \in J$, then $f(x)$ belongs to the closure of $W(x)$ for all $x \in X$, and by Lemma 3.2, $f \in \bar{W} = W$. Hence $J = W$, and the proof is complete.

Remark. For the purpose of proving Theorem 2.1 the above Theorem 3.3 is sufficient. However one would like to characterize *all* closed right (resp., left) ideals in $\mathscr{C}V_\infty(X; E)$ and not only those that are regular. Our first task is then to find conditions on $\mathscr{C}V_\infty(X; E)$ implying that all closed right (resp., left) ideals are $\mathscr{C}_b(X; \mathbb{K})$-modules. Following Kaplansky [3] let us examine the following property that for a locally convex algebra A

(*) *every $u \in A$ lies in the closure of uA.*

We have then the following:

LEMMA 3.4. *Let $\mathscr{C}V_\infty(X; E)$ be a Nachbin algebra satisfying property (*). Then every closed right (resp., left) ideal in $\mathscr{C}V_\infty(X; E)$ is a $\mathscr{C}_b(X; \mathbb{K})$-module.*

Proof. Let $W \subset \mathscr{C}V_\infty(X; E)$ be a closed right ideal. Let $\varphi \in \mathscr{C}_b(X; \mathbb{K})$ and $h \in W$ be given. Then φh belongs to $\mathscr{C}V_\infty(X; E)$, and by property (*) φh

lies in the closure of the space $\varphi h\mathscr{C}V_\infty(X; E)$. Now

$$\varphi h\mathscr{C}V_\infty(X; E) \subset h\mathscr{C}V_\infty(X; E) \subset W.$$

Hence φh lies in the closure of W. Since W is closed, $\varphi h \in W$. That is, W is a $\mathscr{C}_b(X; \mathbb{K})$-module, and the proof is complete.

When does $\mathscr{C}V_\infty(X; E)$ satisfy property $(*)$? We shall see that it is sufficient that E satisfies property $(*)$, when $\mathscr{C}V_\infty(X; E)$ is essential.

PROPOSITION 3.5. *Let E be a locally convex algebra satisfying property $(*)$. Then $\mathscr{C}V_\infty(X; E)$ satisfies property $(*)$ too, if it is essential.*

Proof. Let $f \in \mathscr{C}V_\infty(X; E)$ be given, and define $W = f\mathscr{C}V_\infty(X; E)$. Since $\mathscr{C}V_\infty(X; E)$ is a $\mathscr{C}_b(X; \mathbb{K})$-module, so is W. We claim that $f \in \overline{W}$. By Lemma 3.0, it is enough to prove that $f(x) \in \overline{W(x)}$ for each $x \in X$. Since $\mathscr{C}V_\infty(X; E)$ is essential, $W(x) = f(x)E$, for every $x \in X$. By property $(*)$ applied to $u = f(x)$, we see that $f(x)$ lies in the closure of $W(x)$, as desired.

THEOREM 3.6. *Let $\mathscr{C}V_\infty(X; E)$ be a Nachbin algebra satisfying property $(*)$. Then every closed right (resp., left) ideal W in $\mathscr{C}V_\infty(X; E)$ has the following form: For every $x \in X$ a closed right (resp., left) ideal W_x in E is given, and W consists of all those $f \in \mathscr{C}V_\infty(X; E)$ such that $f(x) \in W_x$ for each $x \in X$.*

Proof. As in the proof of Theorem 3.3, any W of the above form is a closed right (resp., left) ideal in $\mathscr{C}V_\infty(X; E)$.

Conversely, let W be a closed right (resp., left) ideal in $\mathscr{C}V_\infty(X; E)$, and let W_x be the closure of $W(x)$ in E, where

$$W(x) = \{f(x); f \in W\}.$$

Then W_x is a closed right (resp., left) ideal in E. Define

$$J = \{f \in \mathscr{C}V_\infty(X; E); f(x) \in W_x \text{ for all } x \in X\}.$$

Clearly, $W \subset J$. Conversely, if $f \in J$, then $f(x) \in W_x$ for all $x \in X$. By Lemma 3.4, W is a $\mathscr{C}_b(X; \mathbb{K})$-module. Therefore, by Lemma 3.0, $f \in \overline{W} = W$.

COROLLARY 3.7. *Under the hypothesis of Theorem 3.6 assume that the algebra E is simple (i.e., E has no proper closed two-sided ideals). Then a subset $W \subset \mathscr{C}V_\infty(X; E)$ is a closed two-sided ideal in $\mathscr{C}V_\infty(X; E)$ if, and only if, there exists a closed subset $N \subset X$ such that*

$$W = \{f \in \mathscr{C}V_\infty(X; E); f(x) = 0 \textit{ for all } x \in N\}.$$

Proof. If $N \subset X$ is any closed subset of X, clearly the set

$$Z(N) = \{f \in \mathscr{C}V_\infty(X; E); f(x) = 0 \text{ for all } x \in N\}$$

is a closed two-sided ideal.

Conversely, let W be a closed two-sided ideal in $\mathscr{C}V_\infty(X; E)$ and let

$$N = \{x \in X; f(x) = 0 \text{ for all } f \in W\}.$$

Clearly, N is closed in X and we have $W \subset Z(N)$. Let now $f \in Z(N)$ and assume, by contradiction, that $f \notin W$. By Theorem 3.6, there is some point $x \in X$, such that $f(x) \notin W_x = \overline{W(x)}$. Since W_x is a closed two-sided ideal and E is simple, $W_x = \{0\}$. Hence $f(x) \neq 0$, i.e., $x \notin N$. However, $W(x) \subset W_x$, and therefore $x \in N$. Hence $f \in W$.

COROLLARY 3.8. *Assume that X is not a singleton and that $\mathscr{C}V_\infty(X; \mathbb{K})$ is strongly separating. Then a subset $M \subset \mathscr{C}V_\infty(X; \mathbb{K})$ is a proper closed maximal ideal if, and only if,*

$$M = \{f \in \mathscr{C}V_\infty(X; \mathbb{K}); f(x) = 0\}$$

for some $x \in X$.

Proof. Let $x \in X$ be given and define $M = \{f \in \mathscr{C}V_\infty(X; \mathbb{K}); f(x) = 0\}$. Now $M = \delta_x^{-1}(0)$. Since X is not a singleton and $\mathscr{C}V_\infty(X; \mathbb{K})$ is strongly separating,

$$0 \subsetneq M \subsetneq \mathscr{C}V_\infty(X; \mathbb{K}).$$

Hence M is the kernel of a nontrivial homomorphism of $\mathscr{C}V_\infty(X; \mathbb{K})$ onto its underlying field $\mathbb{K}$, and therefore it is a maximal ideal. Since δ_x is continuous, M is closed.

Conversely, let $M \subset \mathscr{C}V_\infty(X; \mathbb{K})$ be a proper closed maximal ideal. Since M is proper, there is some $x \in X$ such that $M(x) = \{0\}$. Let $I = \{f \in \mathscr{C}V_\infty(X; \mathbb{C}); f(x) = 0\}$. Then I is an ideal and $M \subset I$. Since M is maximal, either $M = I$ or else $I = \mathscr{C}V_\infty(X; \mathbb{K})$. Since X is not a singleton and $\mathscr{C}V_\infty(X; \mathbb{K})$ is strongly separating $I \subsetneq \mathscr{C}V_\infty(X; \mathbb{K})$.

Remark. The Nachbin algebra of Examples 1, 2, and 3, in the case $E = \mathbb{K}$, are strongly separating, and $\mathbb{K}$ being a field trivially satisfies property (*) and it is simple. Therefore we can apply to them both Corollaries 3.7 and 3.8, thus obtaining the familiar representation of closed ideals of $\mathscr{C}(X; \mathbb{K})$, $\mathscr{C}_\infty(X; \mathbb{K})$, and $\mathscr{C}_b(X; \mathbb{K})$ with their respective topologies κ, σ, and β.

4. PROOF OF THEOREM 2.1

Step 1. *G is onto $\Delta(\mathscr{C}V_\infty(X; E))$.*

Proof. Let $H \in \Delta(\mathscr{C}V_\infty(X; E))$. Its kernel $I = H^{-1}(0)$ is a proper closed maximal ideal in $\mathscr{C}V_\infty(X; E)$ which is two sided and regular. By Theorem 3.3, there exists some point $x \in X$ such that the ideal $I_x = \overline{I(x)}$ is proper. Choose a function $f \in \mathscr{C}V_\infty(X; E)$ such that $H(f) = 1$. Then f is an identity modulo I. For each $u \in E$, let $u^* = uf \in \mathscr{C}V_\infty(X; E)$. Define $h: E \to \mathbb{K}$ by setting

$$h(u) = H(u^*)$$

for all $u \in E$. Clearly $h \in E'$, the dual of E. Now, if $u, t \in E$, then

$$h(ut) = H((ut)^*) = H(utf) = H(futf) = H(fu)h(t) = H(fuf)h(t) = h(u)h(t).$$

Hence h is multiplicative. Let J be its kernel. We claim that $J \subset I(x)$. Indeed, if $u \in J$, then $u^* \in I$. Choose $g \in \mathscr{C}V_\infty(X; E)$ with $g(x) = u$. Then $gf - g$ belongs to I, and therefore $uf(x) - u$ belongs to $I(x)$. Now $uf(x) = u^*(x) \in I(x)$. Since I_x is proper, it now follows that $h \neq 0$, i.e., $h \in \Delta(E)$. Let W be the kernel of $h \circ \delta_x$. We claim that $I \subset W$. Indeed, let $g \in \mathscr{C}V_\infty(X; E)$ be such that $g \notin W$. Then $g(x) \notin J$. On the other hand, J is a maximal ideal and $J \subset I(x) \subset I_x$. Since I_x is proper, it follows that $J = I_x$. Hence $g(x) \notin I_x$. By Theorem 3.3, $g \notin I$. Since I is maximal and W is closed and proper, $I = W$. This shows that H and $h \circ \delta_x$ have the same kernel. Since both are multiplicative,

$$H = h \circ \delta_x = G(x, h).$$

Step 2. *G is continuous.*

Proof. Let $(x_0, h_0) \in X \times \Delta(E)$, $\varepsilon > 0$, and $g \in \mathscr{C}V_\infty(X; E)$ be given. Choose an equicontinuous neighborhood N of h_0 in $\Delta(E)$ such that

$$N \subset \{h \in \Delta(E); |(h - h_0)(g(x_0))| < \varepsilon/2.$$

Let W be a neighborhood of $g(x_0)$ in E such that $|h(w - g(x_0))| < \varepsilon/2$ for all $w \in W$ and $h \in N$. Finally, choose a neighborhood U of x_0 in X such that $g(x) \in W$ for all $x \in U$. Then $(x, h) \in U \times N$ implies $g(x) \in W$ and $h \in N$. Therefore $|h(g(x) - g(x_0))| < \varepsilon/2$ and $|(h - h_0)(g(x_0))| < \varepsilon/2$. It follows that

$$\begin{aligned}|G(x, h)(g) - G(x_0, h_0)(g)| &= |h(g(x)) - h_0(g(x_0))| \\ &\leq |(h - h_0)(g(x_0))| + |h(g(x) - g(x_0))| < \varepsilon\end{aligned}$$

for all $(x, h) \in U \times N$.

Step 3. *G^{-1} is continuous.*

Proof. Let $H_\alpha \to H$ in $\Delta(\mathscr{C}V_\infty(X; E))$. Since G is onto $\Delta(\mathscr{C}V_\infty(X; E))$, there exists nets $\{x_\alpha\}$ in X and $\{h_\alpha\}$ in $\Delta(E)$ such that $H_\alpha = G(x_\alpha, h_\alpha)$, and points $x \in X$ and $h \in \Delta(E)$ such that $H = G(x, h)$. Since $H \neq 0$, there is some $f \in \mathscr{C}V_\infty(X; E)$ such that $H(f) = 1$. Choose α_0 such that $\alpha \geq \alpha_0$ implies $H_\alpha(f) \neq 0$. Let $g \in \mathscr{C}_b(X; \mathbb{K})$ be given. Then gf belongs to $\mathscr{C}V_\infty(X; E)$ and for all $\alpha \geq \alpha_0$

$$g(x_\alpha) = \frac{g(x_\alpha)h_\alpha(f(x_\alpha))}{h_\alpha(f(x_\alpha))} = \frac{H_\alpha(gf)}{H_\alpha(f)}.$$

Therefore $g(x_\alpha) \to g(x)$. Indeed,

$$g(x_\alpha) \to H(gf) = h(g(x)f(x)) = g(x)h(f(x)) = g(x)H(f) = g(x).$$

Since X is a completely regular Hausdorff space and $g \in \mathscr{C}_b(X; \mathbb{K})$ was arbitrary, $x_\alpha \to x$ in X.

On the other hand, given $u \in E$, then $uf \in \mathscr{C}V_\infty(X; E)$, and for all $\alpha \geq \alpha_0$ we have

$$h_\alpha(u) = \frac{h_\alpha(u)h_\alpha(f(x_\alpha))}{h_\alpha(f(x_\alpha))} = \frac{H_\alpha(uf)}{H_\alpha(f)}.$$

Therefore $h_\alpha(u) \to h(u)$. Indeed,

$$h_\alpha(u) \to H(uf) = h(uf(x)) = h(u)h(f(x)) = h(u)H(f) = h(u).$$

Hence $h_\alpha \to h$ in the relative weak topology of $\Delta(E)$.

5. Properties of Nachbin Algebras

Definition 5.1. A locally convex algebra E is said to be *semisimple* if the relation $\hat{u} = 0$ implies $u = 0$, for every $u \in E$. The function $\hat{u}$ is the *Gelfand transform* of u, defined on the spectrum $\Delta(E)$ by

$$\hat{u}(h) = h(u)$$

for all $h \in \Delta(E)$.

Proposition 5.2. *If E is semisimple, then $\mathscr{C}V_\infty(X; E)$ is semisimple.*

Proof. Let $f \in \mathscr{C}V_\infty(X; E)$ be such that $\hat{f} = 0$. Let $x \in X$ be given. For each $h \in \Delta(E)$, $h \circ \delta_x$ belongs to $\Delta(\mathscr{C}V_\infty(X; E))$. Hence $h(f(x)) = \hat{f}(h \circ \delta_x) = 0$. Therefore $(f(x))\hat{} = 0$. Since E is semisimple, $f(x) = 0$. Hence $f = 0$, and $\mathscr{C}V_\infty(X; E)$ is semisimple.

Proposition 5.3. *Let V be such that $\mathscr{C}V_\infty(X; \mathbb{K})$ is essential and $\mathscr{C}V_\infty(X; E)$ is semisimple. Then E is semisimple.*

Proof. Let $v \in E$, be such that $h(v) = 0$ for all $h \in \Delta(E)$. Let $g \in \mathscr{C}V_\infty(X; \mathbb{K})$ be given and let $H \in \Delta(\mathscr{C}V_\infty(X; E))$. Since $\mathscr{C}V_\infty(X; \mathbb{K})$ is essential, and $\mathscr{C}V_\infty(X; \mathbb{K}) \otimes E$ is contained in $\mathscr{C}V_\infty(X; E)$, the Nachbin algebra $\mathscr{C}V_\infty(X; E)$ is essential. Therefore the map G is onto the spectrum $\Delta(\mathscr{C}V_\infty(X; E)$, by Step 1, Section 4. Hence $H = h \circ \delta_x$ for some $h \in \Delta(E)$ and $x \in X$. Then $H(g \otimes v) = h(g(x)v) = g(x)h(v) = 0$. Since $\mathscr{C}V_\infty(X; E)$ is semisimple, $g \otimes v = 0$. Now $g \in \mathscr{C}V_\infty(X; \mathbb{K})$ was arbitrary, so $v = 0$ and E is semisimple.

Definition 5.4. A locally convex algebra E is said to be *regular* if, for every closed subset $A \subset \Delta(E)$ and every $h \in \Delta(E)$ with $h \notin A$, there is $u \in E$ such that $\hat{u}(h) = 1$ and $\hat{u}(k) = 0$ for every $k \in A$.

PROPOSITION 5.5. *Let V be such that $\mathscr{C}V_\infty(X;\mathbb{K})$ is essential, and let E be a locally convex algebra such that $\Delta(E)$ is equicontinuous. Then, E is regular if, and only if, $\mathscr{C}V_\infty(X;E)$ is regular.*

Proof. We saw in the proof of Proposition 5.3 that, if $\mathscr{C}V_\infty(X;\mathbb{K})$ is essential, then $\mathscr{C}V_\infty(X;E)$ is essential too. Hence G is a homeomorphism between $X \times \Delta(E)$ and $\Delta(\mathscr{C}V_\infty(X;E))$. Let now $A \subset \Delta(\mathscr{C}V_\infty(X;E))$ be a closed subset and $H \in \Delta(\mathscr{C}V_\infty(X;E))$ be an element with $H \notin A$. Then $H = G(x,h)$ for some $x \in X$ and $h \in \Delta(E)$. Choose two closed subsets Y in X and B in $\Delta(E)$, with $(x,h) \in \complement Y \times \complement B \subset \complement(G^{-1}(A))$. Since $\mathscr{C}V_\infty(X;\mathbb{K})$ is essential, choose $g \in \mathscr{C}V_\infty(X;\mathbb{K})$ with $g(x) = 1$. The space X being completely regular, choose $\varphi \in \mathscr{C}_b(X;\mathbb{R})$, with $0 \le \varphi \le 1$, $\varphi(x) = 1$ and $\varphi(y) = 0$ for all $y \in Y$. Then $f = \varphi g \in \mathscr{C}V_\infty(X;\mathbb{K})$, $f(x) = 1$ and $f(y) = 0$ for all $y \in Y$. Notice, since we know from Corollary 2.3 that X is the spectrum of $\mathscr{C}V_\infty(X;\mathbb{K})$, that this shows that the algebra $\mathscr{C}V_\infty(X;\mathbb{K})$ is regular. Now, if E is regular, choose $u \in E$ such that $\hat{u}(h) = 1$ and $\hat{u}(k) = 0$ for all $k \in B$. Let us define $w \in \mathscr{C}V_\infty(X;E)$ to be the function $x \mapsto uf(x)$. Then

$$\hat{w}(H) = G(x,h)(w) = h(uf(x)) = f(x)h(u) = 1.$$

On the other hand, if $K \in A$ and $K = G(y,k)$, then $(y,k) \in G^{-1}(A) \subset \complement(\complement Y \times \complement B)$. Therefore, either $y \in Y$ or $k \in B$. Hence

$$\hat{w}(K) = G(y,k)(w) = k(uf(y)) = f(y)k(u) = 0,$$

because $y \in Y$ implies $f(y) = 0$ and $u \in B$ implies $k(u) = 0$. This proves that $\mathscr{C}V_\infty(X;E)$ is regular, if E is regular.

Conversely, assume that $\mathscr{C}V_\infty(X;E)$ is regular. Let $A \subset \Delta(E)$ be a closed subset and let $h \in \Delta(E)$ be such that $h \notin A$. Choose $x \in X$. Then $\{x\} \times A$ is closed in $X \times \Delta(E)$, and its image $B = G(\{x\} \times A)$ is closed in the spectrum of $\mathscr{C}V_\infty(X;E)$. Now $G(x,h) \notin B$, and there exists $f \in \mathscr{C}V_\infty(X;E)$ such that

$$\hat{f}(G(x,h)) = h(f(x)) = 1, \qquad \text{and} \qquad \hat{f}(g(x,k)) = k(f(x)) = 0,$$

for all $k \in A$. Then $u = f(x) \in E$ satisfies $\hat{u}(h) = 1$ and $\hat{u}(k) = 0$ for all $k \in A$; that is, E is regular.

DEFINITION 5.6. A locally convex algebra E is said to satisfy the *Wiener–Tauber condition* if the set of elements $u \in E$ such that $\hat{u}$ has compact support is dense in E.

PROPOSITION 5.7. *Let X be a locally compact Hausdorff space. Then $\mathscr{C}V_\infty(X;\mathbb{K})$ satisfies the Wiener–Tauber condition.*

Proof. Let $\mathscr{K}(X;\mathbb{K})$ be the set of all elements in $\mathscr{C}(X;\mathbb{K})$ which have compact support. Then $\mathscr{K}(X;\mathbb{K}) \subset \mathscr{C}V_\infty(X;\mathbb{K})$ and it is dense.

PROPOSITION 5.8. *Let X be a locally compact Hausdorff space, and let E be a locally convex algebra such that $\Delta(E)$ is equicontinuous. If E satisfies the Wiener–Tauber condition, then $\mathscr{C}V_\infty(X; E)$ also satisfies it.*

Proof. Since $\mathscr{K}(X; \mathbb{K}) \subset \mathscr{C}V_\infty(X; \mathbb{K})$, then $\mathscr{C}V_\infty(X; \mathbb{K})$ is essential and Theorem 2.1 can be applied. Now $\mathscr{K}(X; \mathbb{K}) \otimes E$ is dense in $\mathscr{C}V_\infty(X; E)$. On the other hand, if we denote by $k(E)$ the set $\{u \in E; \hat{u} \text{ has compact support}\}$, then $k(E)$ is dense in E, since E satisfies the Wiener–Tauber condition. Now any element in $\mathscr{K}(X; \mathbb{K}) \otimes k(E)$ has compact support, and this set is dense in $\mathscr{K}(X; \mathbb{K}) \otimes E$, hence dense in $\mathscr{C}V_\infty(X; E)$.

COROLLARY 5.9. *Let X and E be as in Proposition 5.8. Furthermore, assume that E is a complete regular semisimple commutative locally m-convex algebra satisfying the Wiener–Tauber condition. Then every proper closed ideal in $(\mathscr{C}(X; E), \kappa)$ is contained in a closed regular maximal ideal.*

Proof. Since X is locally compact, $(\mathscr{C}(X; E), \kappa)$ is a complete commutative locally m-convex algebra. Since E is regular, semisimple and satisfies the Wiener–Tauber condition, by Propositions 5.2, 5.5, and 5.8, the same is true of $(\mathscr{C}(X; E), \kappa)$. By Mallios [5, Lemma 2.3], every proper closed ideal in $\mathscr{C}(X; E)$ is contained in a closed regular maximal ideal.

Remark. Corollary 5.9 is also true for $\mathscr{C}_\infty(X; E)$ with the topology of uniform convergence on X.

REFERENCES

1. W. E. DIETRICH, JR., The maximal ideal space of the topological algebra $\mathscr{C}(X; E)$, *Maths. Ann.* **183** (1969), 201–212.
2. A. HAUSNER, Ideals in a certain Banach algebra, *Proc. Amer. Math. Soc.* **8** (1957), 246–249.
3. I. KAPLANSKY, The structure of certain operator algebras, *Trans. Amer. Math. Soc.* **70** (1951), 219–255.
4. A. MALLIOS, Heredity of tensor products of topological algebras, *Math. Ann.* **162** (1966), 246–257.
5. A. MALLIOS, Tensor products and Harmonic Analysis, *Math. Ann.* **158** (1965), 46–56.
6. E. A. MICHAEL, Locally multiplicatively-convex topological algebras, *Mem. Amer. Math. Soc.* **11** (1952).
7. J. B. PROLLA, "Approximation of Vector-Valued Functions," North-Holland Publ., Amsterdam, 1977.
8. J. B. PROLLA, On the spectra of non-archimedean function algebras, *in* "Advances in Functional Analysis, Holomorphy and Approximation Theory" (S. Machado, ed.), Lecture Notes in Mathematics, Springer-Verlag, Berlin and New York, to appear.
9. B. YOOD, Banach algebras of continuous functions, *Amer. J. Math.* **73** (1951), 30–42.

MATHEMATICAL ANALYSIS AND APPLICATIONS, PART B
ADVANCES IN MATHEMATICS SUPPLEMENTARY STUDIES, VOL. 7B

Lebesgue's First Theorem

WALTER RUDIN[†]

Department of Mathematics
University of Wisconsin
Madison, Wisconsin

TO LAURENT SCHWARTZ, ON THE OCCASION OF HIS 65TH BIRTHDAY

If a real-valued function f, defined in the plane, is continuous in x for each y and is continuous in y for each x, must f be measurable?

Several years ago, I used to pose this question to randomly selected analysts. The typical answer was something like this: "Hmm—well—probably not—why should it be?" The only group that did a little better were the probabilists. And there was just one person who said: "Let's see—yes, it is—and it is of Baire class1—and . . .". He knew.

In Lebesgue's first published paper [7] the question is answered affirmatively: For each natural number n, draw vertical lines in the plane, each at distance $1/n$ from its left and right neighbors. Define $f_n(x, y)$ to be $f(x, y)$ on the union of these lines, and determine $f_n(x, y)$ on the rest of the plane by linear interpolation in the x variable. Since f is a continuous function of y, for each fixed x, each f_n is a continuous function on R^2. Since f is a continuous function of x, for each fixed y,

$$\lim_{n\to\infty} f_n(x, y) = f(x, y)$$

for all $(x, y) \in R^2$. Done.

When f is a separately continuous function on R^k, an easy induction shows that f is then of Baire class $k - 1$. In a later paper [8, p. 202] Lebesgue proved this, as well as the much more surprising result that $k - 1$ cannot be replaced by any smaller integer.

To return to Lebesgue's proof, let us define

$$h_j(x) = \max(0, 1 - |nx - j|),$$

[†] This research was partially supported by NSF Grant MCS 78-06860, and by the William F. Vilas Trust Estate.

ISBN 0-12-512802-9

for fixed n, and for $j = 0, \pm 1, \pm 2, \ldots$. Then $\{h_j\}$ is a partition of unity, and our approximating functions f_n can be written in the form

$$f_n(x, y) = \sum_{j=-\infty}^{\infty} h_j(x) f\left(\frac{j}{n}, y\right)$$

in which f_n appears as a convex combination of continuous functions of y, with coefficients that are continuous functions of x. This suggests that the proof should work in much more general situations, namely, for functions

$$f: X \times Y \to E,$$

where X and Y are topological spaces (subject to some hypotheses) and E is a *locally convex topological vector space.*

Such an f induces "slice functions"

$$f_x: Y \to E \qquad \text{and} \qquad f^y: X \to E,$$

defined for $x \in X$ and $y \in Y$ by

$$f_x(y) = f(x, y) = f^y(x).$$

To say that f is *separately continuous* means that all slice functions f_x and f^y are continuous.

The question thus arises whether separately continuous functions are always Borel measurable in this setting, i.e., whether $f^{-1}(\Omega)$ must be a Borel set in $X \times Y$ for every open set $\Omega \subset E$.

Lebesgue's theorem gives an affirmative answer when $X = Y = E = R$. Other affirmative answers have been obtained by Hahn [4, p. 327] (X and Y separable metric, $E = R$), by Kuratowski [6, p. 181] (he removed the separability assumption), by Montgomery [9] (for $f: X \times Y \to Z$, where X, Y, Z are arbitrary metric spaces), by Johnson [5] (if $E = R$ and X, Y are locally compact Hausdorff spaces that carry finite positive Borel measures which vanish on no nonempty open set), and by Moran [10] (if $E = R$ and X, Y are compact Hausdorff spaces of which at least one carries a measure as in Johnson's theorem). Moran showed also that real-valued separately continuous functions on $X \times Y$ need not be Baire functions if neither of the compact Hausdorff spaces X and Y is the support of a finite Borel measure.

The work of Johnson and Moran was in response to a question raised by Glicksberg [2], where the measurability problem arose in a natural way in studying convolutions on separately continuous semigroups: one needs measurability every time that Fubini's theorem is invoked.

Some other recent contributions to this question were made by Gowrisankaran [3], Fremlin [1], and Wong [12].

The proof of Theorem 1 below is basically just as easy as the one given by Lebesgue, because of the (by now well-known) fact that metric spaces are paracompact; a very short proof of this was given by M. E. Rudin [11]. To be more precise, we need the following consequence of paracompactness:

If X is a metric space and n is a natural number, then there exists a locally finite partition of unity on X, of mesh $1/n$; i.e., there are continuous functions $h_{\alpha,n}: X \to [0, 1]$, such that

$$\sum_{\alpha} h_{\alpha,n}(x) = 1, \qquad x \in X,$$

such that every point of X has a neighborhood in which all but finitely many of the functions $h_{\alpha,n}$ vanish identically, and such that the support of each $h_{\alpha,n}$ has diameter at most $1/n$.

THEOREM 1. *Suppose X is a metric space, Y is a topological space, and E is a locally convex topological vector space.*

For $n = 1, 2, 3, \ldots$, let $\{h_{\alpha,n}\}$ be a locally finite partition of unity on X, of mesh $1/n$, let D be a dense subset of X, and choose $x_{\alpha,n} \in D$ so that $h_{\alpha,n}(x_{\alpha,n}) > 0$.

If $f: X \times Y \to E$ satisfies

(a) *$f^y: X \to E$ is continuous for each $y \in Y$,*

(b) *$f_x: Y \to E$ is continuous for each $x \in D$, and $F_n: X \times Y \to E$ is defined by*

$$F_n(x, y) = \sum_{\alpha} h_{\alpha,n}(x) f(x_{\alpha,n}, y) \tag{$*$}$$

then each F_n is continuous on $X \times Y$, and

$$\lim_{n \to \infty} F_n(x, y) = f(x, y)$$

at every point $(x, y) \in X \times Y$.

If E is metrizable, it follows that f is a Borel function on $X \times Y$.

Proof. The continuity of F_n is an obvious consequence of the local finiteness of $\{h_{\alpha,n}\}$.

Fix $(x, y) \in X \times Y$, let V be a convex neighborhood of O in E. By (a) there exists n_0 such that

$$f(\xi, y) - f(x, y) \in V$$

for all $\xi \in X$ with $d(x, \xi) < 1/n_0$, where d denotes the metric of X. If $n > n_0$ and α is an index for which $h_{\alpha,n}(x) > 0$, then our assumptions on $\{h_{\alpha,n}\}$ show that $d(x, x_{\alpha,n}) < 1/n_0$, so that

$$f(x_{\alpha,n}, y) \in f(x, y) + V.$$

It now follows from $(*)$ and the convexity of V that

$$F_n(x, y) \in f(x, y) + V$$

for all $n > n_0$. Thus $F_n(x, y)$ converges to $f(x, y)$ as $n \to \infty$.

To finish, assume E is metrizable. To every open $\Omega \subset E$ corresponds then a continuous function $p: E \to [0, 1]$ such that $p > 0$ exactly in Ω. Since

$$(p \circ f)(x, y) = \lim_{n \to \infty} (p \circ F_n)(x, y),$$

and since each $p \circ F_n$ is a real-valued continuous function on $X \times Y$, $p \circ f$ is a Borel function, hence

$$f^{-1}(\Omega) = \{(x, y) : (p \circ f)(x, y) > 0\}$$

is a Borel set in $X \times Y$.

Remarks. Note that the approximating functions F_n depend linearly on f, and that the process by which they are constructed does not involve Y or E in any significant way. The range of F_n lies in the convex hull of $f(D \times Y)$—not merely in the *closed* convex hull. The target space E is not assumed to be complete. As to the Borel measurability of f, here is an example in which it fails, even though $X = Y = R$.

EXAMPLE. Let E be the vector space of all bounded real Borel functions on R, with the topology of pointwise convergence. Define $g: R \times R \to R$ by

$$g(x, y) = \frac{2xy}{x^2 + y^2} \qquad \text{if} \quad (x, y) \neq (0, 0),$$

put $g(0, 0) = 0$, and define $f: R \times R \to E$ by

$$f(x, y)(t) = g(x - t, y - t).$$

Since g is separately continuous, and since the topology of E is that of pointwise convergence, f is separately continuous.

For each $t \in R$, define

$$V_t = \{\varphi \in E : |\varphi(t)| < \tfrac{1}{2}\}.$$

Then V_t is open in E, and $f(x, y) \in V_t$ if and only if $|g(x - t, y - t)| < \frac{1}{2}$. Let Δ be the diagonal of $R \times R$. Since $g = 1$ at all points of Δ except at the origin, it follows that

$$\Delta \cap f^{-1}(V_t) = \{(t, t)\},$$

a single point of Δ.

Now let T be a non-Borel subset of R, and let V be the union of all V_t, for $t \in T$. Then V is open in E, and

$$\Delta \cap f^{-1}(V) = \{(t, t) : t \in T\},$$

which is not a Borel set. Hence $f^{-1}(V)$ is not a Borel set in $R \times R$, so that f is not a Borel function.

The main purpose of this paper is to stress the simplicity of the proof of Theorem 1, and the linearity of the approximation that it furnishes, rather than to try for the greatest possible generality. Nevertheless, we shall present one more result (Theorem 2), which can be derived from Theorem 1, although neither X nor Y will be assumed to be metrizable. Theorem 2 is related to Moran's work [10], but the hypotheses are a little weaker, and the proof is perhaps a bit more elementary.

We begin with some lemmas. They are known and simple, but their explicit statement will make the proof of Theorem 2 more transparent.

If T is any topological space, $C(T)$ will denote the set of all continuous real-valued functions on T, and for any $A \subset C(T)$, the symbol (A, p) will denote A equipped with the topology of pointwise convergence on T.

LEMMA 1. *If K is compact and if some countable set $\{f_n\} \subset C(K)$ separates points on K, then K is metrizable, hence separable.*

Proof. $$d(x, y) = \sum_n \frac{2^{-n}|f_n(x) - f_n(y)|}{(1 + |f_n(x) - f_n(y)|)}.$$

LEMMA 2. *If T is a topological space with a dense σ-compact subset S, and if some countable subset of $C(T)$ separates points on S, then T is separable.*

Proof. S is the union of countably many compact sets to which Lemma 1 applies, and which are therefore separable. Their union is dense in T.

LEMMA 3. *If T is a separable topological space then every compact subset of $(C(T), p)$ is metrizable.*

Proof. Suppose $A \subset C(T)$ and (A, p) is compact. There is a countable set S that is dense in T. Define $\varphi_s(f) = f(s)$ for $f \in A$, $s \in S$. Each φ_s is continuous on (A, p). Since S is dense in T, the collection $\{\varphi_s : s \in S\}$ separates points on (A, p). Lemma 1 can now be applied to (A, p) in place of K.

LEMMA 4. *If X is a topological space with a dense σ-compact subset, then every separable compact subset of $(C(X), p)$ is metrizable.*

Proof. Suppose $E \subset C(X)$ and (E, p) is compact and separable. Associate to every $x \in X$ the closed set $j(x)$ consisting of all $x' \in X$ that satisfy $f(x') = f(x)$ for every $f \in E$. Let T be the corresponding identification space: T is the set of all equivalence classes $j(x)$, and $\Omega \subset T$ is open if and only if $j^{-1}(\Omega)$ is open in X. To each $f \in E$ corresponds a unique f^* on T such that $f = f^* \circ j$. This f^* is continuous on T. Moreover, $E^* = \{f^* : f \in E\}$ separates points on T, and the map $f \to f^*$ is a homeomorphism of (E, p) onto (E^*, p).

Since (E, p) is separable, it has a countable subset A that is *dense* in (E, p). Therefore A and E induce the same equivalence relation in X. Thus $A^* = \{f^*: f \in A\}$ separates points on T. Since j is continuous, T satisfies the hypotheses of Lemma 2, so that T is separable. By Lemma 3, (E^*, p) is metrizable, hence so is its homeomorph (E, p).

THEOREM 2. *Suppose that X is a topological space with a dense σ-compact subset, and that there is a Borel probability measure μ on X such that $\mu(\Omega) > 0$ for every nonempty open $\Omega \subset X$.*

If Y is compact, then every separately continuous $f: X \times Y \to R$ is a pointwise limit of a sequence of continuous functions.

Note that *every* separable X carries a measure with the required properties.

Proof. Let $E \subset C(X)$ satisfy $|g(x)| \leq 1$ for all $g \in E$, $x \in X$, and assume that (E, p) is compact. Our first objective is to prove that (E, p) is metrizable.

Let A be any infinite countable subset of E, let $\bar{A}$ be its closure in $(C(X), p)$. Then $\bar{A} \subset E$. By Lemma 4, $(\bar{A}, p)$ is compact metric. If $g \in (E, p)$ is a limit point of A, it follows that some sequence of distinct members of A converges pointwise to g. The dominated convergence theorem shows therefore that A has a limit point in (E, μ), i.e., in the set E equipped with the norm topology of $L^1(\mu)$. Since this topology is metric, we have shown that (E, μ) is compact. Hence there is a countable set $B \subset E$ such that B is dense in (E, μ). If $h \in E$, there exists $\{b_n\} \subset B$ such that $\int |b_n - h| \, d\mu \to 0$. By Lemma 4, applied to the closure $\bar{B}$ of B in $(C(X), p)$, some subsequence $\{b_{n_i}\}$ converges pointwise to some $g \in \bar{B}$. Since $\mu(\Omega) > 0$ for every nonempty open Ω, it follows that $g = h$. Thus $h \in \bar{B}$. We have now shown that B is dense in (E, p), i.e., that (E, p) is separable. Another application of Lemma 4 shows that (E, p) is metrizable.

Assume now, without loss of generality, that $|f(x, y)| \leq 1$ for all (x, y), and put

$$E_0 = \{f^y: y \in Y\}.$$

Then $E_0 \subset C(X)$. We claim that the map

$$\pi: y \to f^y$$

is continuous from Y onto (E_0, p). To see this, fix $y \in Y$, and let V be a neighborhood of f^y in (E_0, p). Then there exist $x_1, \ldots, x_n \in X$ and $\varepsilon > 0$ such that V contains all f^z for which

$$|f^z(x_i) - f^y(x_i)| < \varepsilon, \qquad i = 1, \ldots, n.$$

If Ω is the set of all $z \in Y$ for which these inequalities hold, then $\pi(\Omega) \subset V$. The continuity of the functions f_x shows that Ω is open in Y. Thus π is continuous, as claimed.

It follows that (E_0, p) is compact, and hence metrizable, by the first part of the proof.

To finish, define $\psi: X \times (E_0, p) \to R$ by

$$\psi(x, f^y) = f(x, y).$$

Since ψ is separately continuous and (E_0, p) is metric, Theorem 1 shows that ψ is the pointwise limit of a sequence of continuous functions ψ_n of the form

$$\psi_n(x, f^y) = \sum_{\alpha} h_\alpha(f^y) f(x, y_\alpha).$$

Put $F_n(x, y) = \psi_n(x, \pi(y))$. Then $F_n \in C(X \times Y)$, and

$$\lim_{n \to \infty} F_n(x, y) = \lim_{n \to \infty} \psi_n(x, f^y) = \psi(x, f^y) = f(x, y)$$

for all $(x, y) \in X \times Y$.

References

1. D. H. Fremlin, Pointwise compact sets of measurable functions, *Manuscripta Math.* **15** (1975), 219–242.
2. I. Glicksberg, Weak compactness and separate continuity, *Pacific J. Math.* **11** (1961), 205–214.
3. K. Gowrisankaran, Measurability of functions in product spaces, *Proc. Amer. Math. Soc.* **31** (1972), 485–488.
4. H. Hahn, "Reelle Funktionen," Akad. Verlag, Leipzig, 1932.
5. B. E. Johnson, Separate continuity and measurability, *Proc. Amer. Math. Soc.* **20** (1969), 420–422.
6. C. Kuratowski, "Topologie I," Monografje Math., Vol. 3, Polskie Towarzystwo Matematyczne, Warsaw, 1933.
7. H. Lebesgue, Sur l'approximation des fonctions, *Bull. Sci. Math.* **22** (1898), 278–287.
8. H. Lebesgue, Sur les fonctions representables analytiquement, *J. Math. Pures Appl.* **1** (1905), 139–215.
9. D. Montgomery, Non-separable metric spaces, *Fund. Math.* **25** (1935), 527–533.
10. W. Moran, Separate continuity and supports of measures, *J. London Math. Soc.* **44** (1969), 320–324.
11. M. E. Rudin, A new proof that metric spaces are paracompact, *Proc. Amer. Math. Soc.* **20** (1969), 603.
12. J. C. S. Wong, Convolution and separate continuity, *Pacific J. Math.* **75** (1978), 601–611.

MATHEMATICAL ANALYSIS AND APPLICATIONS, PART B
ADVANCES IN MATHEMATICS SUPPLEMENTARY STUDIES, VOL. 7B

Quantization of Symplectic Transformations[†]

IRVING SEGAL[‡]

The Institute for Advanced Study
Princeton, New Jersey

DEDICATED TO LAURENT SCHWARTZ ON THE OCCASION OF HIS 65TH BIRTHDAY

A sympletic transformation on a linear space is quantizable in the sense of having a corresponding induced action on an associated quantum field with vacuum if and only if it is unitarizable. The quantization is unique if and only if the corresponding complex Hilbertian structure is unique. This unicity holds in turn if and only if the transformation is unitarizable with spectrum disjoint (measure theoretically, as appropriate for operators in Hilbert space) from that of its inverse.

INTRODUCTION

The heuristic process of quantization may be analyzed in terms of two successive processes: (1) the formulation of the commutation relations and of generalized operators satisfying these relations and a prescribed differential equation; (2) the determination of a vacuum state, and thereby a concrete Hilbert space on which the quantized field operators act appropriately. Process (1) is simpler by virtue of being local, and is also more general, as regards the class of equations to which it applies. Process (2) is nonlocal, requiring knowledge of the field throughout space and time, and is possible only under stability-type restrictions. For this reason, the quantization of general wave equations, even when linear, remains incomplete; process (2) has remained basically undeveloped for some time.

Here we consider this problem in the algebraic spirit exemplified so usefully and elegantly by the work of L. Schwartz in physical mathematics. This is applied to the proposal in [5] that the vacuum be determined as the presumptively unique regular state that is invariant under the scattering automorphism. Mathematical theory is developed for the quantization of a

[†] Research supported in part by the National Science Foundation.

[‡] Present address: Department of Mathematics, Massachussets Institute of Technology, Cambridge, Massachussetts.

ISBN 0-12-512802-9

given linear symplectic space. This space may be interpreted physically as the phase space of a system described by a linear wave equation, in a curved space–time that is asymptotically flat, or subject to a time dependent perturbation in a flat space–time; however, no specific interpretation is required in the present purely mathematical treatment.

Technical Preliminaries and Notations

By a *symplectic Hilbertizable* space we mean the pair consisting of a real linear topological space $\mathbf{L}$ together with a real bilinear antisymmetric form A on $\mathbf{L}$, having the properties that $\mathbf{L}$ is isomorphic to a complex Hilbert space $\mathbf{H}$ in such a way that A is carried into the imaginary part of the inner product. Thus, if T denotes the isomorphism of $\mathbf{L}$ with $\mathbf{H}$, $A(z, z') = \operatorname{Im}\langle Tz, Tz'\rangle$ for arbitrary $z, z' \in \mathbf{L}$. The real part of the inner product in $\mathbf{H}$, and the complex structure in $\mathbf{H}$, are consequently not at all unique.

The group of all bounded invertible real linear transformations on $\mathbf{L}$ that preserve the form A will be denoted as $\mathrm{Sp}(\mathbf{L}, A)$, or when $\mathbf{L}$ has a given complex Hilbertian structure, as $\mathrm{Sp}(\mathbf{L})$. An element R of $\mathrm{Sp}(\mathbf{L}, A)$ is *elliptic* if it is unitary in some representation of $\mathrm{Sp}(\mathbf{L}, A)$ in the form $\mathrm{Sp}(\mathbf{H})$ just described; it is *regular* if in this representation it has simple spectrum (or equivalently, the W^*-algebra it generates is maximal abelian). In particular, an element of $\mathrm{Sp}(\mathbf{H})$, $\mathbf{H}$ being a given complex Hilbert space, is elliptic if it is conjugate to a unitary transformation on $\mathbf{H}$ and regular if it commutes with no bounded linear operators other than functions of itself.

Quantization will involve the concept of *Weyl system* over $(\mathbf{L}, A)$: This is defined here as a pair $(\mathbf{K}, W)$ in which $\mathbf{K}$ is a complex Hilbert space and W is a strongly continuous map from $\mathbf{H}$ into $U(\mathbf{K})$, which notation signifies the unitary group of $\mathbf{K}$, satisfying the relations

$$W(z)W(z') = e^{iA(z,z')/2}W(z + z').$$

Definition. The system $(S, \mathbf{L}, A)$ is unitary quantizable if there exists $(\mathbf{K}, W, \Sigma, v)$, where $(\mathbf{K}, W)$ is a Weyl system over $(\mathbf{L}, A)$, $\Sigma \in U(\mathbf{K})$, $\Sigma v = v$, and $\Sigma W(z)\Sigma^{-1} = W(Sz)$ for arbitrary z, and the only (closed) subspace of $\mathbf{K}$ containing v and invariant under all the $W(z)$ and Σ is all of $\mathbf{K}$ (i.e., z is *cyclic* in $\mathbf{K}$ for $[W(z): z \in \mathbf{L}]$ together with Σ).

An alternative notion of quantization will involve the concepts of the Weyl algebra of $(\mathbf{L}, A)$ and of a regular state of this algebra.

Definition. If $(\mathbf{K}, W)$ is a Weyl system over $(\mathbf{L}, A)$ and if $\mathbf{M}$ is an arbitrary finite dimensional "nondegenerate" subspace, where nondegenerate means that the restriction of A to $\mathbf{M}$ is nondegenerate, let $\mathbf{A}_{\mathbf{M}}$ denote the W^*-algebra generated by the $W(z)$ with $z \in \mathbf{M}$, and let $\mathbf{A}_W$ denote the C^*- (i.e., uniform) closure of $\bigcup_{\mathbf{M}} \mathbf{A}_{\mathbf{M}}$.

Then the following holds (but would be completely false were the W^*-closure used in place of the C^*-closure, except in the finite dimensional case in which these closures coincide):

Generalized Stone–Von Neumann Theorem

For any two Weyl systems W and W' over $(\mathbf{L}, A)$, there exists a unique algebraic $*$-isomorphism θ of $\mathbf{A}_W$ onto $\mathbf{A}_{W'}$ such that $\theta(W(z)) = W'(z)$, for all $z \in \mathbf{L}$.

The correspondingly unique algebraic equivalence class of the algebras $\mathbf{A}_W$ is called the *Weyl algebra* over $\mathbf{H}$. The theorem implies in particular that for any $S \in \mathrm{Sp}(\mathbf{L}, \mathbf{H})$ there exists a unique automorphism $\theta(S)$ of the Weyl algebra such that $\theta(S)$ carries $W(z)$ into $W(Sz)$ in any concrete representation of the Weyl algebra. A *regular* state of this algebra is defined as a state E in the usual sense applicable to C^*-algebras (i.e., a normalized positive linear functional) with the property that for any finite dimensional nondegenerate subspace $\mathbf{M}$ of $\mathbf{L}$, there exists a relative trace class operator $D_{\mathbf{M}}$ in $\mathbf{A}_{\mathbf{M}}$—which by the Stone–von Neumann theorem is $*$-algebraic isomorphic to the algebra of all bounded linear operators on a Hilbert space, and so possesses a unique *relative* trace, irrespective of the multiplicity with which it acts on $\mathbf{K}$—such that $E(B) = \mathrm{tr}(BD_{\mathbf{M}})$ for all $B \in \mathbf{A}_{\mathbf{M}}$.

DEFINITION. The system $(S, \mathbf{L}, A)$ is C^*-quantizable if there exists a regular state E of the Weyl algebra that is invariant under the C^*-automorphism σ induced by S.

Existence Criteria for Quantizations

The following theorem and in part its proof resembles [8, Theorem 5.1], but deals with a single transformation rather than a one-parameter group, and the proof here eliminates the need to refer to an unpublished work of mine cited in [8].

THEOREM 1. *Let S be an automorphism of the symplectic Hilbertizable space $(\mathbf{L}, A)$. The following conditions on S are all equivalent.*

(i) *S is unitary quantizable.*
(ii) *S is C^*-quantizable.*
(iii) *S is elliptic.*

It is evident that (i) implies (ii) and that (iii) implies (i), so it suffices to show that (ii) implies (iii).

Lemma 1.1. *If S is a symplectic on a complex Hilbert space* **H** *and if there exists a real positive-definite form on* **H** *that is equivalent to $Re\langle\cdot,\cdot\rangle$ and invariant under S, then S is elliptic.*

Proof. Define a new complex structure i' as the orthogonal constituent of the polar decomposition of the operator T for which $A(z, z') = Q(Tz, z')$, where Q is the invariant form whose existence is assumed; then i' is S-invariant and defines a new complex Hilbert space with the same imaginary part for its inner product as the original space **H** and is topologically equivalent to it.

Lemma 1.2. *If S is C^*-quantizable, then $\|S^n\|$ is bounded.*

Proof. By the general theory of representations of C^*-algebras with given states, corresponding to the given state E on the Weyl algebra **A**, invariant under the automorphism σ, there exists a complex Hilbert space **K**, a self-adjoint representation ϕ of **A** by bounded linear operators on **K**, operators $W'(z) = \phi(W(z))$ for arbitrary $z \in \mathbf{L}$, satisfying the Weyl relations and unitary. There is a unitary operator Σ on **K** and a unit vector v in **K** such that $\Sigma v = v$, $\Sigma W(z)\Sigma^{-1} = W(Sz)$ for arbitrary z, such that x is cyclic in **K** for the $W(z)$ together with Σ, and related to E and σ by the equations

$$E(B) = \langle\phi(B)v, v\rangle, \qquad \phi(\sigma(B)) = \Sigma\phi(B)\Sigma^{-1}, \qquad B \in \mathbf{A}.$$

The continuity of $W(\cdot)$ does not follow from general theory, but the regularity of the state E permits this deduction [6]. Thus S is C^*-quantizable only if it is unitary quantizable.

As the basis of an indirect argument, assume that S^n is unbounded; since $\{S^n; n = 0, \pm 1, \ldots\}$, forms a group, this means that $\|S^n x\|/\|x\|$ cannot be bounded away from 0; hence there exists a sequence $\{x_n\}$ of vectors in **L** such that $S^n x_n \to 0$, while $\|x_n\| = 1$. Setting $c_n = \|S^n x_n\|^{1/2}$, $y_n = c_n^{-1} x_n$, and $z_n = ic_n x_n$, and $\phi(u) = \langle W(u)v, v\rangle$ for arbitrary $u \in \mathbf{H}$, then $\phi(z_n) \to 1$ since $z_n \to 0$ and $W'(\cdot)$ is strongly continuous, while $\phi(y_n) = \phi(S^n y_n) \to 1$ since $\|S^n y_n\| = \|S^n x_n\| \to 0$. On the other hand, the Weyl relations imply that $W'(y_n)W'(z_n) = -W'(z_n)W'(y_n)$ whence

$$\langle W'(y_n)W'(z_n)v, v\rangle = \langle W'(z_n)W'(y_n)v, v\rangle,$$

which in passing to the limit $n \to \infty$ and noting that $\phi(x) = \overline{\phi(-x)}$ for arbitrary x gives the requisite contradiction.

Lemma 1.3. *If S is an invertible real linear operator on* **H** *for which $\|S^n z\|$ is bounded ($\pm n = 0, 1, 2, \ldots$), then there exists a real symmetric form on* **H** *that is equivalent to* $\mathrm{Re}\langle x, x'\rangle$ *and also S-invariant.*

Proof. (This result is well known, but for completeness its brief proof is included.) Define $F(z, z') = M[\langle S^n z, S^n z' \rangle : n \in \mathbf{Z}]$, where $M[\cdot]$ is any invariant mean on the infinite cyclic group $\mathbf{Z}$. Then F is an S-invariant real symmetric form that is continuous relative to the form $\operatorname{Re}\langle z, z' \rangle$, and, conversely, the latter form is continuous relative to F since for arbitrary $z \in \mathbf{H}$:

$$\langle z, z \rangle = \langle S^n z, S^{*-n} z \rangle \leq \|S^n z\| \, \|S^{*-n} z\|,$$

whence, noting that by duality the S^{*-n} are also uniformly bounded,

$$M[\langle z, z \rangle] \leq M[\|S^n z\| \, \|S^{*-n} z\|] \leq M[\|S^n z\|^2]^{1/2} M[\|S^{*-n} z\|^2]^{1/2},$$

i.e., $\langle z, z \rangle \leq \text{const} \cdot F(z, z)^{1/2} \langle z, z \rangle^{1/2}$, whence $\langle z, z \rangle \leq \text{const.} \cdot F\langle z, z \rangle$.

Completion of proof. Lemmas 1.2 and 1.3 imply that the hypothesis of Lemma 1.1 is satisfied, implying that S is elliptic.

Remark 1. An arbitrary automorphism of the Weyl algebra will by fixed point theory always admit an invariant state. But in general this state is mathematically quite irregular and physically appears unobservable. A simple example is provided by the case in which S is the transformation $\left(\begin{smallmatrix} \lambda & 0 \\ 0 & \lambda^{-1} \end{smallmatrix}\right)$ on $R^1 \oplus R^1$, with $A(x \oplus y, x' \oplus y') = xy' - x'y$. By standard analysis (so-called diagonalization of the canonical q, otherwise known as the real wave representation), the corresponding Weyl algebra may be identified with the algebra $\mathbf{B}$ of all bounded linear operators on $L_2(R^1)$ in such a way that the corresponding automorphism takes the form

$$\sigma(B) = U^{-1} B U; \qquad U : f(x) \to \lambda^{1/2} f(\lambda x), \qquad f \in L_2(R^1).$$

It may be seen that if E is any invariant state of $\mathbf{B}$ under σ, then E vanishes on all compact operators, as well as on all operators of the form $f(q)$ or $f(p)$, where f is an arbitrary continuous function of compact support, and q and p are the operators, $q : f(x) \to x f(x)$, $p : f(x) \to -i(d/dx) f(x)$ in their usual self-adjoint formulations.

Specifically, if $\mathbf{C}$ denotes the algebra of all compact operators in $\mathbf{B}$, then $E | \mathbf{C}$ must have the form $E(B) = \operatorname{tr} BD$ for some trace class operator D, in view of the known form of the dual of $\mathbf{C}$. For E to be invariant under σ it is necessary and sufficient that D commute with U. This would require U to have finite-dimensional invariant subspaces, but it has none. In the case of the $f(q)$ the restriction of E would define a finite measure on R^1 that is invariant under the transformation $x \to \lambda x$, but there is no nonvanishing such measure, and by Fourier transformation the vanishing on the $f(q)$ implies the same for the $f(p)$.

Physically, this could be interpreted as implying that for any σ-invariant state of $\mathbf{B}$, q and p are expected to be at infinity on the extended real line with probability 1, and so correspond to an unobservable state.

Unicity Criteria for Quantizations

The unicity issue is somewhat parallel to that of the unicity of a ground state, but differs in that semiboundedness is not necessarily involved. For this reason it is essential to restrict the quantization involved.

DEFINITION. A *free-field quantization* of $(S, \mathbf{L}, A)$ is a unitary quantization for which $\langle W(z)v, v \rangle = \exp(-\langle z, z \rangle/4)$, for all $z \in \mathbf{L}$, where $\langle \cdot, \cdot \rangle$ is the inner product in a Hilbertization of $(\mathbf{L}, A)$, i.e., a complex Hilbert space structure on $\mathbf{L}$ that is a topological isomorphism as regards the Hilbert space and given topologies on $\mathbf{L}$, and such that $A(z, z') = \operatorname{Im}\langle z, z' \rangle$, that is, S-invariant.

The existence of "universally" invariant quantizations [5] shows that without further restriction, quantization is far from being unique. The restriction to the free-field quantization appears mathematically natural and physically appropriate.

It is interesting that the free-field quantization of a symplectic, if it exists at all, is generically unique, but does not require regularity. Rather it requires a spectral condition that is invariant not only under the symplectic group but also under antisymplectics (which reverse the fundamental antisymmetric form); this has been treated in the present connection in [3, Section 32]; see the comment there on essentially equivalent concepts involved in work by Gelfand and Lidskii [2] and also, in infinite-dimensional cases, in work of Daleckii and Krein [1]).

In order to avoid an extensive digression concerning the spectral theory of symplectics, I here define the spectrum of an elliptic symplectic as that of any unitarization of it. An elliptic symplectic will be called *asymmetric* if the symplectic and its inverse have disjoint spectra, in a simultaneous unitarization. Recall in this connection that two normal operators A and A' in a complex Hilbert space $\mathbf{H}$, of spectral resolutions $E(\cdot)$ and $E'(\cdot)$, are said to have disjoint spectra in case the measures m_x and m'_y defined by the equations $m_x(G) = \langle E(G)x, x \rangle$ and $m'_y(G) = \langle E'(G)y, y \rangle$, where G is an arbitrary Borel subset of C^1, are mutually singular for arbitrary x and y in $\mathbf{H}$. To avoid ideationally extraneous considerations, it will be assumed that $\mathbf{L}$ is separable.

THEOREM 2. *An elliptic S in* $\mathrm{Sp}(\mathbf{L}, A)$, $(\mathbf{L}, A)$ *being a symplectic Hilbertizable linear space, has a unique free-field quantization if and only if it is asymmetric.*

LEMMA 2.1. *An arbitrary elliptic S in* $\mathrm{Sp}(\mathbf{L}, A)$ *has a unique free-field quantization if and only if* $(\mathbf{L}, A)$ *has a unique S-invariant Hilbertization.*

Proof. It is known [5–7] that an arbitrary quantization is determined within unitary equivalence by the "generating functional" $\phi(z) = \langle W(z)v, v \rangle$.

Thus there is unicity if and only if $\|z\|^2$ is uniquely determined by the given constraints, which means that $(\mathbf{L}, A)$ has a unique S-invariant Hilbertization. If on the other hand this is the case, then $\phi(z)$ and hence the free-field quantization is uniquely determined.

LEMMA 2.2. *If S is an elliptic in* $\mathrm{Sp}(\mathbf{L}, A)$ *whose spectrum is contained in the set* $[e^{i\theta}: \theta \in (0, \pi - \varepsilon)]$, *where* $\varepsilon > 0$, *it has a unique free-field quantization.*

Proof. By virtue of Lemma 2.1, it suffices to show that $(\mathbf{L}, A)$ has a unique S-invariant Hilbertization, and it is no essential loss of generality to assume that S is unitary on the Hilbertization $\mathbf{H}$ of $(\mathbf{L}, A)$. Since $z^a = e^{a \log z}$, with $\log z$ defined via a cut along the negative real axis and $|\arg \log z|$, correspondingly bounded by π, is complex analytic when z ranges over a neighborhood of the subset C_ε of the unit circle $[e^{i\theta}: |\theta| < -\varepsilon]$, there exists a sequence $\{p_n(z)\}$ of polynomials that converges uniformly to z^a on C_ε. By reflection in the real axis it follows that the sequence $\{\overline{p}_n(z)\}$, obtained from the original sequence by complex conjugation of all coefficients, does the same. Adding these two sequences, it follows in turn that a sequence $\{q_n(z)\}$ having the same property as $\{p_n(z)\}$, and in addition having real coefficients, exists.

If j is the complex structure of an alternative S-invariant Hilbertization, then $jS = Sj$, whence $jq_n(s) = q_n(S)j$, and hence $je^{aB} = e^{aB}j$ for $a \in (0, 1)$ where B is the self-adjoint operator with spectrum contained in $(0, -\varepsilon)$ such that $S = e^B$. We now turn to [8, Theorem 1.2] according to which a real linear bounded transformation that commutes with a one-parameter unitary group with positive generator (i.e., nonnegative and annihilating no vector $\neq 0$) is necessarily complex linear. But a complex linear symplectic on a Hilbert space is automatically unitary, and since $j^2 = -1$ and $\mathrm{Im}\langle jz, z\rangle \geq 0$ for all z, it follows, e.g., from the diagonalization of j, that $j = i$.

LEMMA 2.3. *Let j be a positive symplectic complex structure that commutes with the unitary operator S on the complex Hilbert space* $\mathbf{H}$. *Then j maps the spectral manifold of S corresponding to any given subset of the unit circle into the union of this manifold with that corresponding to the complex conjugate of the subset.*

Proof. Since j induces a real ring automorphism of the algebra of all bounded real linear transformations on $\mathbf{H}$, it induces a continuous automorphism of the center, i.e., the scalars. Hence j either commutes or anticommutes with i.

If j commutes with i, it is unitary, and so leaves invariant the spectrum of any complex linear transformation. If j anticommutes with i, then conjugation by j of the spectral resolution $S = \int \lambda \, dE(\lambda)$ shows that $S = \int \lambda^{-1} \, djE(\lambda)j^{-1} = \int \lambda \, djE(\lambda^{-1})j^{-1}$. It follows from the unicity of the spectral representation of S that $E(\lambda) = jE(\lambda^{-1})j^{-1}$.

Completion of proof. The conclusion of Lemma 2.2 holds equally if the spectrum is merely asymmetric, since the unit circle may be represented as the union of countably many intervals, to each of which Lemma 2.2 is applicable. The range of the spectral projection corresponding to each such interval is j-invariant by Lemma 2.3 and the hypothesized asymmetry.

Conversely, if the complex structure is unique, then the spectrum must be asymmetric. For otherwise there exists a spectral manifold for S, the restriction of S to which is unitarily equivalent to the action of a matrix-valued function on $L_2(M, \mathbb{C}^2)$ of the form

$$\begin{pmatrix} e^{i\theta(x)} & 0 \\ 0 & e^{-i\theta(x)} \end{pmatrix}.$$

On the other hand, when M here consists of a single point, there exist alternative complex structures, e.g., as an operator on R^4,

$$j = \begin{pmatrix} a\sigma & b\sigma' \\ b\sigma' & a\sigma \end{pmatrix}, \qquad \text{where} \quad a = (1 + b^2)^{1/2},$$

and

$$\sigma = \begin{pmatrix} 0 & -1 \\ 1 & 0 \end{pmatrix} \qquad \text{and} \qquad \sigma' = \begin{pmatrix} 0 & 1 \\ 1 & 0 \end{pmatrix}.$$

(See [3] for details and a determination of all such.) For general M, the j whose value for each point $x \in M$ takes this form provides an alternative complex structure satisfying the conditions of the theorem.

Remarks 2. The differential equation

$$(\Box + m^2 + V)\phi = 0; \qquad \Box = (\partial/\partial t)^2 - \Delta,$$

where V is a given bounded C^∞-function on space–time on Minkowski space defines a transformation $R(t, t')$ from the Cauchy data at time t' to that at time t, in the space of all initial data, $f \oplus g$, for $\phi(\cdot, t')$ and $\partial\phi(\cdot, t')/\partial t'$ respectively, at the given time t' (see, e.g. [4]). The transformations $R(t, t')$ are symplectic with respect to the form

$$A(f \oplus g, f' \oplus g') = \langle f, g' \rangle - \langle f', g \rangle,$$

where $\langle \cdot, \cdot \rangle$ denotes the usual inner product in $L_2(R^3)$. The Cauchy data space is appropriately Hilbertizable, indeed in a unique Lorentz–invariant fashion, as shown elsewhere.

The problem of quantizing this equation resolves itself into two processes, earlier noted. The first, applicable without any further restriction on V, consists in the construction of operator-valued, rather than numerically-valued, functions satisfying the given differential equation and the canonical commutation relations associated with the given form A (i.e.,

$$[\phi(x,t), (\partial/\partial t)\phi(y,t)] = i\delta(x - y),$$

the $\phi(x,t)$ and $\partial/\partial t\, \phi(x,t)$ being separately mutually commutative for fixed t as x varies). This problem is soluble in terms of the Weyl algebra A, with $R(t,t')$ corresponding not to a unitary operator on a Hilbert space $\mathbf{K}$ but to an automorphism of $\mathbf{A}$, and the regularized version of the quantized fields, $\exp[i\int(\phi(x,t)f(x) + (\partial\phi/\partial t)(x,t)g(x))\,dx]$ by the abstract Weyl operator $W(z)$, for a certain vector z depending linearly on f and g. In general, there is no Hilbert space on which the operators $R(t,t')$ can be simultaneously represented by unitary operators and on which a concrete Weyl system, representing the abstract Weyl operators in $\mathbf{A}$ and intertwining appropriately with the temporal evolution corresponding to $R(t,t')$, also act.

In special cases such a space exists, but the representation of the field and its dynamics on it are not even then at all unique within unitary equivalence, nor does the representation determine the physically important vacuum expectation values. When there is invariance under a temporal translation group, the vacuum may typically be specified uniquely, when it exists, by invariance and positivity properties, which may be interpreted as a generalization to the case of dynamics that is specified by an automorphism group of a C^*-algebra of the concept of the vacuum state vector as the lowest eigenstate of the Hamiltonian. Without temporal invariance, however, this treatment of the vacuum is entirely inapplicable.

Therefore, process (2), which specifies within unitary equivalence a vacuum and corresponding concrete field operators on a Hilbert space, satisfying the commutation relations and the given differential equation, does not follow at all automatically. For the equation indicated, the present Theorem 2 is applicable if $V \geq 0$ and is appropriately bounded; compare the forthcoming work with Paneitz, which also makes applications to nonlinear equations and also see [3].

Acknowledgment

I thank S. M. Paneitz for useful discussion.

References

1. J. L. Daleckii and M. G. Krein, "Stability of Solutions of Differential Equations in Banach Spaces," Translations of Mathematical Monographs, Vol. 43, Amer. Math. Soc., Providence, Rhode Island, 1974.

2. I. M. GELFAND AND V. B. LIDSKII, On the structure of regions of stability of linear canonical systems of differential equations with periodic coefficients, *Usp. Mat. Nauk* (*N.S.*) **10**, No. 1 (63) (1955), 3–40.
3. S. M. PANEITZ, Causal structures in lie groups and applications to stability of differential equations, Ph.D. Thesis, MIT, 1980.
4. L. SCHWARTZ, "Methodes mathématiques pour les sciences physique," Enseignement des Sciences, Hermann, Paris, 1961.
5. I. SEGAL, Foundations of the theory of dynamical systems of infinitely many degrees of freedom, I, *Mat.-Fys. Medd. Danske Vid. Selsk*. **31**, No. 12 (1959), 1–39.
6. I. SEGAL, Foundations of the theory of dynamical systems of infinitely many degrees of freedom, II, *Canad. J. Math.* **13** (1961), 1–18.
7. I. SEGAL, Foundations of the theory of dynamical systems of infinitely many degrees of freedom, III, *Illinois J. Math.* **6** (1962), 500–523.
8. M. WEINLESS, Existence and uniqueness of the vacuum for linear quantized fields, *J, Funct. Anal.* **4** (1969), 350–379.

MATHEMATICAL ANALYSIS AND APPLICATIONS, PART B
ADVANCES IN MATHEMATICS SUPPLEMENTARY STUDIES, VOL. 7B

On the Space $\mathscr{D}_{L^p}$

M. VALDIVIA

Facultad de Matemáticas
Universidad de Valencia
Valencia, Spain

DEDICATED TO LAURENT SCHWARTZ

In this paper we prove that the space $\mathscr{D}_{L^p}$, $1 \leq p < \infty$, is isomorphic to $s \hat{\otimes} l^p$.

If P is a set in a topological space, $\mathring{P}$ is the interior of P. We denote by s the space of all complex rapidly decreasing sequences provided with the usual topology of Fréchet space.

If E and F are two Hausdorff topological vector spaces defined over the field C of complex numbers, we denote by $E \hat{\otimes} F$ the completion of $E \otimes F$ endowed with the projective topology. If E and F are isomorphic we put $E \simeq F$.

If $\alpha = (\alpha_1, \alpha_2, \ldots, \alpha_n)$ is a multi-index, with α_j a nonnegative integer, $j = 1, 2, \ldots, n$, we write, as usual,

$$|\alpha| = \alpha_1 + \alpha_2 + \cdots + \alpha_m,$$

and, since f is an infinitely differentiable complex function defined in the n-dimensional Euclidean space R^n, $D^\alpha f$ is the partial derivative of f of order α. If $\beta = (\beta_1, \beta_2, \ldots, \beta_n)$ is another multi-index, we write $\beta \leq \alpha$ if $\beta_j \leq \alpha_j$, $j = 1, 2, \ldots, n$.

$\mathscr{D}_{L^p}$, also $\mathscr{D}_{L^p}(R^n)$, $1 \leq p < l$, is the L. Schwartz classical space of all infinitely differentiable complex functions f defined in R^n, with $D^\alpha f \in L^p(R^n)$ for every multi-index α. The space $\mathscr{D}_{L^p}$ is supposed to be endowed with the usual topology of Fréchet space; thus, a sequence (f_r) in $\mathscr{D}_{L^p}$ converges to the origin if and only if $(D^\alpha f_r)$ converges to the origin in $L^p(R^n)$ for every multi-index α.

If K is a compact set in R^n, $\mathscr{D}(K)$ is the vector space of all the infinitely differentiable complex functions defined in R^n with supports in K, provided with the ordinary topology of Fréchet space.

Let

$$\{A_r : r = 1, 2, \ldots\} \qquad \text{and} \qquad \{B_r : r = 1, 2, \ldots\}$$

ISBN 0-12-512802-9

be the families of all cubes of the form

$$\{(x_1, x_2, \ldots, x_n): a_j - \tfrac{1}{4} \le x_j \le a_j + \tfrac{5}{4},\ a_j \text{ integer}, j = 1, 2, \ldots, n\}$$

and

$$\{(x_1, x_2, \ldots, x_n): a_j \le x_j \le a_j + 1,\ a_j \text{ integer}, j = 1, 2, \ldots, n\},$$

respectively.

Each A_r intersects 3^n elements of $\{A_r : r = 1, 2, \ldots\}$. Furthermore

$$\bigcup_{r=1}^{\infty} \mathring{A}_r = R^n.$$

For every positive integer n, let us suppose

$$A_r = \{(x_1, x_2, \ldots, x_n): a_j(r) - \tfrac{1}{4} \le x_j \le a_j(r) + \tfrac{5}{4}, j = 1, 2, \ldots, n\}$$

and

$$B_r = \{(x_1, x_2, \ldots, x_n): a_j(r) \le x_j \le a_j(r) + 1, j = 1, 2, \ldots, n\}.$$

Let ψ_r be the bijective mapping from R^n onto R^n defined by

$$\psi_r(x_1, x_2, \ldots, x_n) = (a_1(r) + \tfrac{1}{2} + x_1,\ a_2(r) + \tfrac{1}{2} + x_2, \ldots, a_m(r) + \tfrac{1}{2} + x_n).$$

Therefore, ψ_r maps the sets

$$I = \{(x_1, x_2, \ldots, x_n): -\tfrac{3}{4} \le x_j \le \tfrac{3}{4}, j = 1, 2, \ldots, n\}$$

and

$$J = \{(x_1, x_2, \ldots, x_n): -\tfrac{1}{2} \le x_j \le \tfrac{1}{2}, j = 1, 2, \ldots, n\}$$

onto A_r and B_r, respectively.

Let φ be an infinitely differentiable function defined in R^n such that

$$\varphi(x) > 0, \quad \text{if} \quad x \in \mathring{I},$$
$$\varphi(x) = 0, \quad \text{if} \quad x \in R^n \sim \mathring{I}.$$

If

$$\mu_r = \frac{\varphi \circ \psi_r^{-1}}{\sum_{h=1}^{\infty} \varphi \circ \psi_h^{-1}}$$

then $\{\mu_1, \mu_2, \ldots, \mu_r, \ldots\}$ is a partition of the unity in R^n of class $\mathscr{C}^{\infty}$.

PROPOSITION 1. *Let G be a complemented subspace of $s \hat{\otimes} l^p$, $1 \le p < \infty$. Let H be a complemented subspace of G. If*

$$H \simeq s \hat{\otimes} l^p$$

it follows that

$$G \simeq s \hat{\otimes} l^p.$$

Proof. Let G_1 be a topological complement of G in $s \hat{\otimes} l^p$. Let H_1 be a topological complement of H in G. Then

$$\begin{aligned} s \hat{\otimes} l^p &\simeq (s \hat{\otimes} s) \hat{\otimes} l^p \simeq s \hat{\otimes} (s \hat{\otimes} l^p) \simeq s \hat{\otimes} (G \times G_1) \simeq (s \hat{\otimes} G) \times (s \hat{\otimes} G_1) \\ &\simeq ((s \times C) \hat{\otimes} G) \times (s \hat{\otimes} G_1) \simeq (s \hat{\otimes} G) \times (C \hat{\otimes} G) \times (s \hat{\otimes} G_1) \\ &\simeq (s \hat{\otimes} G) \times G \times (s \hat{\otimes} G_1) \simeq G \times (s \hat{\otimes} (G \times G_1)) \simeq G \times (s \hat{\otimes} l^p). \end{aligned}$$

On the other hand,

$$\begin{aligned} G &\simeq H_1 \times H \simeq H_1 \times (s \hat{\otimes} l^p) \simeq H_1 \times (s \times s) \hat{\otimes} l^p \simeq H_1 \times (s \hat{\otimes} l^p) \times (s \hat{\otimes} l^p) \\ &\simeq H_1 \times H \times (s \hat{\otimes} l^p) \simeq G \times (s \hat{\otimes} l^p). \end{aligned}$$

Hence

$$G \simeq s \hat{\otimes} l^p. \quad \blacksquare$$

Representation of $\mathscr{D}_{L^p}$

Let E be the space of sequences (f_r) of $\mathscr{D}(I)$ such that, for every multi-index α

$$\sum_{r=1}^{\infty} \int_I |D^\alpha f_r(x)|^p \, dx < \infty.$$

If

$$x = (x_1, x_2, \ldots, x_n), \qquad y = (y_1, y_2, \ldots, y_n) \in I,$$
$$\alpha = (\alpha_1, \alpha_2, \ldots, \alpha_n), \qquad \beta = (\alpha_1 + 1, \alpha_2 + 1, \ldots, \alpha_n + 1), \qquad q^{-1} = 1 - p^{-1},$$

then

$$\begin{aligned} |D^\alpha = f_r(y)| &= \left| \int_{-3/4}^{y_1} \int_{-3/4}^{y_2} \cdots \int_{-3/4}^{y_n} (D^\beta f(x)) \, dx_1 \, dx_2 \cdots dx_n \right| \\ &\le \int_I |D^\beta f(x)| \, dx \le \left(\frac{3}{2}\right)^{n/q} \left(\int_I |D^\beta f(x)|^p \, dx \right)^{1/p}, \qquad r = 1, 2, \ldots, \end{aligned}$$

hence

$$\sum_{r=1}^{\infty} \left(\sup_{x \in I} |D^\alpha f_r(x)| \right)^p \le \left(\frac{3}{2}\right)^{np/q} \sum_{r=1}^{\infty} \int_I |D^\beta f(x)|^p \, dx < \infty. \tag{1}$$

On the other hand, if (g_r) is a sequence in $\mathscr{D}(I)$ such that, for every multi-index α,

$$\sum_{r=1}^{\infty} \left(\sup_{x \in I} |D^\alpha g_r(x)| \right)^p < \infty,$$

then

$$\sum_{r=1}^{\infty} \int_I |D^\alpha g_r(x)|^p \, dx \le \sum_{r=1}^{\infty} \left(\frac{3}{2}\right)^n \left(\sup_{x\in I} |D^\alpha g_r(x)|\right)^p < \infty. \tag{2}$$

and thus, $(g_r) \in E$.

If $(f_r) \in E$, let

$$p_m((f_r)) = \sum_{|\alpha|\le m} \left(\sum_{r=1}^{\infty} \int_I |D^\alpha f_r(x)|^p \, dx\right)^{1/p}, \qquad m = 1, 2, \ldots.$$

We have that $\{p_m : m = 1, 2, \ldots\}$ is a family of norms on E. We suppose E endowed with the locally convex topology defined by this family of norms.

If

$$q_m((f_r)) = \sum_{|\alpha|\le m} \left(\sum_{r=1}^{\infty} \left(\sup_{x\in I} |D^\alpha f_r(x)|\right)^p\right)^{1/p}, \qquad m = 1, 2, \ldots,$$

it follows from (1) and (2) that the topology of E is also defined by the family of norms $\{q_m : m = 1, 2, \ldots\}$.

PROPOSITION 2. *The space E is isomorphic to $s \hat{\otimes} l^p$.*

Proof. Since $\mathscr{D}(I)$ is nuclear, the projective and bi-equicontinuous convergence topologies on $\mathscr{D}(I) \otimes l^p$ coincide. Let us suppose $\mathscr{D}(I) \otimes l^p$ endowed with this topology.

If $f_j \in \mathscr{D}(I)$, $(a_r^{(j)}) \in l^p$, $j = 1, 2, \ldots, k$, let

$$S\left(\sum_{j=1}^{k} f_j \otimes (a_r^{(j)})\right) = \left(\sum_{j=1}^{k} a_r^{(j)} f_j\right)$$

It follows easily that S is a linear one-to-one mapping from $\mathscr{D}(I) \otimes l^p$ into E. Furthermore,

$$\begin{aligned} q_m\left(\sum_{j=1}^{k} a_r^{(j)} f_j\right) &= \sum_{|\alpha|\le m} \left(\sum_{r=1}^{\infty} \left(\sup_{x\in I} \left| D^\alpha \sum_{j=1}^{k} a_r^{(j)} f_j(x)\right|\right)^p\right)^{1/p} \\ &\le \sum_{|\alpha|\le m} \sum_{j=1}^{k} \left(\sum_{r=1}^{\infty} |a_r^{(j)}|^p \left(\sup_{x\in I} |D^\alpha f_j(x)|^p\right)\right)^{1/p} \\ &= \sum_{j=1}^{k} \left(\sum_{|\alpha|\le m} \sup_{x\in I} |D^\alpha f_j(x)|\right) \left(\sum_{r=1}^{\infty} |a_r^{(j)}|^p\right)^{1/p}, \end{aligned}$$

whence it follows that S is continuous.

Let

$$U = \left\{g \in \mathscr{D}(I) : \sum_{|\alpha|\le m} \sup_{x\in I} |D^\alpha g(x)| \le 1\right\}, \qquad B = \left\{(a_r) \in l^p : \sum_{r=1}^{\infty} |a_r|^p \le 1\right\}.$$

Let U° and B° be the polar sets of U and B in the topological dual spaces of $\mathscr{D}(I)$ and l^p, respectively. If $u \in U^\circ$ and $(b_r) \in B^\circ$, then

$$\begin{aligned}
\left|\sum_{j=1}^{k} \langle f_j, u\rangle \langle (a_r^{(j)}), (b_r)\rangle\right| &= \left|\sum_{j=1}^{k} \langle f_j, u\rangle \sum_{r=1}^{\infty} a_r^{(j)} b_r\right| \\
&= \left|\sum_{r=1}^{\infty} b_r \left\langle \sum_{j=1}^{k} a_r^{(j)} f_j, u\right\rangle\right| \\
&\leq \left(\sum_{r=1}^{\infty} |b_r|^q\right)^{1/q} \left(\sum_{r=1}^{\infty} \left|\left\langle \sum_{j=1}^{k} a_r^{(j)} f_j, u\right\rangle\right|^p\right)^{1/p} \\
&\leq \left(\sum_{r=1}^{\infty} \left|\left\langle \sum_{j=1}^{k} a_r^{(j)} f_j, u\right\rangle\right|^p\right)^{1/p} \\
&\leq \left(\sum_{r=1}^{\infty} \left(\sum_{|\alpha| \leq m} \sup_{x \in I} \left|D^\alpha \sum_{j=1}^{k} a_r^{(j)} f_j(x)\right|\right)^p\right)^{1/p} \\
&\leq \sum_{|\alpha| \leq m} \left(\sum_{r=1}^{\infty} \left(\sup_{x \in I} \left|D^\alpha \sum_{j=1}^{h} a_r^{(j)} f_j(x)\right|\right)^p\right)^{1/p} \\
&= q_m\left(\sum_{j=1}^{h} a_r^{(j)} f_j\right),
\end{aligned}$$

hence S is an open mapping from $\mathscr{D}(I) \otimes l^p$ onto the subspace $S(\mathscr{D}(I) \otimes l^p)$ of E.

Finally, it is easy to prove that $S(\mathscr{D}(I) \otimes l^p)$ is dense in E and since $\mathscr{D}(I) \simeq s$, [3, p. 210], we have that

$$E \simeq \mathscr{D}(I) \hat{\otimes} l^p \simeq s \hat{\otimes} l^p. \quad \blacksquare$$

Let $A_{r(h)}$ be $h = 1, 2, \ldots, 3^n$, the elements of $A_j : j = 1, 2, \ldots$, which intersects A_r. If (f_j) belongs to E,

$$\begin{aligned}
\left(\int_{B_r} \left|D^\alpha\left(\sum_{j=1}^{\infty} f_j \circ \psi_j^{-1}\right)(x)\right|^p dx\right)^{1/p} &\leq \sum_{h=1}^{3^n} \left(\int_{A_{r(h)}} |D^\alpha(f_{r(h)} \circ \psi_{r(h)}^{-1})(x)|^p \, dx\right)^{1/p} \\
&= \sum_{h=1}^{3^n} \left(\int_I |D^\alpha f_{r(h)}(x)|^p \, dx\right)^{1/p} \\
&\leq \left(\frac{3}{2}\right)^{n/p} \sum_{h=1}^{3^n} \sup_{x \in I} |D^\alpha f_{r(h)}(x)|,
\end{aligned}$$

hence

$$\int_{B_r} \left|D^\alpha\left(\sum_{j=1}^{\infty} f_j \circ \psi_j^{-1}\right)(x)\right|^p dx \leq \frac{3^{np}}{2^n} \sum_{h=1}^{3^n} (\sup_{x \in I} |D^\alpha f_{r(h)}(x)|)^p.$$

Thus

$$\int_{R^n}\left|D^\alpha\left(\sum_{j=1}^{\infty} f_j\circ\psi_j^{-1}\right)(x)\right|^p dx = \sum_{r=1}^{\infty}\int_{B_r}\left|D^\alpha\left(\sum_{j=1}^{\infty} f_j\circ\psi_j^{-1}\right)(x)\right|^p dx$$

$$\leq \frac{3^{np}}{2^n}\sum_{r=1}^{\infty}\sum_{h=1}^{3^n}\left(\sup_{x\in I}|D^\alpha f_{r(h)}(x)|\right)^p$$

$$= \frac{3^{n(p+1)}}{2^n}\sum_{r=1}^{\infty}\left(\sup_{x\in I}|D^\alpha f_r(x)|\right)^p,$$

and so the mapping V from E into $\mathscr{D}_{L^p}$ defined by

$$V((f_r)) = \sum_{r=1}^{\infty} f_r\circ\psi_r^{-1},$$

obviously linear, is also continuous.

PROPOSITION 3. *The mapping V is onto.*

Proof. Let f be an element of $\mathscr{D}_{L^p}$. Let

$$f_r = (f\mu_r)\circ\psi_r, \qquad r = 1, 2, \ldots.$$

Since J is a compact set contained in $\mathring{I}$, it follows that

$$\inf\{\varphi(x): x\in J\} = b > 0.$$

If $x\in R^n$, we can find a positive integer h such that $x\in B_h$. Then

$$\left(\sum_{r=1}^{\infty}\varphi\circ\psi_r^{-1}\right)(x) \geq (\varphi\circ\psi_h^{-1})(x) = \varphi(\psi_h^{-1}(x)) \geq b.$$

Therefore, given any multi-index α, we can find a number Q, which does not depend upon r, such that

$$\left(\int_I |D^\alpha f_r(x)|^p\,dx\right)^{1/p}$$

$$= \left(\int_I\left|D^\alpha\left(f(\psi_r(x))\left[\left(\sum_{j=1}^{\infty}\varphi\circ\psi_j^{-1}\right)(\psi_r(x))\right]^{-1}\varphi(x)\right)\right|^p dx\right)^{1/p}$$

$$\leq Q\left(\int_I\left(\sum_{\beta\leq\alpha}|D^\beta f(\psi_r(x))|\right)^p dx\right)^{1/p}$$

$$\leq Q\sum_{\beta\leq\alpha}\left(\int_{A_r}|D^\beta f(x)|^p\,dx\right)^{1/p},$$

hence

$$\int_I |D^\alpha f_r(x)|^p \, dx \le (Q(|\alpha| + 1)^n)^p \sum_{\beta \le \alpha} \int_{A_r} |D^\beta f(x)|^p \, dx.$$

Then

$$\sum_{n=1}^{\infty} \int_I |D^\alpha f_r(x)|^p \, dx \le (Q(|\alpha| + 1)^n)^p \sum_{\beta \le \alpha} \sum_{v=1}^{\infty} \int_{A_r} |D^\beta f(x)|^p \, dx$$

$$\le 3^n (Q(|\alpha| + 1)^n)^p \sum_{\beta \le \alpha} \int_{R^n} |D^\beta f(x)|^p \, dx. \tag{3}$$

Therefore $(f_r) \in E$.

Finally,

$$V((f_r)) = \sum_{r=1}^{\infty} f_r \circ \psi_r^{-1} = \sum_{r=1}^{\infty} f\mu_r = f \sum_{r=1}^{\infty} \mu_r = f. \quad ∎$$

PROPOSITION 4. *$\mathscr{D}_{L^p}$ is isomorphic to a complemented subspace of* $s \hat{\otimes} l^p$

Proof. Let W be the mapping from $\mathscr{D}_{L^p}$ into E defined by

$$W(f) = ((f\mu_r) \circ \psi_r),$$

for every $f \in \mathscr{D}_{L^p}$.

Obviously, W is linear and one to one. From (3) it follows that W is continuous. On the other hand,

$$V(W(f)) = f,$$

hence W is an isomorphism from the Fréchet space $\mathscr{D}_{L^p}$ into the Fréchet space E.

Finally, if D is the kernel of V, then E is the topological direct sum of D and $W(\mathscr{D}_{L^p})$, hence, by Proposition 2, the result follows. ∎

Let H be the vector space of the sequences (f_r) in $\mathscr{C}^\infty(J)$ such that, for every multi-index α,

$$\sum_{r=1}^{\infty} \left(\sup_{x \in J} |D^\alpha f_r(x)| \right)^p < \infty.$$

Let H be endowed with the locally convex topology defined by the system of norms

$$q_m((f_r)) = \sum_{|\alpha| \le m} \left(\sum_{r=1}^{\infty} \left(\sup_{x \in J} |D^\alpha f_r(x)| \right)^p \right)^{1/p}, \qquad m = 1, 2, \ldots .$$

PROPOSITION 5. *H is isomorphic to* $s \hat{\otimes} l^p$.

Proof. Since $\mathscr{C}^\infty(J) \simeq s$, [3, p. 207], it follows in the same way as Proposition 2. ■

Let T be a continuous linear operator from $\mathscr{C}^\infty(J)$ into $\mathscr{D}(I)$ such that, if $f \in \mathscr{C}^\infty(J)$, the restriction of Tf to J coincides with f, [2] or [4]. We choose a sequence (r_p) of positive integers such that the cubes $\{A_{r_p}: p = 1, 2, \ldots\}$ are mutually disjoint.

If (f_p) belongs to H, it is straightforward that (Tf_p) belongs to E. Therefore, if

$$Z((f_p)) = \sum_{p=1}^{\infty} (Tf_p) \circ \psi_{r_p}^{-1},$$

then Z is an isomorphism from H into $\mathscr{D}_{L^p}$.

Let M be the subspace of $\mathscr{D}_{L^p}$ of all functions vanishing on B_{r_p}, $p = 1, 2, \ldots$.

PROPOSITION 6. *$\mathscr{D}_{L^p}$ has a complemented subspace isomorphic to* $s \hat{\otimes} l^p$.

Proof. $Z(H)$ is isomorphic to H, so that it is enough to prove that

$$\mathscr{D}_{L^p} = Z(H) \oplus M.$$

Obviously,

$$Z(H) \cap M = \{0\}.$$

If $g \in \mathscr{D}_{L^p}$, let k_p be the restriction of g to B_{r_p} and let g_p be the element of $\mathscr{C}^\infty(J)$ such that, for every $x \in J$,

$$g_p(x) = (k_p \circ \psi_{r_p})(x).$$

Since g is the image of an element of E by V, it follows, for every multi-index α, that

$$\sum_{p=1}^{\infty} \left(\sup_{x \in B_{r_p}} |D^\alpha k_p(x)| \right)^p < \infty,$$

hence (g_p) belongs to H. On the other hand, $Z((g_p))$ and g coincide on B_{r_p}, $p = 1, 2, \ldots$, thus

$$g = Z((g_p)) + (g - Z((g_p))), \qquad Z((g_p)) \in Z(H), \qquad g - Z((g_p)) \in M. \quad ■$$

THEOREM 1. *The space $\mathscr{D}_{L^p}$ is isomorphic to* $s \hat{\otimes} l^p$.

Proof. It follows from Propositions 1, 4, and 6. ■

REFERENCES

1. A. GROTHENDIECK, Produits tensoriels topologiques et espaces nucléaires, *Mem. Amer. Math. Soc.* No. 16 (1966).
2. Z. OGRODZKA, On simultaneous extension of infinitely differentiable functions, *Studia Math.* **28** (1967), 193–207.
3. S. ROLEWICZ, "Metric Linear Spaces," PWN-Polish Sci. Publ., Warsaw, 1972.
4. M. VALDIVIA, Representaciones de los espacios $\mathscr{D}(\Omega) y \mathscr{D}'(\Omega)$, *Rev. Real Acad. Cienc. Exact. Fís. Natur. Madrid* **72** (1978), 385–414.

MATHEMATICAL ANALYSIS AND APPLICATIONS, PART B
ADVANCES IN MATHEMATICS SUPPLEMENTARY STUDIES, VOL. 7B

The Choice of Test Functions in Quantum Field Theory

A. S. WIGHTMAN

Joseph Henry Laboratories of Physics
Princeton University
Princeton, New Jersey

DEDICATED TO LAURENT SCHWARTZ ON THE OCCASION OF HIS 65TH BIRTHDAY

1. INTRODUCTION

The main point of this note is metaphysical; it is to provide an answer to the question: In what space is it reasonable to look for solutions of relativistic quantum field theories? To pose the problem in its historical setting, we recall a little of the standard theory, in which the solutions are tempered operator-valued distributions, and some of the examples which have led to generalizations of the formalism to more singular generalized functions. In particular, Schwinger's model for the quantum electrodynamics of massless fermions in two dimensions is discussed. In some gauges, it has solutions which are not tempered distributions because of the growth of the matrix elements of fields at large distances. On the other hand, examples are discussed for which the local behavior is too singular for the field to be a distribution. The solutions may then be regarded as hyperfunction quantum fields.

2. FIELDS AS OPERATOR-VALUED DISTRIBUTIONS

In classical field theory, the basic fields are functions on space–time, and it is customary in physics to manipulate the equations for these functions without specifying in detail how smooth they are. Similarly, in the quantum field theory of the 1920s, the fields were taken to be functions on space–time with values linear operators in the Hilbert space of states. These functions again had unspecified smoothness properties but it was already evident from the first examples (the quantized vibrating string [1] and the quantized Maxwell theory in the absence of charges [2]) that the functions involved

ISBN 0-12-512802-9

are very singular in their space–time dependence. This was explicitly recognized by Bohr and Rosenfeld, who discussed the behavior of the expectation values of the electric and magnetic field strengths in quantum electrodynamics and concluded that one can hope to measure only the averages of these operators over space–time regions, not their values at a point: $\vec{\mathscr{E}}_\Delta = (1/|\Delta|)\int_\Delta d^4x\vec{\mathscr{E}}(x)$ where Δ is some space–time region rather than $\vec{\mathscr{E}}(x)$ itself [3]. Heisenberg studied the corresponding problem for the electromagnetic charge and current operators j_μ, $\mu = 0, 1, 2, 3$, and found that only averages $j(f)$ with a sufficiently smooth function f made sense in that case: $j(f) = \int d^4x f^\mu(x) j_\mu(x)$ [4]. So it was already evident in the 1930s that the field operators of quantum field theory, if they were to be mathematically well defined at all, had to be regarded as some singular kind of functions. However, little systematic work was done to determine what kind of singularities should be expected; people were then unaccustomed to choosing a fixed space in advance and looking for solutions in it.

Thus it was providential for workers attempting to construct a mathematically rigorous quantum theory of fields in the early 1950s, that they had the model of Laurent Schwartz's theory of distributions before their eyes [5]. That theory provided a convenient language as well as a systematic body of results which had only to be adapted to the special circumstances of relativistic quantum field theory.

More explicitly, the description of a field is as follows. Let $\mathscr{H}$ be the Hilbert space whose vectors describe the quantum mechanical states of the system under study. Then a field ϕ is defined as a linear mapping of a test function space $\mathscr{T}_\phi$ into the operators of $\mathscr{H}$:

$$f \mapsto \phi(f). \tag{2.1}$$

The test functions were chosen, following Schwartz, as elements of $\mathbb{C}^N \times \mathscr{D}(\mathbb{R}^\nu)$ where $\mathscr{D}(\mathbb{R}^\nu)$ is the space of infinitely differentiable functions of compact support or of $\mathbb{C}^N \times \mathscr{S}(\mathbb{R}^\nu)$ where $\mathscr{S}(\mathbb{R}^\nu)$ is the space of infinitely differentiable functions of fast decrease. In any case, ν is the dimension of space–time and N is the number of components of the field, e.g., for the electric field strength $\vec{\mathscr{E}}$ mentioned above, $N = 3$. This definition gives meaning to the expression

$$\phi(f) = \sum_{j=1}^{N} \int d^\nu x\, f_j(x)\phi_j(x).$$

The very first relativistic fields constructed, the quantized electric and magnetic fields of Maxwell's theory in the absence of charges, have the property that the $\mathscr{E}(f)$ and $B(g)$ for $f, g \in \mathbb{C}^3 \times \mathscr{S}(\mathbb{R}^4)$ are unbounded operators. Thus, it is natural to admit unbounded operators in the general definition of fields. Then one expects that it will be necessary to make some assumptions about the domains of the unbounded operators. The standard proposal for dealing with this situation is as follows. One assumes that in specifying a quantum field, one should give a basic set $\{\phi\}$ of fields ϕ, as well as test function spaces $\mathscr{T}_\phi$ for each of them. Then one requires that there exist a dense linear subset D of the Hilbert space $\mathscr{H}$ on which all operators

$$\phi(f) \quad \text{and} \quad \phi(f)^*, \qquad f \in \mathscr{T}_\phi$$

are defined, and such that on vectors of D the relations

$$\begin{aligned} \phi(\alpha f) &= \alpha\phi(f), & \phi(f_1 + f_2) &= \phi(f_1) + \phi(f_2), \\ \phi(\alpha f)^* &= \bar{\alpha}\phi(f)^*, & \phi(f_1 + f_2)^* &= \phi(f_1)^* + \phi(f_2)^* \end{aligned} \tag{2.2}$$

hold for all $\alpha \in \mathbb{C}$ and f, f_1 and $f_2 \in \mathscr{T}_\phi$. Finally, for Φ, Ψ vectors in D we assume that the linear functional

$$(\Phi, \phi(f)\Psi) \tag{2.3}$$

of f is continuous in f for $f \in \mathscr{T}_\phi$. The requirements (2.2) and (2.3) on the mappings (2.1) give a precise meaning to the statement that the fields ϕ are operator-valued distributions. We now turn to matters concerned with relativistic invariance.

In general quantum mechanics, it is shown that the relativistic invariance of a theory under the inhomogeneous Lorentz (= Poincaré) group implies the existence of a continuous unitary representation of its covering group ISL(2, $\mathbb{C}$). If the elements of ISL(2, $\mathbb{C}$) are denoted $\{a, A\}$, where a is a space–time translation and A is a 2×2 matrix of determinants 1, then the group multiplication law in ISL(2, $\mathbb{C}$) is given by

$$\{a, A\}\{b, B\} = \{a + \Lambda(A)b, AB\} \tag{2.4}$$

The representation $\{a, A\} \mapsto U(a, A)$ expresses the law of Poincaré transformation of states; the unitarity of U guarantees the Poincaré invariance of transition probabilities

$$|(U(a, A)\Phi, U(a, A)\Psi)|^2 = |(\Phi, \Psi)|^2. \tag{2.5}$$

The general theory of the representations of the translation group assures us that

$$U(a, 1) = \exp iP^\mu a_\mu, \tag{2.6}$$

where the $P = P^0, P^1, P^2, P^3$ are commuting self-adjoint operators. Their joint spectrum is interpreted as the set of possible energy momentum vectors of the theory, and it is a fundamental physical stability requirement that this joint spectrum lie in the future cone

$$\mathrm{sp}(P) \subset \bar{V}_+, \tag{2.7}$$

where $\bar{V}_+$ is the closed cone

$$\bar{V}_+ = \{p|p \cdot p \geq 0; p^0 \geq 0\}.$$

Furthermore, we will require that there is a unique vector, Ψ_0, invariant under U, where

$$U(a, A)\Psi_0 = \Psi_0.$$

This vector is called the *vacuum*. These requirements will be referred to as the *spectral condition*.

To obtain a link between the domain D of the fields and the spectral condition, one assumes

$$\Psi_0 \in D$$

and

$$\phi(f)D \subseteq D, \qquad \phi(f)^*D \subseteq D,$$

for all $f \in \mathscr{T}_\phi$.

In a relativistic field theory it is natural to consider as basic fields, only those which have a definite transformation law under the Poincaré group or its covering group. Such a transformation law can be partly given in terms of representations on the test function spaces $\mathscr{T}_\phi$.

$$f \mapsto \{a, A\}f.$$

Typically,

$$(\{a, A\}f)_j(x) = \sum_{k=1}^{N} S(A^{-1})_{kj} f_k(\Lambda(A^{-1})(x - a)), \tag{2.8}$$

where

$$A \mapsto S(A)$$

is an $N \times N$ matrix representation of SL(2, $\mathbb{C}$). It is worth noting that the representations (2.8) are automatically continuous on the test function spaces $\mathscr{T}_\phi$, if we choose either of the spaces mentioned above. As we will see, the same is true for all the other test function spaces considered later. Thus the requirement of relativistic invariance does not impose a restriction on the choice of test function space.

The full transformation law of the fields expresses the compatibility of the transformation law $\{a, A\} \mapsto U(a, A)$ of states and that $f \mapsto \{a, A\}f$ of test functions. It says

$$U(a, A)\phi(f)U(a, A)^{-1} = \phi(\{a, A\}f). \tag{2.9}$$

In addition to the requirement of Poincaré invariance it is traditional to impose that of *local commutativity*. For observable fields, this axiom expresses the independence of field measurements carried out at points in space–time separated by a spacelike interval. Explicitly,

$$[\phi(f), \psi(g)] = 0 \tag{2.10}$$

for any observable fields ϕ and ψ if every point of the support of f is separated from every point of the support of g by a spacelike interval. For unobservable fields which do not satisfy $\phi(\bar{f}) = \phi(f)^*$, one imposes also

$$[\phi(f), \psi(g)^*] = 0. \tag{2.11}$$

Furthermore, it turned out (that was the striking discovery of Jordan and Wigner [6]) that it is fruitful to consider the possibility that the fields anticommute, i.e., the commutator in (2.10) and (2.11) is replaced by the anticommutator

$$[\phi(f), \psi(g)^*]_+ = \phi(f)\psi(g)^* + \psi(g)^*\phi(f).$$

It should be emphasized that for the test function spaces mentioned above (Schwartz's $\mathscr{D}$ and $\mathscr{S}$) there is no lack of test functions of compact support to use in these formulas. On the other hand, as will be seen shortly, for other interesting choices of test function space there are no test functions of compact support and the formulation of local commutativity requires reexamination.

To complete the above axiomatic framework, it is customary to add the requirement of asymptotic completeness, i.e., the completeness of the scattering states of the theory. Since this assumption will only play a peripheral role in what follows, we will replace it by a weaker assumption, the cyclicity of the vacuum vector with respect to the polynomial algebra of the smeared fields: vectors of the form $\mathscr{P}(\phi(f), \psi(g), \ldots)\Psi_0$, where $\mathscr{P}$ is an arbitrary polynomial, are dense in the Hilbert space of states, $\mathscr{H}$. This latter assumption is known to imply the irreducibility of the polynomial algebra [7] and so is natural for a theory in which the fields are supposed to provide a complete description.

In summary, standard field theory is a triple $\{\mathscr{H}, U, \{\phi\}\}$, consisting of a separable Hilbert space $\mathscr{H}$, a unitary representation U of ISL(2, $\mathbb{C}$), and a denumerable set of fields $\{\phi\}$, satisfying the five axioms:

(I) spectral condition,
(II) domain axiom for fields,
(III) relativistic invariance,
(IV) local commutativity,
(V) asymptotic completeness;

but we replace (V) by

(V′) cyclicity of the vacuum.

With this specification of the framework of axiomatic field theory in hand, we can now begin the discussion of the implications of the choice of test function space.

Most work on quantum field theory assumes tempered distribution fields, i.e., $\mathscr{T}_\phi = \mathbb{C}^N \times \mathscr{S}(\mathbb{R}^\nu)$. There are several reasons for this. First of all, there are the reasons of simplicity and convenience; the theory of tempered distributions is elegant and the theory of the Fourier transformation, which plays an important role in field theory, is symmetrical and neat. Second, perturbation theory yields tempered distributions. Until recently, the principal source of information about quantum field theory lay in the renormalized perturbation series for Lagrangean field theories with polynomial interactions. These are formal power series in some coupling constant, and it is an elementary consequence of the polynomial character of the interaction that the terms of the series for the basic quantities of the theory (vacuum expectation values of products of fields $(\Psi_0, \prod_i \phi^{(i)}(x_i)\Psi_0)$) are tempered distributions in the space–time variables. Needless to say, this does not imply that the solutions to which the formal power series are asymptotic, supposing they exist, are also tempered distributions. A third reason for the assumption of temperedness is dispersion theory. In general, it presupposes that scattering amplitudes and vacuum expectation values are polynomially bounded distributions and therefore are tempered. The argument from dispersion theory is not airtight, however. It is known that certain theories, for which the vacuum expectation values are not distributions at all, have scattering amplitudes which are tempered distributions. We shall return to this point later.

The question of the distinction between distributions with test functions in $\mathscr{D}$ and $\mathscr{S}$ actually arose very early in the history of axiomatic quantum field theory. It was realized that the Bochner–Schwartz theorem on the structure of distributions of positive type implies that a two-point function which is a distribution test functions in $\mathscr{D}$ can be uniquely extended to become a tempered distribution with test functions in $\mathscr{S}$ [8]. The point is that the translation invariance of the vacuum implies that

$$\|\phi(f)\Psi_0\|^2 = (\Psi_0, \phi(f)^*\phi(f)\Psi_0) \tag{2.12}$$

defines a distribution in the difference variable $x_1 - x_2$ which is of positive type and therefore the Fourier transform of a positive measure of slow increase. Consequently, it defines a tempered distribution. This argument extends easily to quantities of the form

$$(\Psi_0, \phi(f)\psi(g)\Psi_0). \tag{2.13}$$

They can be bounded by Schwarz's inequality in terms of quantities of the form (2.12) and it was natural to ask whether it extends to the three-point function

$$(\Psi_0, \phi(f)\psi(g)\chi(h)\Psi_0) \tag{2.14}$$

and higher. A partial answer was provided by Borchers who pointed out that a counterexample is provided by the distribution obtained from the analytic function [9]

$$\exp z_1 z_2 z_3 [z_1 z_2 + z_2 z_3 + z_1 z_3]^{-1}$$

by writing

$$z_1 = (x_2 - x_3 - i\eta_1)^2, \qquad z_2 = (x_3 - x_1 - i\eta_2)^2, \qquad z_3 = (x_1 - x_2 - i\eta_3)^2,$$

and passing to boundary $\eta_1, \eta_2, \eta_3 \to 0; \eta_1, \eta_2, \eta_3 \in V_+$. This function is analytic in the product of the cut planes $\{z_j \neq \rho \geq 0\}$, $j = 1, 2, 3$, and is polynomially bounded there in each variable when the other two are held fixed. It is not polynomially bounded when $z_1 = z_2 = z_3$ and therefore its boundary value is not a tempered distribution. It is only a partial answer because although it satisfies the requirements of the spectrum condition of relativistic invariance and of local commutativity it is not known to arise from a field theory. Thus, although a large fraction of the general theory of fields can be carried out for distributions which are not tempered, it does not seem likely that every such theory is in fact tempered.

In the next section of this chapter I will discuss examples in which even the two-point function is a nontempered distribution. That is possible because the theory generalizes the above notion of quantized field by introducing an indefinite sesquilinear form for expectation values. This example shows that if one considers quantum field theory in the enlarged framework with an indefinite metric there can be a real distinction between the test function spaces $\mathscr{D}$ and $\mathscr{S}$ even in elementary examples.

The first serious proposals for the use of test function spaces other than $\mathscr{D}$ or $\mathscr{S}$ were made for quite a different class of examples: nonrenormalizable Lagrangean field theories. The history is rather complicated but merits retelling in brief outline because it explains to some extent the notation adopted later.

In the late 1930s Heisenberg became convinced that there must be a fundamental length in nature which would play a role analogous to Planck's constant h and the velocity of light c [10]. h was regarded as fundamental because its being strictly positive reflects the limitations on classical mechanics implied by quantum mechanics. c was regarded as fundamental because its finiteness reflects analogous limitations arising from relativity theory. Heisenberg found, in the then existing situation in particle physics, hints of a third such limitation. He noted that in the collisions of cosmic ray particles with matter there appeared to be multiple production of particles.

On the other hand, to obtain such production from a theory of fields, he suggested an interaction in which the coupling constant has the dimensions of a length to a positive power. In such theories, multiple production becomes more important the higher the energy of a collision, at least when it is calculated in the primitive manner then available. He found the appearance of such a length l_0 very satisfactory because, using appropriate powers of $\hbar$, c, and l_0, one can reduce all physical constants to dimensionless form.

When, later on, the refined study of Lagrangian field theories led to a classification into super-renormalizable, renormalizable, and nonrenormalizable, it turned out that the theories proposed by Heisenberg more or less coincided with the nonrenormalizable class. The coupling constants of super-renormalizable theories have the dimensions of a length to a negative power, while those of renormalizable theories are dimensionless. In a formal perturbation expansion, the vacuum expectation values of nonrenormalizable theories are characterized by requiring an infinite number of types of regularizing terms whereas for the other two classes the number of types is finite.

When, in the middle 1950s, nonperturbative (but still formal) methods became available, it was argued that this last feature of nonrenormalizable theories was an artifact produced by perturbation expansions; with appropriate nonperturbative methods the models studied appeared to be labeled by a finite number of parameters. In addition, the models showed "essential singularities" in their vacuum expectation values at points of coincidence in coordinate space [11, 12]. At first, little effort was made to link these essential singularities with generalized functions which are not distributions, but by the end of the 1950s such a procedure was made easy by the appearance of the treatise by Gelfand and Shilov [13]. In some pioneering works of this kind, there was a lingering effect of Heisenberg's fundamental length, in that efforts were made to prove that the generalized functionals associated with nonrenormalizable theories are in some sense "nonlocal" [14]. In one obvious sense, that can be the case because some of the test function spaces used do not contain any functions of compact support, consisting as they do of analytic functions. However, it turns out that in such cases there is another definition of support (sometimes called the *carrier*) for a generalized function which serves equally well so it makes sense to talk about a commutator vanishing outside the light cone also in these theories using analytic functions as test functions.

On the other hand, it was shown that when the Fourier transforms of vacuum expectation values grow sufficiently fast it can happen that a breakdown of local commutativity occurs; commutators become nonvanishing at spacelike separations. Schroer, in a basic paper [15], distinguished such really nonlocal nonrenormalizable theories by calling them of the second kind, the ones for which axioms I–IV, V′ are of the first kind. Later Jaffe distinguished a subclass of theories of the first kind in which the generalized functions in momentum space have less than exponential growth [16]. There is a whole class of these spaces for each of which the test function has plenty of functions of compact support. Theories using such test function spaces are called *strictly localizable*. Their further development can be traced from [17, 18]. Perhaps the most striking physical result obtained in these theories is the proof that the scattering amplitude is a tempered distribution even though the vacuum expectation values themselves are much more singular generalized functions [16, 19].

The most general class of generalized functions for which axiomatic quantum field theory has been developed is that of hyperfunctions [20, 21]. One advantage of this theory is that it gives a very symmetrical relation between Euclidean field theory and the Minkowski quantum field theory described above. The Osterwalder–Schrader theorem that connects the two was originally developed for tempered distributions and for them yields a somewhat asymmetrical relation, in the sense that the practical conditions for the recovery of a Minkowski quantum field theory from a Euclidean field theory are sufficient but not necessary.

This sketchy review will have to serve for our purposes. The reader interested in filling some of its gaps may consult [22]. We now turn to the evidence from concrete models which forces us to generalize the conventional formalism.

3. Examples Requiring Nontempered Distributions for Infrared Reasons

The first example of this section shows the necessity of introducing an indefinite metric in the theory of a massless scalar field in a space–time of two dimensions, or, alternatively, constraining the test functions. That infrared divergences cause surprises in this model was first pointed out by Schroer [23] and the introduction of an infrared indefinite metric was advocated by the author [24]. However, it is only recently that it has been shown that this can be done in such a way that the theory takes the form proposed for quantum electrodynamics by Gupta and Bleuler [25–27]. We will repeat some of the elementary steps in the construction, because some instructive points arise.

Since ϕ is supposed to be a free field, it has an n-point vacuum expectation value expressible in terms of two-point vacuum expectation values as follows

$$(\Psi_0, \phi(x_1)\ldots\phi(x_n)\Psi_0) = \begin{cases} 0, & n \text{ odd}, \\ [1\ldots n], & n \text{ even}, \end{cases} \tag{3.1}$$

where the hafnian $[1\ldots n]$ is defined recursively

$$[1\ldots n] = \sum_{j=2}^{n} [1j][1\ldots\hat{j}\ldots n]. \tag{3.2}$$

Here

$$[jk] = (\Psi_0, \phi(x_j)\phi(x_k)\Psi_0). \tag{3.3}$$

The two-point function itself is invariant under translations so it depends only on the difference of its arguments

$$(\Psi_0, \phi(x_1)\phi(x_2)\Psi_0) = F(x_1 - x_2).$$

Since ϕ is assumed to satisfy

$$\Box\phi(x) = 0, \tag{3.4}$$

F satisfies

$$\Box F(x) = 0. \tag{3.5}$$

Furthermore, because Ψ_0 is invariant under U, the unitary representation of the Poincaré group and ϕ has the transformation law of a scalar field

$$F(\Lambda x) = F(x) \tag{3.6}$$

for all Lorentz transformations Λ in the connected component of the identity. Since the Fourier transform of F satisfies

$$p^2\hat{F}(p) = 0 \tag{3.7}$$

the support of F lies on the light cone $p^2 = 0$. The light cone $p^2 = 0$ consists of five pieces on each of which the Lorentz transformations act transitively. They are characterized by

$$p^0 \gtrless 0, \qquad p^1 \gtrless 0 \tag{3.8}$$

and

$$p^0 = p^1 = 0. \tag{3.9}$$

On each of the first four, $F(p)$ is uniquely determined up to a constant factor by Lorentz invariance. (The easiest way to see this is to introduce locally the variable $\tau = x^\mu x_\mu$ in place of one of the other variables that does not vanish, say x^j. Then an invariant distribution is locally independent of all variables but τ, because differentiation with respect to the other variables is equivalent to $x_j^{-1}[x_j\partial/\partial x^k - x_k\partial/\partial x^j]$ and the square brackets indicate an infinitesimal Lorentz transformation. (See [28] for a detailed account.) If we introduce light cone coordinates $u = p^0 + p^1$ and $v = p^0 - p^1$, it means that away from $u = 0 = v$, the invariant distributions are all linear combinations of

$$\theta(u)u^{-1}\delta(v), \quad \theta(-u)u^{-1}\delta(v), \quad \theta(v)v^{-1}\delta(u), \quad \theta(-v)v^{-1}\delta(u), \tag{3.10}$$

where θ is the Heaviside function

$$\theta(x) = \begin{cases} 1, & x > 0, \\ \frac{1}{2}, & x = 0, \\ 0, & x < 0. \end{cases}$$

The expressions (3.10) define distributions on the half-planes with the lines $u = 0$ or $v = 0$ omitted. When one attempts to extend them to the whole plane one encounters a striking phenomenon. Off the line $u = 0$, $\theta(\pm u)u^{-1}\delta(v)$ coincides with

$$\frac{d}{du}[\theta(\pm u)\ln|u|]\delta(v), \tag{3.11}$$

and off $v = 0$, $\theta(\pm v)v^{-1}\delta(u)$ coincides with

$$\frac{d}{dv}[\theta(\pm v)\ln|v|]\delta(u). \tag{3.12}$$

These expressions define distributions in the whole place, and they are unique up to addition of a multiple of

$$\delta(u)\delta(v). \tag{3.13}$$

However, (3.11) and (3.12) are not Lorentz invariant because, for example, if $\lambda > 0$,

$$\frac{d}{d\lambda u}[\theta(\pm\lambda u)\ln|\lambda u|]\delta(\lambda^{-1}v) = \frac{d}{du}[\theta(\pm u)\ln|u|]\delta(v) + \ln\lambda\,\delta(u)\delta(v). \tag{3.14}$$

Only if the four functions are combined in pairs so that the $\ln\lambda$ terms cancel does one get a Lorentz invariant result. Thus, the most general solution of (3.7) is a linear combination of (3.13) and the following six possibilities:

\ $$\frac{d}{du}[\theta(u)\ln u]\delta(v) - \frac{d}{du}[\theta(-u)\ln|u|]\delta(v), \tag{3.15a}$$

/ $$\frac{d}{dv}[\theta(v)\ln v]\delta(u) - \frac{d}{dv}[\theta(-v)\ln|v|]\delta(u), \tag{3.15b}$$

V $$\frac{d}{du}[\theta(u)\ln u]\delta(v) + \frac{d}{dv}[\theta(v)\ln v]\delta(u), \tag{3.15c}$$

$$\wedge \qquad \frac{d}{du}[\theta(-u)\ln|u|]\delta(v) + \frac{d}{dv}[\theta(-v)\ln|v|]\delta(u), \tag{3.15d}$$

$$> \qquad \frac{d}{du}[\theta(u)\ln u]\delta(v) + \frac{d}{dv}[\theta(-v)\ln|v|]\delta(u), \tag{3.15e}$$

$$< \qquad \frac{d}{du}[\theta(-u)\ln|u|]\delta(v) + \frac{d}{dv}[\theta(v)\ln v]\delta(u), \tag{3.15f}$$

where the symbols to the left of the expressions indicate the location of the support of the distribution. Evidently, the spectral condition is compatible only with one of these, (3.15c), apart from the trivial contribution (3.13). (In particular, there is no Lorentz invariant solution of (3.5) whose Fourier transform has support on a single one of the closed sets

$$p^0 \gtreqless 0, \qquad p^1 \gtreqless 0.$$

Such a two-point function would be needed to construct a scalar field theory of massless particles which can travel only to the right or only to the left.) Thus, the most general Lorentz two-point distribution satisfying the wave equation is a multiple of

$$\begin{aligned}\frac{1}{i}\,\Delta^{(+)}(0,x) &= \frac{1}{4\pi}\iint du\,dv\,\exp\{-i/2[u(x^0+x')+v(x^0-x')]\}\\ &\quad\times\left[\frac{d}{du}(\theta(n)\ln u)\delta(v)+\frac{d}{dv}(\theta(v)\ln v)\delta(u)\right]\\ &= \lim_{\varepsilon\to 0^+}\frac{1}{4\pi}[2\Gamma'(1)-\ln(-x^2+i\varepsilon x^0)]\end{aligned} \tag{3.16}$$

plus a constant. Now the striking feature of (3.16) is that $\Delta^{(+)}$ is not a distribution of positive type because (3.15c) is not a positive measure. Evaluated on a test function $f(u,v)$, (3.15c) yields

$$\int_0^\infty -\left(\frac{\partial f}{\partial u}\right)(u,0)\ln u\,du + \int_0^\infty -\left(\frac{\partial f}{\partial v}\right)(0,v)\ln v\,dv, \tag{3.17}$$

and it is easy to choose a positive f so that this expression is negative. For example, take $f(u,v) = \exp - \alpha(u+v)$ with α sufficiently large. No addition of a finite amount of (3.13) can cure this negativity. Thus the simple example of a massless field in two dimensional space–time shows the necessity of enlarging the framework of the theory to include indefinite metric.

A failure in positivity also occurs in quantum electrodynamics (in four dimensional space–time) and Gupta and Bleuler's proposal for handling it goes roughly as follows: Embed the space of states in a larger Hilbert space,

$\mathscr{H}$, with scalar product $(\cdot, \cdot)$ so that the given indefinite sesquilinear form $\langle \cdot, \cdot \rangle$ is expressible

$$\langle \Phi, \Psi \rangle = (\Phi, \eta\Psi), \tag{3.18}$$

where η is a self-adjoint bounded operator. For the massless scalar field in two-dimensional space–time this can be done in several ways [26]. For example, one can define

$$\begin{aligned}(\phi(f)\Psi_0, \phi(g)\Psi_0) \\ = \frac{1}{2\pi}\bigg\{\int_0^\infty du(1 + |\log u|)\Big[\overline{f(u,0)}g(u,0) + \overline{\frac{\partial f}{\partial u}(u,0)}\frac{\partial g}{\partial u}(u,0)\Big] \\ + \int_0^\infty dv(1 + |\log v|)\Big[\overline{f(0,v)}g(0,v) + \overline{\frac{\partial f}{\partial v}(0,v)}\frac{\partial g}{\partial v}(0,v)\Big]\bigg\},\end{aligned} \tag{3.19}$$

and then using the formula (3.1) express the n-point vacuum expectation values in terms of the two point. This is equivalent to constructing a symmetric Fock space

$$\mathfrak{F}_s(\mathscr{H}^{(1)}) = \bigoplus_{n=0}^{\infty} \mathscr{H}^{(n)};$$

$$\mathscr{H}^{(0)} = \mathbb{C}; \qquad \mathscr{H}^{(n)} = (\mathscr{H}^{(1)})_s^{\oplus n}, \qquad n \geq 1.$$

Here $\mathscr{H}^{(1)}$ is the Hilbert space obtained by completion in the scalar product (3.19) from $\mathscr{S}(\mathbb{R}^2)$ restricted to the light cone. Then, the sesquilinear form defined by

$$\langle \phi(f)\Psi_0, \phi(g)\Psi_0 \rangle = \iint dx\, dy\, \overline{f(x)}\, \frac{1}{i}\, \Delta^{(+)}(x - y)g(y) \tag{3.20}$$

satisfies

$$|\langle \phi(f)\Psi_0, \phi(g)\Psi_0 \rangle|^2 \leq (\phi(f)\Psi_0, \phi(f)\Psi_0)(\phi(g)\Psi_0, \phi(g)\Psi_0). \tag{3.21}$$

This inequality guarantees the existence of the required operator η. From this point on, to keep clear that the two-point vacuum expectation value is indefinite, we shall use angular brackets to denote it.

As always in the Gupta–Bleuler formalism, there is a maximal subspace $\mathscr{H}'$, of $\mathscr{H}$ on which the form $\langle \cdot, \cdot \rangle$ is positive and the physical states are identified with the elements of

$$\mathscr{H}_{\text{phys}} = \overline{(\mathscr{H}'/\mathscr{H}'')}.$$

Here $\mathscr{H}''$ is the subspace of $\mathscr{H}'$ consisting of the vectors of zero length; the bar denotes the Hilbert space closure in the metric determined by $\langle \cdot, \cdot \rangle$.

At first sight, the Gupta–Bleuler formalism seems somewhat arbitrary and contrived, but it has so far been the only direct method for defining covariant and local gauge fields, so we propose to follow it where it leads. One of the directions is to the construction of functions of free fields.

The first of these is an entire function

$$:\exp i\lambda\phi:(x) = \sum_{n=0}^{\infty} \frac{(i\lambda)^n}{n!} :\phi^n:(x), \tag{3.22}$$

where $:\phi^n:$ is the so-called Wick ordered nth power of ϕ. It can be defined by

$$\lim_{x_1 \cdots x_n \to x} :\phi(x_1)\ldots\phi(x_n):, \tag{3.23}$$

where

$$:\phi(x_1)\ldots\phi(x_n): = \sum_{r=0}^{[n/2]} (-1)^r \sum_{C_r} [j_1\ldots j_{2r}]\phi(x_{k_1})\cdots\phi(x_{k_{n-2r}}), \tag{3.24}$$

$[n/2]$ is the largest integer less than or equal to $n/2$, and the sum $\sum_{C_r}$ is over all partitions of the integers $12\ldots n$ into two subsets $j_1\ldots j_{2r}$ and $k_1\ldots k_{n-2r}$ that satisfy $j_1 < j_2 < \cdots < j_{2r}$, $k_1 < k_2 < \cdots < k_{n-2r}$. The hafnian is defined as before but with our new notation

$$[pq] = \langle\Psi_0, \phi(x_p)\phi(x_q)\Psi_0\rangle. \tag{3.25}$$

The proof that (3.22) actually defines a set of vacuum expectation values can be carried out following [15] and [24] and will not be discussed here. The main point for present purposes is that the two-point vacuum expectation value turns out to be

$$\langle\Psi_0, :\exp i\lambda\phi:(x_1) :\exp i\mu\phi:(x_2)\Psi_0\rangle = \exp\left[-\lambda\mu \frac{1}{i} D^{(+)}(x_1 - x_2)\right] \tag{3.26}$$

If we insert the expression (3.16) for $D^{(+)}(x)$ we get

$$\lim_{\varepsilon\to 0^+} [-(x_1 - x_2)^2 + i\varepsilon(x_1^0 - x_2^0)]^{+\lambda\mu/4\pi}. \tag{3.27}$$

Thus, when $\lambda = -\mu$ the two-point function falls off at large spacelike $(x_1 - x_2)$ as a power $-\lambda^2/4\pi$. On the other hand, for $\lambda = \mu$, it grows with a power $\lambda^2/4\pi$, which is arbitrarily large for sufficiently large values of λ^2. Note that this behavior would be reversed if for some reason the sign of $D^{(+)}$ were changed. As will be discussed shortly, just such changes in sign are made in the definition of auxiliary fields for the solution of quantum electrodynamics. The significance of such growth for $\lambda = -\mu$ is that it is absolutely forbidden in quantum field theories with positive metric. There the cluster decomposition theorem says that a two-point function of the form $[(\Psi_0, \phi(x)\phi(y)\Psi_0) - (\Psi_0, \phi(x)\Psi_0)^2]$ must behave at infinity no worse than the theory of a free field, i.e., in our case has no worse than logarithmic growth. The moral is that in theories with indefinite metric infrared singularities can be appreciably worse.

Whatever the choice of sign in the preceding example, the result is a tempered distribution quantized field. To get nontempered fields one must make further constructions.

One such appears as an auxiliary construct in the solution of the quantum electrodynamics of massless fermions in two-dimensional space–time. It has a two-point function F_1 that is related to (3.5) by

$$\Box F_1(x) = F(x), \tag{3.28}$$

i.e., in terms of light cone coordinates

$$uv\hat{F}_1(u, v) = \frac{d}{du}[\theta(u)\ln u]\delta(v) + \frac{d}{dv}[\theta(v)\ln v]\delta(u).$$

$\hat{F}_1$ is thereby uniquely determined up to addition of a multiple of $\delta(u)\delta(v)$. A particular solution of the homogeneous equation is

$$\hat{F}_1(u, v) = \frac{d^2}{du^2}[\theta(u)\ln u]\delta'(v) + \frac{d^2}{dv^2}[\theta(v)\ln v]\delta'(u), \tag{3.29}$$

whose Fourier transform yields the two-point expectation value

$$\langle\Psi_0, \phi_1(x_1)\phi_1(x_2)\Psi_0\rangle = -\frac{(x_1 - x_2)^2}{4}\frac{1}{i}\Delta^{(+)}(x_1 - x_2). \tag{3.30}$$

The free field ϕ_1 with this two-point function is what is needed for the following; it is obtained by a standard Fock space construction following the ideas of [26].

The analog of the construction of $:\exp i\lambda\phi:$ from ϕ yields here $:\exp i\lambda\phi_1:$ with a two-point vacuum expectation value

$$\langle\Psi_0, :\exp i\lambda\phi_1:(x_1) :\exp i\mu\phi_1:(x_2)\Psi_0\rangle = \exp[-\lambda\mu\langle\Psi_0, \phi_1(x_1)\phi_1(x_2)\Psi_0\rangle]. \tag{3.31}$$

The behavior of this expression is quite analogous to (3.26). The main change is that the expectation value $\langle\Psi_0, \phi_1(x_1)\phi_1(x_2)\Psi_0\rangle$ grows like $(x_1 - x_2)^2 \log(x_1 - x_2)^2$ in spacelike directions. Thus, if $\lambda = -\mu$ and the overall sign is changed as before (3.31) grows in spacelike directions like

$$\exp(\lambda^2/16\pi)(-(x_1 - x_2)^2)\ln[-(x_1 - x_2)^2]. \tag{3.32}$$

This implies that (3.31) is not a tempered distribution. [There are tempered distributions which grow this fast, e.g., $\partial/\partial x_1^\mu \exp\{i\exp[\lambda^2(x_1 - x_2)^2]\}$ but they oscillate. The direct application of the criterion for temperedness

$$|T(\phi(\cdot + a))| \leq C(|a|^N + 1),$$

for some C and N, easily yields that (3.31) is nontempered.]

Although the preceding construction yields a field defined for test functions in $\mathscr{D}$ but not tempered, i.e., with no extension to test functions in $\mathscr{S}$, it could seem rather artificial especially in the last step in which the sign of the two-point vacuum expectation value of the ϕ_1 field is apparently arbitrarily changed. A more convincing justification is to be found in the construction of the solution of the quantum electrodynamics of a fermion field in two-dimensional space–time. There one finds (in all covariant gauges but the Landau gauge) that auxiliary fields of precisely the type described above are necessary to construct the solution. It is known that one cannot construct gauge theories in a local and covariant gauge except with indefinite metric. Thus, after all, it may not be so surprising that the indefinite metric forces one to the above choice.

To give an idea where the field ϕ_1 comes from, we shall sketch the construction of the solution of the classical coupled field equations. The quantum field theory is then obtained from it, by replacing classical fields by appropriate quantized fields. However, there is one important additional effect of a nonclassical nature, a mass shift.

The Maxwell–Dirac equations to be solved are

$$-i\gamma^{\mu}(\partial_{\mu} + qiA_{\mu})\psi(x) = 0, \tag{3.33}$$

$$A_{\mu}(x) + (\alpha - 1)\partial_{\mu}(\partial \cdot A) = q\psi^{+}\gamma_{\mu}\psi, \tag{3.34}$$

where

$$\gamma^0 = \begin{pmatrix} 0 & 1 \\ 1 & 0 \end{pmatrix}, \qquad \gamma' = \begin{pmatrix} 0 & -1 \\ 1 & 0 \end{pmatrix}, \qquad \gamma^5 = \gamma^0\gamma^1 = \begin{pmatrix} 1 & 0 \\ 0 & -1 \end{pmatrix},$$

and ψ is a two-component quantity on which the γ matrices act by matrix multiplication. The adjoint ψ^{+} is defined as $\bar{\psi}\gamma^0$ where bar stands for complex conjugation. Following the now traditional approach we consider the Ansatz

$$\psi(x) = \exp[-iq\Omega(x)]\psi^{(0)}(x), \qquad \Omega(x) = c(x) + \gamma^5 d(x), \tag{3.35}$$

and $\psi^{(0)}$ is supposed to satisfy the free Dirac equation

$$-i\gamma^{\mu}\partial_{\mu}\psi^{(0)}(x) = 0. \tag{3.36}$$

Now

$$\gamma^{\mu}\exp[-iq\Omega] = \exp[-iq\Omega']\gamma^{\mu},$$

where Ω' is Ω with d replaced by $-d$, so

$$-i\gamma^{\mu}\partial_{\mu}\exp[-iq\Omega]\psi^{(0)} = -i\gamma^{\mu}(-iq)[\partial_{\mu}c + \gamma^5\varepsilon_{\mu\nu}\partial^{\nu}d]\exp[iq\Omega]\psi^{(0)}.$$

Now $\gamma^\mu\gamma^5\partial_\mu = \gamma_\nu(\varepsilon^{\nu\mu}\partial_\mu)$ so we see that to compensate the A_μ in the Dirac equation (3.33) we need only choose

$$A^\mu = \partial^\mu c + \varepsilon^{\mu\nu}\partial_\nu d. \tag{3.37}$$

This implies

$$\partial \cdot A = \Box c. \tag{3.38}$$

Now note

$$\begin{aligned}\psi^+(x)\gamma^\mu\psi(x) &= \psi^{(0)+}(x)\exp[iq\Omega']\gamma^\mu\exp[-iq\Omega]\psi^{(0)}(x)\\ &= \psi^{(0)+}(x)\gamma^\mu\psi^{(0)}(x)\end{aligned} \tag{3.39}$$

so insertion of the expression for A_μ in (3.34) yields

$$\alpha\partial^\mu\Box c + \varepsilon^{\mu\nu}\partial_\nu\Box d = j^{(0)\mu}(x) \tag{3.40}$$

where $j^{(0)\mu}$ is the free Dirac current given by the right-hand side of (3.39).

Now the right-hand side of (3.40) satisfies both

$$\partial^\mu j^{(0)}_\mu = 0 \qquad \text{and} \qquad \operatorname{curl} j^{(0)} = 0$$

by virtue of the free Dirac equation. Thus there exist potentials ρ and σ such that

$$j^{(0)\mu} = \frac{1}{\sqrt{\pi}}\partial^\mu\rho = \frac{1}{\sqrt{\pi}}\varepsilon^{\mu\nu}\partial_\nu\sigma \tag{3.41}$$

(the $\sqrt{\pi}$s are traditional). Thus, if we choose

$$\Box c = \frac{\alpha_1}{\alpha}\frac{q}{\sqrt{\pi}}\rho, \qquad \Box d = (1-\alpha)\frac{q}{\sqrt{\pi}}\sigma, \tag{3.42}$$

(3.34) is satisfied and, modulo the solution of (3.41) and (3.42) for ρ and σ and c and d, a family of solutions of the coupled Maxwell–Dirac equations is given in terms of a family of solutions of the free Dirac and wave equations. We see that for the quantum field theory problem we would be led naturally to introduce quantized fields satisfying

$$\Box^2 c(x) = \Box^2 d(x) = 0$$

as we have described above. Where then is the quantum field theory surprise? It is in the definition of the current. When full attention is paid to the singular character of the operators $\psi(x)$, the current has to be given a definition for which (3.39) does *not* hold: the current of the coupled theory is not proportional to that of the free theory. In fact, for the quantized j^μ

$$\operatorname{curl} j = -(e^2/\pi)F_{01} \tag{3.43}$$

where

$$F_{\mu\nu} = \partial_\mu A_\nu - \partial_\nu A_\mu$$

so

$$(\square + e^2/\pi)F_{01}(x) = 0, \tag{3.44}$$

i.e., F_{01} satisfies the free field equation for a massive particle with $m^2 = e^2/\pi$. Nevertheless, the vacuum expectation values of the theory contain precisely the kind of factors discussed above and so, as first shown by Capri and Ferrari [29], the Schwinger model in any gauge except one for which $\alpha = \infty$ (the Landau gauge), is a theory of nontempered quantized fields.

4. The Evidence for Hyperfunction Field Theories

We have cited above some evidence for the existence of nonrenormalizable field theories in which the vacuum-expectation values may be generalized functions which are not distributions. Here the possibility shall be discussed that this phenomenon appears for one range of coupling constant and not for another.

The first clear indication of a bound on coupling constants in a tempered model quantum field theory came in the work of Glimm and Jaffe on the so-called $\lambda\phi_2^4$ model [30]. This is the theory of a scalar hermitean field in two-dimensional space–time satisfying a quantized version of the classical field equation

$$(\square + m_0^2)\phi(x) = \lambda\phi^3(x). \tag{4.1}$$

They showed that in this model there is an absolute upper bound on the renormalized coupling constant g. The definition of renormalized coupling constant used here is (roughly) defined as the Fourier transform of the four-point vacuum-expectation value evaluated at the origin, divided by the square of the Fourier transform of the two-point vacuum-expectation value and a power of the mass of the lowest massive state sufficient to give a dimensionless result. In lowest order perturbation theory in λ, the renormalized coupling constant for $\lambda\phi_\nu^4$ is $g = \lambda/m_0^{4-\nu}$. The idea of using g rather than $\lambda/m_0^{4-\nu}$ to parametrize solutions of a theory arises rather naturally in perturbative renormalization theory and is usually accepted without question in the heuristic literature of quantum field theory, but its validity will only be assured when one has actually established the existence and basic properties of the theories to be parametrized; the task has barely been begun. Thus any discussion of the general situation necessarily involves a certain amount of poking around in the dark. Glimm and Jaffe's own conjecture for the $\lambda\phi_2^4$ theory is shown in Fig. 1 and encapsulated in the phrase *critical point dominance.* [30] actually establishes a bound for any integer space–time dimension provided the two-point vacuum-expectation value satisfies a

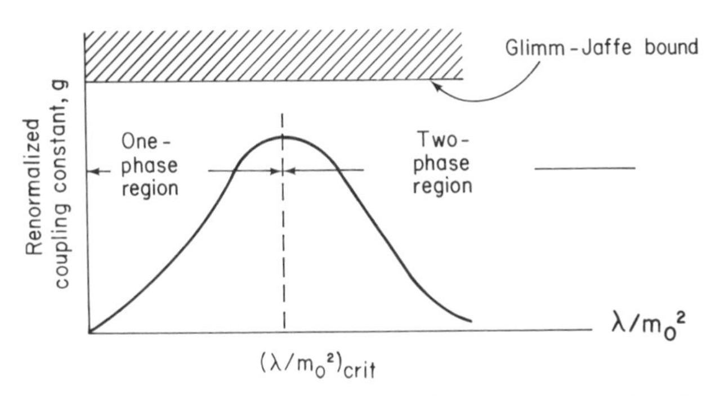

FIGURE 1. Behavior of the renormalized coupling constant g as function of λ/m_0^2 in the $\lambda\phi_2^4$ model. The behavior in a neighborhood of $(\lambda/m_0^2)_{\text{crit}}$ is conjectural.

certain ultraviolet boundedness condition. To state it, one needs the spectral representation of the two-point vacuum-expectation value

$$(\psi_0, \phi(x_1)\phi(x_2)\psi_0) = \int d\rho(a)\,\frac{1}{i}\,\Delta^{(+)}(a, x_1 - x_2).$$

Here

$$\Delta^{(+)}(x) = i[2(2\pi)^{d-1}]^{-1} \int \exp(-ikx)\, d\Omega_m(k),$$

where

$$d\Omega_m(k) = d^{n-1}k/[\mathbf{k}^2 + m^2]^{1/2}$$

is the Lorentz invariant measure on the hyperboloid

$$p^2 = m^2, \qquad p^0 > 0.$$

The basic hypothesis of Glimm and Jaffe is

$$\int \frac{d\rho(a)}{a} < \infty.$$

This bound is known to hold for the solutions of $\lambda\phi_2^4$ obtained in constructive quantum field theory and in the analogous theory $\lambda\phi_3^4$ in three-dimensional space–time. In $\lambda\phi_4^4$, the theory in four-dimensional space–time, it holds to all orders in perturbation theory but nothing is yet known about the non-perturbative solutions if there are any, except for the free field for which $g = 0$.

To gain an overview of this situation, it is helpful to adopt the point of view of the renormalization group and to interpolate and extrapolate these hard-won results in space–time dimension ν. (To prove that an analytic interpolation is possible is a program which has made some headway lately [31].) The results are displayed in Fig. 2.

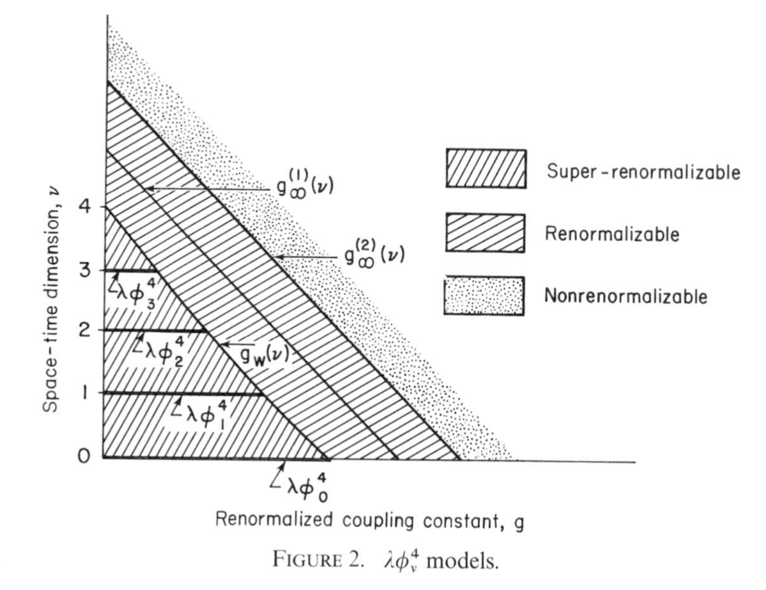

FIGURE 2. $\lambda\phi_\nu^4$ models.

The heavy horizontal lines in integer dimension correspond to the solutions whose existence has been established by the methods of constructive field theory. The interpolation indicated by diagonal lines //// is conjectured. (The curve $g_w(\nu)$, which is the boundary of the region, has a subscript which stands for Wilson, who was a pioneer in the analysis of the properties of the $\lambda\phi_\nu^4$ models near this boundary.) To understand the significance of this curve and the others $g_\infty^{(1)}(\nu)$ and $g_\infty^{(2)}(\nu)$ it is necessary to introduce the function β which is believed to govern the high energy and low energy behavior of the vacuum expectation values of the theory. See Fig. 3. A zero g_0 of β is an

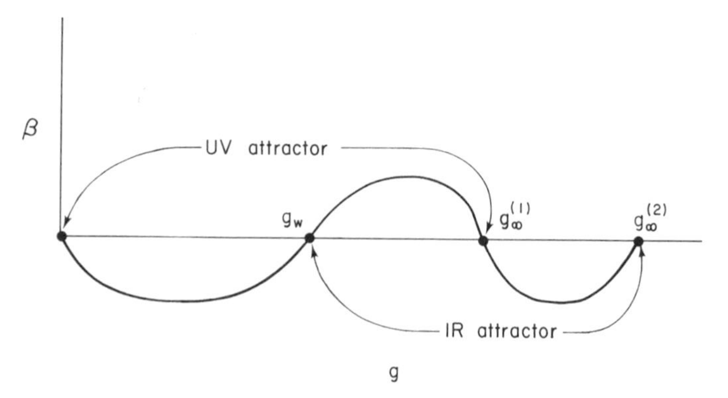

FIGURE 3. Qualitative behavior of the function β for $\lambda\phi_\nu^4$ with $\nu < 4$. The existence of the zeros $g_\infty^{(1)}(\nu)$ and $g_\infty^{(2)}(\nu)$ is conjectured.

ultraviolet (uv) *attractor* if $\beta'(g_0) < 0$ and an *infrared* (ir) *attractor* if $\beta'(g_0) > 0$. Theories whose renormalized coupling constants lie in an interval between two zeros have their asymptotic behavior determined by the theories associated with coupling constant at those adjacent zeros.

There is one obvious feature of this diagram which should be mentioned first of all. If it is really true that interpolation of the known super-renormalizable solutions in integer dimensions results in a family smoothly parametrized by the space–time dimension ν, the physical mass m and the renormalized coupling constant g, then it seems plausible that there should be other bands lying between the curves $g(\nu)$, $g_\infty^{(1)}(\nu)$, $g_\infty^{(2)}(\nu)$, ... which are similarly parametrized. That would mean that there might be solutions of $\lambda\phi_1^4$ (the anharmonic oscillator) which are not super-renormalizable and that such more singular solutions would provide more guidance about $\lambda\phi_4^4$ than any super-renormalizable solution.

A striking simple argument which provides some support for this point of view has been given by Newman [32], who has shown directly that in dimension 0 and in dimension 1,

$$g \leq g_{\text{crit}},$$

where g_{crit} is the theory that corresponds to an Ising model. For dimension 0, the argument is so simple and enlightening that we borrow it and recapitulate it here. In space–time of zero dimensions there is only one point and the vacuum-expectation values reduce to moments of a measure

$$S_n = \int x^n \, d\mu(x)$$

with $\int d\mu(x) = 1$ and $\int x \, d\mu(x) = 0$, $\int x^2 \, d\mu(x) > 0$. The renormalized coupling constant is

$$g = -[(S_4 - 3S_2^2)/S_2^2] = 3 - (S_4/S_2^2).$$

Now Schwarz's inequality implies

$$\left| \int x^2 1 \, d\mu(x) \right|^2 \leq \int x^4 \, d\mu(x) \int d\mu(x)$$

with equality only when x^2 and 1 are proportional almost everywhere relative to μ, i.e., $S_2^2 \leq S_4$ or

$$g \leq 2$$

with $g = 2$ only when

$$\tfrac{1}{2}[\delta(x - c) + \delta(x + c)] \, dx$$

for some $c > 0$. Thus, $g_{\text{w}}(0) = 2$.

It is not clear that enlarging the framework by considering hyperfunction fields will make it possible to evade Newman's argument and to find solutions of $\lambda\phi_v^4$ with $g > g_w(v)$. However, if one again appeals to "continuity," the studies of the perturbation series for $v > 4$ suggest this.

There is another argument which is indicative if not wholly convincing. The study of the quantized version of the so-called sine–Gordon equation

$$\Box\phi + \frac{\alpha_0}{\beta}\sin\beta\phi = 0$$

shows there are several distinguished ranges of the parameters. For $0 \leq \beta \leq 4\pi$ the theory is determined by its perturbation series, and a solution can be constructed nonperturbatively, but for $4\pi < \beta < 8\pi$, an extra renormalization is necessary [33]. For $\beta \geq 8\pi$, opinions differ, Coleman arguing that the theory loses its vacuum while Schroer and Truong hold out the possibility that there may be nonrenormalized solutions. Now it is admittedly the case that the sine–Gordon theory is exceedingly special in structure. For one thing, the above-mentioned function β vanishes identically. Nevertheless, it does provide evidence for what one may call an *ultraviolet phase transition*—a qualitative change in the ultraviolet behavior of a quantum field theory as a parameter passes through a critical value. This phenomenon will have to be understood before the status of such models as $\lambda\phi_4^4$ can be regarded as clarified.

References

1. M. Born, W. Heisenberg, and P. Jordan, Zur Quantenmechanik II, *Z. Phys.* **35** (1925–1926), 557–615.
2. P. Jordan and W. Pauli, Zur Quantenmechanik ladungsfreier Felder, *Z. Phys.* **47** (1928), 151–173.
3. N. Bohr and L. Rosenfeld, Zur Frage der Messbarkeit der electromagnetischen Feldgrössen, *Dansk Vid. Selsk. Mat. Fys. Medd.* **12**, No. 8 (1933), 1–65.
4. W. Heisenberg, Über die mit der Entstehung von Materie aus Strahlung verknüpften Ladungsschwankungen, *Ber. Verh. Sächs. Akad. Wiss. Leipzig Math. Phys. Kl.* **86** (1934), 317–322.
5. L. Schwartz, "Theorie des Distributions, I, II," Hermann, Paris, 1950, 1951.
6. P. Jordan and E. Wigner, Über das Paulische Äquivalenzverbot, *Z. Phys.* **47** (1928), 631–651.
7. D. Ruelle, On the asymptotic condition in quantum field theory, *Helv. Phys. Acta* **35** (1962), 147–163.
8. A. S. Wightman, Quantum field theory in terms of vacuum expectation values, *Phys. Rev.* **101** (1956), 860–866.
9. H. J. Borchers, unpublished observations.
10. W. Heisenberg, Über die in der Theorie der Elementarteilchen auftretende universelle Länge, *Ann, Physik* **32** (1938), 20–33.

11. R. Arnowitt and S. Deser, Renormalization of derivative coupling theories, *Phys. Rev.* **100** (1955), 349–361.
12. L. Cooper, Some notes on non-renormalizable field theory, *Phys. Rev.* **100** (1955), 362–370.
13. I. M. Gelfand and G. E. Shilov, "Generalized Functions," Vols. 1 and 2, Academic Press, New York, 1964, 1968.
14. W. Güttinger, Non-local structure of field theories with non-renormalizable interaction, *Nuovo Cimento* **10** (1958), 1–36.
15. B. Schroer, The concept of non-localizable fields and its connection with non-renormalizable field theory, *J. Math. Phys.* **5** (1964), 1361–1367.
16. A. M. Jaffe, High energy behavior in quantum field theory I, strictly localizable fields, *Phys. Rev.* **158** (1967), 1454–1461.
17. F. Constantinescu, Analytic properties of non-strictly localizable fields, *J. Math. Phys.* **12** (1971), 293–298.
18. F. Constantinescu and W. Thalheimer, Euclidean Green's functions for Jaffe fields, *Comm. Math. Phys.* **38** (1974), 299–316.
19. H. Epstein, V. Glaser, and A. Martin, Polynomial behavior of scattering amplitudes at fixed momentum transfer in theories with local observables, *Comm. Math. Phys.* **13** (1969), 257–316.
20. S. Nagamachi and N. Mugibayashi, Hyperfunction quantum field theory, *Comm. Math. Phys.* **46** (1976), 119–134.
21. S. Nagamachi and N. Mugibayashi, II Euclidean Green's functions, *Comm. Math. Phys.* **49** (1976), 257–275.
22. K. Pohlmeyer, Non-renormalizable quantum field theories, *in* "Renormalization Theory" (G. Velo and A. S. Wightman, eds.), pp. 461–482, Reidel Publ., Dordrecht Netherlands, 1976.
23. B. Schroer, Infrateilchen in der Quantenfeldtheorie, *Fortschr. Phys.* **11** (1963), 1–32.
24. A. S. Wightman, Introduction to some aspects of the relativistic dynamics of quantized fields, *in* "High Energy Electromagnetic Interactions and Field Theory," Cargèse Lectures in Physics, 1964 (M. Levy, ed.), pp. 171–291, Gordon & Breach, New York, 1967.
25. N. Nakanishi, Free massless scalar field in two dimensional space–time, *Progr. Theoret. Phys.* **57** (1977), 269–278.
26. G. Morchio and F. Strocchi, Infra-red singularities, vacuum structure and pure phases in local quantum field theory, preprint, Univ. of Pisa, May 1979.
27. O. W. Greenberg, J. S. Kang, and C. H. Woo, Infra-red regularization of the massless scalar free field in two dimensional space–time via Lorentz expansion, preprint, Univ. of Maryland, 1980.
28. L. Gårding and J. L. Lions, Functional analysis, *in* "Mathematical Problems of the Quantum Theory of Particles and Fields," *Nuovo Cimento Suppl.* **14** (1959), 9–66 (esp. pp. 45–53).
29. A. Capri and R. Ferrari, Schwinger model chiral symmetry, anomaly and θ-vacuua preprint, Univ. of Alberta, Edmonton, 1980.
30. J. Glimm and A. Jaffe, Absolute bounds on vertices and couplings; on the approach to the critical point, *Ann. Inst. H. Poincaré Sect. A* **22**, (1975), 97–107.
31. V. Rivasseau and E. Speer, The Borel transform in Euclidean ϕ_v^4 local existence for Re $v < 4$, *Comm. Math. Phys.* **72** (1980), 293–302.
32. C. Newman, to appear.
33. B. Schroer and T. Truong, Equivalence of the sine-Gordon and Thirring models and cumulative mass effects, *Phys. Rev. D* **15** (1977), 1684–1692.